D1693489

Compendium of Polymer Terminology and Nomenclature
IUPAC RECOMMENDATIONS 2008

International Union of Pure and Applied Chemistry

Compendium of Polymer Terminology and Nomenclature
IUPAC RECOMMENDATIONS 2008

Issued by the Polymer Division

Prepared for publication by

Richard G. Jones
University of Kent,
Kent, UK

Jaroslav Kahovec
Academy of Sciences,
Czech Republic

Robert Stepto
Manchester University,
Manchester, UK

Edward S. Wilks
Hockessin,
Delaware, USA

Michael Hess
University of Essen,
Germany

Tatsuki Kitayama
Osaka University,
Japan

W. Val Metanomski
CAS, Columbus,
Ohio, USA

With advice from

Aubrey Jenkins
Hassocks,
Sussex, UK

Pavel Kratochvil
Academy of Sciences,
Czech Republic.

RSCPublishing

ISBN: 978-0-85404-491-7

A catalogue record for this book is available from the British Library

© International Union of Pure and Applied Chemistry 2009

All rights reserved

Apart from fair dealing for the purposes of research for non-commercial purposes or for private study, criticism or review, as permitted under the Copyright, Designs and Patents Act 1988 and the Copyright and Related Rights Regulations 2003, this publication may not be reproduced, stored or transmitted, in any form or by any means, without the prior permission in writing of The Royal Society of Chemistry or the copyright owner, or in the case of reproduction in accordance with the terms of licences issued by the Copyright Licensing Agency in the UK, or in accordance with the terms of the licences issued by the appropriate Reproduction Rights Organization outside the UK. Enquiries concerning reproduction outside the terms stated here should be sent to The Royal Society of Chemistry at the address printed on this page.

Published by The Royal Society of Chemistry,
Thomas Graham House, Science Park, Milton Road,
Cambridge CB4 0WF, UK

Registered Charity Number 207890

For further information see our web site at www.rsc.org

Preface

This Compendium of Polymer Terminology and Nomenclature is an expansion and revision of the Compendium of Macromolecular Nomenclature published in 1991 (the so-called Purple Book) under the auspices of the Commission on Macromolecular Nomenclature of the Macromolecular Division of the International Union of Pure and Applied Chemistry (IUPAC). The present compendium contains twenty-two chapters, compared with the nine chapters and introduction of the 1991 edition, and only four of the chapters of the 1991 edition are retained in their original form. One chapter included in the 1991 edition (Nomenclature for regular single-strand and quasi-single-strand inorganic and coordination polymers) has been excluded from this edition pending major revision in accord with the 2005 edition of the Nomenclature of Inorganic Chemistry (the Red Book).

This edition is published under the auspices of the Commission on Macromolecular Nomenclature of the Macromolecular Division of the IUPAC, the Sub-committee on Macromolecular Terminology and the Sub-committee on Polymer Terminology of the Polymer (previously Macromolecular) Division of the IUPAC. The changes of title of the bodies concerned reflect both the growth in the activities of the former Macromolecular Division and the reorganisation of the IUPAC to a project-based rather than a commission-based organisation. At the beginning of 2002, the Division of Chemical Nomenclature and Structure Representation of the IUPAC was formed with the task of overseeing all the chemical nomenclature projects of the IUPAC. At the same time, the Commission on Macromolecular Nomenclature was discontinued and the Sub-committee on Macromolecular Terminology formed with the same scientists actively involved. Finally, in 2004, the Macromolecular Division voted to change its name to Polymer Division and, in 2005, the sub-committee changed its name to the Sub-committee on Polymer Terminology. These changes followed the gradual expansion of the division's and sub-committee's activities that had occurred to cover the materials as well as the molecular aspects of the subject. The sub-committee has responsibility for projects on polymer terminology and works on macromolecular nomenclature in collaboration with the Division of Chemical Nomenclature and Structure Representation.

The synchronisation of the formation of the Sub-committee on Macromolecular Terminology with the formation of the Division of Chemical Nomenclature and Structure Representation was essential to ensuring the smooth continuation and expansion of the IUPAC's work on polymer terminology and nomenclature. Except for Chapter 14, which is an introduction to polymer nomenclature, and Chapter 23, dealing with abbreviations for names of polymeric substances, the chapters of the present compendium are reproductions, with some editorial corrections in line with current IUPAC practice, of official IUPAC recommendations published in *Pure and Applied Chemistry*. It can be seen from the range of dates of the original articles constituting the chapters that the restructuring of the IUPAC and the creation of a polymer terminology sub-committee from a macromolecular nomenclature commission did not cause a hiatus in publications.

The first, 1991 edition of the Purple Book contained two main sections, namely Definitions, having four chapters, and Nomenclature, having an introduction and five

PREFACE

chapters. In keeping with the expansion of the Commission's and Sub-committees' work into topics related to polymeric and polymer-based materials, the first section of the present edition is renamed Terminology and increased to thirteen chapters, and the second, Nomenclature section has ten chapters. In addition, the present edition retains a bibliography of biopolymer-related IUPAC-IUBMB nomenclature.

Following this preface are a copy of the preface to the first edition of the Purple Book, a history of the work of the IUPAC Polymer Division on polymer terminology and nomenclature by the past secretaries of the Commission and Sub-committees, and comprehensive lists of members of the several bodies involved in the work (Sub-commission (1952-1967), Commission (1968-2001) and Sub-committees (from 2002)).

This compendium is also available in electronic form via the Polymer Division's website, www.iupac.org/divisions/IV.

Finally, we would like to thank our co-editors, Val Metanomski, Ted Wilks, Jaroslav Kahovec and Tatsuki Kitayama, for their sterling efforts over the long period of time needed to prepare this book for publication, and also the help of colleagues Aubrey Jenkins and Pavel Kratochvíl, past Chairmen of the Commission on Macromolecular Nomenclature, for their invaluable contributions during the final stages of the editorial labours.

We hope that this, new edition of the 'Purple Book' continues the significant contribution to clear and precise communication in polymer science made by its predecessor.

ROBERT STEPTO
Chairman, IUPAC Commission on Macromolecular Nomenclature, 1992 – 1999

MICHAEL HESS
Chairman, IUPAC Commission on Macromolecular Nomenclature, 2000 – 2001
Chairman, IUPAC Sub-committee on Macromolecular Terminology, 2002 - 2005
Chairman, IUPAC Sub-committee on Polymer Terminology, 2005

RICHARD JONES
Chairman, IUPAC Sub-committee on Polymer Terminology, from 2006

Contents

Preface, v

IUPAC Polymer Division (IV): History of IUPAC Polymer Terminology and Nomenclature, ix

Membership of the Sub-commission on Nomenclature (1952 to 1967), xiii

Membership of the Commission on Macromolecular Nomenclature (1968 to 2001), xiii

Membership of the Sub-committee on Macromolecular Terminology (2002 to 2005) and of the Sub-committee on Polymer Terminology (from 2005), xv

Meeting Locations of the Commission on Macromolecular Nomenclature (1968-2001) and of the Sub-committees on Macromolecular Terminology and Polymer Terminology (2002-2006), xvi

References, xvii

TERMINOLOGY, 1

1. Glossary of Basic Terms in Polymer Science (1996), 3

2. Stereochemical Definitions and Notations Relating to Polymers (1980), 22

3. Definitions of Terms Relating to Individual Macromolecules, their Assemblies, and Dilute Polymer Solutions (1988), 44

4. Basic Classification and Definitions of Polymerization Reactions (1994), 70

5. Definitions Relating to Stereochemically Asymmetric Polymerizations (2002), 73

6. Definitions of Terms Relating to Crystalline Polymers (1988), 80

7. Definitions of Terms Relating to Low-Molar-Mass and Polymer Liquid Crystals (2001), 93

8. Definitions of Terms Relating to the Non-Ultimate Mechanical Properties of polymers (1997), 146

9   Definitions of Terms Related to Polymer Blends, Composites, and Multiphase Polymeric Materials (2004), 186

10 Terminology of Polymers Containing Ionizable or Ionic Groups and of Polymers Containing Ions (2006), 205

11 Definitions of Terms Relating to the Structure and Processing of Sols, Gels, Networks and Inorganic-Organic Hybrid Materials (2007), 211

12 Definitions of Terms Relating to Reactions of Polymers and to Functional Polymeric Materials (2003), 237

13 Definitions of Terms Relating to Degradation, Aging, and Related Chemical Transformations of Polymers (1996), 251

NOMENCLATURE, 259

14 Introduction to Polymer Nomenclature, 261

15 Nomenclature of Regular Single-Strand Organic Polymers (2002), 280

16 Nomenclature of Regular Double-Strand (Ladder and Spiro) Organic Polymers (1993), 318

17 Structure-Based Nomenclature for Irregular Single-Strand Organic Polymers (1994), 336

18 Graphic Representations (Chemical Formulae) of Macromolecules (1994), 350

19 Source-Based Nomenclature for Copolymers (1985), 368

20 Source-Based Nomenclature for Non-Linear Macromolecules and Macromolecular Assemblies (1997), 382

21 Generic Source-Based Nomenclature for Polymers (2001), 394

22 Abbreviations, 403

APPENDIX

Bibliography of Biopolymer-Related IUPAC-IUBMB Nomenclature Recommendations, 408

INDEX, 411

IUPAC Polymer Division (IV):
History of IUPAC Polymer Terminology and Nomenclature

From the 1920s, as polymer science developed and came of age, so too a common language came into being through the efforts of individuals who recognized the need for such a language. They formed committees to consider issues that included not only systematic nomenclature, but terminology and definitions, symbols, and other matters that might affect communication. All of this effort forms a part of the prehistory of the work of Division IV on polymer terminology and nomenclature.

SUB-COMMISSION ON NOMENCLATURE

The first publication of the IUPAC in the area of macromolecular nomenclature was in 1952 by the Sub-commission on Nomenclature of the then IUPAC Commission on Macromolecules, which drew on the talents of such remarkable individuals as J. J. Hermans, M. L. Huggins, O. Kratky, and H. F. Mark. That report [1] was a landmark in that, for the first time, it systematized the naming of macromolecules and certain symbols and terms commonly used in polymer science. It introduced the use of parentheses in source-based polymer names when the monomer from which the polymer is derived consists of more than one word, a practice that is now widely followed, and it recommended an entirely new way of naming polymers based on their structure that included the suffix "amer", a recommendation that has been almost totally ignored. After ten years, the Sub-commission issued its second report [2], which dealt with the then-burgeoning field of stereoregular polymers. A revision [3] of definitions in the original report appeared four years later. In 1968, a summary report [4] of the activities of the Sub-commission was published.

COMMISSION ON MACROMOLECULAR NOMENCLATURE (COMMISSION IV.1)

In 1968, the Commission on Macromolecular Nomenclature of the Macromolecular Division (Division IV)* was established under the Chairmanship of Kurt L. Loening with first Lionel C. Cross and later Robert B. Fox as Secretary. A series of major documents was produced that shaped modern polymer language. Most noteworthy was one that defined basic terms [5,6] and another on structure-based nomenclature for regular single-strand polymers [7,8]. The latter, originally developed by the Nomenclature Committee of the Polymer Division of the American Chemical Society and refined by the Commission, revolutionized polymer nomenclature by providing a systematic, consistent scheme particularly well-adapted to indexing; it became the standard for *Chemical Abstracts* and

---

* The name of Division IV was changed in 2004 to 'Polymer Division.'

IUPAC POLYMER DIVISION

major polymer journals throughout the world. A list of standard abbreviations was published [9-11] and later revised [12].

As the 1970s came to a close, Aubrey Jenkins assumed the Chairmanship, with Robert B. Fox continuing as Secretary through 1979, to be succeeded by Norbert Bikales, who served as Secretary until 1987. There followed in the 1980s a complete revision of the stereochemical definitions [13,14], terminology for molar masses in polymer science [15], an extension of structure-based nomenclature to inorganic and coordination polymers [16,17], the systemization of source-based nomenclature for copolymers [18] and key documents dealing with physicochemical terminology in the polymer field, covering definitions for individual macromolecules, their assemblies, and dilute solutions [19], crystalline polymers [20], a new method of classifying polymers [21] and a basic classification and definitions of polymerization reactions [22]. These documents were completed under the Chairmanship of Pavel Kratochvíl, who assumed this post in 1985; William J. Work was elected Secretary in 1987. In 1991, the first edition of the Purple Book (the Compendium of Macromolecular Nomenclature) [23] was published. The Compendium was the first major compilation of the Commission and consisted of an introduction to macromolecular nomenclature and nine chapters corresponding to the then valid IUPAC recommendations.

The final decade of the 20th century saw the activity of the Commission unabated. Robert Stepto succeeded to the Chairmanship in 1991, Máximo Barón became Secretary in 1998 and Michael Hess became Chairman in 2000. In the course of this decade, the graphical representation of polymer structures was addressed for the first time in 1994 [24], a revised and enlarged glossary of basic terms was published in 1996 [25], along with definitive documents dealing with the terminology of polymer aging and degradation (1998) [26] and non-ultimate mechanical properties (1998) [27]. Terminology concerned with liquid-crystal polymers was also published (2001) [28,29]. In the field of structure-based nomenclature, the Commission published recommendations covering regular double-strand polymers (1993) [30] and irregular single-strand polymers (1994) [31], and a revision of the Commission's 1975 rules for structure-based nomenclature [8] was completed (2002) [32]. In 1997, a document concerned with a new area, source-based nomenclature for non-linear macromolecules and macromolecular assemblies [33], was published. Documents on definitions relating to stereochemically asymmetric polymerizations [34] and the source-based generic nomenclature for polymers [35] were prepared in 2001.

SUB-COMMITTEE ON MACROMOLECULAR TERMINOLOGY and SUB-COMMITTEE ON POLYMER TERMINOLOGY

With effect from 1st January, 2002, the Bureau and Council of the IUPAC decided to form a new Division of Chemical Nomenclature and Structure Representation (Division VIII) to deal with nomenclature in an integrated manner across all branches of chemistry. In keeping with this change and the change of the IUPAC to project-based funding, the Commission on Macromolecular Nomenclature decided to become the Sub-committee on Macromolecular Terminology of Division IV. Michael Hess and Máximo Barón remained as Chairman and Secretary, respectively. The development was timely as it reflected the change in the emphasis of the work the Commission was carrying out. Under the chairmanship of Robert Stepto and Michael Hess, the majority of projects had been

HISTORY OF TERMINOLOGY AND NOMENCLATURE

concerned with terminology, related particularly to polymer and polymer-based materials. The trend continued, with recommendations on the terminology of polymer reactions and functional polymeric materials [36], and of polymer blends and composites [37] being published in 2004.

Richard Jones became Secretary in 2004. In the same year, the Macromolecular Division changed its name to Polymer Division and, in 2005, the subcommittee changed its name to the Subcommittee on Polymer Terminology. Richard Jones became Chairman in 2006 and Michael Hess became Secretary.

In keeping with the Subcommittee's mission, projects are planned that keep pace with the ever increasing variety of new polymer processes and polymeric materials and new classes of polymers. In 2006, recommendations on the terminology of polymers containing ionizable or ionic groups and of polymers containing ions [38] and, in 2007, jointly with the Inorganic Chemistry Division of the IUPAC, recommendations on definitions of terms relating to the structure and processing of sols, gels, networks and inorganic-organic hybrid materials [39] were published. Presently, in various stages of preparation are recommendations on kinetics and thermodynamics of polymerization, radical polymerizations with minimal termination – the so-called 'controlled' or 'controlled/living' radical polymerizations, terms relating to polymers in dispersed systems, biopolymers and biomedical polymers, and the thermal properties of polymers. In the field of nomenclature, working parties are preparing and completing recommendations concerned with the nomenclature of cyclic, hyperbranched, dendritic and rotaxane macromolecules, and with definitions of polymer class names. Finally, a general guide to polymer terminology and nomenclature is being prepared for publication on the IUPAC web site.

TRANSLATIONS OF NOMENCLATURE AND TERMINOLOGY RECOMMENDATIONS

Although the nomenclature and terminology recommendations have all been published in English, those involved have purposefully pursued their further dissemination and discussion in other languages, including Chinese, Croatian, Czech, French, German, Italian, Japanese, Polish, Portuguese, Russian, and Spanish. Over the years, there has been no doubt about the global influence of the recommendations on the language of chemistry. A list of translations can be obtained from:
<http://www.chem.qmul.ac.uk/iupac/bibliog/macro.html>.

ROBERT B. FOX
Secretary, IUPAC Commission on Macromolecular Nomenclature, 1973 – 1979

NORBERT M. BIKALES
Secretary, IUPAC Commission on Macromolecular Nomenclature, 1979 – 1987

WILLIAM J. WORK
Secretary, IUPAC Commission on Macromolecular Nomenclature, 1987 – 1997

IUPAC POLYMER DIVISION

MÁXIMO BARÓN
Secretary, IUPAC Commission on Macromolecular Nomenclature, 1998 – 2001
Secretary, IUPAC Subcommittee on Macromolecular Terminology, 2002 – 2003

RICHARD JONES
Secretary, IUPAC Subcommittee on Macromolecular Terminology, 2004 – 2005
Secretary, IUPAC Subcommittee on Polymer Terminology, 2005

MICHAEL HESS
Secretary, IUPAC Subcommittee on Polymer Terminology, from 2006

April 2007

HISTORY OF TERMINOLOGY AND NOMENCLATURE

MEMBERSHIP OF THE SUB-COMMISSION ON NOMENCLATURE (1952 to 1967)

Paolo Corradini (Italy)
Victor Desreux (Belgium)
Jan J. Hermans (The Netherlands/USA)
Maurice L. Huggins (USA)
Otto Kratky (Austria)
Herman F. Mark (USA)
Giulio Natta (Italy)

MEMBERSHIP OF THE COMMISSION ON MACROMOLECULAR NOMENCLATURE (1968 to 2001)

Members of the Commission were either Titular Members (TM), Associate Members (AM), or National Representatives (NR). Until and including 1989, commission memberships commenced and terminated (in odd-numbered years) immediately following a biennial IUPAC General Assembly; from 1991 they commenced (in even-numbered years) on 1st January immediately following a General Assembly and terminated at the end of the (odd-numbered) year of a General Assembly. CC denotes Commission Chairman and CS Commission Secretary.

Tae-Oan Ahn (South Korea)	NR (1989-1991)
José V. Alemán (Spain)	AM (1987-1995)
Giuseppe Allegra (Italy)	TM (1977-1989)
Devrim Balköse (Turkey)	NR (1987-1991)
Rolf E. Bareiss (FRG; Germany)[a]	TM (1983-1993)
Máximo Barón (Argentina)	NR (1987-1991); AM (1992-1995); TM (1996-2001, CS 1998-2001)
Norbert Bikales (USA)	AM (1975-1977); TM (1977-1987, CS 1979-1987)
Dietrich Braun (FRG)[a]	AM (1981-1989)
Anthony B. Brennan (USA)	TM (1994-1997)
Paolo Corradini (Italy)	TM (1968-1977)
John ("Ian") M. G. Cowie (UK)	AM (1983-1987)
Lionel C. Cross (UK)	TM (1968-1975, CS 1968-1973); AM (1975-1979)
Dinkar D. Deshpande (India)	NR (1985-1987)
L. Guy Donaruma (USA)	AM (1977-1985)
Robert B. Fox (USA)	TM (1968-1979, CS 1973-1979); AM (1979-1983)
Koichi Hatada (Japan)	AM (1987-1989); TM (1989-1997)
Jisaong He (China)	AM (2000-2001)
Kazuyuki Horie (Japan)	AM (1996-1997); TM (1998-2001)
Yadong Hu (China)	NR (1981-1983)

IUPAC POLYMER DIVISION

Marly A. M. Jacobi (Brazil)	NR (1994-1995)
Aubrey Jenkins (UK)	TM (1975-1985, CC 1977-1985); AM (1985-1987)
Jung-Il Jin (South Korea)	NR (1992-1993); AM (1994-2001)
Richard Jones (UK)	TM (1998-2001);
Jaroslav Kahovec (Czechoslovakia; Czech Republic)[b]	AM (1987-1991); TM (1992-1999)
Kaushal Kishore (India)	NR (1989-1991, 1998-2001)
Tatsuki Kitayama (Japan)	TM (2000-2001)
Vasilii V. Korshak (USSR)[c]	TM (1968-1971)
Ole Kramer (Denmark)	AM (1996-2001)
Pavel Kratochvíl (Czechoslovakia; Czech Republic)[b]	AM (1977-1979); TM (1979-1991, CC 1985-1991)
Przemysław Kubisa (Poland)	AM (1996-1999); TM (2000-2001)
Gordon J. Leary (New Zealand)	NR (1981-1987)
Kurt L. Loening (USA)	TM (1968-1977, CC 1968-1977); AM (1977-1985); NR (1985-1987)
Edgar M. Macchi (Argentina)	NR (1983-1987)
Eloisa B. Mano (Brazil)	NR (1979-1985)
Robert H. Marchessault (Canada)	AM (1977-1981, 1983-1987)
Ernest Maréchal (France)	AM (1992-1993); TM (1994-2001)
Krzysztof Matyjaszewski (USA)	AM (1992-1995)
Ingrid Meisel (Germany)	AM (1998-1999); TM (2000-2001)
W. Val Metanomski (USA)	AM (1987-1991); TM (1992-1999)
Itaru Mita (Japan)	AM (1977-1979); TM (1979-1989); AM (1989-1991)
Graeme Moad (Australia)	NR (2000-2001)
Werner Mormann (Germany)	NR (1998-2001)
N. Nhlapo (Republic of South Africa)	NR (1996-2001)
Claudine Noël (France)	TM (1985-1993)
Ivan M. Papisov (USSR; Russia)[c]	TM (1979-1987); AM (1987-1991)
Stanisław Penczek (Poland)	AM (1979-1983, 1994-2001)
Nikolai A. Platé (USSR; Russia)[c]	TM (1971-1979); AM (1979-1983); NR (1983-1991)
Roderic P. Quirk (USA)	AM (1983-1987)
Marguerite Rinaudo (France)	AM (1981-1985)
Wolfhardt Ring (FRG)[a]	TM (1971-1981)
Takeo Saegusa (Japan)	AM (1989-1993)
Raquel Santos Mauler (Brazil)	NR (1996-1999)
Lianghe Shi (China)	NR (1983-1987); AM (1987-1995)
Valerii P. Shibaev (USSR; Russia)[c]	TM (1987-1995); AM (1996-1999)
Pierre Sigwalt (France)	TM (1975-1983); AM (1983-1987)
Augusto Sirigu (Italy)	NR (1992-2001)

HISTORY OF TERMINOLOGY AND NOMENCLATURE

Stanisław Słomkowski (Poland)	AM (2000-2001)
George J. Smets (Belgium)	TM (1968-1975)
Robert Stepto (UK)	AM (1987-1989); TM (1989-1999, CC 1992-1999)
Claus Suhr (FRG)[a]	TM (1968-1971)
Lars-Oluf Sundelöf (Sweden)	AM (1981-1985)
Ulrich W. Suter (Switzerland)	AM (1979-1981); TM (1981-1991); AM (1992-1993)
Graham Swift (USA)	AM (2000-2001)
David Tabak (Brazil)	NR (2000-2001)
A. S. Tan (Malaysia)	NR (1983-1987)
Sukant K. Tripathy (USA)	AM (1987-1991)
Teiji Tsuruta (Japan)	TM (1968-1979); AM (1979-1983)
Ferenc Tüdős (Hungary)	NR (1985-1987)
Marcel van Beylen (Belgium)	NR (1985-1987)
J. K. Varna (India)	NR (1994-1997)
Jiří Vohlídal (Czech Republic)	AM (2000-2001)
Raymond E. Wetton (UK)	AM (1977-1981)
Edward ("Ted") S. Wilks (USA)	AM (1998-1999); TM (2000-2001)
Hans Wilski (FRG)[a]	AM (1975-1979)
William J. Work (USA)	AM (1985-1987); TM (1987-1997, CS 1987-1997)

[a] Prior to 1990, Germany was divided into West Germany (FRG) and East Germany (GDR).
[b] Prior to 1993, Czechoslovakia included the Czech Republic and Slovakia.
[c] Prior to 1991, Russia was part of the USSR.

MEMBERSHIP OF THE SUB-COMMITTEE ON MACROMOLECULAR TERMINOLOGY (2002 to 2005)

and

MEMBERSHIP OF THE SUB-COMMITTEE ON POLYMER TERMINOLOGY (from 2005)

Giuseppe Allegra (Italy), Máximo Barón (Argentina, Secretary, 2002-2003), Taihyun Chang (Korea), Alain Fradet (France), Jiasong He (China), Koichi Hatada (Japan), Karl-Heinz Hellwich Germany), Michael Hess (Germany, Chairman, 2002-2005; Secretary, from 2006), Roger C. Hiorns (France), Philip Hodge (UK), Kazuyuki Horie (Japan), Aubrey Jenkins (UK), Jung-Il Jin (Korea), Richard G. Jones (UK, Secretary, 2004-2005; Chairman, from 2006), Jaroslav Kahovec (Czech Republic), Tatsuki Kitayama (Japan), Pavel Kratochvíl (Czech Republic), Przemysław Kubisa (Poland), Ernest Maréchal (France), Ingrid Meisel (Germany), W. Val Metanomski (USA), Itaru Mita (Japan), Valdo Meille (Italy), Graeme Moad (Australia), Werner Mormann (Germany), Nabuo

IUPAC POLYMER DIVISION

Nakabayashi (Japan), Christopher K. Ober (USA), Stanisław Penczek (Poland), Luis P. Rebelo (Portugal), Marguerite Rinaudo (France), Claudio dos-Santos (Brazil), Ivan Schopov (Bulgaria), Francois Schué (France), Mark Schubert (USA), Valerii P. Shibaev (Russia), Stanisław Słomkowski (Poland), Robert F. T. Stepto (UK), David Tabak (Brazil), Jean-Pierre Vairon (France), Michel Vert (France), Jiří Vohlídal (Czech Republic), Edward S. Wilks (USA), and William J. Work (USA).

MEETING LOCATIONS of the COMMISSION ON MACROMOLECULAR NOMENCLATURE (1968-2001) and of the SUB-COMMITTEES ON MACROMOLECULAR TERMINOLOGY and POLYMER TERMINOLOGY (2002-2006)

YEAR	LOCATION
1968	Toronto, Canada
1969	Oberursel, Germany
1970	Ravello, Italy
1971	Washington, DC, USA
1972	Knokke Zoute, Belgium
1973	München (Munich), Germany
1974	Santiago de Compostela, Spain
1975	Madrid, Spain
1976	Dorking, UK
1977	Warszawa (Warsaw), Poland
1978	Moskva (Moscow), USSR
1979	Davos, Switzerland
1980	Arco Felice, Italy
1981	Leuven/Louvain, Belgium
1982	Amherst, Massachusetts, USA
1983	Lyngby, Denmark
1984	Praha (Prague), Czechoslovakia
1985	Lyon (Lyons), France
1986	Höhr-Grenzhausen, Germany
1987	Boston, Massachusetts, USA
1988	Tokyo, Japan
1989	Lund, Sweden
1990	Montréal/Montreal, Canada
1991	Hamburg, Germany
1992	Zürich, Switzerland
1993	Lisboa (Lisbon), Portugal
1994	Columbus, Ohio, USA
1995	Guildford, UK
1996	Seoul, South Korea
1997	Genève (Geneva), Switzerland
1998	Sydney, Australia
1999	Berlin, Germany
2000	Warszawa (Warsaw), Poland
2001	Brisbane, Australia
2002	Beijing, China
2003	Ottawa, Canada
2004	Bordeaux, France
2005	Beijing, China
2006	Rio de Janeiro, Brazil
2007	Torino, Italy
2008	Taipei, Taiwan

HISTORY OF TERMINOLOGY AND NOMENCLATURE

REFERENCES

1. Report on nomenclature in the field of macromolecules, *J. Polym. Sci.* 8, 257-277 (1952). Obsolete.
2. Report on nomenclature dealing with steric regularity in high polymers, *J. Polym. Sci.* 56, 153-161 (1962). Superseded by Ref. 14.
3. Report on nomenclature dealing with steric regularity in high polymers, *Pure Appl. Chem.* 12, 645-656 (1966). Superseded by Ref. 14.
4. Report of the Committee on Nomenclature of the International Commission on Macromolecules, *J. Polym. Sci., Part B: Polym. Lett.* 6, 257-260 (1968). Obsolete.
5. Basic definitions of terms relating to polymers, *IUPAC Inf. Bull. Append.* No. 13 (1971). Superseded by Ref. 25.
6. Basic definitions of terms relating to polymers 1974. *Pure Appl. Chem.* 40, 477-491 (1974). Superseded by Ref. 25.
7. Nomenclature of regular single-strand organic polymers, *IUPAC Inf. Bull. Append.* No. 29, (1972); *Macromolecules* 6, 149-158 (1973); *J. Polym. Sci., Polym. Lett. Ed.* 11, 389-414 (1973). Superseded by Ref. 32.
8. Nomenclature of regular single-strand organic polymers (Rules Approved 1975), *Pure Appl. Chem.* 48, 373-385 (1976). Superseded by Ref. 32.
9. Recommendations for abbreviations of terms relating to plastics and elastomers, *Pure Appl. Chem.* 18, 581-589 (1969). Obsolete.
10. List of abbreviations for synthetic polymers and polymer materials, *Inf. Bull. Append.* No. 12, (1971). Superseded by Ref. 12.
11. List of standard abbreviations (symbols) for synthetic polymers and polymer materials 1974, *Pure Appl. Chem.* 40, 473-476 (1974). Superseded by Ref. 12.
12. Use of abbreviations for names of polymeric substances (IUPAC Recommendations 1986), *Pure Appl. Chem.* 59, 691-693 (1987). Revised version reprinted as Chapter 22, this edition.
13. Stereochemical definitions and notations relating to polymers (Provisional), *Pure Appl. Chem.* 51, 1101-1121 (1979). Superseded by Ref. 14.
14. Stereochemical definitions and notations relating to polymers (IUPAC Recommendations 1980), *Pure Appl. Chem.* 53, 733-752 (1981). Reprinted as Chapter 2, this edition. See also Ref. 34.
15. Note on the terminology of the molar masses in polymer science, *Macromol. Chem.* 185, Appendix to No. 1 (1984); *J. Polym. Sci., Polym. Lett. Ed.* 22, 57 (1984); *J. Colloid Interface Sci.* 101, 277 (1984); *J. Macromol. Sci., Chem.* A21, 903-904 (1984); *Br. Polym. J.* 17, 92 (1985).
16. Nomenclature for regular single-strand and quasi-single-strand inorganic and coordination polymers (Provisional), *Pure Appl. Chem.* 53, 2283-2302 (1981). Superseded by Ref. 17.
17. Nomenclature for regular single-strand and quasi-single-strand inorganic and coordination polymers (IUPAC Recommendations 1984), *Pure Appl. Chem.* 57, 149-168 (1985). Reprinted as Chapter II-7 in *Nomenclature of Inorganic Chemistry II – Recommendations 2000*, Royal Society of Chemistry, Cambridge, 2001.
18. Source-based nomenclature for copolymers (IUPAC Recommendations 1985), *Pure Appl. Chem.* 57, 1427-1440 (1985). Reprinted as Chapter 19, this edition.

IUPAC POLYMER DIVISION

19. Definitions of terms relating to individual macromolecules, their assemblies, and dilute polymer solutions (IUPAC Recommendations 1988), *Pure Appl. Chem.* 61, 211-241 (1989). Reprinted as Chapter 3, this edition.
20. Definitions of terms relating to crystalline polymers (IUPAC Recommendations 1988), *Pure Appl. Chem.* 61, 769-785 (1989). Reprinted as Chapter 6, this edition.
21. A classification of linear single-strand polymers (IUPAC Recommendations 1988), *Pure Appl. Chem.* 61, 243-254 (1989).
22. Basic classification and definitions of polymerization reactions (IUPAC Recommendations 1994), *Pure Appl. Chem.* 66, 2483-2486 (1994). Reprinted as Chapter 4, this edition.
23. *Compendium of Macromolecular Nomenclature* (the "Purple Book"), prepared for publication by W. V. Metanomski, Blackwell Scientific Publications, Oxford (1991). (This compendium contains 9 chapters that are, in order, reprints of Refs. 6, 14, 19, 20, 8, 17, 18, 21, and 12.)
24. Graphic representations (chemical formulae) of macromolecules (IUPAC Recommendations 1994), *Pure Appl. Chem.* 66, 2469-2482 (1994). Reprinted as Chapter 18, this edition.
25. Glossary of basic terms in polymer science (IUPAC Recommendations 1996), *Pure Appl. Chem.* 68, 2287-2311 (1996). Reprinted as Chapter 1, this edition.
26. Definition of terms relating to degradation, aging, and related chemical transformations of polymers (IUPAC Recommendations 1996)", *Pure Appl. Chem.* 68, 2313-2323 (1996). Reprinted as Chapter 13, this edition.
27. Definitions of terms relating to the non-ultimate mechanical properties of polymers (IUPAC Recommendations 1998), *Pure Appl. Chem.* 70, 701-754 (1998). Reprinted as Chapter 8, this edition.
28. Definitions of basic terms relating to low-molar-mass and polymer liquid crystals (IUPAC Recommendations 2001), *Pure Appl. Chem.* 73, 845-895 (2001). Reprinted as Chapter 7, this edition.
29. Definitions of basic terms relating to polymer liquid crystals (IUPAC Recommendations 2001), *Pure Appl. Chem.* 74, 493-509 (2002).
30. Nomenclature of regular double-strand (ladder and spiro) organic polymers (IUPAC Recommendations 1993), *Pure Appl. Chem.* 65, 1561-1580 (1993). Reprinted as Chapter 16, this edition.
31. Structure-based nomenclature for irregular single-strand organic polymers (IUPAC Recommendations 1994), *Pure Appl. Chem.* 66, 873-889 (1994). Reprinted as Chapter 17, this edition.
32. Nomenclature of regular single-strand organic polymers (IUPAC Recommendations 2002), *Pure Appl. Chem.* 74, 1921-1956 (2002). Reprinted as Chapter 15, this edition.
33. Source-based nomenclature for non-linear macromolecules and macromolecular assemblies (IUPAC Recommendations 1997), *Pure Appl. Chem.* 69, 2511-2521 (1997). Reprinted as Chapter 20, this edition.
34. Definitions relating to stereochemically asymmetric polymerizations (IUPAC Recommendations 2001), *Pure Appl. Chem.* 74, 915-922 (2002). Reprinted as Chapter 5, this edition.

HISTORY OF TERMINOLOGY AND NOMENCLATURE

35. Generic source-based nomenclature for polymers (IUPAC Recommendations 2001), *Pure Appl. Chem.* 73, 1511-1519 (2001). Errata, *Pure Appl. Chem.* 74, 2019 (2002). Reprinted as Chapter 21, this edition.
36. Definitions of terms relating to reactions of polymers and to functional polymeric materials (IUPAC Recommendations 2003), *Pure Appl. Chem.* 76, 889-906 (2004). Reprinted as Chapter 12, this edition.
37. Definitions of terms related to polymer blends, composites, and multiphase polymeric materials (IUPAC Recommendations 2004), *Pure Appl. Chem.* 76, 1985-2007 (2004). Reprinted as Chapter 9, this edition.
38. Terminology of polymers containing ionizable or ionic groups and of polymers containing ions (IUPAC Recommendations 2006), *Pure Appl. Chem.* 78, 2067-2074 (2006). Reprinted as Chapter 10, this edition.
39. Definitions of terms relating to the structure and processing of sols, gels, networks and inorganic-organic hybrid materials (IUPAC Recommendations 2007), *Pure Appl. Chem.* 79, 1801-1829 (2007). Reprinted as Chapter 11, this edition.

TERMINOLOGY

In the earlier of the original manuscripts, first published in *Pure and Applied Chemistry*, which make up the chapters of this book, it was the practice to cross reference definitions within parentheses as follows: (see Definition xx). Whilst beneficial, this led to extensive interruption to the flow of definitions, so throughout this edition of the *Compendium of Polymer Terminology and Nomenclature*, a practice that is now adopted as standard in the preparation of documents under the auspices of the Subcommittee on Polymer Terminology is used, i.e., terms defined elsewhere in the same chapter are italicized.

A substantial number of definitions in the terminology section are either of physical quantities or are expressed mathematically. In such cases, there are recommended symbols for the quantities and, when appropriate, corresponding SI units. Other terms have common abbreviations. The following format is used to indicate these essential characteristics: **name of term (abbreviation)**, symbol, SI unit: unit. Typical examples are: **tensile stress**, σ, SI unit: Pa; **interpenetrating polymer network (IPN)**. If there are any, alternative names or synonyms follow on the next line, and the definition on the succeeding lines.

1: Glossary of Basic Terms in Polymer Science

CONTENTS

Preamble
1. Molecules and molecular structure
2. Substances
3. Reactions
Alphabetical index of terms

PREAMBLE

In order to present clear concepts it is necessary that idealized definitions be adopted but it is recognized that the realities of polymer science must be faced. Deviations from ideality arise with polymers at both molecular and bulk levels in ways that have no parallel with the ordinary small molecules of organic or inorganic chemistry. Although such deviations are not explicitly taken into account in the definitions below, the terms recommended can usefully be applied to the *predominant* structural features of real polymer molecules, if necessary with self-explanatory, if imprecise, qualifications such as 'essentially....', 'almost completely....', or 'highly....'. Although such expressions lack the rigour beloved by the purist, every experienced polymer scientist knows that communication in this discipline is impossible without them.

Conventionally, the word *polymer* used as a noun is ambiguous; it is commonly employed to refer to both polymer substances and polymer molecules. Henceforth, *macromolecule* is used for individual molecules and *polymer* is used to denote a substance composed of *macromolecules*. *Polymer* may also be employed unambiguously as an adjective, according to accepted usage, e.g., *polymer blend, polymer molecule*.

1 MOLECULES AND MOLECULAR STRUCTURE

1.1 macromolecule
polymer molecule

Molecule of high relative molecular mass, the structure of which essentially comprises the multiple repetition of units derived, actually or conceptually, from molecules of low relative molecular mass.

Note 1: In many cases, especially for synthetic *polymers*, a molecule can be regarded as having a high relative molecular mass if the addition or removal of one or a few of the units has a negligible effect on the molecular properties. This statement fails in the case of certain properties of *macromolecules* which may be critically dependent on fine details of the molecular structure, e.g., the enzymatic properties of polypeptides.

Originally prepared by a working group consisting of A. D. Jenkins (UK), P. Kratochvíl (Czech Republic), R. F. T. Stepto (UK), and U. W. Suter (Switzerland). Reprinted from *Pure Appl. Chem.* **68**, 2287-2311 (1996).

TERMINOLOGY

Note 2: If a part or the whole of the molecule has a high relative molecular mass and essentially comprises the multiple repetition of units derived, actually or conceptually, from molecules of low relative molecular mass, it may be described as either macromolecular or polymeric, or by polymer used adjectivally.

Note 3: In most cases, the polymer can actually be made by direct *polymerization* of its parent monomer but in other cases, e.g., poly(vinyl alcohol), the description 'conceptual' denotes that an indirect route is used because the nominal monomer does not exist.

1.2 oligomer molecule

Molecule of intermediate relative molecular mass, the structure of which essentially comprises a small plurality of units derived, actually or conceptually, from molecules of lower relative molecular mass.

Note 1: A molecule is regarded as having an intermediate relative molecular mass if it has properties which do vary significantly with the removal of one or a few of the units.

Note 2: If a part or the whole of the molecule has an intermediate relative molecular mass and essentially comprises a small plurality of units derived, actually or conceptually, from molecules of lower relative molecular mass, it may be described as oligomeric, or by oligomer used adjectivally.

1.3 monomer molecule

Molecule which can undergo *polymerization*, thereby contributing *constitutional units* to the essential structure of a *macromolecule*.

1.4 regular macromolecule

Macromolecule, the structure of which essentially comprises the repetition of a single *constitutional unit* with all units connected identically with respect to directional sense.

1.5 irregular macromolecule

Macromolecule, the structure of which essentially comprises the repetition of more than one type of *constitutional unit*, or a macromolecule the structure of which comprises *constitutional units* not all connected identically with respect to directional sense.

1.6 linear macromolecule

Macromolecule, the structure of which essentially comprises the multiple repetition in linear sequence of units derived, actually or conceptually, from molecules of low relative molecular mass.

1.7 regular oligomer molecule

Oligomer molecule, the structure of which essentially comprises the repetition of a single *constitutional unit* with all units connected identically with respect to directional sense.

1.8 monomeric unit
     monomer unit
     mer

Largest *constitutional unit* contributed by a single *monomer molecule* to the structure of a *macromolecule* or *oligomer molecule*.

Note: The largest *constitutional unit* contributed by a single *monomer molecule* to the structure of a *macromolecule* or *oligomer molecule* may be described as either monomeric, or by monomer used adjectivally.

GLOSSARY OF BASIC TERMS IN POLYMER SCIENCE

1.9 macromonomer molecule
Macromolecule or *oligomer molecule* that has one end-group which enables it to act as a *monomer molecule*, contributing only a single *monomeric unit* to a *chain* of the final *macromolecule*.

1.10 macroradical
Macromolecule which is also a radical.

1.11 prepolymer molecule
Macromolecule or *oligomer molecule* capable of entering, through reactive groups, into further *polymerization*, thereby contributing more than one *constitutional unit* to at least one type of *chain* of the final *macromolecules*.

Note: A prepolymer molecule capable of entering into further *polymerization* through reactive *end-groups*, often deliberately introduced, is known as a telechelic molecule.

1.12 macromonomeric unit
macromonomer unit
Largest *constitutional unit* contributed by a single *macromonomer molecule* to the structure of a *macromolecule*.

1.13 degree of polymerization
Number of *monomeric units* in a *macromolecule* an *oligomer molecule*, a *block*, or a *chain*.

1.14 constitutional unit
Atom or group of atoms (with pendant atoms or groups, if any) comprising a part of the essential structure of a *macromolecule*, an *oligomer molecule*, a *block*, or a *chain*.

1.15 constitutional repeating unit (CRU)
Smallest *constitutional unit*, the repetition of which constitutes a regular *macromolecule*, a regular *oligomer molecule*, a regular *block*, or a regular *chain*.

1.16 configurational unit
Constitutional unit having at least one site of defined stereoisomerism.

1.17 configurational base unit
Constitutional repeating unit in a regular *macromolecule*, a regular *oligomer molecule*, a regular *block*, or a regular *chain*, the configuration of which is defined at least at one site of stereoisomerism in the *main chain*.

1.18 configurational repeating unit
Smallest set of successive *configurational base units* that prescribes configurational repetition at one or more sites of stereoisomerism in the *main chain* of a regular *macromolecule*, a regular *oligomer molecule*, a regular *block*, or a regular *chain*.

1.19 stereorepeating unit
Configurational repeating unit having defined configurations at all sites of stereoisomerism in the *main chain* of a regular *macromolecule*, a regular *oligomer molecule*, a regular *block*, or a regular *chain*.

1.20 tacticity
Orderliness of the succession of *configurational repeating units* in the *main chain* of a regular *macromolecule*, a regular *oligomer molecule*, a regular *block*, or a regular *chain*.

1.21 tactic macromolecule
Regular *macromolecule* in which essentially all the *configurational (repeating) units* are identical.

1.22 stereoregular macromolecule
Regular *macromolecule* essentially comprising only one species of *stereorepeating unit*.

1.23 isotactic macromolecule
Tactic *macromolecule*, essentially comprising only one species of *configurational base unit*, which has chiral or prochiral atoms in the *main chain* in a unique arrangement with respect to its adjacent *constitutional units*.
Note: In an isotactic *macromolecule*, the *configurational repeating unit* is identical with the *configurational base unit*.

1.24 syndiotactic macromolecule
Tactic *macromolecule*, essentially comprising alternating enantiomeric *configurational base units*, which have chiral or prochiral atoms in the *main chain* in a unique arrangement with respect to their adjacent *constitutional units*.
Note: In a syndiotactic *macromolecule*, the *configurational repeating unit* consists of two *configurational base units* that are enantiomeric.

1.25 atactic macromolecule
Regular *macromolecule* that has an equal number of the possible *configurational base units* in a random sequence distribution.

1.26 block macromolecule
Macromolecule which is composed of *blocks* in linear sequence.

1.27 junction unit
Non-repeating atom or non-repeating group of atoms between *blocks* in a *block macromolecule*.

1.28 graft macromolecule
Macromolecule with one or more species of *block* connected to the *main chain* as side-*chains*, these side-*chains* having constitutional or configurational features that differ from those in the *main chain*.

1.29 stereoblock macromolecule
Block macromolecule composed of stereoregular, and possibly some non-stereoregular, *blocks*.

GLOSSARY OF BASIC TERMS IN POLYMER SCIENCE

1.30 chain
Whole or part of a *macromolecule*, an *oligomer molecule*, or a *block*, comprising a linear or branched sequence of *constitutional units* between two boundary *constitutional units*, each of which may be either an *end-group*, a *branch point*, or an otherwise-designated characteristic feature of the *macromolecule*.
Note 1: Except in linear single-strand *macromolecules*, the definition of a *chain* may be somewhat arbitrary.
Note 2: A cyclic *macromolecule* has no *end-groups* but may nevertheless be regarded as a *chain*.
Note 3: Any number of *branch points* may be present between the boundary units.
Note 4: Where appropriate, definitions relating to *macromolecule* may also be applied to *chain*.

1.31 subchain
Arbitrarily chosen contiguous sequence of *constitutional units*, in a *chain*.
Note: The term *subchain* may be used to define designated subsets of the *constitutional units* in a *chain*.

1.32 linear chain
Chain with no *branch points* between the boundary units.

1.33 branched chain
Chain with at least one *branch point* between the boundary units.

1.34 main chain
     backbone
That linear *chain* to which all other *chains*, *long* or *short* or both, may be regarded as being pendant.
Note: Where two or more *chains* could equally be considered to be the *main chain*, that one is selected which leads to the simplest representation of the molecule.

1.35 end-group
Constitutional unit that is an extremity of a *macromolecule* or *oligomer molecule*.
Note: An *end-group* is attached to only one *constitutional unit* of a *macromolecule* or *oligomer molecule*.

1.36 long chain
Chain of high relative molecular mass.
Note: See Note 1 to Definition 1.1.

1.37 short chain
Chain of low relative molecular mass.
Note: See Note 1 to Definition 1.2.

1.38 single-strand chain
Chain that comprises *constitutional units* connected in such a way that adjacent *constitutional units* are joined to each other through two atoms, one on each *constitutional unit*.

TERMINOLOGY

1.39 single-strand macromolecule
Macromolecule that comprises *constitutional units* connected in such a way that adjacent *constitutional units* are joined to each other through two atoms, one on each *constitutional unit*.

1.40 double-strand chain
Chain consisting of an uninterrupted sequence of rings with adjacent rings having one atom in common (spiro chain) or two or more atoms in common (ladder chain).

1.41 double-strand macromolecule
Macromolecule consisting of an uninterrupted sequence of rings with adjacent rings having one atom in common (spiro macromolecule) or two or more atoms in common (ladder macromolecule).

1.42 spiro chain
Double-strand chain consisting of an uninterrupted sequence of rings, with adjacent rings having only one atom in common.
Note: A spiro chain is a *double-strand chain* with adjacent *constitutional units* joined to each other through three atoms, two on one side and one on the other side of each *constitutional unit*.

1.43 spiro macromolecule
Double-strand macromolecule consisting of an uninterrupted sequence of rings, with adjacent rings having only one atom in common.
Note: A spiro *macromolecule* is a *double-strand macromolecule* with adjacent *constitutional units* joined to each other through three atoms, two on one side and one on the other side of each *constitutional unit*.

1.44 ladder chain
Double-strand chain consisting of an uninterrupted sequence of rings, with adjacent rings having two or more atoms in common.
Note: A ladder chain is a *double-strand chain* with adjacent *constitutional units* joined to each other through four atoms, two on one side and two on the other side of each *constitutional unit*.

1.45 ladder macromolecule
Double-strand macromolecule consisting of an uninterrupted sequence of rings, with adjacent rings having two or more atoms in common.
Note: A ladder macromolecule is a *double-strand macromolecule* with adjacent *constitutional units* joined to each other through four atoms, two on one side and two on the other side of each *constitutional unit*.

1.46 multi-strand chain
Chain that comprises *constitutional units* connected in such a way that adjacent *constitutional units* are joined to each other through more than four atoms, more than two on at least one side of each *constitutional unit*.
Note: A *chain* that comprises *constitutional units* joined to each other through n atoms on at least one side of each *constitutional unit* is termed an n-strand *chain*, e.g., three-strand *chain*. If an uncertainty exists in defining n, the highest possible number is selected.

1.47 multi-strand macromolecule
Macromolecule that comprises *constitutional units* connected in such a way that adjacent *constitutional units* are joined to each other through more than four atoms, more than two on at least one side of each *constitutional unit*.
Note: A *macromolecule* that comprises *constitutional units* joined to each other through n atoms on at least one side of each *constitutional unit* is termed an n-strand *macromolecule*, e.g., three-strand *macromolecule*. If an ambiguity exists in defining n, the highest possible number is selected.

1.48 skeletal structure
Sequence of atoms in the *constitutional unit(s)* of a *macromolecule*, an *oligomer molecule*, a *block*, or a *chain*, which defines the essential topological representation.

1.49 skeletal atom
Atom in a *skeletal structure*.

1.50 skeletal bond
Bond connecting two *skeletal atoms*.

1.51 star macromolecule
Macromolecule containing a single *branch point* from which linear *chains* (arms) emanate.
Note 1: A star macromolecule with n linear *chains* (arms) attached to the *branch point* is termed an n-star macromolecule, e.g., five-star macromolecule.
Note 2: If the arms of a star macromolecule are identical with respect to constitution and *degree of polymerization*, the *macromolecule* is termed a regular star macromolecule.
Note 3: If different arms of a star macromolecule are composed of different *monomeric units*, the *macromolecule* is termed a variegated star macromolecule.

1.52 comb macromolecule
Macromolecule comprising a *main chain* with multiple trifunctional *branch points* from each of which a linear side-*chain* emanates.
Note 1: If the sub*chains* between the *branch points* of the *main chain* and the terminal sub*chains* of the *main chain* are identical with respect to constitution and *degree of polymerization*, and the side-*chains* are identical with respect to constitution and *degree of polymerization*, the *macromolecule* is termed a regular comb macromolecule.
Note 2: If at least some of the *branch points* are of functionality greater than three, the *macromolecule* may be termed a brush *macromolecule*.

1.53 branch
side-chain
pendant chain
Oligomeric or polymeric offshoot from a macromolecular *chain*.
Note 1: An oligomeric branch may be termed a short-chain branch.
Note 2: A polymeric branch may be termed a long-chain branch.

1.54 branch point
Point on a *chain* at which a *branch* is attached.
Note 1: A *branch point* from which f linear *chains* emanate may be termed an f-functional *branch point*, e.g., five-functional *branch point*. Alternatively, the terms trifunctional, tetrafunctional, pentafunctional, etc. may be used, e.g., pentafunctional *branch point*.

TERMINOLOGY

Note 2: A *branch point* in a *network* may be termed a junction point.

1.55 branch unit
Constitutional unit containing a *branch point*.
Note: A *branch* unit from which f linear *chains* emanate may be termed an f-functional *branch* unit, e.g., five-functional *branch* unit. Alternatively, the terms trifunctional, tetrafunctional, pentafunctional, etc. may be used, e.g., pentafunctional *branch* unit.

1.56 pendant group
side-group
Offshoot, neither oligomeric nor polymeric, from a *chain*.

1.57 macrocycle
Cyclic *macromolecule* or a macromolecular cyclic portion of a *macromolecule*.
Note 1: See Note 2 to Definition 1.30.
Note 2: In the literature, the term macrocycle is sometimes used for molecules of low relative molecular mass that would not be considered *macromolecules* as specified in Definition 1.1.

1.58 network
Highly ramified *macromolecule* in which essentially each *constitutional unit* is connected to each other *constitutional unit* and to the macroscopic phase boundary by many permanent paths through the *macromolecule*, the number of such paths increasing with the average number of intervening bonds; the paths must on the average be co-extensive with the *macromolecule*.
Note 1: Usually, and in all systems that exhibit rubber elasticity, the number of distinct paths is very high, but, in most cases, some *constitutional units* exist which are connected by a single path only.
Note 2: If the permanent paths through the structure of a network are all formed by covalent bonds, the term covalent network may be used.
Note 3: The term physical network may be used if the permanent paths through the structure of a network are not all formed by covalent bonds but, at least in part, by physical interactions, such that removal of the interactions leaves individual *macromolecules* or a *macromolecule* that is not a network.

1.59 crosslink
Small region in a *macromolecule* from which at least four *chains* emanate, and formed by reactions involving sites or groups on existing *macromolecules* or by interactions between existing *macromolecules*.
Note 1: The small region may be an atom, a group of atoms, or a number of *branch points* connected by bonds, groups of atoms, or oligomeric *chains*.
Note 2: In the majority of cases, a crosslink is a covalent structure but the term is also used to describe sites of weaker chemical interactions, portions of crystallites, and even physical interactions and entanglements.

1.60 micronetwork
Polymer network that has dimensions of the order 1 nm to 1 μm.

1.61 loose end
Polymer chain within a *network* which is connected by a *junction point* at one end only.

GLOSSARY OF BASIC TERMS IN POLYMER SCIENCE

1.62 block
Portion of a *macromolecule*, comprising many *constitutional units*, that has at least one constitutional or configurational feature which is not present in the adjacent portions.
Note: Where appropriate, definitions relating to *macromolecule* may also be applied to *block*.

1.63 constitutional sequence
Whole or part of a *chain* comprising one or more species of *constitutional unit(s)* in defined sequence.
Note: Constitutional sequences comprising two *constitutional units* are termed diads, those comprising three *constitutional units* triads, and so on. In order of increasing sequence lengths they are called tetrads, pentads, hexads, heptads, octads, nonads, decads, undecads, etc.

1.64 configurational sequence
Whole or part of a *chain* comprising one or more species of *configurational unit(s)* in defined sequence.
Note: Configurational sequences comprising two configurational units are termed diads, those with three such configurational units triads, and so on. In order of increasing sequence lengths they are called tetrads, pentads, hexads, heptads, octads, nonads, decads, undecads, etc.

1.65 polyelectrolyte molecule
Macromolecule in which a substantial portion of the *constitutional units* has ionizable or ionic groups, or both.

1.66 ionomer molecule
Macromolecule in which a small but significant proportion of the *constitutional units* has ionizable or ionic groups, or both.
Note: Some protein molecules may be classified as ionomer molecules

2 SUBSTANCES

2.1 monomer
Substance composed of *monomer molecules*.

2.2 polymer
Substance composed of *macromolecules*.

2.3 oligomer
Substance composed of *oligomer molecules*.
Note: An oligomer obtained by telomerization is often termed a telomer.

2.4 homopolymer
Polymer derived from one species of (real, implicit or hypothetical) *monomer*.
Note 1: Many *polymer*s are made by the mutual reaction of complementary *monomers*. These *monomers* can readily be visualized as reacting to give an 'implicit monomer' or 'hypothetical monomer', the homo*polymerization* of which would give the actual product,

TERMINOLOGY

which can be regarded as a homopolymer. Well-known examples are poly(ethylene terephthalate) and poly(N,N'-hexane-1,6-diyladipamide).

Note 2: Some *polymer*s are obtained by the chemical modification of other *polymer*s such that the structure of the *macromolecules* that constitute the resulting *polymer* can be thought of as having been formed by the homo*polymerization* of a hypothetical *monomer*. These *polymer*s can be regarded as homopolymers. A well-known example is poly(vinyl alcohol).

2.5 copolymer

Polymer derived from more than one species of *monomer*.

Note: Copolymers that are obtained by *copolymerization* of two *monomer* species are sometimes termed bipolymers, those obtained from three *monomers* terpolymers, those obtained from four *monomers* quaterpolymers, etc.

2.6 pseudo-copolymer

Irregular *polymer*, the molecules of which are derived from only one species of *monomer* but which display a variety of structural features more appropriate for description in *copolymer* terms.

Note: Where appropriate, adjectives specifying the types of *copolymer* may be applied to *pseudo-copolymer*. The term statistical pseudo-*copolymer*, for instance, may be used to describe an irregular *polymer* in the molecules of which the sequential distribution of configurational units obeys known statistical laws.

2.7 co-oligomer

Oligomer derived from more than one species of *monomer*.

2.8 pseudo-co-oligomer

Irregular *oligomer*, the molecules of which are derived from only one species of *monomer* but which display a variety of structural features more appropriate for description in co-oligomer terms.

2.9 statistical copolymer

Copolymer consisting of *macromolecules* in which the sequential distribution of the *monomeric units* obeys known statistical laws.

Note: An example of a statistical *copolymer* is one consisting of *macromolecules* in which the sequential distribution of *monomeric units* follows Markovian statistics.

2.10 random copolymer

Copolymer consisting of *macromolecules* in which the probability of finding a given *monomeric unit* at any given site in the *chain* is independent of the nature of the adjacent units.

Note: In a random *copolymer*, the sequence distribution of *monomeric units* follows Bernoullian statistics.

2.11 alternating copolymer

Copolymer consisting of *macromolecules* comprising two species of *monomeric units* in alternating sequence.

Note: An alternating *copolymer* may be considered as a *homopolymer* derived from an implicit or hypothetical *monomer*; see Note 1 to Definition 2.4.

2.12 periodic copolymer
Copolymer consisting of *macromolecules* comprising more than two species of *monomeric units* in regular sequence.

2.13 uniform polymer
monodisperse polymer
Polymer composed of molecules uniform with respect to relative molecular mass and constitution.

Note 1: A *polymer* comprising a mixture of *linear* and *branched chains*, all of uniform relative molecular mass, is not uniform.

Note 2: A *copolymer* comprising linear molecules of uniform relative molecular mass and uniform elemental composition but different sequential arrangements of the various types of *monomeric units*, is not uniform (e.g., a *copolymer* comprising molecules with a random arrangement as well as a *block* arrangement of *monomeric units*).

Note 3: A *polymer* uniform with respect only to either relative molecular mass or constitution may be termed uniform, provided a suitable qualifier is used (e.g., 'a *polymer* uniform with respect to relative molecular mass').

Note 4: The adjectives monodisperse and polydisperse are deeply rooted in the literature, despite the former being non-descriptive and self-contradictory. They are in common usage and it is recognized that they will continue to be used for some time; nevertheless, more satisfactory terms are clearly desirable. After an extensive search for possible replacements, the terms uniform and non-uniform have been selected and they are now the preferred adjectives.

2.14 non-uniform polymer
polydisperse polymer
Polymer comprising molecules non-uniform with respect to relative molecular mass or constitution or both.

Note: See notes 2 and 4 to Definition 2.13.

2.15 regular polymer
Polymer composed of regular *macromolecules*, regular *star macromolecules*, or regular *comb macromolecules*.

Note: A *polymer* consisting of *star macromolecules* with arms identical with respect to constitution and *degree of polymerization* is considered to be regular; see Note 2 to Definition 1.51. Analogously, a *polymer* consisting of *comb macromolecules* with the sub-*chains* between the *branch points* of the *main chain* and the terminal subchains of the *main chain* identical with respect to constitution and *degree of polymerization* and the side-*chains* identical with respect to constitution and *degree of polymerization* is considered to be regular; see Note 1 to Definition 1.52.

2.16 irregular polymer
Polymer composed of irregular *macromolecules*.

2.17 tactic polymer
Polymer composed of tactic *macromolecules*.

2.18 isotactic polymer
Polymer composed of isotactic *macromolecules*.

TERMINOLOGY

2.19 syndiotactic polymer
Polymer composed of syndiotactic *macromolecules*.

2.20 stereoregular polymer
Polymer composed of *stereoregular macromolecules*.

2.21 atactic polymer
Polymer composed of *atactic macromolecules*.

2.22 block polymer
Polymer composed of *block macromolecules*.

2.23 graft polymer
Polymer composed of *graft macromolecules*.

2.24 block copolymer
Copolymer that is a block *polymer*.
Note: In the constituent *macromolecules* of a block copolymer, adjacent *blocks* are constitutionally different, i.e., adjacent *blocks* comprise *constitutional* derived from different species of *monomer* or from the same species of *monomer* but with a different composition or sequence distribution of *constitutional units*.

2.25 graft copolymer
Copolymer that is a *graft polymer*.
Note: In the constituent *macromolecules* of a graft *copolymer*, adjacent *blocks* in the *main chain* or side-*chains*, or both, are constitutionally different, i.e., adjacent *blocks* comprise *constitutional units* derived from different species of *monomer* or from the same species of *monomer* but with a different composition or sequence distribution of *constitutional units*.

2.26 stereoblock polymer
Polymer composed of *stereoblock macromolecules*.

2.27 linear polymer
Polymer composed of linear *macromolecules*.

2.28 linear copolymer
Copolymer composed of linear *macromolecules*.

2.29 single-strand polymer
Polymer, the *macromolecules* of which are single-strand *macromolecules*.

2.30 double-strand polymer
Polymer, the *macromolecules* of which are *double-strand macromolecules*.
Note 1: A *polymer*, the *macromolecules* of which are *spiro macromolecules*, is termed a spiro polymer.
Note 2: A *polymer*, the *macromolecules* of which are *ladder macromolecules*, is termed a ladder polymer.

2.31 double-strand copolymer
Copolymer, the *macromolecules* of which are *double-strand macromolecules*.

GLOSSARY OF BASIC TERMS IN POLYMER SCIENCE

2.32 star polymer
Polymer composed of *star macromolecules*.

2.33 comb polymer
Polymer composed of *comb macromolecules*.
Note: See the Notes to Definitions 1.52 and 2.15.

2.34 branched polymer
Polymer, the molecules of which are *branched chains*.

2.35 macromonomer
Polymer or *oligomer* composed of *macromonomer molecules*.

2.36 mesogenic monomer
Monomer which can impart the properties of liquid crystals to the *polymers* formed by its *polymerization*.

2.37 prepolymer
Polymer or *oligomer* composed of *prepolymer molecules*.

2.38 polyelectrolyte
Polymer composed of *polyelectrolyte molecules*.

2.39 ionomer
Polymer composed of *ionomer molecules*.

2.40 polymer blend
Macroscopically homogeneous mixture of two or more different species of *polymer*.
Note 1: In most cases, blends are homogeneous on scales smaller than several times visual optical wavelengths.
Note 2: For polymer blends, no account is taken of the miscibility or immiscibility of the constituent *polymers*, i.e., no assumption is made regarding the number of phases present.
Note 3: The use of the term polymer alloy for a polymer blend is discouraged.

2.41 network polymer
 polymer network
Polymer composed of one or more *networks*.

2.42 semi-interpenetrating polymer network (SIPN)
Polymer comprising one or more *networks* and one or more *linear* or *branched polymer(s)* characterized by the penetration on a molecular scale of at least one of the *networks* by at least some of the linear or *branched macromolecules*.
Note: Semi-interpenetrating polymer *networks* are distinguished from interpenetrating polymer *networks* because the constituent *linear* or *branched polymers* can, in principle, be separated from the constituent *polymer network(s)* without breaking chemical bonds; they are *polymer blends*.

TERMINOLOGY

2.43 interpenetrating polymer network (IPN)
Polymer comprising two or more *networks* which are at least partially interlaced on a molecular scale but not covalently bonded to each other and cannot be separated unless chemical bonds are broken.
Note: A mixture of two or more pre-formed *polymer networks* is not an IPN.

2.44 polymer-polymer complex
Complex, at least two components of which are different *polymer*s.

3 REACTIONS

3.1 polymerization
Process of converting a *monomer* or a mixture of *monomers* into a *polymer*.

3.2 oligomerization
Process of converting a *monomer* or a mixture of *monomers* into an *oligomer*.
Note: An oligomerization by chain reaction carried out in the presence of a large amount of chain-transfer agent, so that the *end-groups* are essentially fragments of the chain-transfer agent, is termed telomerization.

3.3 homopolymerization
Polymerization in which a *homopolymer* is formed.

3.4 copolymerization
Polymerization in which a *copolymer* is formed.

3.5 co-oligomerization
Oligomerization in which a co-*oligomer* is formed.

3.6 chain polymerization
Chain reaction in which the growth of a *polymer chain* proceeds exclusively by the reaction or reactions between a monomer or *monomers* and a reactive site or reactive sites on the *polymer chain* with regeneration of the reactive site or reactive sites at the end of each growth step.
Note 1: A chain polymerization consists of initiation and propagation reactions, and may also include termination and chain-transfer reactions.
Note 2: The adjective *chain* in chain polymerization denotes a chain reaction rather than a *polymer chain*.
Note 3: Propagation in chain polymerization usually occurs without the formation of small molecules. However, cases exist where a low molar-mass by-product is formed, as in the *polymerization* of oxazolidine-2,5-diones derived from amino acids (commonly termed amino acid *N*-carboxy anhydrides). When a low-molar-mass by-product is formed, the adjective *condensative* is recommended to give the term condensative chain polymerization.
Note 4: The growth steps are expressed by

$$P_x + M \rightarrow P_{x+1} (+ L) \ \{x\} \in \{1, 2, ...\infty \}$$

where P_x denotes the growing *chain* of *degree of polymerization x*, M a *monomer*, and L a low-molar-mass by-product formed in the case of condensative chain polymerization.
Note 5: The term chain polymerization may be qualified further, if necessary, to specify the type of chemical reactions involved in the growth step, e.g., ring-opening chain polymerization, cationic chain polymerization.
Note 6: There exist, exceptionally, some *polymerizations* that proceed *via* chain reactions that, according to the definition, are not chain polymerizations. For example, the *polymerization*

$$HS-X-SH + CH_2=CH-Y-CH=CH_2 \rightarrow -(S-X-S-CH_2-CH_2-Y-CH_2-CH_2)_n-$$

proceeds *via* a radical *chain* reaction with intermolecular transfer of the radical center. The growth step, however, involves reactions between molecules of all *degrees of polymerization* and, hence, the *polymerization* is classified as a polyaddition. If required, the classification can be made more precise and the *polymerization* described as a chain-reaction polyaddition.

3.7 polycondensation
Polymerization in which the growth of *polymer chains* proceeds by condensation reactions between molecules of all *degrees of polymerization*.
Note 1: The growth steps are expressed by
$P_x + P_y \rightarrow P_{x+y} + L$ $\{x\} \in \{1, 2, ...\infty\}; \{y\} \in \{1, 2, ...\infty\}$
where P_x and P_y denote *chains* of *degree of polymerization x* and *y*, respectively, and L denotes a low-molar-mass by-product.
Note 2: The earlier term polycondensation was synonymous with condensation polymerization. It should be noted that the current definitions of polycondensation and condensative *chain polymerization* were both embraced by the earlier term polycondensation.

3.8 polyaddition
Polymerization in which the growth of *polymer chains* proceeds by addition reactions between molecules of all *degrees of polymerization*.
Note 1: The growth steps are expressed by
$P_x + P_y \rightarrow P_{x+y}$ $\{x\} \in \{1, 2, ...\infty\}; \{y\} \in \{1, 2, ...\infty\}$
where P_x and P_y denote *chains* of *degrees of polymerization x* and *y*, respectively.
Note 2: The earlier term addition polymerization embraced both the current concepts of polyaddition and *chain polymerization*, but did not include *condensative chain polymerization*.

3.9 statistical copolymerization
Copolymerization in which a *statistical copolymer* is formed.

3.10 random copolymerization
Copolymerization in which a *random copolymer* is formed.

3.11 alternating copolymerization
Copolymerization in which an *alternating copolymer* is formed.

TERMINOLOGY

3.12 periodic copolymerization
Copolymerization in which a *periodic copolymer* is formed.

3.13 ring-opening polymerization
Polymerization in which a cyclic *monomer* yields a *monomeric unit* which is acyclic or contains fewer cycles than the *monomer*.
Note: If the *monomer* is polycyclic, the opening of a single ring is sufficient to classify the reaction as a ring-opening polymerization.

3.14 ring-opening copolymerization
Copolymerization which is a *ring-opening polymerization* with respect to at least one *monomer*.

3.15 radical polymerization
Chain polymerization in which the kinetic-chain carriers are radicals.
Note: Usually, the growing *chain* end bears an unpaired electron.

3.16 radical copolymerization
Copolymerization which is a radical *polymerization*.

3.17 ionic polymerization
Chain polymerization in which the kinetic-chain carriers are ions or ion-pairs.
Note: Usually, the growing *chain ends* are ions.

3.18 ionic copolymerization
Copolymerization which is an ionic *polymerization*.

3.19 anionic polymerization
Ionic polymerization in which the kinetic-chain carriers are anions.

3.20 cationic polymerization
Ionic polymerization in which the kinetic-chain carriers are cations.

3.21 living polymerization
Chain polymerization from which chain transfer and chain termination are absent.
Note: In many cases, the rate of chain initiation is fast compared with the rate of chain propagation, so that the number of kinetic-chain carriers is essentially constant throughout the *polymerization*.

3.22 living copolymerization
Copolymerization which is a living *polymerization*.

3.23 cyclopolymerization
Polymerization in which the number of cyclic structures in the *constitutional units* of the resulting *macromolecules* is larger than in the *monomer molecules*.

3.24 chain scission
Chemical reaction resulting in the breaking of skeletal bonds.

3.25 depolymerization
Process of converting a *polymer* into its *monomer* or a mixture of *monomers*.

Note: Unzipping is depolymerization occurring by a sequence of reactions, progressing along a *macromolecule* and yielding products, usually *monomer molecules*, at each reaction step, from which *macromolecules* similar to the original can be regenerated.

ALPHABETICAL INDEX OF TERMS

Term	Ref	Term	Ref
alternating copolymer	2.11	decad	1.63, 1.64
alternating copolymerization	3.11	degree of polymerization	1.13
anionic polymerization	3.19	depolymerization	3.25
atactic macromolecule	1.25	diad	1.63, 1.64
atactic polymer	2.21	double-strand chain	1.40
backbone	1.34	double-strand copolymer	2.31
bipolymer	2.5	double-strand macromolecule	1.41
block	1.62	double-strand polymer	2.30
block copolymer	2.24	end-group	1.35
block macromolecule	1.26	*f*-functional branch point	1.54
block polymer	2.22	*f*-functional branch unit	1.55
branch	1.53	graft copolymer	2.25
branch point	1.54	graft macromolecule	1.28
branch unit	1.55	graft polymer	2.23
branched chain	1.33	heptad	1.63, 1.64
branched polymer	2.34	hexad	1.63, 1.64
brush macromolecule	1.52	homopolymer	2.4
cationic polymerization	3.20	homopolymerization	3.3
chain	1.30	interpenetrating polymer network	2.43
chain polymerization	3.6	ionic copolymerization	3.18
chain scission	3.24	ionic polymerization	3.17
co-oligomer	2.7	ionomer	2.39
co-oligomerization	3.5	ionomer molecule	1.66
comb macromolecule	1.52	irregular macromolecule	1.5
comb polymer	2.33	irregular polymer	2.16
condensative chain polymerization	3.6	isotactic macromolecule	1.23
configurational base unit	1.17	isotactic polymer	2.18
configurational repeating unit	1.18	junction point	1.54
configurational sequence	1.64	junction unit	1.27
configurational unit	1.16	ladder chain	1.44
constitutional repeating unit	1.15	ladder macromolecule	1.45
constitutional sequence	1.63	ladder polymer	2.30
constitutional unit	1.14	linear chain	1.32
copolymer	2.5	linear copolymer	2.28
copolymerization	3.4	linear macromolecule	1.6
covalent network	1.58	linear polymer	2.27
crosslink	1.59	living copolymerization	3.22
cyclopolymerization	3.23	living polymerization	3.21

long-chain branch	1.53	polymer-polymer complex	2.44
long chain	1.36	polymeric	1.1
loose end	1.61	polymerization	3.1
macrocycle	1.57	prepolymer molecule	1.11
macromolecular	1.1	pseudo-co-oligomer	2.8
macromolecule	1.1	pseudo-copolymer	2.6
macromonomer	2.35	quaterpolymer	2.5
macromonomer molecule	1.9	radical copolymerization	3.16
macromonomer unit	1.12	radical polymerization	3.15
macromonomeric unit	1.12	random copolymer	2.10
macroradical	1.10	random copolymerization	3.10
main chain	1.34	regular comb macromolecule	1.52
mer	1.8	regular macromolecule	1.4
mesogenic monomer	2.36		
micronetwork	1.60	regular oligomer molecule	1.7
monodisperse polymer	2.13	regular polymer	2.15
monomer	1.8, 2.1	regular star macromolecule	1.51
monomer molecule	1.3	ring-opening copolymerization	3.14
monomer unit	1.8	ring-opening polymerization	3.13
monomeric unit	1.8	segregated star macromolecule	1.51
multi-strand chain	1.46	semi-interpenetrating polymer	
multi-strand macromolecule	1.47	network	2.42
network	1.58	short-chain branch	1.53
network polymer	2.41	short chain	1.37
non-uniform polymer	2.14	side-chain	1.53
nonad	1.63, 1.64	side-group	1.56
n-star macromolecule	1.51	single-strand chain	1.38
n-strand chain	1.46	single-strand macromolecule	1.39
n-strand macromolecule	1.47	single-strand polymer	2.29
octad	1.63, 1.64	skeletal atom	1.49
oligomer	1.2, 2.3	skeletal bond	1.50
oligomer molecule	1.2	skeletal structure	1.48
oligomeric	1.2	spiro chain	1.42
oligomerization	3.2	spiro macromolecule	1.43
pendant chain	1.53	spiro polymer	2.30
pendant group	1.56	star macromolecule	1.51
pentad	1.63, 1.64	star polymer	2.32
pentafunctional	1.54, 1.55	statistical copolymer	2.9
periodic copolymer	2.12	statistical copolymerization	3.9
periodic copolymerization	3.12	statistical pseudo-copolymer	2.6
physical network	1.58	stereoblock macromolecule	1.29
polyaddition	3.8	stereoblock polymer	2.26
polycondensation	3.7	stereoregular macromolecule	1.22
polydisperse polymer	2.14	stereoregular polymer	2.20
polyelectrolyte	2.38	stereorepeating unit	1.19
polyelectrolyte molecule	1.65	subchain	1.31
polymer	1.1, 2.2	syndiotactic macromolecule	1.24
polymer blend	2.40	syndiotactic polymer	2.19
polymer molecule	1.1	tactic macromolecule	1.21
polymer network	2.41	tactic polymer	2.17

tacticity	1.20
telechelic molecule	1.11
telomer	2.3
telomerization	3.2
terpolymer	2.5
tetrad	1.63, 1.64
tetrafunctional	1.54, 1.55
triad	1.63, 1.64
trifunctional	1.54, 1.55
undecad	1.63, 1.64
uniform polymer	2.13
unzipping	3.25
variegated star macromolecule	1.51

2: Stereochemical Definitions and Notations Relating to Polymers

CONTENTS

Preamble
1. Basic definitions
2. Sequences
3. Conformations
4. Supplementary definitions
References

PREAMBLE

A report entitled 'Nomenclature Dealing with Steric Regularity in High Polymers' was issued in 1962 by the Committee on Nomenclature of the Commission on Macromolecules of the IUPAC [1]. Since then, the development of increasingly sophisticated techniques for structure determination has greatly enlarged the field of polymer stereochemistry and this, in turn, has revealed the need for a detailed knowledge of molecular conformations in order to correlate chemical structure with physical properties. The terminology and nomenclature relating to the constitution and configuration of macromolecules has been refined, using structure-based concepts, in documents of this Commission [2,3], while an IUPAC paper [4] on the graphical representation of stereochemical configuration of organic molecules and an IUPAC-IUB document [5] on abbreviations and symbols to be used for the description of the conformations of polypeptide chains have appeared as definitive publications. The present statement is intended to bring up to date the nomenclature of features corresponding to stereoregularity in polymers; it employs the definitions prescribed in [2] and takes into account all the previously elaborated material cited above; it also introduces new concepts dealing with the microstructure of polymer chains, and it proposes a set of definitions and notations for the description of the conformations of polymer molecules. Consistency with documents [4] and [5] has been maintained as far as is possible.

Throughout this document, stereochemical formulae for polymer chains are shown as Fischer projections rotated through 90°, i.e., displayed horizontally rather than vertically, (at variance with [1]) or as hypothetical extended zigzag chains.

The use of rotated Fischer projections corresponds to the practice of using horizontal lines to denote polymer backbone bonds, but it is most important to note that this does not give an immediately visual impression of the zigzag chain. In the projections as used in this document, *at each individual backbone carbon atom* the horizontal lines represent bonds directed below the plane of the paper from the carbon atom while the vertical lines project above the plane of the paper from the carbon atom. Thus, the rotated Fischer projection

---

Originally prepared for publication by A. D. Jenkins (UK). Reprinted from *Pure Appl. Chem.* **53**, 733-752 (1981).

The use of rotated Fischer projections has been retained in the present edition in order to provide a link with, and an explanation of, the bulk of existing published polymer literature, although the present common practice [4] is to depict main-chain bonds in planar, extended zigzag (*all-trans*) conformations, together with a stereochemical representation of side-groups at tetrahedrally-bonded atoms.

Unless otherwise stated, the drawings of configurational base units, configurational repeating units, stereorepeating units, etc., provide information regarding *relative* configurations.

In a polymer molecule, the two portions of the main chain attached to any constitutional unit are, in general, non-identical; consequently, a backbone carbon atom that also bears two different side-groups is considered to be a chiral centre.

The absence from a formula of any one of the horizontal or vertical lines at a chiral or prochiral carbon atom (as in examples on pages 27 and 32), or of E or Z designations[*] at double bonds, indicates that the configuration of that stereoisomeric centre is not known. Also, as in our previous document [2], the convention of orienting polymer structures (and the corresponding constitutional and configurational units) from left to right is used. Thus, the two bracketted constitutional units in

are regarded as different, even though the repetition of either one of them would give the same regular polymer. Some of the definitions presented also appear in a previous paper of this Commission [2], but they are repeated here (with minor grammatical improvement) in order to provide a complete set of stereochemical definitions in a single document.

In order to present clear concepts it is necessary that idealized definitions be adopted but it is recognized that the realities of polymer science must be faced. Deviations from ideality arise with polymers at both molecular and bulk levels in ways that have no parallel with the ordinary small molecules of organic or inorganic chemistry. Although such deviations are not explicitly taken into account in the definitions below, the nomenclature recommended can usefully be applied to the predominant structural features of real polymer molecules, if necessary with self-explanatory, if imprecise, qualifications such as 'almost completely isotactic' or 'highly syndiotactic'. Although such expressions lack the rigour beloved by the purist, every experienced polymer scientist knows that communication in this discipline is impossible without them.

---

[*] With regard to the arrangement of substituents at double bonds (double-bond configuration), it is recommended that E and Z stereodescriptors rather than *cis* and *trans* stereodescriptors be used throughout.

TERMINOLOGY

1 BASIC DEFINITIONS

1.1 configurational unit
Constitutional unit having one or more sites of defined stereoisomerism.

1.2 configurational base unit
Constitutional repeating unit in a regular macromolecule, a regular oligomer molecule, a regular block, or a regular chain, the configuration of which is defined at least at one site of stereoisomerism in the main chain.
Note: In a regular polymer, a configurational base unit corresponds to the constitutional repeating unit.

1.3 configurational repeating unit
Smallest set of successive *configurational base units* that prescribes configurational repetition at one or more sites of stereoisomerism in the main chain of a regular macromolecule, a regular oligomer molecule, a regular block, or a regular chain.

1.4 stereorepeating unit
Configurational repeating unit having defined configurations at all sites of stereoisomerism in the main chain of a regular macromolecule, a regular oligomer molecule, a regular block, or a regular chain.
Note: Two *configurational units* that correspond to the same constitutional unit are considered to be *enantiomeric* if they are non-superposable mirror images. Two non-superposable *configurational units* that correspond to the same constitutional unit are considered to be *diastereomeric* if they are *not* mirror images.

Examples:
In the regular polymer molecule $-[CH(CH_3)-CH_2-]_n$, polypropene, the constitutional repeating unit is $-CH(CH_3)-CH_2-$ and the corresponding *configurational base units* are

$$\begin{array}{cc} \text{H} & \text{CH}_3 \\ | & | \\ -\text{C}-\text{CH}_2- \quad \text{and} \quad -\text{C}-\text{CH}_2- \\ | & | \\ \text{CH}_3 & \text{H} \end{array}$$

(1) (2)

The *configurational base units* (1) and (2) are enantiomeric, while the *configurational units* (1) and (3) cannot be enantiomeric because the constitutional units are different species, according to this nomenclature.

$$\begin{array}{cc} \text{H} & \text{H} \\ | & | \\ -\text{C}-\text{CH}_2- & -\text{CH}_2-\text{C}- \\ | & | \\ \text{CH}_3 & \text{CH}_3 \end{array}$$

(1) (3)

It is immaterial whether (1) or (2) is taken as the *configurational repeating unit* and stercorepeating unit of isotactic polypropene (see Definition 1.7); this is so because the two infinite chains, one built up of identical *configurational units* (1) and the other built up of identical *configurational units* (2), are *not* enantiomeric and differ only in the chain orientation. Within each pair of units,

$$\text{such as} \quad \begin{array}{c} H \\ | \\ -C- \\ | \\ CH_3 \end{array} \;,\; \begin{array}{c} CH_3 \\ | \\ -C- \\ | \\ H \end{array} \quad \text{or} \quad \begin{array}{c} H \\ | \\ -C-CH_2-C- \\ | \quad\quad\quad | \\ R \quad\quad\quad R \end{array} \begin{array}{c} H \\ | \\ \\ | \\ \end{array} \;,\; \begin{array}{c} R \\ | \\ -C-CH_2-C- \\ | \quad\quad\quad | \\ H \quad\quad\quad H \end{array} \begin{array}{c} R \\ | \\ \\ | \\ \end{array}$$

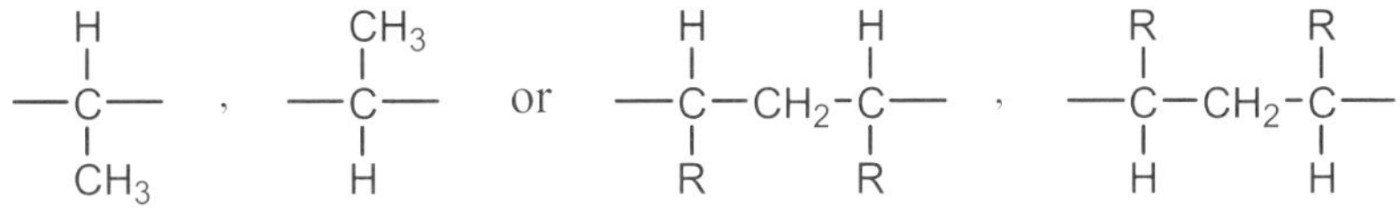

the components are enantiomeric since they are non-superposable mirror images, as defined above. However, with the constitutional unit –CHR–CHR'–CH$_2$–CH$_2$–, the two corresponding *configurational units*

$$\begin{array}{c} H \quad H \\ | \quad | \\ -C-C-CH_2-CH_2- \\ | \quad | \\ R \quad R' \end{array}$$

and

$$\begin{array}{c} H \quad R' \\ | \quad | \\ -C-C-CH_2-CH_2- \\ | \quad | \\ R \quad H \end{array}$$

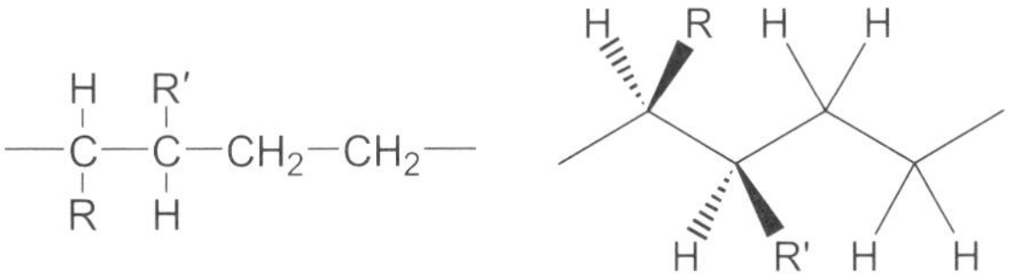

are diastereomeric. The units

$$-CH=CH-\underset{CH_3}{\overset{H}{\underset{|}{C}}}-CH_2- \quad \text{and} \quad -CH=CH-\underset{H}{\overset{CH_3}{\underset{|}{C}}}-CH_2-$$
$$\quad (E) \quad\quad\quad\quad\quad\quad\quad\quad\quad\quad (E)$$

are enantiomeric. The units

$$-CH=CH-\underset{CH_3}{\overset{H}{\underset{|}{C}}}-CH_2- \quad \text{and} \quad -CH=CH-\underset{H}{\overset{CH_3}{\underset{|}{C}}}-CH_2-$$
$$\quad (E) \quad\quad\quad\quad\quad\quad\quad\quad\quad\quad (Z)$$

are not enantiomeric, but diastereomeric. The simplest possible stereorepeating units in a stereoregular polypropene are

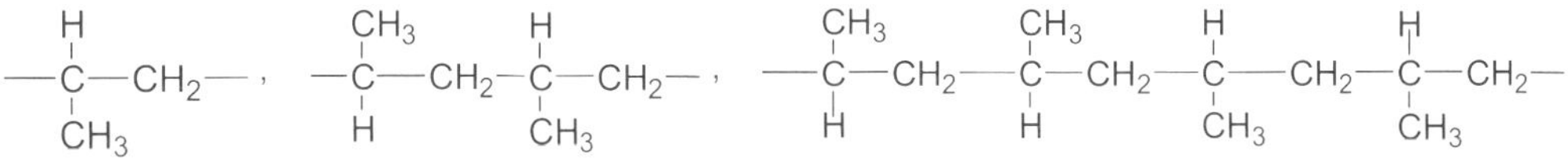

$$\begin{array}{c} H \\ | \\ -C-CH_2- \\ | \\ CH_3 \end{array} ,\; \begin{array}{cc} CH_3 & H \\ | & | \\ -C-CH_2-C-CH_2- \\ | & | \\ H & CH_3 \end{array} ,\; \begin{array}{cccc} CH_3 & CH_3 & H & H \\ | & | & | & | \\ -C-CH_2-C-CH_2-C-CH_2-C-CH_2- \\ | & | & | & | \\ H & H & CH_3 & CH_3 \end{array}$$

and the corresponding *stereoregular polymers* are

TERMINOLOGY

$$\left[\begin{array}{c} H \\ -C-CH_2- \\ CH_3 \end{array}\right]_n , \quad \left[\begin{array}{ccc} CH_3 & & H \\ -C-CH_2-C-CH_2- \\ H & & CH_3 \end{array}\right]_n , \quad \left[\begin{array}{cccc} CH_3 & CH_3 & H & H \\ -C-CH_2-C-CH_2-C-CH_2-C-CH_2- \\ H & H & CH_3 & CH_3 \end{array}\right]_n$$

(4)   (5)   (6)

an isotactic polymer, see Definition 1.7   a syndiotactic polymer, see Definition 1.8   a hypothetical heterotactic polymer, see Section 2.2

1.5 tactic polymer
Regular polymer, the molecules of which have essentially all identical *configurational repeating units*.

1.6 tacticity
Orderliness of the succession of *configurational repeating units* in the *main chain* of a regular *macromolecule*, a regular *oligomer molecule*, a regular *block*, or a regular *chain*..
Note: For the definition of *degree of tacticity*, see Section 4.

1.7 isotactic polymer
Regular polymer, comprising tactic macromolecules having essentially only one species of *configurational base unit* which has chiral or prochiral atoms in the main chain in a unique arrangement with respect to its adjacent constitutional units.
Note 1: In an isotactic polymer, the *configurational repeating unit* is identical with the *configurational base unit*.
Note 2: An isotactic polymer consists of *meso* diads.

1.8 syndiotactic polymer
Regular polymer, comprising tactic macromolecules having alternating enantiomeric *configurational base units*, which have chiral or prochiral atoms in the main chain in a unique arrangement with respect to their adjacent constitutional units.
Note 1: In a syndiotactic macromolecule, the *configurational repeating unit* consists of two *configurational base units* that are enantiomeric.
Note 2: A syndiotactic macromolecule consists of racemo diads.

1.9 stereoregular polymer
Regular polymer, the molecules of which essentially comprise only one species of stereorepeating unit

1.10 atactic polymer
Regular polymer, the molecules of which have equal numbers of the possible *configurational base units* in a random sequence distribution.

Examples:
For the polymer $-[CH(COOR)CH(CH_3)]_n-$, if only the ester-bearing main-chain site in each constitutional repeating unit has defined stereochemistry, the *configurational repeating unit* is (7) and the corresponding *isotactic polymer* is (8).

STEREOCHEMICAL DEFINITIONS

$$-\overset{\overset{H}{|}}{\underset{\underset{COOR}{|}}{C}}-CH(CH_3)-$$

(7)

$$\left[-\overset{\overset{H}{|}}{\underset{\underset{COOR}{|}}{C}}-CH(CH_3)-\right]_n$$

(8)

In the corresponding syndiotactic case, the *configurational repeating unit* is (9) and the *syndiotactic polymer* is (10):

$$-\overset{\overset{H}{|}}{\underset{\underset{COOR}{|}}{C}}-CH(CH_3)-\overset{\overset{COOR}{|}}{\underset{\underset{H}{|}}{C}}-CH(CH_3)-$$

(9)

$$\left[-\overset{\overset{H}{|}}{\underset{\underset{COOR}{|}}{C}}-CH(CH_3)-\overset{\overset{COOR}{|}}{\underset{\underset{H}{|}}{C}}-CH(CH_3)-\right]_n$$

(10)

As the definition of a *stereoregular polymer* (see Definitions 1.4 and 1.9) requires that the configuration be defined at all sites of stereoisomerism, structures (8) and (10) do not represent *stereoregular polymers*. The same is true of (11) and (12), which differ from (8) and (10) in that the sites of specified and unspecified configuration have been interchanged.

$$\left[-CH(COOR)-\overset{\overset{H}{|}}{\underset{\underset{CH_3}{|}}{C}}-\right]_n$$

(11)

$$\left[-CH(COOR)-\overset{\overset{H}{|}}{\underset{\underset{CH_3}{|}}{C}}-CH(COOR)-\overset{\overset{CH_3}{|}}{\underset{\underset{H}{|}}{C}}-\right]_n$$

(12)

Examples (4), (5), (6), (8), (10), (11) and (12) are *tactic polymers*. A *stereoregular polymer* is always a *tactic polymer*, but a *tactic polymer* is not always stereoregular because a *tactic polymer* need not have all sites of stereoisomerism defined.

Further examples of *tactic polymers* are:

$$\left[-O-\overset{\overset{CH_3}{|}}{\underset{\underset{H}{|}}{C}}-CH_2-\right]_n$$

isotactic poly[oxy(1-methylethane-1,2-diyl)]*

$$\left[-\overset{\overset{CH_3}{|}}{\underset{\underset{H}{|}}{C}}-\right]_n$$

isotactic poly(methylmethylene)

---

* Formerly, the structure-based name, poly[oxy(1-methylethylene)], would have been acceptable for this molecule but longer is (see page 257 for an explanation of traditional names, structure-based names and source-based names of polymers).

TERMINOLOGY

syndiotactic poly(methylmethylene)

Note: Structure-based names of *tactic polymers* are formed before the application of adjectives designating *tacticity*; thus, 'syndiotactic poly(methylmethylene)' is preferred to 'syndiotactic poly(1,2-dimethylethane-1,2-diyl)' because a shorter repeating unit is identified, in conformity with the rules in Ref. 3.

On atactic polymers:
Note: As the definition above indicates, a regular polymer, the *configurational base units* of which contain one site of stereoisomerism only, is atactic if it has equal numbers of the possible types of *configurational base units* arranged in a random distribution. If the constitutional repeating unit contains more than one site of stereoisomerism, the polymer may be atactic with respect to only one type of site if there are equal numbers of the possible configurations of that site arranged in a random distribution.

Examples:

constitutional repeating unit

—CH(CH₃)CH₂—

configurational base units (randomly distributed in an *atactic polymer*)

constitutional repeating unit

—CH=CH–CH₂–CH₂—

configurational base units (randomly distributed in an *atactic polymer*)

A polymer such as $\pm$CH=CH–CH(CH₃)–CH₂$\pm_n$, which has two main-chain sites of stereoisomerism, may be atactic with respect to the double bond only, with respect to the chiral atom only or with respect to both centres of stereoisomerism. If there is a random distribution of equal numbers of units in which the double bond is *cis* and *trans*, the polymer is atactic with respect to the double bond, and if there is a random distribution of equal numbers of units containing the chiral atom in the two possible configurations, the polymer is atactic with respect to the chiral atom. The polymer is completely atactic when it contains, in a random distribution, equal numbers of the four possible *configurational base units* which have defined stereochemistry at both sites of stereoisomerism.

In addition to *isotactic, syndiotactic* and *atactic polymers* (and other well-defined types of *tactic polymers*), there exists the whole range of possible arrangements between the completely ordered and the completely random distributions of *configurational base units*, and it is necessary to employ the concept of degree of *tacticity* (see Section 4) to describe such systems.

1.11 stereospecific polymerization
Polymerization in which a *tactic polymer* is formed. However, polymerization in which stereoisomerism present in the monomer is merely retained in the polymer is not to be regarded as stereospecific. For example, the polymerization of a chiral monomer, e.g., (R)-propylene oxide ((R)-methyloxirane), with retention of configuration is not considered to be a stereospecific reaction; however, selective polymerization, with retention, of one of the enantiomers present in a mixture of (R)- and (S)-propylene oxide molecules is so classified.

1.12 ditactic polymer
Tactic polymer that contains two sites of defined stereoisomerism in the main chain of the *configurational base unit*.

Examples:

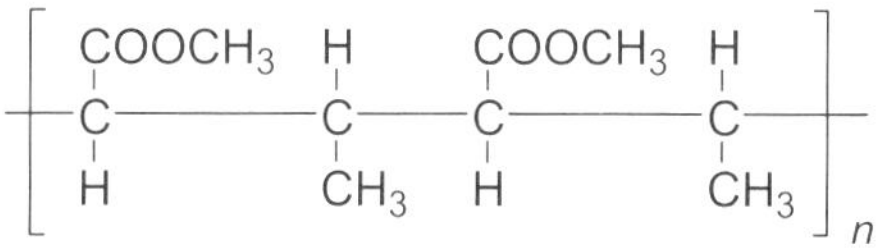

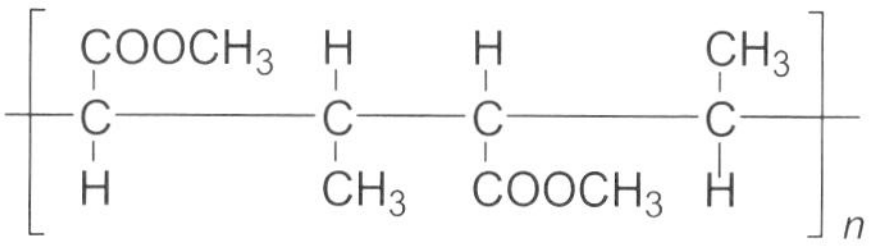

$$\left[\begin{array}{cccc} \text{COOCH}_3 & \text{H} & \text{COOCH}_3 & \text{H} \\ -\text{C} & -\text{C} & -\text{C} & -\text{C}- \\ \text{H} & \text{CH}_3 & \text{H} & \text{CH}_3 \end{array}\right]_n \;,\; \left[\begin{array}{cccc} \text{COOCH}_3 & \text{H} & \text{H} & \text{CH}_3 \\ -\text{C} & -\text{C} & -\text{C} & -\text{C}- \\ \text{H} & \text{CH}_3 & \text{COOCH}_3 & \text{H} \end{array}\right]_n$$

are both ditactic.

1.13 tritactic polymer
Tactic polymer that contains three sites of defined stereoisomerism in the main chain of the *configurational base unit*.

Example:

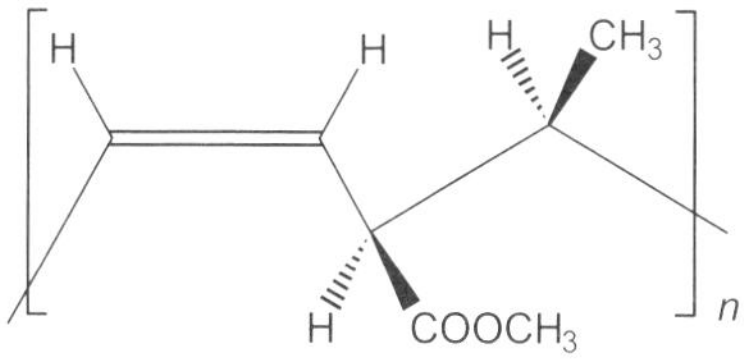

poly[(Z)-(3S)-3-(methoxycarbonyl)-(4S)-4-methylbut-1-ene-1,4-diyl]

1.14 diisotactic polymer
Isotactic polymer that contains two chiral or prochiral atoms with defined stereochemistry in the main chain of the *configurational base unit*.

TERMINOLOGY

1.15 disyndiotactic polymer
Syndiotactic polymer that contains two chiral or prochiral atoms with defined stereochemistry in the main chain of the *configurational base unit*.

Examples:

$$\left[\begin{array}{cc} \text{COOCH}_3 & \text{H} \\ -\text{C}- & -\text{C}- \\ \text{H} & \text{CH}_3 \end{array} \right]_n \quad , \quad \left[\begin{array}{cc} \text{COOCH}_3 & \text{CH}_3 \\ -\text{C}- & -\text{C}- \\ \text{H} & \text{H} \end{array} \right]_n$$

diisotactic

$$\left[\begin{array}{cccc} \text{COOCH}_3 & \text{H} & \text{H} & \text{CH}_3 \\ -\text{C}- & -\text{C}- & & \\ \text{H} & \text{CH}_3 & \text{COOCH}_3 & \text{H} \end{array} \right]_n$$

disyndiotactic*

A polymer with the repeating unit

$$-\text{CH}=\text{CH}-\underset{\underset{\text{H}}{|}}{\overset{\overset{\text{CH}_3}{|}}{\text{C}}}-\text{CH}_2-\text{CH}=\text{CH}-\underset{\underset{\text{CH}_3}{|}}{\overset{\overset{\text{H}}{|}}{\text{C}}}-\text{CH}_2-$$
(Z)   (Z)

is ditactic and may be described as syndiotactic (see Definition 1.8), but it is not disyndiotactic.

The relative configuration of adjacent, constitutionally non-equivalent, carbon atoms can be specified as erythro or threo, as appropriate, by adding the required prefix to the terms 'diisotactic' and 'disyndiotactic', as necessary (see Section 2.2).

1.16 cistactic polymer
Tactic polymer in which the main-chain double bonds of the *configurational base units* are entirely in the Z configuration.

1.17 transtactic polymer
Tactic polymer in which the main-chain double bonds of the *configurational base units* are entirely in the E configuration.
Note 1: Terms referring to the *tacticity* of polymers (tactic, ditactic, tritactic, isotactic, cistactic, etc.) can also be applied with similar significance to chains, sequences, blocks, etc.

---

*

$$\left[\begin{array}{cccc} \text{COOCH}_3 & \text{CH}_3 & \text{H} & \text{H} \\ -\text{C}- & -\text{C}- & -\text{C}- & -\text{C}- \\ \text{H} & \text{H} & \text{COOCH}_3 & \text{CH}_3 \end{array} \right]_n$$

does not represent a different disyndiotactic polymer.

Note 2: Terms defining stereochemical arrangements are to be italicized only when they form part of the name of a polymer; the use of such terms as adjectives, even when immediately preceding names, does not require italics. This practice is illustrated in the examples below.

Examples:

isotactic poly[(*E*)-3-methylbut-1-ene-1,4-diyl];
transisotactic poly(3-methylbut-1-ene-1,4-diyl)[*]

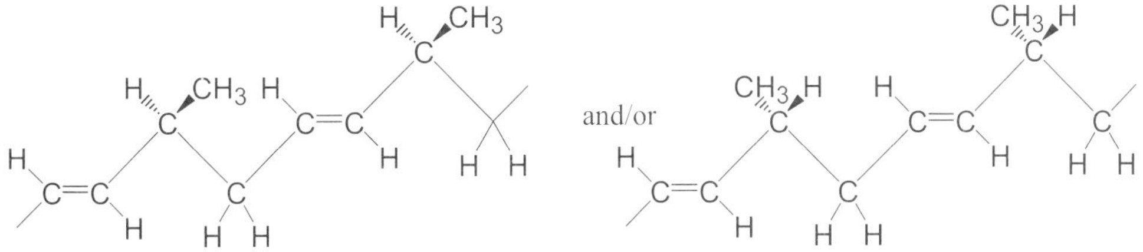

diisotactic poly[*threo*-(*E*)-3-(methoxycarbonyl)-4-methylbut-1-ene-1,4-diyl];
transthreodiisotactic poly[3-(methoxycarbonyl)-4-methylbut-1-ene-1,4-diyl][*]

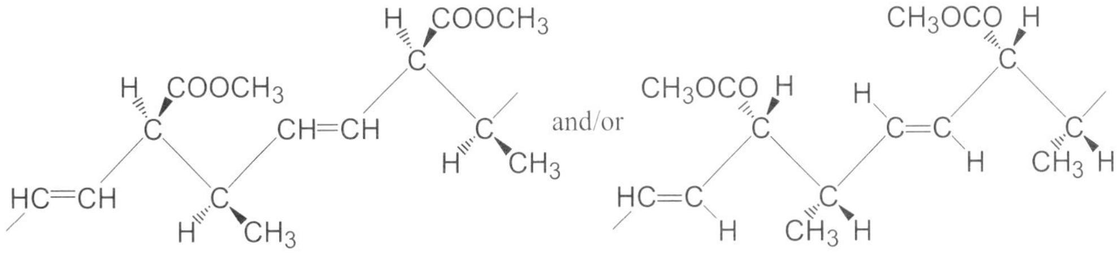

1.18   block
Portion of a *macromolecule*, comprising many *constitutional units*, that has at least one constitutional or configurational feature which is not present in the adjacent portions.

1.19   tactic block
Regular block that can be described by only one species of *configurational repeating unit* in a single sequential arrangement.

1.20   atactic block
Regular block that has equal numbers of the possible *configurational base units* in a random sequence distribution.

1.21   stereoblock
Regular block that can be described by one species of stereorepeating unit in a single sequential arrangement.

1.22   tactic block polymer
Polymer, the molecules of which consist of tactic blocks connected linearly.

---

[*] Structure-based name; either may be used.

TERMINOLOGY

1.23 stereoblock polymer
Polymer, the molecules of which consist of stereoblocks, and possibly some non-stereoregular blocks.

Examples:
Tactic block polymer
$-A_k-B_l-A_m-B_n-$
where A and B are, for example,

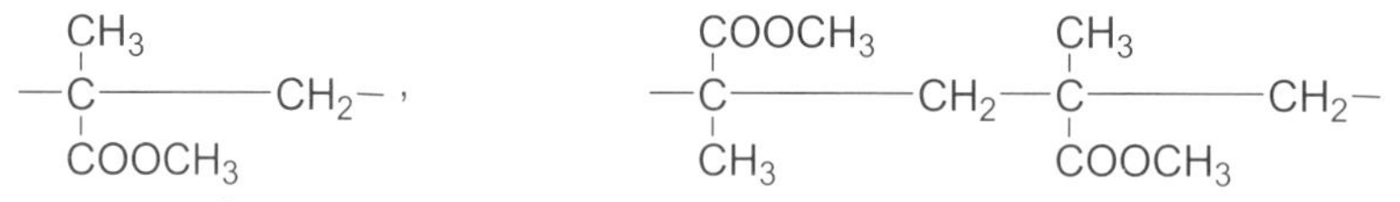

$$-\underset{\underset{\text{COOCH}_3}{|}}{\overset{\overset{\text{CH}_3}{|}}{C}}-CH_2- \text{ (A)} , \quad -\underset{\underset{\text{CH}_3}{|}}{\overset{\overset{\text{COOCH}_3}{|}}{C}}-CH_2-\underset{\underset{\text{COOCH}_3}{|}}{\overset{\overset{\text{CH}_3}{|}}{C}}-CH_2- \text{ (B)}$$

In this case the blocks are stereoblocks but the block polymer is not a block copolymer because all the units derive from a single monomer.

In the folowing example of a regular polypropene chain, the stereoblocks are denoted by |____| . Here, the sequence of identical relative configurations of adjacent units that characterizes the stereoblock is terminated at each end of the block. Note that ┆----┆ represents a *configurational sequence*, which may or may not be identical with a *stereoblock*.

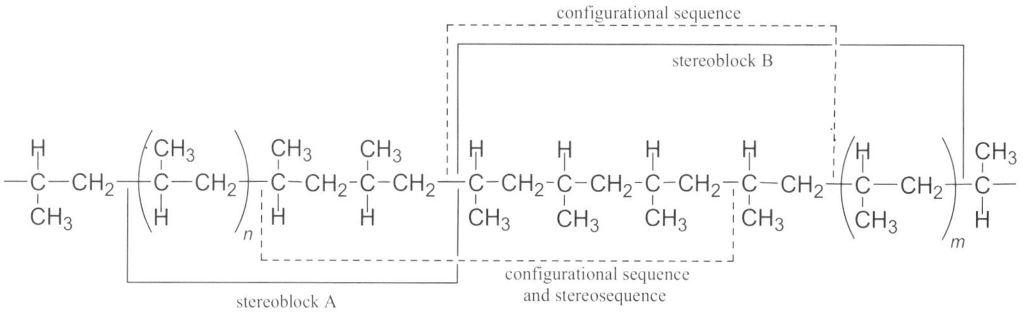

The *configurational sequence* and *stereosequence* coincide in this particular case because there is only one site of stereoisomerism in each constitutional repeating unit (compare Definitions 2.1.3 and 2.1.4).

2 SEQUENCES

2.1 Constitutional and Configurational Sequences
Descriptions of polymer structures revealed by studies of physical properties focus attention on the distribution of local arrangements present in the molecules, and terms useful in this context are defined in this section. (The terms defined here in relation to complete polymer molecules can also be applied to sequences and to blocks, as in Ref. 2, Definition 3.14.)

2.1.1 constitutional sequence
Whole or part of a chain comprising one or more species of constitutional unit(s) in defined sequence.

Example:

$$-CH_2-CH_2-CH_2-CH(CH_3)- \quad -CH_2-CH(CH_3)-CH_2-CH(CH_3)-$$

Note: Constitutional sequences comprising two constitutional units are termed diads, those comprising three constitutional units triads, and so on. In order of increasing sequence lengths they are called tetrads, pentads, hexads, heptads, octads, nonads, decads, undecads, etc.

2.1.2 constitutional homosequence
Constitutional sequence which contains constitutional units of only one species and in one sequential arrangement.

Example:

$$CH(CH_3)-CH_2-CH(CH_3)-CH_2- \quad -[CH(CH_3)-CH_2]_6$$

In these two cases, the constitutional unit $-CH(CH_3)-CH_2-$ can be called the constitutional repeating unit of the homosequence.

2.1.3 configurational sequence
Whole or part of a chain comprising one or more species of *configurational unit(s)* in defined sequence.
Note 1: See examples following Definition 1.23.
Note 2: Configurational sequences comprising two configurational units are termed diads, those with three such configurational units triads, and so on. In order of increasing sequence lengths they are called tetrads, pentads, hexads, heptads, octads, nonads, decads, undecads, etc.

2.1.4 stereosequence
Configurational sequence in which the relative or absolute configuration is defined at all sites of stereoisomerism in the main chain of a polymer molecule.

2.1.5 configurational homosequence
Constitutional homosequence in which the relative or absolute configuration is defined at one or more sites of stereoisomerism in each constitutional unit in the main chain of a polymer molecule.

2.1.6 stereohomosequence
Configurational homosequence in which the relative or absolute configuration is defined at all sites of stereoisomerism in the main chain of a polymer molecule.

2.2 Description of Relative Configurations

erythro and threo structures
Relative configurations at two contiguous carbon atoms in main chains bearing, respectively, substituents a and b (a ≠ b), are designated by the prefix *erythro or threo,* as appropriate, by analogy with the terminology for carbohydrate systems [6] in which the substituents are –OH.

TERMINOLOGY

Examples:

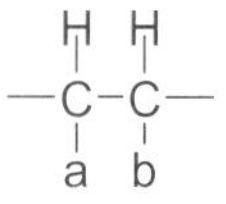

$$\begin{array}{cc} \text{H} & \text{H} \\ -\text{C}-\text{C}- \\ \text{a} & \text{b} \end{array} \qquad \begin{array}{cc} \text{H} & \text{b} \\ -\text{C}-\text{C}- \\ \text{a} & \text{H} \end{array}$$

  erythro 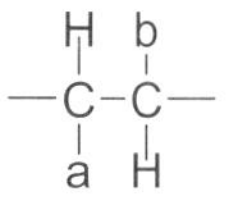   *threo*

Similar systems in which a higher level of substitution exists may be treated analogously if the *erythro* or *threo* designation is employed to denote the relative placements of those two substituents, one for each backbone carbon atom, which rank highest according to the Sequence Rule. Thus, the following hypothetical examples would be designated as indicated:

Examples:

erythro

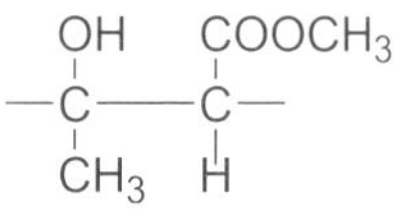

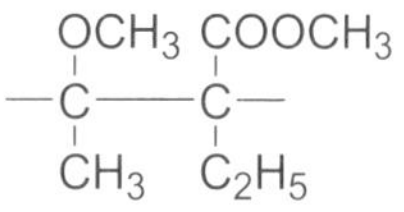

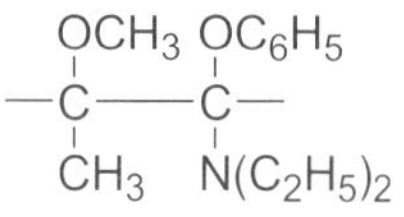

$$\begin{array}{cc} \text{OH} & \text{COOCH}_3 \\ -\text{C}-\!\!-\!\!-\text{C}- \\ \text{CH}_3 & \text{H} \end{array}$$

$$\begin{array}{cc} \text{OCH}_3 & \text{COOCH}_3 \\ -\text{C}-\!\!-\!\!-\text{C}- \\ \text{CH}_3 & \text{C}_2\text{H}_5 \end{array}$$

$$\begin{array}{cc} \text{OCH}_3 & \text{OC}_6\text{H}_5 \\ -\text{C}-\!\!-\!\!-\text{C}- \\ \text{CH}_3 & \text{N(C}_2\text{H}_5)_2 \end{array}$$

threo

$$\begin{array}{cc} \text{OH} & \text{H} \\ -\text{C}-\!\!-\!\!-\text{C}- \\ \text{CH}_3 & \text{COOCH}_3 \end{array}$$

$$\begin{array}{cc} \text{OCH}_3 & \text{C}_2\text{H}_5 \\ -\text{C}-\!\!-\!\!-\text{C}- \\ \text{CH}_3 & \text{COOCH}_3 \end{array}$$

$$\begin{array}{cc} \text{OCH}_3 & \text{N(C}_2\text{H}_5)_2 \\ -\text{C}-\!\!-\!\!-\text{C}- \\ \text{CH}_3 & \text{OC}_6\text{H}_5 \end{array}$$

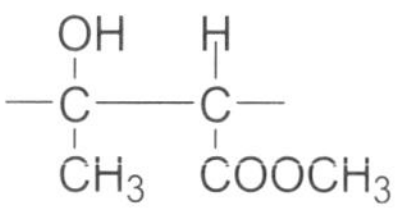

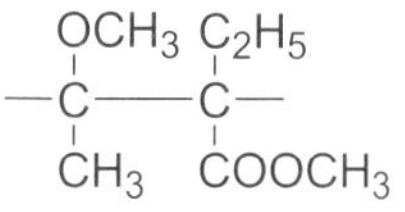

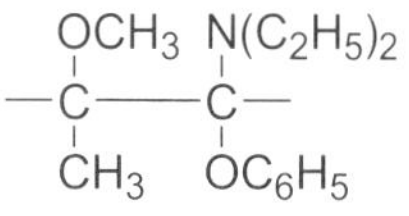

This novel extension of the *erythro/threo* terminology, especially its conjunction with the Sequence Rule, is specifically proposed solely to cope with the problems incurred in describing the steric structures of macromolecules.

STEREOCHEMICAL DEFINITIONS

meso and racemo structures

Relative configurations of consecutive, but not necessarily contiguous, constitutionally equivalent carbon atoms that have a symmetrically constituted connecting group (if any) are designated as *meso* or *racemo*, as appropriate.

abbreviation *m*

meso

abbreviation *r*

racemo

(The symbol $-\!\!\!\diagup\!\!\!\diagdown\!\!\!-$ represents a symmetrically constituted connecting group, such as $-CH_2-$, $-CH_2-CH_2-$, or $-CR_2-CH_2-CR_2-$.)

Note: The structures

and

both have the *meso* relative configuration but the boldly printed carbon atoms in each of the formulae below cannot be considered as in a *meso* arrangement because the connecting group lacks the necessary symmetry.

The term '*racemo*' is introduced here as the logical prefix for the designation of an arrangement that is analogous to racemic, in the sense defined above. It is unfortunate that the meaning of the term 'racemic' current in organic chemistry is not directly applicable to polymers, but the use of the prefix '*racemo*' proposed here should not cause confusion because of the special context. To achieve a full configurational description, it may be necessary to preface the name of a polymer with a compound adjective that combines a term such as *erythro*, *threo*, *meso* or *racemo* with a term such as 'diisotactic' or 'disyndiotactic'.

Examples:

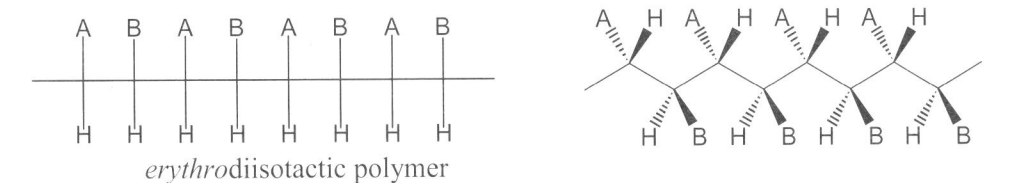

*erythro*diisotactic polymer

TERMINOLOGY

*threo*diisotactic polymer

*di*syndiotactic polymer[*]

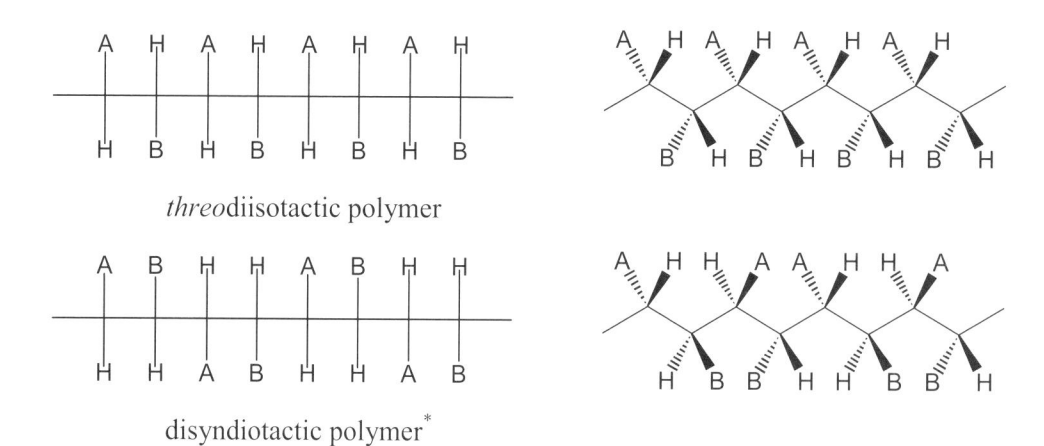

Polymers with chiral centres arising from rings linking adjacent main-chain carbon atoms can be included in this nomenclature:

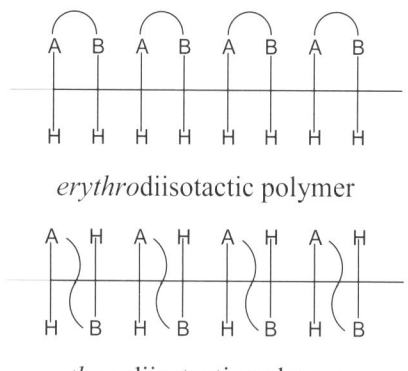

*erythro*diisotactic polymer

*threo*diisotactic polymer

In the last two cases, the chiralities of the asymmetric centres should be designated *R*- or *S*-, if known.

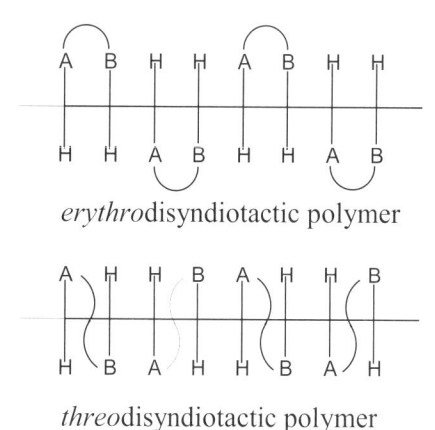

*erythro*disyndiotactic polymer

*threo*disyndiotactic polymer

---

[*] This polymer cannot be expressed as erythrodisyndiotactic or as threodisyndiotactic. Instead:

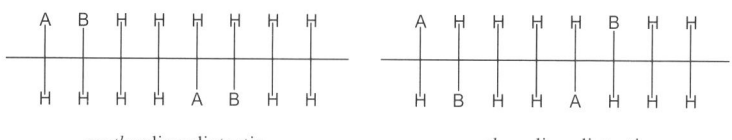

*erythro*disyndiotactic   *threo*disyndiotactic

If the rings are symmetrical:

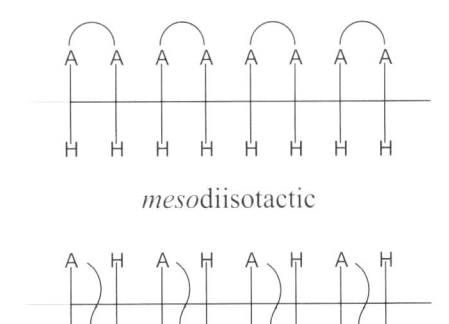

*meso*diisotactic

*racemo*diisotactic

Stereosequences
Stereosequences terminating in tetrahedral stereoisomeric centres at both ends, and which comprise two, three, four, five, etc., consecutive centres of that type, may be called diads, triads, tetrads, pentads, etc., respectively.
   Typical diads are:

   When it is necessary to specify the internal stereochemistry of the group, a prefix is required. In vinyl polymers there are meso *(m)* and racemic *(r)* diads and *mm, mr, rr* triads. The latter may be called isotactic, heterotactic and syndiotactic triads, respectively. *Stereoregular* vinyl polymers can be defined in terms of the regular sequences of diads; thus an isotactic vinyl polymer consists entirely of *m* diads, i.e., it corresponds to the following succession of relative configuration -*mmmmmm*-, whereas a syndiotactic vinyl polymer consists entirely of *r* diads, corresponding to the sequence –*rrrrrrr*-. Similarly, a vinyl polymer consisting entirely of *mr* (= *rm*) triads is called a heterotactic polymer.

3   CONFORMATIONS

3.1   Designation of conformation of polymer molecules

Bond lengths
If a specific A–B bond is denoted as A_i–B_j, the bond length is written $b(A_i, B_j)$. Abbreviated notations, such as b_i, may be used if this meaning is clarified by a diagram.

Bond angles
The bond angle formed by three consecutive atoms

A_i — B_j — C_k

is written τ(A$_i$, B$_j$, C$_k$) which may be abbreviated, if there is no ambiguity, to τ(B$_j$), $τ_j^B$, τB(j), or τ$_j$.

Torsion angles
If a system of four consecutive atoms

$$\begin{array}{c} A \diagdown \diagup D \\ B-C \end{array}$$

is projected onto a plane normal to bond B–C, the angle between the projection of A–B and the projection of C–D is described as the torsion angle of A and D about bond B–C; this angle may also be described as the angle between the plane containing A, B and C and the plane containing B, C and D. The torsion angle is written in full as $θ$(A$_i$, B$_j$, C$_k$, D$_l$,), which may be abbreviated, if there is no ambiguity, to $θ$(B$_j$, C$_k$), $θ$(B$_j$) or $θ_j^B$, etc. In the eclipsed conformation in which the projections of A–B and C–D coincide, $θ$ is given the value 0° (synperiplanar conformation). A torsion angle is considered positive *(+θ)* or negative *(-θ)* depending on whether, when the system is viewed along the central bond in the direction B–C (or C–B), the bond to the front atom A (or D) requires the smaller angle of rotation to the right or to the left, respectively, in order that it may eclipse the bond to the rear atom D (or A); note that it is immaterial whether the system be viewed from one end or the other. According to this definition, a sequence of consecutive positive torsion angles generates a right-handed helix (Helix sense, p. 43). It is to be noted that:

1. torsion angles are measured in the range -180° < $θ$ < +180° rather than from 0° to 360°, so that the relationship between enantiomeric configurations or conformations can be readily appreciated;
2. any Greek letter from the end of the alphabet, except τ, can be used to denote torsion angles; $θ$ or $ω$ of are recommended;
3. abbreviated notations are preferably restricted to bond lengths, bond angles and torsion angles related to main-chain atoms.

Conformations referring to torsion angles $θ$ (A, B, C, D), where A, B, C, D are main-chain atoms, can be described as: *cis* or *synperiplanar* (C); *gauche* or *synclinal* (G); *anticlinal* (A); and *trans* or *antiperiplanar* (T), corresponding to torsion angles within ±30° of, respectively, 0°, ±60°, ± 120° and ± 180°. The letters shown in parentheses (upper case C, G, A, T) are the recommended abbreviations.[*]
The symbols G$^+$, G$^-$ (or A$^+$, A$^-$, for example) refer to torsion angles of similar type but opposite known sign, i.e., ~ + 60°, ~ − 60° (or ~ + 120°, ~ −120°). The notation G, $\overline{G}$; A, $\overline{A}$; (and T, $\overline{T}$; C, $\overline{C}$ whenever the torsion angles are not exactly equal to 180° and 0° respectively) is reserved for the designation of enantiomorph conformations, i.e., conformations of opposite but unspecified sign. Where necessary, a deviation from the proper value of the torsion angle can be indicated by the sign (~), as in the following examples: G(~); $\overline{G}$(~); G$^+$(~); G$^-$(~).

Examples:

The chain conformation of isotactic polypropene in the crystalline state is:
... TGTGTGTG ...

---

[*] Different authors variously use upper and lower case letters in this context. The desire for uniformity necessitates an arbitrary choice between the alternatives, and the upper case has been selected in the belief that it conflicts less with other designations, for example, the use of *c* and *t* in Section 3.2

STEREOCHEMICAL DEFINITIONS

The chain conformation of syndiotactic polypropene in the crystalline state is:
... TTGGTTGG ... or ... TTTTTTTT ...
The chain conformation of a right-handed α-helix is:
... G⁻G⁻(*trans*) G⁻G⁻(*trans*) ...
or ... G⁻(~)G⁻(~)(*trans*)G⁻(~)G⁻(~)(*trans*) ...

The *cis* and *trans* notation may be used to designate rigid dihedral angles such as those occurring with double bonds.
The chain conformation of crystalline poly(1,1-difluoroethene), modification 2, is:
... TGTḠTGT ...
The chain conformation of crystalline poly[(Z)-1-methylbut-1-ene-1,4-diyl] in the α-form is:
... (*trans*) CTA(*trans*) CTĀ ...
The chain conformation of isotactic vinyl polymers in the crystalline state is:
... T(~)G(~)T(~)G(~) ...

A possible conformation of isotactic polypropene in the melt can be described as:

$$\cdots (TG)_{n1}(TT)(\overline{G}T)_{m1} \begin{matrix} (\overline{A}G) \\ \\ (\overline{G}A) \end{matrix} (TG)_{n2}(TT)(\overline{G}T)_{m2} \begin{matrix} (\overline{A}G) \\ \\ (\overline{G}A) \end{matrix} (TG)_{n3} \cdots$$

3.2 Specific terminology for crystalline polymers

The crystallographic identity period parallel to the chain axis should preferably be designated c in descriptions of macromolecular crystallography.

In the description of helices, the following parameters and symbols should be employed: n signifies the number of conformational repeating units per turn (the conformational repeating unit in a crystalline polymer is the smallest unit of given conformation that is repeated through symmetry operations which comprise a translation. In most cases it corresponds to the *configurational repeating unit*); h signifies the unit height, i.e., the translation along the helix axis per conformational repeating unit; t signifies the unit twist, i.e., the angle of rotation about the helix axis per conformational repeating unit.

Examples:
If the number of conformational repeating units along the identity period c is M and the number of turns is N, then:

$n = M/N$
$h = c/M$
$t = 2\pi N/M$

For isotactic polypropene, since $M = 3$, $N = 1$ and $c = 6.50$ Å,

$n = 3$
$h = 2.17$ Å
$t = 2\pi/3$

For poly(oxymethylene), since $M = 9$, $N = 5$ and $c = 17.39$ Å,

TERMINOLOGY

$n = 1.8$
$h = 1.93$ Å
$t = 2\pi(5/9)$.

Helix sense
The right-handed sense of a helix traces out a clockwise rotation moving away from the observer; the left-handed sense of a helix traces out an anticlockwise rotation moving away from the observer, e.g., the ...TG⁺TG⁺TG⁺... helix of isotactic polypropene is left-handed.

Isomorphous and enantiomorphous structures
In the crystalline state, polymer chains are generally parallel to one another but neighbouring chains of equivalent conformation may differ in chirality and/or orientation.
Chains of identical chirality and conformation are *isomorphous*. Chains of opposite chirality but equivalent conformation are *enantiomorphous*.
Examples:
Two ...TG⁺TG⁺TG⁺... helices of isotactic polypropene are isomorphous.
Isotactic polypropene chains of the ...TG⁺TG⁺TG⁺... and ...G⁻TG⁻TG⁻T... types are mutually enantiomorphous.

Isoclined and anticlined structures
With regard to orientation, consider a repeating side-group originating at atom A_1^i, the first atom of the side-group being B_α^i. For certain chain symmetries (helical, for instance) the bond vectors

$\vec{b}(A_1^i, B_\alpha^i)$

have the same components (positive or negative)

$\vec{b} \cdot \vec{c} / |\vec{c}|$

along the c axis for every i.

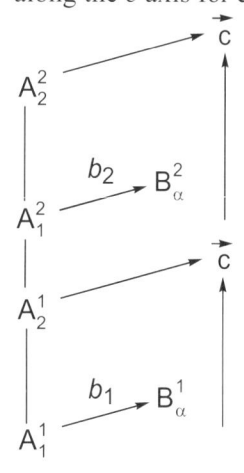

Two equivalent (isomorphous or enantiomorphous) chains in the crystal lattice, having identical components of the bond vectors along c, both positive or both negative, are designated *isoclined*; two equivalent chains having bond vectors along c of the same magnitude but opposite sign are designated *anticlined*.

Examples:
1. Isotactic poly[(*Z*)-3-methylbut-1-ene-1,4-diyl]

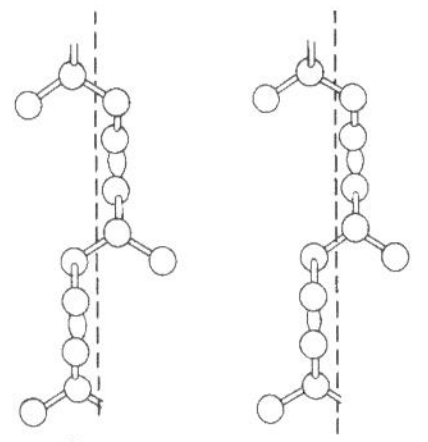

Isoclined isomorphous chains: the two chains have parallel axes and the same orientation of the pendant methyl groups.

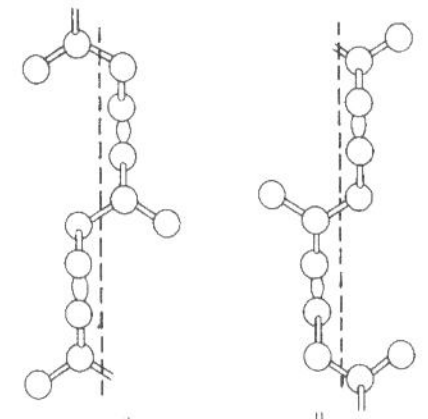

Anticlined isomorphous chains: the two chains have parallel axes and opposite orientation of the pendant methyl groups.

2. Isotactic polypropene

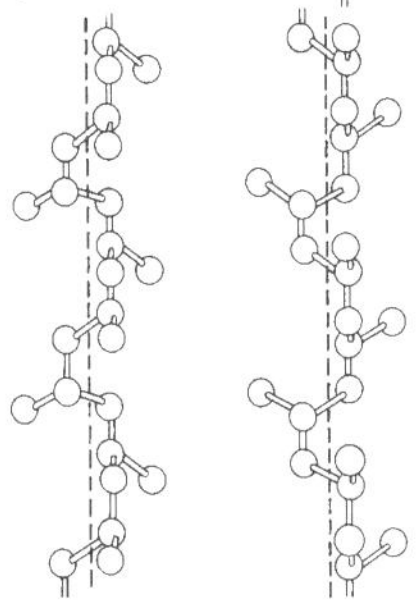

Anticlined enantiomorphous chains: the conformation of A corresponds to a $(TG^-)_n$ bond succession (right-handed helix). The conformation of B corresponds to a $(G^+T)_n$ bond succession (left-handed helix).

Line repetition groups and symmetry elements
To designate linear chain conformations in the crystalline state, the use of line repetition groups [7] is recommended.

First symbol   t   translation
               s   screw repetition

[In this case of screw repetition, the number of conformational repeating residues per turn is included in parentheses, i.e.: $s(11/3)$ or $s(3.67 \pm 0.02)$.]

Second and further symbols. The symmetry elements required to define the line repetition group are suggested in Reference [8]. Possible symmetry elements are:

i centre of symmetry
m plane of symmetry perpendicular to the chain axis
c glide plane parallel to the chain axis
d plane of symmetry parallel to the chain axis
2   two-fold axis of symmetry perpendicular to the chain axis

The possible line repetition groups are listed below, with examples. (The structure-based name is given first, the source-based name second, in each case.)

TERMINOLOGY

Table 1. Chain symmetry of some crystalline polymers [9]

Line repetition group [a]	Polymer (source-based name, structure-based name)
t*l*	*trans*-1,4-polyisoprene, poly[(*E*)-2-methylbut-2-ene-1,4-diyl]
s(A*M/N)	isotactic polypropene, poly(1-methylethane-1,2-diyl) (M/N=3/1, A=2)
s(A*M/N)2	syndiotactic polypropene, poly(1-methylethane-1,2-diyl) (M/N=2/1, A=4, helical modification)
t*m*	poly(heptane-1,7-diylheptanediamide), poly(iminoheptanedioyliminoheptane-1,7-diyl)
t*c*	poly(1,1-difluoroethene), poly(1,1-difluoroethane-1,2-diyl) (modification 2)
t*i*	diisotactic poly[ethene-*alt*-(*E*)-but-2-ene], diisotactic poly(1,2-dimethylbutane-1,4-diyl)
s(5*2/1)*m*	poly(cyclopentene), poly(cyclopentane-1,2-diyl)
s(14*2/1)*d*	poly(hexamethylene adipamide), poly(iminoadipoyliminohexane-1,6-diyl)
t*cm*	syndiotactic 1,2-polybuta-1,3-diene, poly(1-vinylethane-1,2-diyl)
s(1*2/1)*dm*	polyethene, poly(methylene), PE

[a] See the note, Definition 2.9 for explanation of symbols.

4 SUPPLEMENTARY DEFINITIONS

4.1 degrees of triad isotacticity, syndiotacticity and heterotacticity
Fractions of triads in a regular vinyl polymer that are of the *mm*, *rr* and *mr* = *rm* types, respectively.
Note: In cases where triad analysis is not attainable, the diad isotacticity and diad syndiotacticity may be defined as the fractions of diads in a regular-vinyl polymer that are of the *m* and *r* types, respectively.

4.2 degrees of cistacticity and transtacticity
For a regular polymer containing double bonds in the main chain of the constitutional repeating units, these are the fractions of such double bonds that are in the *cis* and *trans* configurations, respectively.

4.3 degree of crystallinity
Fractional amount of crystallinity in a polymer sample.

4.4 lateral order
Order in the side-by-side packing of the molecules of a linear polymer.

4.5 longitudinal order
Order in the atomic positions along the chains of a linear polymer.

REFERENCES

1. M. L. Huggins, G. Natta, V. Desreux, H. Mark (for IUPAC Commission on Macromolecules). 'Report on nomenclature dealing with steric regularity in high polymers', *J. Polym. Sci.* **56**, 153-161 (1962); *Pure Appl. Chem.* **12**, 643-656 (1966).
2. IUPAC. 'Glossary of basic terms in polymer science (IUPAC Recommendations 1996)', *Pure Appl. Chem.* **68**, 2287-2311 (1996). Reprinted as Chapter 1 this volume.
3. IUPAC. 'Nomenclature of regular single-strand organic polymers (IUPAC Recommendations 2002)', *Pure Appl. Chem.* **74**, 1921-1956 (2002). Reprinted as Chapter 15 this volume.
4. IUPAC. 'Graphical representation of stereochemical configuration (IUPAC Recommendations 2006)', *Pure Appl. Chem.* **78**, 1897-1970 (2006).
5. IUPAC-IUB. 'Abbreviations and symbols for the description of the conformation of polypeptide chains (Rules Approved 1974)', *Pure Appl. Chem.* **40**, 291-308 (1974).
6. IUPAC. 'Nomenclature of Carbohydrates (Recommendations 1996)', *Pure Appl. Chem.* **68**, 1919-2008 (1996)
7. A. Klug, F. H. C. Crick, H. W. Wyckoff. *Acta Crystallogr.* **11**, 199 (1958).
8. P. Corradini. 'Chain conformation and crystallinity' in *Stereochemistry of Macromolecules* (Ed. A. Ketley). Marcel Dekker, New York (1968), Vol. III, pp. 1-60.
9. IUPAC, 'Definitions of terms relating to crystalline polymers (IUPAC Recommendations 1989)', *Pure Appl. Chem.* **61**, 769-785 (1989). Reprinted as Chapter 6 this volume.

3: Definitions of Terms Relating to Individual Macromolecules, Their Assemblies, and Dilute Polymer Solutions

CONTENTS

Preamble
1. Individual macromolecules
2. Assemblies of macromolecules
3. Dilute polymer solutions
References

PREAMBLE

This document is part of a series published by the Commission on Macromolecular Nomenclature dealing with definitions for the important terms in polymer science [1-3]. The recommendations presented here deal with such key areas of the physical chemistry of macromolecules as individual macromolecules, their assemblies and dilute polymer solutions; they include recommended terminology for molecular weight, molecular-weight averages, distribution functions, radius of gyration, the Flory-Huggins theory, viscosity of solutions, scattering of radiation by polymers, fractionation, etc.

The reader's attention is especially directed to the new terms 'uniform polymer' and 'nonuniform polymer' which denote polymers composed of molecules that are uniform or nonuniform, respectively, with respect to relative molecular mass and constitution. These terms replaced the widely used, but non-descriptive and self-contradictory terms 'monodisperse polymer' and 'polydisperse polymer'.

1 INDIVIDUAL MACROMOLECULES

1.1 relative molecular mass, M_r
     molecular weight
Ratio of the average mass per formula unit of a substance to 1/12 of the mass of an atom of nuclide ^{12}C.
Note: See Definition 1.2.

1.2 molar mass, M
Mass divided by amount of substance.
Note 1: Molar mass is usually expressed in g mol^{-1} or kg mol^{-1} units. The g mol^{-1} unit is recommended in polymer science, since then the numerical values of the molar mass and the *relative molecular mass* of a substance are equal.

Originally prepared by a working group consisting of P. Kratochvíl (Czech Republic) and U. W. Suter (Switzerland). Reprinted from *Pure Appl. Chem.* **61**, 211-241 (1989), with changes and corrections (2006).

INDIVIDUAL MACROMOLECULES

Note 2: *Relative molecular mass* (relative molar mass, *molecular weight*) is a pure number and must not be associated with any units.
Note 3: The dalton, symbol Da, is an alternative name for the unified atomic mass unit, m_u, symbol u, i.e., $m_u = 1$ u.
Note 4: The terms 'molar' and 'molecular' may also be used for particles consisting of more than one molecule, such as complexes, aggregates, micelles, etc.
Note 5: If there is no danger of confusion, the subscript r in the recommended symbol for the relative molecular mass, M_r, may be omitted.

1.3 degree of polymerization (DP), X
Number of monomeric units in a macromolecule or an oligomer molecule, a block, or a chain.

1.4 thermodynamically equivalent sphere
Sphere, impenetrable to other spheres, displaying the same *excluded volume* as an actual polymer molecule.

1.5 short-range intramolecular interaction
Steric or other interaction involving atoms or groups or both situated nearby in sequence along a polymer chain.
Note 1: The interacting atoms or groups are typically separated by fewer than ten consecutive bonds in a chain.
Note 2: If no confusion can occur, the word 'intramolecular' may be omitted.

1.6 long-range intramolecular interaction
Interaction between segments, widely separated in sequence along a polymer chain, that occasionally approach one another during molecular flexing.
Note 1: This type of interaction is closely related to the *excluded volume of a segment*, the latter quantity reflecting interactions involving segments and solvent molecules.
Note 2: If no confusion can occur, the word 'intramolecular' may be omitted.

1.7 unperturbed dimensions
Dimensions of an actual polymer *random coil* in a *theta state*.

1.8 perturbed dimensions
Dimensions of an actual polymer *random coil* not in a *theta state*.

1.9 radius of gyration, s, $\langle s^2 \rangle^{1/2}$
Parameter characterizing the size of a particle of any shape.
For a rigid particle consisting of mass elements of mass m_i, each located at a distance r_i from the centre of mass, the radius of gyration, s, is defined as the square root of the mass-average of r_i^2 for all the mass elements, i.e.,

$$s = \left(\frac{\sum_i m_i r_i^2}{\sum_i m_i} \right)^{1/2}$$

For a non-rigid particle, an average over all conformations is considered, i.e.,

TERMINOLOGY

$$\langle s^2 \rangle^{1/2} = \langle \sum_i m_i r_i^2 \rangle^{1/2} \Big/ \left(\sum_i m_i \right)^{1/2}$$

Note: The subscript zero is used to indicate *unperturbed dimensions* as in $\langle s^2 \rangle_0^{1/2}$.

1.10 end-to-end vector, $r \rightarrow \underline{r}$

Vector connecting the two ends of a linear macromolecular chain in a particular conformation.

1.11 end-to-end distance, r
Length of the *end-to-end vector*.

1.12 root-mean-square end-to-end distance, $\langle r^2 \rangle^{1/2}$

Square root of the mean-square *end-to-end distance* of a linear macromolecular chain averaged over all conformations of the chain. For a *freely jointed chain* consisting of N segments each of length L, $\langle r^2 \rangle^{1/2} = N^{1/2} L$.

Note 1: The subscript zero is used to indicate *unperturbed dimensions*, as in $\langle r^2 \rangle_0^{1/2}$.

Note 2: If this term is used repeatedly, and if it is not confusing, the abbreviated name 'end-to-end distance' may be used.

1.13 characteristic ratio, C_N (C_∞ when $N \rightarrow \infty$)

Ratio of the mean-square *end-to-end distance*, $\langle r^2 \rangle_0$, of a linear macromolecular chain in a *theta state* to NL^2, where N is the number of rigid sections in the main chain, each of length L; if all of the rigid sections are not of equal length, the mean-square value of L is used, i.e.,

$$L^2 = \sum_i \overline{L_i^2} \Big/ N$$

Note: In simple single-strand chains, the bonds are taken as the rigid sections.

1.14 contour length
Maximum *end-to-end distance* of a linear macromolecular chain.

Note 1: For a single-strand polymer molecule with skeletal bonds all joined by the same value of valence angle, the contour length is equal to the *end-to-end distance* of the chain extended to the all-trans conformation. For chains of complex structures, only approximate values of the contour length may be accessible.

Note 2: The sum of the lengths of all skeletal bonds of a single-strand polymer molecule is occasionally termed 'contour length'. This use of the term in this sense is discouraged.

1.15 random coil, statistical coil
Complete set of spatial arrangements of a chain molecule with a large number of segments that randomly change mutual orientation with time, under conditions in which it is free from external constraints that would affect its conformation.

Note: If the solution of the chain molecules is not in a *theta state*, the segments change mutual orientation only approximately randomly.

1.16 freely jointed chain
Hypothetical linear chain consisting of infinitely thin rectilinear segments uniform in length, each of which can take all orientations in space with equal probability, independently of its neighbours.
Note 1: For models in which the segments are not all uniform in length, the name 'random-walk chain' has been used.
Note 2: In the freely-jointed-chain approach, two or more segments can occupy the same volume simultaneously.

1.17 equivalent chain
Hypothetical *freely jointed chain* with the same *mean-square end-to-end distance* and *contour length* as an actual macromolecular chain in a *theta state*.

1.18 statistical segment
Segment of an actual macromolecular chain which behaves, with respect to some property, virtually as a segment of a *freely jointed chain*.

1.19 freely rotating chain
Hypothetical linear chain molecule, free from *short-range* and *long-range interactions*, consisting of infinitely thin rectilinear segments (bonds) of fixed length, jointed at fixed bond angles; the torsion angles of the bonds can assume all values with equal probability.

1.20 steric factor, σ
Ratio of the *root-mean-square end-to-end distance* of a macromolecular chain with *unperturbed dimensions*, $\langle r^2 \rangle_0^{1/2}$, to that of a *freely rotating chain* with the same structure, $\langle r^2 \rangle_{0,f}^{1/2}$, i.e., $\left(\langle r^2 \rangle_0 / \langle r^2 \rangle_{0,f} \right)^{1/2}$ in the limit of infinite chain length.
Note: The steric factor reflects the effect of hindrance to free rotation.

1.21 worm-like chain
       continuously curved chain
Hypothetical linear macromolecule consisting of an infinitely thin chain of continuous curvature; the direction of curvature at any point is random.
Note 1: The model describes the whole spectrum of chains with different degrees of chain stiffness from rigid rods to *random coils*, and is particularly useful for representing stiff chains.
Note 2: In the literature this chain is sometimes referred to as Porod-Kratky chain.

1.22 persistence length, a
Average projection of the *end-to-end vector* on the tangent to the chain contour at a chain end in the limit of infinite chain length.
Note: The persistence length is the basic characteristic of the *worm-like chain*.

1.23 short-chain branch
Oligomeric offshoot from a macromolecular chain.

1.24 long-chain branch
Polymeric offshoot from a macromolecular chain.

TERMINOLOGY

1.25 branching index, g

Parameter characterizing the effect of *long-chain branches* on the size of a branched molecule in solution and defined as the ratio of the mean-square *radius of gyration* of a branched molecule, $\langle s_b^2 \rangle$, to that of an otherwise identical linear molecule $\langle s_l^2 \rangle$, with the same *relative molecular mass* in the same solvent at the same temperature, i.e., $g = \langle s_b^2 \rangle / \langle s_l^2 \rangle$.

1.26 network

Highly ramified polymer structure in which each constitutional unit is connected to each other constitutional unit and to the macroscopic phase boundary by many permanent paths through the structure, their number increasing with the average number of intervening bonds; these paths must on the average be coextensive with this structure.

Note 1: Usually, and in all systems that exhibit rubber elasticity, the number of distinct paths is very high, but some constitutional units exist, in most cases, which are connected by a single path only. Sometimes, a structure without any multiple path has also been called a network.

Note 2: If the permanent paths through the structure of a network are all formed by covalent bonds, the term covalent network may be used.

Note 3: The term physical network may be used if the permanent paths through the structure of a network are not all formed by covalent bonds but, at least in part, by physical interactions, such that removal of the interactions leaves individual macromolecules or a macromolecule that is not a network.

Note 4: See also Definition 4.1.21 in Chapter 11.

1.27 microgel

Particle of a *gel* of any shape with an equivalent diameter of approximately 0.1 to 100 μm.

1.28 copolymer micelle

Micelle formed by one or more block or graft copolymer molecules in a *selective solvent*.

2 ASSEMBLIES OF MACROMOLECULES

2.1 compositional heterogeneity

Variation in elemental composition from molecule to molecule usually found in copolymers.

2.2 constitutional heterogeneity

Variation in constitution from molecule to molecule in polymers with molecules uniform with respect to elemental composition.

Note: An example is a polymer composed of linear and branched molecules; another example is a statistical copolymer comprising two isomeric constitutional units.

2.3 uniform polymer
 monodisperse polymer

Polymer composed of molecules uniform with respect to *relative molecular mass* and constitution.

Note 1: A polymer comprising a mixture of linear and branched chains, all of uniform *relative molecular mass*, is not uniform.

INDIVIDUAL MACROMOLECULES

Note 2: A copolymer comprising linear molecules of uniform relative molecular mass and uniform elemental composition, but different sequence arrangement of the various types of monomeric units, is not uniform (e.g., a copolymer comprising molecules with random arrangement as well as block arrangement of monomeric units).

Note 3: A polymer uniform with respect only to either relative molecular mass or constitution may be termed 'uniform', provided a suitable qualifier is used (e.g., 'a polymer uniform with respect to relative molecular mass').

Note 4: The adjectives 'monodisperse' and 'polydisperse' are deeply rooted in the literature despite being non-descriptive and self-contradictory. They are in common usage and it is recognized that they will continue to be used for a certain time, nevertheless more satisfactory terms are clearly desirable. After an extensive search for possible replacements, the new terms 'uniform' and 'non-uniform' have been selected and they are now the preferred adjectives.

2.4 non-uniform polymer
 polydisperse polymer

Polymer comprising molecules non-uniform with respect to *relative molecular mass* or constitution or both.

Note: See Definition 2.3, Note 3.

2.5 molar-mass average, $\overline{M}_k$,
 relative-molecular-mass average/molecular-weight average, $\overline{M}_{r,k}$

Any average of the *molar mass* or *relative molecular mass* (*molecular weight*) for a non-uniform polymer. In both symbols, k specifies the type of average.

Note 1: An infinite number of molar-mass averages can in principle be defined, but only a few types of averages are directly accessible experimentally. The most important averages are defined by simple moments of the *distribution functions* and are obtained by methods applied to systems in thermodynamic equilibrium, such as osmometry, light scattering and sedimentation equilibrium. Hydrodynamic methods, as a rule, yield more complex molar-mass averages.

Note 2: Any molar-mass average can be defined in terms of mass fractions or mole fractions. In this document only a few of the important molar-mass averages are given in terms of the mass fractions, w_i, of the species with molar mass M_i. These definitions are most closely related to the experimental determination of molar-mass averages.

2.6 number-average molar mass, $\overline{M}_n$

$$\overline{M}_n = \frac{1}{\sum_i (w_i / M_i)}$$

 number-average relative molecular mass
 number-average molecular weight, $\overline{M}_{r,n}$

$$\overline{M}_{r,n} = \frac{1}{\sum_i (w_i / M_{r,i})}$$

For definitions of symbols, see Definition 2.5.

TERMINOLOGY

2.7 mass-average molar mass, $\overline{M}_w$

$$\overline{M}_w = \sum_i w_i M_i$$

mass-average relative molecular mass, $\overline{M}_{r,w}$
weight-average molecular weight

$$\overline{M}_{r,w} = \sum_i w_i M_{r,i}$$

For definitions of symbols, see Definition 2.5.

2.8 z-average molar mass, $\overline{M}_z$

$$\overline{M}_z = \frac{\sum_i w_i M_i^2}{\sum_i w_i M_i}$$

z-average relative molecular mass, $\overline{M}_{r,z}$
z-average molecular weight

$$\overline{M}_{r,z} = \frac{\sum_i w_i M_{r,i}^2}{\sum_i w_i M_{r,i}}$$

For definitions of symbols, see Definition 2.5.

2.9 (z + 1)-average molar mass, $\overline{M}_{z+1}$

$$\overline{M}_{z+1} = \frac{\sum_i w_i M_i^3}{\sum_i w_i M_i^2}$$

(z + 1)-average relative molecular mass, $\overline{M}_{r,z+1}$
(z + 1)-average molecular weight

$$\overline{M}_{r,z+1} = \frac{\sum_i w_i M_{r,i}^3}{\sum_i w_i M_{r,i}^2}$$

For definitions of symbols, see Definition 2.5.

2.10 viscosity-average molar mass, $\overline{M}_v$

$$\overline{M}_v = \left[\sum_i w_i M_i^\alpha \right]^{1/\alpha}$$

viscosity-average relative molecular mass, $\overline{M}_{r,v}$
viscosity-average molecular weight

$$\overline{M}_{r,v} = \left[\sum_i w_i M_{r,i}^a\right]^{1/a}$$

where a is the exponent in the *Mark-Houwink equation*, $[\eta] = KM^a$; For definitions of the other symbols, see Definition 2.5.

Note: The exponent a is not identical with the adjustable parameter of some of the *distribution functions* or with the *persistence length*.

2.11 **apparent molar mass,** M_{app}

apparent relative molecular mass, $M_{r,app}$

apparent molecular weight

Molar mass, *relative molecular mass*, or *molecular weight* calculated from experimental data without the application of appropriate corrections, such as for finite polymer concentration, association, preferential solvation, *compositional heterogeneity*, *constitutional heterogeneity*.

2.12 **average degree of polymerization,** $\overline{X}_k$

Average of the *degree of polymerization* for a polymer, where k specifies the type of average.

Note: Definitions 2.5-2.10 apply directly to averages of the degree of polymerization when X is substituted for M in the formulae.

2.13 **distribution function**

Normalized function giving the relative amount of a portion of a polymeric substance with a specific value, or a range of values, of a random variable or variables.

Note 1: Distribution functions may be discrete, i.e., take on only certain specified values of the random variable(s), or continuous, i.e., take on any intermediate value of the random variable(s), in a given range. Most distributions in polymer science are intrinsically discrete, but it is often convenient to regard them as continuous or to use distribution functions that are inherently continuous.

Note 2: Distribution functions may be integral (or cumulative), i.e., give the proportion of the population for which a random variable is less than or equal to a given value. Alternatively they may be differential distribution functions (or probability density functions), i.e., give the (maybe infinitesimal) proportion of the population for which the random variable(s) is (are) within a (maybe infinitesimal) interval of its (their) range(s).

Note 3: Normalization requires that: (i) for a discrete differential distribution function, the sum of the function values over all possible values of the random variable(s) be unity; (ii) for a continuous differential distribution function, the integral over the entire range of the random variable(s) be unity; (iii) for an integral (cumulative) distribution function, the function value at the upper limit of the random variable(s) be unity.

2.14 **number-distribution function**

Distribution function in which the relative amount of a portion of a substance with a specific value, or a range of values, of the random variable(s) is expressed in terms of mole fraction.

2.15 mass-distribution function
weight-distribution function

Distribution function in which the relative amount of a portion of a substance with a specific value, or a range of values, of the random variable(s) is expressed in terms of mass fraction.

2.16 Schulz-Zimm distribution

Continuous distribution with the *differential mass-distribution function* of the form

$$f_w(x)\,dx = \frac{a^{b+1}}{\Gamma(b+1)} x^b \exp(-ax)\,dx,$$

where x is a parameter characterizing the chain length, such as *relative molecular mass* or *degree of polymerization*, a and b are positive adjustable parameters, and $\Gamma(b+1)$ is the gamma function of argument $(b+1)$.

2.17 most probable distribution

Discrete distribution with the *differential mass-distribution function* of the form

$$f_w(x) = a^2 x (1-x)^{x-1},$$

For definitions of symbols, see Definition 2.16.
Note 1: For large values of x, the *most probable distribution* converges to the particular case of the *Schulz-Zimm distribution* with $b = 1$.
Note 2: In the literature, this distribution is sometimes referred to as the Flory distribution or the Schulz-Flory distribution.

2.18 Poisson distribution

Discrete distribution with the *differential mass-distribution function* of the form

$$f_w(x) = \frac{x\,e^{-a}\,a^{x-1}}{(a+1)(x-1)!},$$

For definitions of symbols, see Definition 2.16.

2.19 Tung distribution

Continuous distribution with the *differential mass-distribution function* of the form

$$f_w(x)\,dx = abx^{b-1} \exp(-ax^b)\,dx,$$

For definitions of symbols, see Definition 2.16.

2.20 logarithmic normal distribution

Continuous distribution with the *differential mass-distribution function* of the form

$$f_w(x)\,dx = \frac{1}{ax\sqrt{\pi}} \exp\left(-\frac{1}{a^2}\left[\ln\left(\frac{x}{b}\right)\right]^2\right) dx$$

For definitions of symbols, see Definition 2.16.

2.21 polymolecularity correction
Correction applied to relationships between a property and the *molar mass* or *relative molecular mass*, obtained from polymers *non-uniform* with respect to *relative molecular mass*, in order to obtain the corresponding relationship for polymers strictly *uniform* with respect to *relative molecular mass*.

3 DILUTE POLYMER SOLUTIONS

3.1 General and Thermodynamic Terms

3.1.1 dilute solution
Solution in which the sum of the volumes of the domains occupied by the solute molecules or particles is substantially less than the total volume of the solution.
Note: The term 'domain' refers to the smallest convex body that contains the molecule or particle in its average shape.

3.1.2 cross-over concentration, c^*
    overlap concentration
Concentration range at which the sum of the volumes of the domains occupied by the solute molecules or particles in solution is approximately equal to the total volume of that solution.
Note 1: For the meaning of the term 'domain', see the note in Definition 3.1.1.
Note 2: Cross-over concentration is defined as a range because different measurement techniques give different values. The symbol c^* refers usually to amount concentration, but in polymer science it is generally used for mass concentration.

3.1.3 polymer-solvent interaction
Sum of the effects of all intermolecular interactions between polymer and solvent molecules in solution that are reflected in the values of the Gibbs and Helmholtz energies of mixing.

3.1.4 thermodynamic quality of solvent
    quality of solvent
Qualitative characterization of the *polymer-solvent interaction*. A solution of a polymer in a 'better' solvent is characterized by a higher value of the *second virial coefficient* than a solution of the same polymer in a 'poorer' solvent.

3.1.5 theta state
    θ state
State of a polymer solution for which the *second virial coefficient* is zero.

TERMINOLOGY

Note 1: In some respects, a polymer solution in the theta state resembles an ideal solution and the theta state may be referred to as a pseudo-ideal state. However, a solution in the theta state must not be identified with an ideal solution.
Note 2: The solvent involved is often referred to as 'theta solvent' or 'θ solvent'.
Note 3: It is assumed that the *degree of polymerization* of the polymer is high.

3.1.6 theta temperature; θ, SI unit: K
    θ temperature
Temperature at which a solution is in the *theta state*.

3.1.7 virial coefficients, A_i where $i = 1, 2$, etc.
    virial coefficients of the chemical potential
Coefficients in the expansion of the chemical potential of the solvent, μ_s, in powers of the mass concentration, c, of the solute, i.e.,

$$\mu_s - \mu_s^o = -\pi V_s = -RTV_s \left(A_1 c + A_2 c^2 + A_3 c^3 + ... \right),$$

where μ_s^o is the chemical potential of the solvent in the reference state at the temperature of the system and ambient pressure, π is the osmotic pressure and V_s is the partial molar volume of the solvent. In solvents comprising more than one component, the definition applies to any solvent component. The first virial coefficient is the reciprocal *number-average molar mass*, i.e., $A_1 = 1/\overline{M_n}$. The second and higher virial coefficients, $A_2, A_3, ...$, respectively, describe *polymer-solvent* and polymer-polymer interactions.
Note: The factor RT is sometimes included in the virial coefficients.

3.1.8 excluded volume of a segment
Volume from which a segment of a macromolecule in solution effectively excludes all other segments, i.e., those belonging to the same macromolecule as well as those belonging to other macromolecules.
Note: The excluded volume of a segment depends on the Gibbs and Helmholtz energies of mixing of solvent and polymer, i.e., on the *thermodynamic quality of the solvent*, and is not a measure of the geometrical volume of that segment.

3.1.9 excluded volume of a macromolecule
Volume from which a macromolecule in a dilute solution effectively excludes all other macromolecules.
Note: The excluded volume of a macromolecule depends on the Gibbs and Helmholtz energies of mixing of solvent and polymer, i.e., on the *thermodynamic quality of the solvent*, and is not a measure of the geometrical volume of that macromolecule.

3.1.10 expansion factor α_r, α_s, α_η
Ratio of a dimensional characteristic of a macromolecule in a given solvent at a given temperature to the same dimensional characteristic in the *theta state* at the same temperature. The most frequently used expansion factors are: expansion factor of the *mean-square end-to-end distance*, $\alpha_r = (\langle r^2 \rangle / \langle r^2 \rangle_0)^{1/2}$; expansion factor of the *radius of gyration*, $\alpha_s = (\langle s^2 \rangle / \langle s^2 \rangle_0)^{1/2}$; *relative viscosity*, $\alpha_\eta = ([\eta]/[\eta]_0)^{1/3}$, where $[\eta]$ and $[\eta]_0$ are the *intrinsic viscosity* in a given solvent and in the *theta state* at the same temperature, respectively.

Note: Expansion factors defined by different dimensional characteristics are not exactly equal, nor need they have a constant ratio as a function of relative molecular mass.

3.1.11 Flory-Huggins theory
Flory-Huggins-Staverman theory

Statistical thermodynamic mean-field theory of polymer solutions, first formulated independently by Flory, Huggins, and Staverman, in which the thermodynamic quantities of the solution are derived from a simple concept of combinatorial entropy of mixing and a reduced Gibbs-energy parameter, the χ *interaction parameter*.

3.1.12 χ parameter, χ
χ interaction parameter

Numerical parameter employed in the *Flory-Huggins theory*, to account for the contribution of the noncombinatorial entropy of mixing and the enthalpy of mixing to the Gibbs energy of mixing.

3.1.13 preferential sorption
selective sorption

Equilibrium phenomenon, operative in polymer solutions, in multicomponent solvents, and in *polymer networks* swollen by multicomponent solvents, that produces differences in solvent composition in the polymer-containing region and in the pure solvent which is in thermodynamic equilibrium with that region.

3.1.14 selective solvent

Medium that is a solvent for at least one component of a mixture of polymers, or for at least one block of a block or graft polymer, but a non-solvent for the other component(s) or block(s).

3.1.15 co-solvency

Dissolution of a polymer in a solvent comprising more than one component, each component of which by itself is a non-solvent for the polymer.

3.1.16 solubility parameter (of a polymer), δ, SI unit: $Pa^{1/2} = J^{1/2}\ m^{-3/2}$

Parameter used in predicting the *solubility* of non-electrolytes (including polymers) in a given *solvent*. For a substance B:

$$\delta_B = \left(\Delta_{vap} E_{m,B} / V_{m,B}\right)^{1/2}$$

where $\Delta_{vap}E_m$ is the molar energy of vaporization at zero pressure and V_m is the molar volume.

Note 1: Alternative units are $\mu Pa^{1/2} = J^{1/2}\ cm^{-3/2}$ and $cal^{1/2}\ cm^{-3/2}$, where $1\ J^{1/2}\ cm^{-3/2} \approx 2.045\ cal^{1/2}\ cm^{-3/2}$.

Note 2: For a polymer, the value of the solubility parameter is usually taken to be the value of the solubility parameter of the solvent producing the solution with maximum *intrinsic viscosity* or maximum swelling of a *network* of the polymer.

Note 3: For a substance of low molecular weight, the value of the solubility parameter can be estimated most reliably from the enthalpy of vaporization and the molar volume.

Note 4: The solubility of a substance B can be related to the square of the difference between the solubility parameters for supercooled liquid B and solvent at a given temperature, with appropriate allowances for entropy of mixing. Thus a value can be estimated from the solubility of the solid in a series of solvents of known solubility parameter.

3.1.17 isopycnic
Adjective describing components of a multicomponent system with equal partial specific volumes.

3.2 Transport Properties

3.2.1 frictional coefficient, f, SI unit: kg s^{-1}
Tensor correlating the frictional force, F, opposing the motion of a particle in a viscous fluid and the velocity u of this particle relative to the fluid.
Note: In the case of an isolated spherical particle in a viscous isotropic fluid, f is a scalar and $F = fu$.

3.2.2 hydrodynamically equivalent sphere
Hypothetical sphere, impenetrable to the surrounding medium, displaying in a hydrodynamic field the same frictional effect as an actual polymer molecule.
Note: The size of a hydrodynamically equivalent sphere may be different for different types of motion of the macromolecule, e.g., for diffusion and for viscous flow.

3.2.3 hydrodynamic volume
Volume of a *hydrodynamically equivalent sphere*.

3.2.4 bead-rod model
Model simulating the hydrodynamic properties of a chain macromolecule consisting of a sequence of beads, each of which offers hydrodynamic resistance to the flow of the surrounding medium and is connected to the next bead by a rigid rod which does not. The mutual orientation of the rods is random.

3.2.5 bead-spring model
Model simulating the hydrodynamic properties of a chain macromolecule consisting of a sequence of beads, each of which offers hydrodynamic resistance to the flow of the surrounding medium and is connected to the next bead by a spring which does not contribute to the frictional interaction but which is responsible for the elastic and deformational properties of the chain. The mutual orientation of the springs is random.

3.2.6 freely draining
Adjective referring to a chain macromolecule the segments of which produce such small frictional effects when moving in a medium that the hydrodynamic field in the vicinity of a given segment is not affected by the presence of other segments. Thus, the solvent can flow virtually undisturbed through the domain occupied by a freely draining macromolecule.

3.2.7 non-draining
Adjective describing a chain macromolecule that behaves in a hydrodynamic field as though the solvent within the domain of the macromolecule were virtually immobilized with respect to the macromolecule.

3.2.8 partially draining
Adjective describing a chain macromolecule that behaves in a hydrodynamic field as though the solvent within the domain of the macromolecule were progressively more immobilized with respect to the macromolecule in the direction from its outer fringes inward.
Note: A *freely draining* macromolecule and a *non-draining* macromolecule are two extremes of the concept of a partially draining macromolecule.

3.2.9 streaming birefringence
flow birefringence
Birefringence induced by flow in liquids, solutions and dispersions of optically anisotropic, anisometric or deformable molecules or particles due to a non-random orientation of the molecules or particles.

3.2.10 rotational diffusion
Process by which the equilibrium statistical distribution of the overall orientation of molecules or particles is maintained or restored.
Note: Rotational diffusion may be compared to translational diffusion, through which the equilibrium statistical distribution of position in space is maintained or restored.

3.2.11 sedimentation coefficient, s, SI unit: s
Parameter characterizing the motion of a particle in a centrifugal field and defined as the velocity of motion u due to unit centrifugal acceleration, i.e., $s = u/(r\omega^2)$, where ω is the angular velocity and r the distance from the centre of rotation.
Note: The unit 10^{-13} s is useful; this unit has been referred to as a 'svedberg' (Sv). 1 Sv = 10^{-13} s =
0.1 ps.

3.2.12 sedimentation equilibrium
Equilibrium established in a centrifugal field when there is no net flux of any component across any plane perpendicular to the centrifugal force.

3.2.13 equilibrium sedimentation (method)
Method by which the distribution of the concentration of the solute or dispersed component in a dilute solution or dispersion along the centrifuge cell is measured at *sedimentation equilibrium*, and the results are interpreted in terms of *molar masses* or their distribution, or both.

3.2.14 sedimentation velocity method
Method by which the velocity of motion of solute component(s) or dispersed particles is measured and the result is expressed in terms of its (their) *sedimentation coefficient(s)*.

3.2.15 Archibald's method
Sedimentation method based on the fact that at the meniscus and at the bottom of the centrifuge cell there is never a flux of the solute across a plane perpendicular to the radial

TERMINOLOGY

direction and the equations characterizing the *sedimentation equilibrium* always apply there, even though the system as a whole may be far from equilibrium.
Note: The use of the term 'approach to sedimentation equilibrium' for Archibald's method is discouraged, since it has a more general meaning.

3.2.16 equilibrium sedimentation in a density gradient
Equilibrium sedimentation technique working with a multi-component solvent forming a density gradient in a centrifugal field

3.2.17 relative viscosity, η_r
viscosity ratio
Ratio of the viscosity of the solution, η, to the viscosity of the solvent, η_s, i.e., $\eta_r = \eta/\eta_s$.

3.2.18 relative viscosity increment, η_i
Ratio of the difference between the viscosities of solution and solvent to the viscosity of the solvent, i.e., $\eta_i = (\eta - \eta_s)/\eta_s$. For definitions of symbols, see Definition 3.2.17.
Note: The use of the term 'specific viscosity' for this quantity is discouraged, since the relative viscosity increment does not have the attributes of a specific quantity.

3.2.19 reduced viscosity, SI unit: $m^3\,kg^{-1}$
viscosity number
Ratio of the *relative viscosity increment* to the mass concentration of the polymer, c, i.e., η_i/c.
Note: This quantity and those in Definitions 3.2.20 and 3.2.21 are neither viscosities nor pure numbers. The terms are to be looked on as traditional names. Any replacement by consistent terminology would produce unnecessary confusion in the polymer literature.

3.2.20 inherent viscosity, η_{inh}, SI unit: $m^3\,kg^{-1}$
logarithmic viscosity number, η_{ln}
Ratio of the natural logarithm of the *relative viscosity* to the mass concentration of the polymer, c, i.e.,

$$\eta_{inh} \equiv \eta_{ln} = (\ln \eta_r)/c.$$

Note: See note under Definition 3.2.19.

3.2.21 intrinsic viscosity, $[\eta]$, SI unit: $m^3\,kg^{-1}$
limiting viscosity number
Staudinger index
Limiting value of the *reduced viscosity* (or the *inherent viscosity*) at infinite dilution of the polymer, i.e.,

$$[\eta] = \lim_{c \to 0} (\eta_i/c) = \lim_{c \to 0} \eta_{inh}$$

Note: See note under Definition 3.2.19.

3.2.22 Huggins equation

Equation describing the dependence of the *reduced viscosity*, η_i/c, on the mass concentration of the polymer, c, for *dilute polymer solutions* of the form

$$\eta_i/c = [\eta] + k_H [\eta]^2 c$$

where k_H is the *Huggins coefficient* and $[\eta]$ is the *intrinsic viscosity*.

3.2.23 Huggins coefficient, k_H

Parameter in the *Huggins equation*.

3.2.24 viscosity function, ϕ
Flory constant

Coefficient connecting the *intrinsic viscosity*, the *radius of gyration* and the *molar mass* of a chain macromolecule, according to the equation

$$[\eta] = \Phi 6^{3/2} \langle s^2 \rangle^{3/2} / M$$

3.2.25 Mark-Houwink equation

Equation describing the dependence of the *intrinsic viscosity* of a polymer on its *relative molecular mass (molecular weight)* and having the form

$$[\eta] = K M_r^a$$

where K and a are constants, the values of which depend on the nature of the polymer and solvent as well as on temperature, and M_r, is usually one of the *relative molecular-mass averages*.

Note 1: The use of this equation with the *relative molecular mass (molecular weight)* is recommended, rather than with *molar mass* (which has the dimension of mass divided by amount of substance), since in the latter case the constant K assumes awkward and variable dimensions owing to the fractional and variable nature of the exponent a.

Note 2: Kuhn and Sakurada have also made important contributions and their names are sometimes included, as, for example, in the Kuhn-Mark-Houwink-Sakurada equation.

3.3 coherent elastic scattering of radiation

A beam of radiation traversing a medium may be attenuated and partially scattered. The definitions below are for those cases in which the attenuation of the incident beam is due only to scattering, the energy of scattering quanta is the same as that of quanta in the primary beam (elastic scattering) and phase relationships between independent scatterers are retained (coherent scattering). This section deals with light scattering (LS), small-angle x-ray scattering (SAXS), and small-angle neutron scattering (SANS). In light scattering the polarization of light is relevant; plane-polarized light is considered here only, and it is called vertically polarized (v) if the electric vector of the beam is perpendicular to the plane containing the source, sample and detector, and horizontally polarized (h) if the electric vector lies in that plane. Unpolarized light is considered to be a mixture of equal parts of v and h light.

TERMINOLOGY

3.3.1 small particle
Particle much smaller than the wavelength of the radiation in the medium. In practice, all dimensions of a particle considered small must be less than about one-twentieth of the wavelength employed.

3.3.2 large particle
Particle with dimensions comparable with the wavelength of the radiation in the medium or larger. In practice a particle must be treated as large, if its largest dimension exceeds about one-twentieth of the wavelength employed.

3.3.3 scattering angle, θ
angle of observation
Angle between the forward direction of the incident beam and a straight line connecting the scattering point and the detector.

3.3.4 scattering vector; q
Vector difference between the wave propagation vectors of the incident and the scattered beam, both of length $2\pi/\lambda$, where λ is the wavelength of the scattered radiation in the medium.

3.3.5 length of the scattering vector, q
Length of the *scattering vector* is $q = (4\pi/\lambda) \sin(\lambda/2)$, where λ is the wavelength of the scattered radiation in the medium and θ is the *scattering angle*.

3.3.6 refractive index increment, $\partial n/\partial c$
Change of the solution refractive index, n, with solute concentration, c.
Note 1: The solute concentration is most frequently expressed in terms of mass concentration, molality or volume fraction. If expressed in terms of mass concentration or molality, the corresponding refractive index increments are referred to as specific or molal refractive index increments, respectively.
Note 2: Following use of the full name, the abbreviated name refractive increment may be used.

3.3.7 Rayleigh ratio, $R(\theta)$ or R_θ, SI unit: m^{-1}
Quantity used to characterize the scattered intensity at the *scattering angle*, θ, defined as $R(\theta) = i_\theta r^2/(I f V)$, where I is the intensity of the incident radiation, i_θ is the total intensity of scattered radiation observed at an angle θ and a distance r from the point of scattering and V is the scattering volume. The factor f takes account of polarization phenomena.
Note 1: The value of f depends upon the type of radiation employed:
(i) for light scattering, depending upon the polarization of the incident beam, $f = 1$ for vertically polarized light, $f = \cos^2 \theta$ for horizontally polarized light, $f = (1 + \cos^2\theta)/2$ for unpolarized light;
(ii) for small-angle neutron scattering $f = 1$;
(iii) for small-angle x-ray scattering $f \approx 1$, if $\theta <$ ca 5°.
Note 2: In physics, the factor, f, may not be included in the definition of the Rayleigh ratio.
Note 3: In small-angle neutron scattering, the term cross-section is often used instead of $R(\theta)$; the two quantities are identical.

3.3.8 excess Rayleigh ratio
Difference between the *Rayleigh ratio* for a dilute solution and for pure solvent.
Note: If the scattering intensity is not reduced to the *Rayleigh ratio*, the difference between the scattering intensities for a dilute solution and that for pure solvent is named 'excess scattering'.

3.3.9 turbidity, τ
Apparent absorbance of the incident radiation due to scattering.
Note: For *small particles* direct proportionality exists between turbidity and the *Rayleigh ratio*.

3.3.10 particle scattering function, $P(\theta)$ or P_θ
particle scattering factor
Ratio of the intensity of radiation scattered at an angle of observation θ to the intensity of radiation scattered at an angle zero, i.e., $P(\theta) \equiv R(\theta)/R(0)$.

3.3.11 Zimm plot
Diagrammatic representation of data on scattering from *large particles*, corresponding to the equation

$$\frac{Kc}{\Delta R(\theta)} = \frac{1}{\overline{M}_w P(\theta)} + 2A_2 c + \ldots$$

and used for the simultaneous evaluation of the *mass-average molar mass*, $\overline{M}_w$, the *second virial coefficient of the chemical potential*, A_2, and (usually) the z-average *radius of gyration*; c is the mass concentration of the solute, $\Delta R(\theta)$ the excess *Rayleigh ratio*, and $P(\theta)$ the *particle scattering function* that comprises (usually) the z-average *radius of gyration*. K depends on the solute, the temperature and the type of radiation employed.
Note: Several modifications of the Zimm plot are in frequent use; the most common one uses the *excess scattering* instead of the *excess Rayleigh ratio*.

3.3.12 Guinier plot
Diagrammatic representation of data on scattering from *large particles*, obtained at different angles but at the same concentration, constructed by plotting $\lg[\Delta R(\theta)]$ or $\lg[P(\theta)]$ versus $\sin^2(\theta/2)$ or q^2, and (usually) used for the evaluation of the *radius of gyration*, $P(\theta)$ the *particle scattering function*, θ, the *scattering angle* and q, the length of the *scattering vector*.

3.3.13 Kratky plot
Diagrammatic representation of scattering data on *large particles*, obtained at different angles but at the same concentration, constructed by plotting $\sin^2(\theta/2)\Delta R(\theta)$ versus $\sin(\theta/2)$, or $q^2\Delta R(\theta)$ versus q, and used for the determination of molecular shape. For definitions of symbols, see Definition 3.3.12.

3.3.14 dissymmetry of scattering, $z(\theta_1, \theta_2)$
Ratio of two *Rayleigh ratios* for different *angles of observation*, i.e.,

TERMINOLOGY

$z(\theta_1,\theta_2) = R(\theta_1)/R(\theta_2)$, $\theta_1 > \tilde{\theta}_2$.

Note: The angles must be specified; in light scattering it is customary to let $\theta_2 = 180° - \theta_1$, and, most frequently, $\theta_1 = 45°$ and $\theta_2 = 135°$.

3.3.15 depolarization of scattered light
Phenomenon consequent upon the electric vectors of the incident and scattered beams being non-coplanar, such that light scattered from a vertically (horizontally) polarized incident beam contains a horizontal (vertical) component.
Note: The phenomenon is due primarily to the anisotropy of the polarizability of the scattering medium.

3.3.16 turbidimetric titration
Process in which a precipitant is added incrementally to a highly dilute polymer solution and the intensity of light scattered by, or the *turbidity* due to, the finely dispersed particles of the *polymer-rich phase* is measured as a function of the amount of precipitant added.

3.3.17 isorefractive
Adjective describing components of a multicomponent system having zero refractive index increments with respect to each other.

3.3.18 Mie scattering
Scattering of light by particles with size larger than approximately one-half of the wavelength of incident light.
Note: For homogeneous spheres, this phenomenon is rigorously described by the theory developed by Mie.

3.4 Separation

3.4.1 fractionation
Process by means of which macromolecular species differing in some characteristic (chemical composition, *relative molecular mass*, branching, stereoregularity, etc.) are separated from each other.

3.4.2 polymer-poor phase
dilute phase
Phase of a two-phase equilibrium system, consisting of a polymer and low-molecular-weight material, in which the polymer concentration is lower.
Note: The use of the name 'sol phase' is discouraged.

3.4.3 polymer-rich phase
concentrated phase
Phase of a two-phase equilibrium system, consisting of a polymer and low-molecular-weight material, in which the polymer concentration is higher.
Note: The use of the name 'gel phase' is discouraged.

3.4.4 precipitation fractionation
Process in which a polymeric material, consisting of macromolecules differing in some characteristic affecting their solubility, is separated from solution into fractions by successively decreasing the solution power of the solvent, resulting in the repeated

formation of a two-phase system in which the less soluble components concentrate in the *polymer-rich phase*.

3.4.5 extraction fractionation
Process in which a polymeric material, consisting of macromolecules differing in some characteristic affecting their solubility, is separated from *a polymer-rich phase* into fractions by successively increasing the solution power of the solvent, resulting in the repeated formation of a two phase system in which the more soluble components concentrate in the *polymer-poor phase*.

3.4.6 size-exclusion chromatography (SEC)
Separation technique in which separation mainly according to the *hydrodynamic volume* of the molecules or particles takes place in a porous non-adsorbing material with pores of approximately the same size as the effective dimensions in solution of the molecules to be separated.

3.4.7 gel-permeation chromatography (GPC)
Size-exclusion chromatography in which the porous non-adsorbing material is a gel.

3.4.8 molar-mass exclusion limit
molecular-weight exclusion limit
Maximum value of the *molar mass* or *molecular weight* of molecules or particles, in a specific polymer-solvent system, that can enter into the pores of the porous non-adsorbing material used in *size-exclusion chromatography*.
Note: For particles with molar mass or molecular weight larger than the exclusion limit the separation effect of the size-exclusion chromatography vanishes.

3.4.9 elution volume, V_{el}, SI unit: m^3
Volume of a solvent passed, since the injection of the sample, through a *size-exclusion chromatography* bed at the time at which a specified signal of the detector has been recorded.

3.4.10 retention volume, V_R
Elution volume at the maximum concentration of an elution peak.

3.4.11 universal calibration
Calibration of *size-exclusion chromatography* columns based on the finding that the *retention volume* of a molecular or particulate species is usually a single-valued function of an appropriate size parameter of this molecule or particle, irrespective of its chemical nature and structure.
Note: The product of the *intrinsic viscosity* and *molar mass*, $[\eta] M$, has been widely used as a size parameter.

3.4.12 spreading function
Normalized signal produced, as a function of *elution volume*, at the outlet of a *size-exclusion chromatography* set-up, by an instantaneous injection of a uniform sample.

3.4.13 plate number, N

Characteristic of the efficiency of a *size-exclusion chromatography* set-up in terms of band broadening, defined as $N = (V_R/\sigma_V)^2$, where V_R is the *retention volume* of an individual low-molecular-weight compound, and σ_V is the corresponding full width at 60.7% peak height of the elution peak.

3.4.14 plate height, H
      height equivalent to a theoretical plate

Length of a part of a *size-exclusion chromatography* bed corresponding to one plate, i.e., the length of the bed, L, divided by its *plate number*, N, or $H = L/N$.

REFERENCES

1. IUPAC. 'Glossary of basic terms in polymer science (IUPAC Recommendations 1996)', *Pure Appl. Chem.* **68**, 2287-2311 (1996). Reprinted as Chapter 1 this volume.
2. IUPAC. 'Stereochemical definitions and notations relating to polymers (Recommendations 1980)', *Pure Appl. Chem.* **53**, 733-752 (1981). Reprinted as Chapter 2 this volume.
3. IUPAC. 'Note on the terminology for molar masses in polymer science', *Makromol. Chem.* **185**, Appendix to No. 1 (1984); *J. Polym. Sci., Polym. Lett. Ed.* **22**, 57 (1984); and in other journals.
4. IUPAC. Compendium of Chemical Terminology, 2nd ed. (the 'Gold Book'). Compiled by A. D. McNaught and A. Wilkinson. Blackwell Scientific Publications, Oxford (1997). XML on-line version: http://goldbook.iupac.org (2006-) created by M. Nic, J. Jirat, B. Kosata; updates compiled by A. Jenkins.

ALPHABETICAL INDEX OF TERMS

Term	Symbol	Definition number
angle of observation	θ	3.3.3
apparent molar mass	M_{app}	2.11
apparent molecular weight	$M_{r,app}$	2.11
apparent relative molecular mass	$M_{r,app}$	2.11
Archibald's method		3.2.15
average degree of polymerization	$\overline{X}_k$	2.1.2
bead-rod model		3.2.4
bead-spring model		3.2.5
branching index	g	1.25
characteristic ratio	C_N, C_∞	1.13
chi parameter	χ	3.1.12
coherent elastic scattering of radiation		3.3
compositional heterogeneity		2.1
concentrated phase		3.4.3
constitutional heterogeneity		2.2
continuous distribution function		2.13
continuously curved chain		1.21
contour length		1.14
copolymer micelle		1.28
co-solvency		3.1.15
cross-over concentration	c^*	3.1.2
cross-section		3.3.7
cumulative distribution function		2.1.3
degree of polymerization	X	1.3
depolarization of scattered light		3.3.15
differential distribution function		2.13
dilute phase		3.4.2
dilute solution		3.1.1
discrete distribution function		2.13
distribution function		2.13
dissymmetry of scattering	$z(\theta_1, \theta_2)$	3.3.14
elution volume		3.4.9
end-to-end distance	r	1.11, 1.12
end-to-end vector	$\mathbf{r}$	1.10
equilibrium sedimentation (method)		3.2.13
equilibrium sedimentation in a density gradient		3.2.16
equivalent chain		1.17
equivalent sphere		1.4, 3.2.2
excess Rayleigh ratio		3.3.8
excess scattering		3.3.8
excluded volume of a macromolecule		3.1.9
excluded volume of a segment		3.1.8
expansion factor	$\alpha_r, \alpha_s, \alpha_\eta$	3.1.10
extraction fractionation		3.4.5
Flory constant	Φ	3.2.24

TERMINOLOGY

Flory distribution		2,17
Flory-Huggins theory		3.1.11
Flory-Huggins-Staverman theory		3.1.11
flow birefringence		3.2.9
fractionation		3.4.1
freely draining		3.2.6
freely jointed chain		1.16
freely rotating chain		1.19
frictional coefficient	f	3.2.1
gel-permeation chromatography		3.4.7
Guinier plot		3.3.12
Height equivalent to a theoretical plate	H	3.4.14
Huggins coefficient	K_H	3.2.23
Huggins equation		3.2.22
hydrodynamic volume		3.2.3
hydrodynamically equivalent sphere		3.2.2
inherent viscosity	η_{inh}	3.2.20
integral distribution function		2.13
intrinsic viscosity	$[\eta]$	3.2.21
isopycnic		3.1.17
isorefractive		3.3.17
Kratky plot		3.3.13
Kuhn-Mark-Houwink-Sakurada equation		3.2.25
large particle		3.3.2
length of the scattering vector	q	3.3.5
limiting viscosity number	$[\eta]$	3.2.21
logarithmic normal distribution		2.20
logarithmic viscosity number	η_{ln}	3.2.20
long-chain branch		1.24
long-range interaction		1.6
long-range intramolecular interaction		1.6
Mark-Houwink equation		3.2.25
mass-average degree of polymerization	$\overline{X}_w$	2.12
mass-average molar mass	$\overline{M}_w$	2.7
mass-average relative molecular mass	$\overline{M}_{r,w}$	2.7
mass-distribution function		2.15
microgel		1.27
Mie scattering		3.3.18
molal refractive index increment		3.3.6
molar mass	M	1.2
molar-mass average	$\overline{M}_k$	2.5
molar-mass exclusion limit		3.4.8
molecular weight	M_r	1.1
molecular-weight average	$\overline{M}_{r,k}$	2.5
molecular-weight exclusion limit		3.4.8
monodisperse polymer		2.3
most probable distribution		2.17
network		1.26

non-draining		3.2.7
non-uniform polymer		2.4
number-average degree of polymerization	$\overline{X}_n$	2.12
number-average molar mass	$\overline{M}_n$	2.6
number-average molecular weight	$\overline{M}_{r,n}$	2.6
number-average relative molecular mass	$\overline{M}_{r,n}$	2.6
number-distribution function		2.14
partially draining		3.2.8
particle scattering factor	$P(\theta), P_\theta$	3.3.10
particle scattering function	$P(\theta), P_\theta$	3.3.10
persistence length	a	1.22
perturbed dimensions		1.8
plate height	H	3.4.14
plate number	N	3.4.13
Poisson distribution		2.18
polydisperse polymer		2.4
polymer-poor phase		3.4.2
polymer-rich phase		3.4.3
polymer-solvent interaction		3.1.3
polymolecularity correction		2.21
Porod-Kratky chain		1.21
precipitation fractionation		3.4.4
preferential sorption		3.1.13
probability density function		2.13
quality of solvent		3.1.4
radius of gyration	$s, \langle s^2 \rangle^{1/2}$	1.9
random coil		1.15
random-walk chain		1.16
Rayleigh ratio	$R(\theta), R_\theta$	3.3.7
reduced viscosity		3.2.19
refractive increment	$\partial n/\partial c$	3.3.6
refractive index increment	$\partial n/\partial c$	3.3.6
relative molecular mass	M	1.1
relative molecular-mass average	$\overline{M}_{r,k}$	2.5
relative viscosity	η_r	3.2.17
relative viscosity increment	η_i	3.2.18
retention volume	V_R	3.4.10
root-mean-square end-to-end distance	$\langle r^2 \rangle^{1/2}$	1.12
rotational diffusion		3.2.10
scattering angle	θ	3.3.3
scattering vector		3.3.4
Schulz-Flory distribution		2.17
Schulz-Zimm distribution		2.16
second virial coefficient	A_2	3.1.7
sedimentation coefficient	s	3.2.11
sedimentation equilibrium		3.2.12

TERMINOLOGY

sedimentation velocity method		3.2.14
selective solvent		3.1.14
selective sorption		3.1.13
short-chain branch		1.23
short-range interaction		1.5
short-range intramolecular interaction		1.5
size-exclusion chromatography		3.4.6
small particle		3.3.1
solubility parameter	δ	3.1.16
specific refractive index increment		3.3.6
spreading function		3.4.12
statistical coil		1.15
statistical segment		1.18
Staudinger index	$[\eta]$	3.2.21
steric factor	σ	1.20
streaming birefringence		3.2.9
thermodynamic quality of solvent		3.1.4
thermodynamically equivalent sphere		1.4
theta solvent		3.1.5
theta state		3.1.5
theta temperature		3.1.6
Tung distribution		2.19
turbidimetric titration		3.3.16
turbidity	τ	3.3.9
uniform polymer		2.3
universal calibration		3.4.11
unperturbed dimensions		1.7
virial coefficients	$A_1, A_2, \ldots$	3.1.7
virial coefficients of the chemical potential	$A_1, A_2, \ldots$	3.1.7
viscosity function	Φ	3.2.24
viscosity number		3.2.19
viscosity ratio	η_r	3.2.17
viscosity-average degree of polymerization	$\overline{X}_v$	2.12
viscosity-average molar mass	$\overline{M}_v$	2.10
viscosity-average molecular weight	$\overline{M}_{r,v}$	2.10
viscosity-average relative molecular mass	$\overline{M}_{r,v}$	2.10
weight-average degree of polymerization	$\overline{X}_w$	2.12
weight-average molecular weight	$\overline{M}_{r,w}$	2.7
weight-distribution function		2.15
worm-like chain		1.21
z-average degree of polymerization	$\overline{X}_z$	2.12
z-average molar mass	$\overline{M}_z$	2.8
z-average molecular weight	$\overline{M}_{r,z}$	2.8
z-average relative molecular mass	$\overline{M}_{r,z}$	2.8

(z + 1)-average molar mass	$\overline{M}_{z+1}$	2.9
(z + 1)-average degree of polymerization	$\overline{X}_{z+1}$	2.12
(z + 1)-average relative molecular mass	$\overline{M}_{r,z+1}$	2.9
(z + 1)-average molecular weight	$\overline{M}_{r,z+1}$	2.9
Zimm plot		3.3.11

4: Basic Classification and Definitions of Polymerization Reactions

CONTENTS

Preamble
Definitions
Summary
References

PREAMBLE

In its report on 'Basic Definitions of Terms Relating to Polymers' published in 1974 [1], the IUPAC Commission on Macromolecular Nomenclature defined the terms 'addition polymerization' (polymerization by repeated addition processes) and 'condensation polymerization' (polymerization by repeated condensation processes, i.e., with the elimination of small molecules). At that time, the terms were intended to classify polymerization reactions according to whether or not small molecules are formed in the growth reaction. Meanwhile, widespread use of the term addition polymerization for polymerizations with growth steps that are chain reactions has resulted in the introduction of alternative terms aimed at a clear distinction in terminology between the chain or non-chain nature of the growth reaction and the stoichiometry associated with the formation or non-formation of a small molecule during that reaction [2]. Most prominent amongst these are the terms 'chain-growth polymerization' and 'step-growth polymerization.' However, conflicting usage in the literature and contradictory explanations of these terms in textbooks indicate the timeliness of a new set of definitions.

The definitions given in this document use a single term for polymerizations in which polymer chains grow via chain-reaction mechanisms, with the stoichiometry specified additionally, if so desired. 'Polyaddition' and 'polycondensation' are invoked only in the naming of polymerizations proceeding through reactions between molecules of all degrees of polymerization. The definitions are in accord with the IUPAC terms for organic chemistry [3] and the basic definitions [1].

DEFINITIONS

1 chain polymerization

Chain reaction in which the growth of a polymer chain proceeds exclusively by the reaction or reactions between a monomer or monomers and a reactive site or reactive sites on the polymer chain with regeneration of the reactive site or reactive sites at the end of each growth step.

Originally prepared by a working group consisting of I. Mita (Japan), R. F. T. Stepto (UK) and U. W. Suter (Switzerland). Reprinted from *Pure Appl. Chem.* **66**, 2483-2486 (1994).

CLASSIFICATION OF POLYMERIZATION REACTIONS

Note 1: A chain polymerization consists of initiation and propagation reactions, and may also include termination and chain-transfer reactions.

Note 2: The adjective 'chain' in 'chain polymerization' denotes 'chain reaction' rather than 'polymer chain'.

Note 3: Propagation in chain polymerization often occurs without the formation of small molecules. However, cases exist where, at each propagation step, a low-molar-mass by-product is formed, as in the polymerization of 1,3-oxazolidine-2,5-diones derived from amino acids (commonly termed amino acid N-carboxyanhydrides). When a low-molar-mass by-product is formed, the adjective *condensative* is recommended to give the term 'condensative chain polymerization'.

Note 4: The growth steps are expressed by

$$P_x + M \rightarrow P_{x+1}(+ L) \quad \{x\} \in \{1, 2, ...\infty\}$$

where P_x denotes the growing chain of degree of polymerization x, M a monomer and L a low-molar-mass by-product formed in the case of condensative chain polymerization.

Note 5: The term 'chain polymerization' may be qualified further, if necessary, to specify the type of chemical reactions involved in the growth step, e.g. ring-opening chain polymerization or cationic chain polymerization.

Note 6: There exist, exceptionally, some polymerizations that can proceed via chain reactions that, according to the definition, are not chain polymerizations. For example, the polymerization

$$HS-X-SH + CH_2=CH-X'-CH=CH_2 \rightarrow -(-S-X-S-CH_2-CH_2-X'-CH_2-CH_2-)_n-$$

proceeds via a radical chain reaction with intermolecular transfer of the radical centre. The growth step, however, involves reactions between molecules of all degrees of polymerization and, hence, the polymerization is classified as a polyaddition. If required, the classification can be made more precise and the polymerization described as a chain-reaction polyaddition.

2   polycondensation

Polymerization in which the growth of polymer chains proceeds by condensation reactions between molecules of all degrees of polymerization.

Note 1: The growth steps may be expressed by

$$P_x + P_y \rightarrow P_{x+y} + L \quad \{x\} \in \{1,2, ...\infty\}; \{y\} \in \{1, 2, ...\infty\}$$

where P_x and P_y denote chains of degrees of polymerization x and y, respectively, and L denotes a low-molar-mass by-product.

Note 2: The earlier term 'polycondensation' (namely, polymerization by a repeated condensation process, i.e., with elimination of simple molecules [1]) was synonymous with 'condensation polymerization'. It should be noted that the current definitions of polycondensation and condensative chain polymerization were *both* embraced by the earlier term 'polycondensation'.

3   polyaddition

Polymerization in which the growth of polymer chains proceeds by addition reactions between molecules of all degrees of polymerization.

TERMINOLOGY

Note 1: The growth steps may be expressed by

$$P_x + P_y \rightarrow P_{x+y} \qquad \{x\} \in \{1, 2, ...\infty\}; \{y\} \in \{1, 2, ...\infty\}$$

where P_x and P_y denote chains of degrees of polymerization x and y, respectively.

Note 2: The earlier term 'addition polymerization' (as defined in [1]) embraced both the current concepts of 'polyaddition' and 'chain polymerization', but did not include 'condensative chain polymerization'.

SUMMARY

The classification of the types of polymerization resulting from the definitions, except for the polymerizations of Definition 1, Note 6, may be summarized in the following table:

GROWTH MECHANISM	REACTION TYPE	STOICHIOMETRY	
		with low molecular mass by-products	without low molecular mass by-products
monomer molecules reacting with active polymer chains	chain reaction	**condensative chain polymerization**	**chain polymerization**
monomer, oligomer and polymer molecules of all sizes reacting together	usually non-chain reaction	**polycondensation**	**polyaddition**

REFERENCES

1. Glossary of basic terms in polymer science (IUPAC Recommendations 1996), *Pure Appl. Chem.* 68, 2287-2311 (1996). Reprinted as Chapter 1 this volume.
2. IUPAC. Compendium of Chemical Terminology, 2nd ed. (the 'Gold Book'). Compiled by A. D. McNaught and A. Wilkinson. Blackwell Scientific Publications, Oxford (1997). XML on-line version: http://goldbook.iupac.org (2006-) created by M. Nic, J. Jirat, B. Kosata; updates compiled by A.D. Jenkins.
3. Glossary of Terms Used in Physical Organic Chemistry, *Pure Appl. Chem.* **66**, 1077-1184 (1994).

5: Definitions Relating to Stereochemically Asymmetric Polymerizations

CONTENTS

Preamble
Definitions
References

PREAMBLE

The contribution of Professor Sigwalt to the conception of this chapter is acknowledged. He started working on it in 1981, but originally there was no agreement within the Commission and with outside specialists in organic chemistry. The current document reflects the continuation of this work with the assistance from the former IUPAC Commission on Nomenclature of Organic Chemistry.

Basic definitions of terms relating to polymerization reactions [1,2] and stereochemical definitions and notations relating to polymers [3] have been published, but no reference was made explicitly to reactions involving the asymmetric synthesis of polymers. It is the aim of the present document to recommend classification and definitions relating to asymmetric polymerizations that may produce optically active polymers.

As in Chapter 2 and in previous IUPAC documents [1,3], the rotated Fischer projection is used to denote a polymer backbone by a horizontal line. Hence, *at each individual backbone carbon atom* the horizontal lines represent the bonds directed below the plane of the paper from the carbon atom while the vertical lines project above the plane of the paper from the carbon atom. Thus the rotated Fischer projection* of

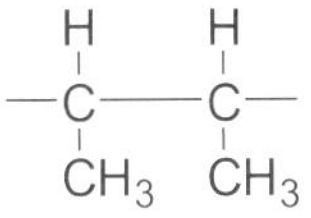

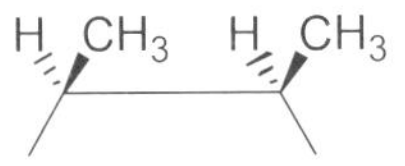

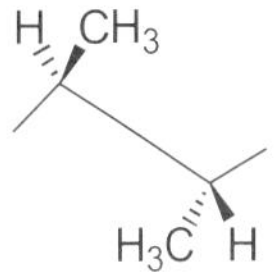

DEFINITIONS

1 asymmetric polymerization

Polymerization that proceeds in an unsymmetrical manner in terms of chirality under the influence of chiral features present in one or more components of the reaction system.

---

* Usually in Fischer projections the carbon atoms of the main chain are omitted [4]. In general, in the area of macromolecular chemistry, element symbols are not omitted in the backbone and are usually shown in the rotated Fischer projection [3].

Originally prepared by a working group consisting of K. Hatada (Japan), J. Kahovec (Czech Republic), M. Barón (Argentina), K. Horie (Japan), T. Kitayama (Japan), P. Kubisa (Poland), G. P. Moss (UK), R. F. T. Stepto (UK), and E. S. Wilks (USA). Reprinted from *Pure Appl. Chem.* **74**, 915-922 (2002)

TERMINOLOGY

Note 1: An asymmetric polymerization generally produces a polymer which contains chirality centres of opposite configuration in unequal amounts.
Note 2: Chiral features may be present in monomers, solvents, initiators, catalysts and supports.
Note 3: Polymerization is defined as the process of converting a monomer or a mixture of monomers into a polymer [1]. Thus the definition of an asymmetric polymerization covers homopolymerization and copolymerization.
Note 4: Some stereospecific polymerizations produce tactic polymers [3] that contain a mixture of pairs of enantiomeric polymer molecules in equal amounts. For example, in the case of a polymerization leading to an isotactic polymer the product consists of

$-(-R-)-_i, -(-R-)-_{i+1}, -(-R-)-_{i+2}, \ldots$

and their corresponding enantiomers

$-(-S-)-_i, -(-S-)-_{i+1}, -(-S-)-_{i+2}, \ldots$

in equal amounts; here, -R- and -S- represent enantiomeric configurational repeating units. The product can be considered to be a mixture of polymer racemates because a racemate is defined as an equimolar mixture of enantiomers [4]. Such polymerizations can be named by using the adjective 'racemate-forming', as in 'racemate-forming chirogenic polymerization' (see Note 4 to Definition 2) and 'racemate-forming enantiomer-differentiating polymerization' (see Note 2 to Definition 3). Polymerizations of prochiral substituted ethenes leading to atactic polymers usually give enantiomeric polymer molecules in equal amounts. However, use of the adjective 'racemate-forming' is discouraged for such cases where a mixture of polymer racemates molecules with the same degree of polymerization consists of a large number of diastereomers.

2 asymmetric chirogenic polymerization

Asymmetric polymerization in which the polymer molecules formed contain one (or more) new type(s) of elements of chirality not existing in the starting monomer(s).
Note 1: The new elements of chirality generated in the course of the polymerization may be new types of chirality centres in the polymer molecules (see Notes 2, 4, and 5 to Definition 2) or may arise from the helicity of the polymer molecules (see Note 3 to Definition 2).
Note 2: In asymmetric chirogenic polymerizations of some prochiral monomers, such as Examples 2.1 and 2.2, at least one new type of chirality centre in the main chain is generated at each propagation step, which lead to polymer molecules having the same configuration (R or S) at each corresponding chirality centre. The resulting polymer is isotactic and optically active.
Note 3: Some asymmetric chirogenic polymerizations give helical polymer molecules of only one screw sense that usually show optical activity due to the helicity (see Examples 2.5, 2.6 and 2.7). These polymerizations are termed 'asymmetric helix-chirogenic polymerizations'.
Note 4: Some polymerizations produce enantiomeric polymer molecules in equal amounts; each polymer molecule contains, in its main chain, a single type of chirality centre that does not exist in the starting monomer. The resulting polymer is optically inactive, and the polymerization is not an asymmetric chirogenic polymerization. Such a polymerization is termed a 'racemate-forming chirogenic polymerization'. The polymerizations described in Examples 2.5, 2.6 and 2.7, carried out using an optically inactive initiator of the corresponding racemate, are examples of racemate-forming chirogenic polymerizations.

STEREOCHEMICALLY ASYMMETRIC POLYMERIZATIONS

Another example is the polymerization of penta-1,3-diene with an optically inactive initiator of the corresponding racemate (see Example 2.1), which leads to an isotactic product.

Note 5: In some polymerizations of vinyl monomers leading to isotactic polymers using an optically active initiator, stereorepeating units [3] of one type of chirality centre are formed at every propagation step. Hence, they give only one type of enantiomeric polymer molecule (A or B), and are asymmetric chirogenic polymerizations. Optical activities of the resulting isotactic polymers are usually very small or not detectable, because the polymer molecules can be regarded as having a plane of symmetry if their degrees of polymerization are so large that the presence of their end-groups is negligible.

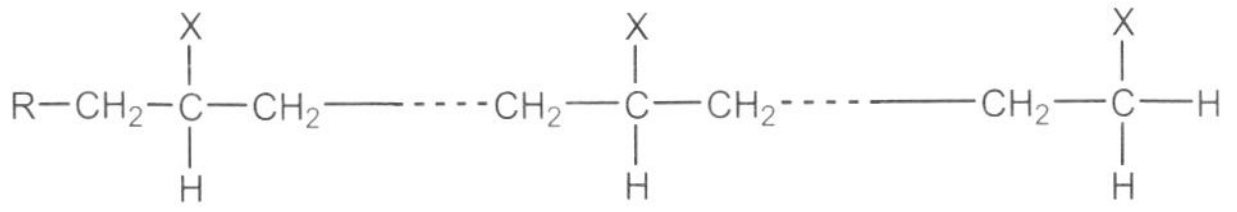

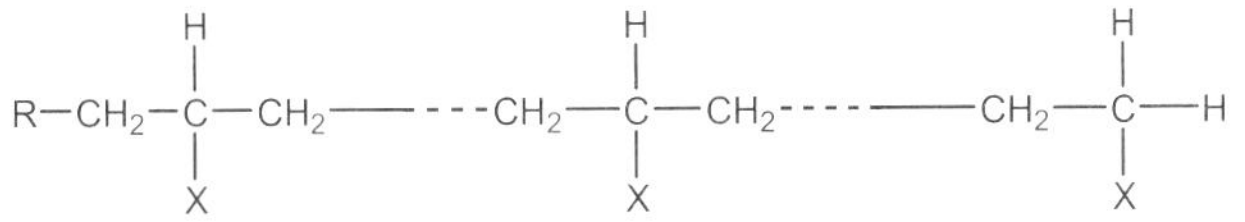

In usual polymerizations of vinyl monomers leading to isotactic polymers, the enantiomeric polymer molecules (A) and (B) are formed in equal amounts, and the polymerizations are not asymmetric but racemate-forming chirogenic polymerizations.

Example 2.1
Polymerization of penta-1,3-diene by 1,4-addition with an optically active catalyst gives an optically active polymer comprising configurational repeating units with predominantly one type of chirality centre.

Example 2.2
Polymerization of 1-benzofuran with an optically active initiator gives the optically active polymer, poly[(2R,3S)-2,3-dihydro-1-benzofuran-2,3-diyl], containing predominantly one type of stereorepeating unit.

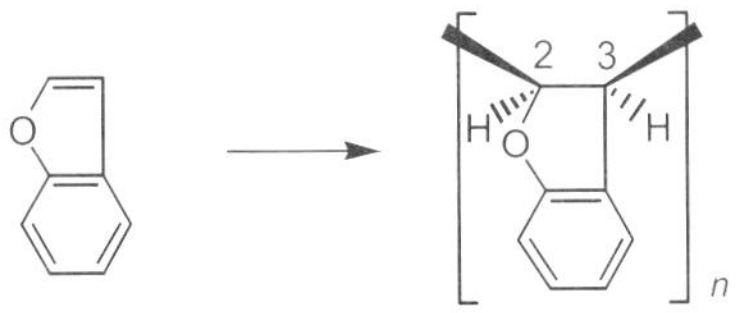

Example 2.3
Polymerization of *cis*-2,3-dimethylthiirane with an optically active initiator results in an optically active polymer, poly{sulfanediyl[(1R,2R)-1,2-dimethylethylene]}, containing predominantly RR configurational repeating units. Inversion of configuration occurs on ring opening, which gives contiguous monomer units with two identical chirality centres.

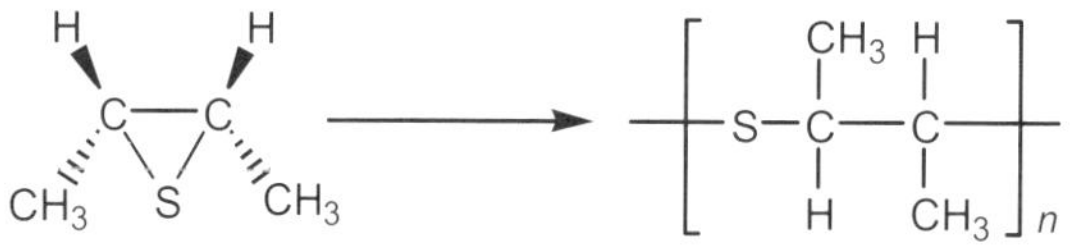

TERMINOLOGY

Example 2.4
Copolymerization of a monomer having two styrene moieties attached to a chiral template molecule with a comonomer (*e.g.*, methyl methacrylate) gives copolymers with strong optical activity after removal of the template molecules. In this case styrene diads of an *S,S* configuration separated from other styrene diads by comonomeric units are responsible for the optical activity.

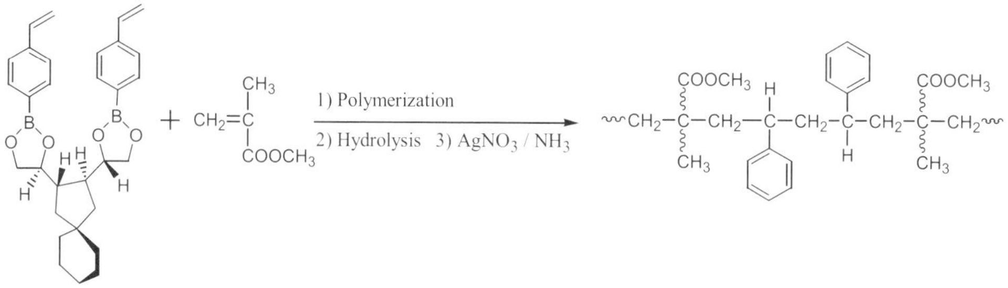

Example 2.5
A polymerization of a bulky methacrylate ester (*e.g.* trityl methacrylate) using an optically active anionic initiator can give an isotactic polymer, poly{1-methyl-1-[(trityloxy)carbonyl]ethylene} of high optical activity owing to the formation of helical polymer molecules with units of predominantly one chirality sense.

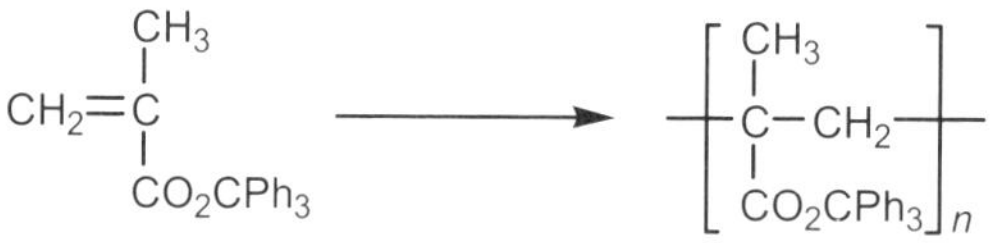

Example 2.6
Polymerization of trichloroacetaldehyde in bulk using an optically active initiator gives an isotactic polymer, poly{oxy[(trichloromethyl)methylene]}, of high optical activity owing to the formation of the helical polymer molecules with units of predominantly one chirality sense.

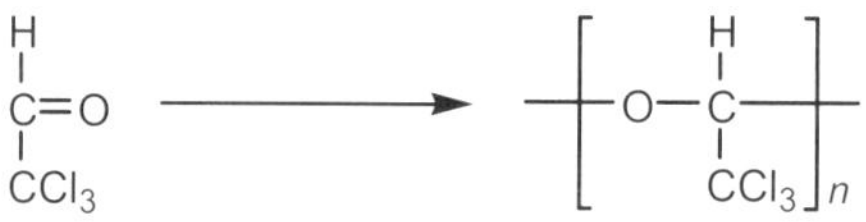

Example 2.7
Polymerization of *tert*-butyl isocyanide using an optically active initiator gives an optically active product comprising helical polymer molecules with units of predominantly one chirality sense.

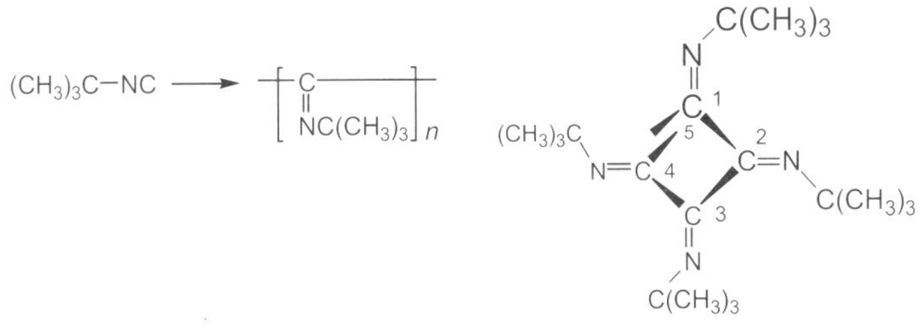

poly[(*tert*-butylimino)methylene] (view of the right-handed helix along the helical axis; monomer unit 5 is below unit 1)

3 asymmetric enantiomer-differentiating polymerization

Asymmetric polymerization in which, starting from a mixture of enantiomeric monomer molecules, only one enantiomer is polymerized.

Note 1: A 'stereoselective polymerization' is defined as 'a polymerization in which a polymer molecule is formed from a mixture of stereoisomeric monomer molecules by the incorporation of only one stereoisomeric species' [1]. Thus, an asymmetric enantiomer-differentiating polymerization is a *stereoselective polymerization*, in which all the polymer molecules are formed by the incorporation of only one type of stereoisomeric species.

Note 2: A polymerization in which, starting from the racemate of a chiral monomer, two types of polymer molecules, each containing monomeric units derived from one of the enantiomers, form in equal amounts is termed 'racemate-forming enantiomer-differentiating polymerization'. The resulting polymer is optically inactive (see Note 4 of Definition 2).

Example 3.1
Polymerization of racemic 3-methylpent-1-ene (MP) using an optically active catalyst may give an optically active polymer by a polymerization that is partially asymmetric; preferential consumption of one of the two enantiomers leaves a monomer mixture having optical activity.

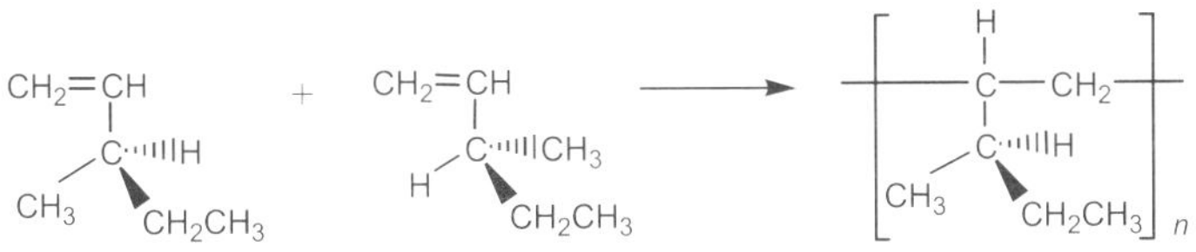

Example 3.2
Polymerization of racemic 1-phenylethyl methacrylate (PEMA) using a chiral complex of a Grignard reagent with a diamine as an initiator may proceed by reaction of only one of the two enantiomers to give an optically active polymer.

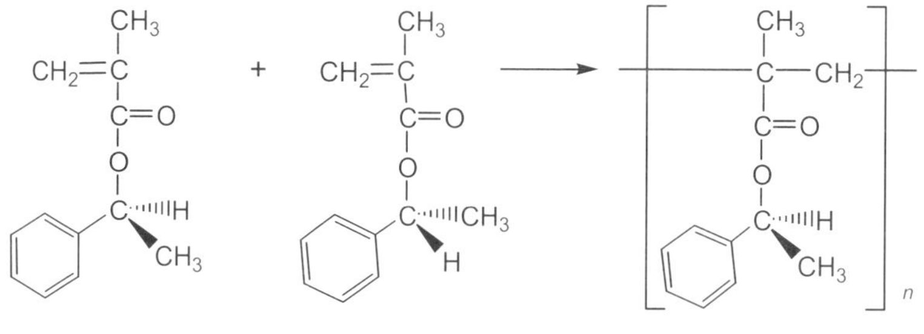

TERMINOLOGY

(S)-PEMA — reacts

(R)-PEMA — does not react

poly[(S)-PEMA]

Example 3.3
Polymerization of racemic methylthiirane (MT) using an optically active initiator may proceed by reaction of only one of the two enantiomers to produce stereoregular polymer molecules, comprising only one type of configurational repeating unit, as a result of either complete retention or complete inversion of configuration at the chirality centre of the monomer. The following reaction scheme represents the case of complete retention of monomer configuration in the polymer formed, which is optically active.

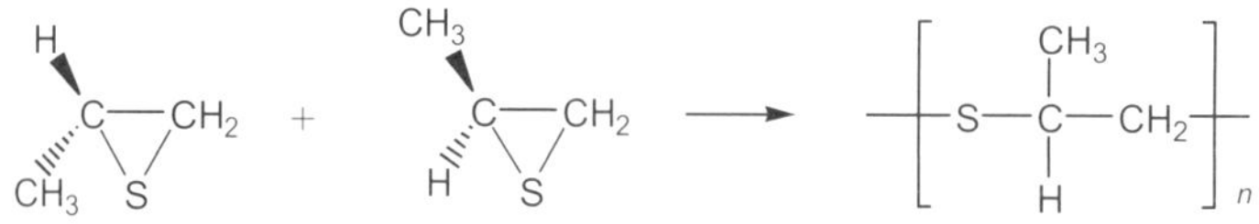

(R)-MT — reacts

(S)-MT — does not react

poly{sulfanediyl[(1R)-1-methylethylene]}

Example 3.4: Polymerization of racemic *trans*-2,3-dimethylthiirane (DMT) using an optically active initiator may proceed by reaction of only one of the two enantiomers to give stereoregular but optically inactive, non-chiral polymer molecules as a result of inversion of the configuration of the attacked carbon atom.

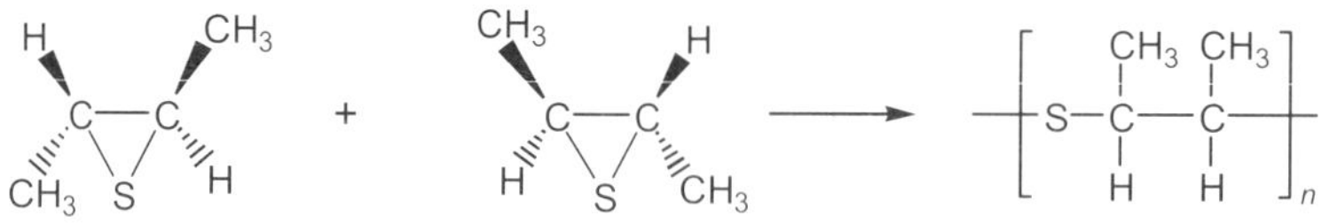

trans-(R,R)-DMT — reacts

trans-(S,S)-DMT — does not react

poly{sulfanediyl[(1R,2S)-1,2-dimethylethylene]}

REFERENCES

1. IUPAC. 'Glossary of basic terms in polymer science (IUPAC Recommendations 1996)', *Pure Appl. Chem.* **68**, 2287-2311 (1996). Reprinted as Chapter 1 this volume.
2. IUPAC. 'Basic classification and definitions of polymerization reactions (IUPAC Recommendations 1994)', *Pure Appl. Chem.* **66**, 2483-2486 (1994). Reprinted as Chapter 4 this volume.
3. IUPAC. 'Stereochemical definitions and notations relating to polymers (Recommendations 1980)', *Pure Appl. Chem.* **53**, 733-752 (1981). Reprinted as Chapter 2 this volume.
4. IUPAC. 'Basic terminology of stereochemistry (IUPAC Recommendations 1996)', *Pure Appl. Chem.* **68**, 2193-2222 (1996).

6: Definitions of Terms Relating to Crystalline Polymers

CONTENTS

Preamble
 1. General definitions
 2. Terminology relating to local conformation and structural aspects
 3. Terminology relating to morphological aspects
 4. Terminology relating to molecular conformation within polymer crystals
 5. Terminology relating to crystallization kinetics
References

PREAMBLE

The recommendations embodied in this document are concerned with the terminology relating to the structure of crystalline polymers and the process of macromolecular crystallization. The document is limited to systems exhibiting crystallinity in the classical sense of three-dimensionally periodic regularity. The recommendations deal primarily with crystal structures that are comprised of essentially rectilinear, parallel-packed polymer chains, and secondarily, with those composed of so-called globular macromolecules. Since the latter are biological in nature, they are not covered in detail here. In general, macromolecular systems with mesophases are also omitted, but crystalline polymers with *conformational disorder* are included.

After a listing of some general definitions relating to crystalline polymers (Section 1), the subject is divided into sections dealing, successively, with local structural arrangements at the scale of a few bond lengths (Section 2), morphological aspects (Section 3), molecular conformation within polymer crystals (Section 4) and, finally, kinetic aspects of crystallization (Section 5). An alphabetical index of terms is provided for the convenience of the reader.

This document relies on the basic definitions of terms in polymer science [1]. It was the second in a series published by the Commission on Macromolecular Nomenclature dealing with definitions of physical and physicochemical terms in the polymer field (for the first in the series, see Reference [2]).

Originally prepared by a Working Group consisting of G. Allegra (Italy), P. Corradini (Italy), H.-G. Elias (USA), P. H. Geil (USA), H. D. Keith (USA) and B. Wunderlich (USA). Reprinted from *Pure Appl. Chem.* **61**, 769-785 (1989).

CRYSTALLINE POLYMERS

1 GENERAL DEFINITIONS

1.1 crystallinity
Presence of three-dimensional order on the level of atomic dimensions.
Note: Crystallinity may be detected by diffraction techniques, heat of fusion measurements, etc. Some amount of disorder within the crystalline region is not incompatible with this concept.

1.2 crystalline polymer
Polymer showing *crystallinity*.
Note 1: One- or two-dimensional order leads to mesophase structure.
Note 2: The range of order may be as small as about 2 nm in one (or more) crystallographic direction(s) and is usually below 50 nm in at least one direction.

1.3 degree of crystallinity
Fractional amount of *crystallinity* in a polymer sample.
Note 1: The assumption is made that the sample can be subdivided into a crystalline phase and an amorphous phase (the so-called two-phase model).
Note 2: Both phases are assumed to have properties identical with those of their ideal states, with no influence of interfaces.
Note 3: The degree of crystallinity may be expressed either as the mass fraction, w_c, or as the volume fraction, ϕ_c, the two quantities being related by

$$w_c = \phi_c \rho_c / \rho$$

where ρ and ρ_c are the densities of the entire sample and of the crystalline fraction, respectively.
Note 4: The degree of crystallinity can be determined by several experimental techniques; among the most commonly used are: (i) X-ray diffraction, (ii) calorimetry, (iii) density measurements, and (iv) infrared spectroscopy (IR). Imperfections in *crystals* are not easily distinguished from the amorphous phase. Also, the various techniques may be affected to different extents by imperfections and interfacial effects. Hence, some disagreement among the results of quantitative measurements of *crystallinity* by different methods is frequently encountered.
Note 5: The following expressions for $w_{c,\alpha}$, are recommended, where the subscript α specifies the particular experimental method used.
(i) By X-ray diffraction: the degree of crystallinity, $w_{c,x}$, is given by

$$w_{c,x} = I_c / (I_c + K_x I_a)$$

where I_c and I_a are the integrated intensities scattered over a suitable angular interval by the crystalline and the amorphous phases, respectively, and K_x is a calibration constant. If the sample is anisotropic, a suitable average of the diffracted intensity in reciprocal space must be obtained.
(ii) By calorimetry: the degree of crystallinity, $w_{c,h}$ is given by

$$w_{c,h} = \Delta_{fus} h / \Delta_{fus,c} h$$

where $\Delta_{fus} h$ is the specific enthalpy of fusion of the sample and $\Delta_{fus,c} h$ is the specific enthalpy of fusion of the completely *crystalline polymer* over the same temperature range.

The value of $\Delta_{\text{fus},c} h$ may be obtained by extrapolating $\Delta_{\text{fus}} h$ to the density of the completely *crystalline polymer*, which in turn may be obtained from x-ray diffraction data. The specific enthalpies of fusion are temperature dependent.

(iii) By density measurements: the degree of crystallinity, $w_{c,d}$, is given by

$$w_{c,d} = \frac{\rho_c (\rho - \rho_a)}{\rho (\rho_c - \rho_a)}$$

where ρ, ρ_c and ρ_a are the densities of the sample, of the completely *crystalline polymer* and of the completely amorphous polymer, respectively.

(iv) By infrared spectroscopy: the degree of crystallinity, $w_{c,i}$, is given by

$$w_{c,i} = \frac{1}{a_c \rho l} \log_{10}(I_0 / I)$$

where I_0 and I are, respectively, the incident and the transmitted intensities at the frequency of the absorption band due to the crystalline portion, a_c is the mass decadic absorption coefficient of the crystalline material and l is the thickness of the sample.

1.4 (polymer) crystal
Crystalline domain usually limited by well-defined boundaries.
Note 1: Polymer crystals frequently do not display the perfection that is usual for low-molar-mass substances.
Note 2: Twinned polymer crystals are, sometimes, erroneously referred to as 'crystals'.
Note 3: Polymer crystals that can be manipulated individually are often called (polymer) single crystals. A single crystal may contain different *fold domains*.

1.5 (polymer) crystallite
Small crystalline domain.
Note 1: A crystallite may have irregular boundaries and parts of its constituent macromolecules may extend beyond its boundaries.
Note 2: This definition is not identical with that used in classical crystallography.

1.6 unit cell
Smallest, regularly repeating material portion contained in a parallelepiped from which a *crystal* is formed by parallel displacements in three dimensions [3].
Note 1: Unlike in the case of low-molar-mass substances, the unit cell of *polymer crystals* usually comprises only parts of the polymer molecules and the regularity of the periodic repetition may be imperfect.
Note 2: In the case of *parallel-chain crystals*, the chain axis is usually denoted by c or, sometimes, b.
Note 3: This definition applies to the so-called primitive unit cell. In practice, the effective unit cell may consist of more than one primitive unit cell.

1.7 molecular conformation
Conformation of the macromolecule as a whole.
Note 1: In the polymer literature, molecular conformation is sometimes referred to as macroconformation.

Note 2: In molecular conformations involving parallel *stems*, the latter may be confined to the same *crystal* or may also extend over several *crystals*.

1.8 local conformation
Conformation of a macromolecule at the scale of the constitutional units.
Note: In the polymer literature, local conformation is sometimes referred to as microconformation.

2 TERMINOLOGY RELATING TO LOCAL CONFORMATION AND STRUCTURAL ASPECTS [4]

2.1 chain axis
Straight line parallel to the direction of chain extension, connecting the centres of mass of successive blocks of chain units, each of which is contained within an *identity period* (see Fig. 1).

2.2 (chain) identity period
(chain) repeating distance
Shortest distance along the *chain axis* for translational repetition of the chain structure.
Note 1: The chain identity period is usually denoted by c.
Note 2: An example is given in Fig. 1.

2.3 (chain) conformational repeating unit
Smallest structural unit of a polymer chain with a given conformation that is repeated along that chain through symmetry operations [5].

2.4 geometrical equivalence
Symmetry correspondence among units belonging to the same chain.
Note: The symmetry elements always bear a special relationship to the *chain axis* (see also the note, Definition 2.9).

2.5 equivalence postulate
Working hypothesis that the chain monomeric units are geometrically equivalent [6].

2.6 helix
Molecular conformation of a spiral nature, generated by regularly repeating rotations around the backbone bonds of a macromolecule.
Note: An example is shown in Fig. 1.

2.7 helix residue
Smallest set of one or more successive configurational base units that generates the whole chain through helical symmetry.

2.8 class of helix
Number of skeletal chain atoms contained within the *helix residue*.

2.9 line repetition groups
Possible symmetries of arrays extending in one direction with a fixed *repeating distance* [3, 5, 7].

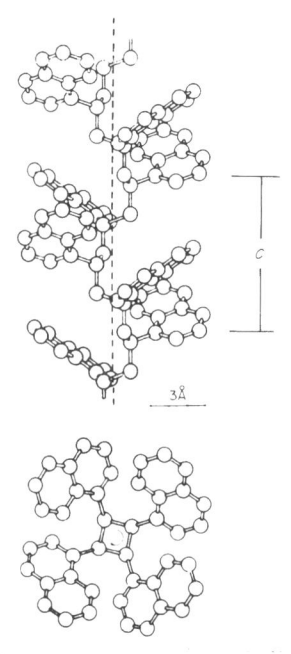

Fig. 1. Side view (above) and end view (below) of the macromolecule of isotactic poly[1-(1-naphthyl)ethane-1,2-diyl] in the crystalline state. The helix symbol is s($2^*4/1$). The *chain axis* is shown by the dashed line, and *c* is the *chain identity period*. Hydrogen atoms are omitted. [From P. Corradini and P. Ganis. *Nuovo Cimento*, Suppl. **15**, 96 (1960)].

Note: Linear polymer chains in the crystalline state must belong to one of the line repetition groups (see Table 1 for some examples). Permitted symmetry elements are: the identity operation (symbol 1); the translation along the *chain axis* (symbol t); the mirror plane orthogonal to the *chain axis* (symbol m) and that containing the *chain axis* (symbol d); the glide plane containing the *chain axis* (symbol c); the inversion centre, placed on the *chain axis* (symbol i); the two-fold axis orthogonal to the *chain axis* (symbol 2); the helical, or screw, symmetry where the axis of the *helix* coincides with the *chain axis*. In the latter case, the symbol is s(A^*M/N), where s stands for the screw axis, A is the class of the *helix*,[*] and / are separators, and *M* is the integral number of residues contained in *N* turns, corresponding to the *identity period* (*M* and *N* must be prime to each other) [8, 9] (see Fig. 1). The class index A may be dropped if deemed unnecessary, so that the *helix* may also be simply denoted as s(*M/N*) [5, 7].

2.10 structural disorder
Any deviation from the ideal three-dimensional regularity of the *polymer crystal* structure.

Note: Examples of structural disorder in *crystalline polymers* are given in Table 2.

2.10.1 lattice distortion
Structural disorder resulting from misalignment of the *unit cells* within the *crystals*.

Table 1. Chain symmetry of some crystalline polymers

Line repetition group [a]	Polymer (source-based name, structure-based name)
t1	trans-1,4-polyisoprene, poly[(E)-1-methylbut-2-ene-1,4-diyl]
s(A*M/N)	isotactic polypropene, poly(1-methylethane-1,2-diyl) (M/N=3/1, A=2)
s(A*M/N)2	syndiotactic polypropene, poly(1-methylethane-1,2-diyl) (M/N=2/l, A=4, helical modification)
tm	poly(N,N'-heptane-1,7-diylheptanediamide), poly(iminoheptanedioyliminoheptane-1,7-diyl)
tc	poly(1,1-difluoroethene), poly(1,1-difluoroethane-1,2-diyl) (modification 2)
ti	diisotactic poly[ethene-alt-(E)-but-2-ene], diisotactic poly(1,2-dimethylbutane-1,4-diyl)
s(5*2/1)m	poly(cyclopentene), poly(cyclopentane-1,2-diyl)
s(14*2/1)d	poly(hexamethylene adipamide), poly(iminohexanedioyliminohexane-1,6-diyl)
tcm	syndiotactic 1,2-polybuta-1,3-diene, poly(1-vinylethane-1,2-diyl)
s(1*2/1)dm	polyethene, poly(methylene), PE

[a] See the note, Definition 2.9 for explanation of symbols.

Table 2. Examples of structural disorder occurring in crystalline polymers

Type of structural disorder	Examples Polymer (source-based name , structure-based name)
(i) lattice distortion	as in usual crystallization (i.e., mechanical strain, lattice dislocation, impurities etc.)
(ii) chain orientation disorder	isotactic polypropene, poly(1-methylethane-1,2-diyl) [12], isotactic polystyrene, poly(1-phenylethane-1,2-diyl) [13], poly(1,1-difluoroethene), poly(1,1-difluoroethane-1,2-diyl)-form II [14]
(iii) configurational disorder	atactic polymers capable of crystallization: poly(vinyl alcohol), poly(1-hydroxyethane-1,2-diyl) [15], poly(vinyl fluoride), poly(1-fluoroethane-1,2-diyl) [16] (see Fig. 2)
(iv) conformational disorder	high-temperature polymorph of trans-1,4-polybuta-1,3-diene, poly[(E)-but-1-ene-1.4-diyl] [17] (Fig. 3); cis-1,4-polyisoprene, poly[(Z)-2-methylbut-2-ene-1,4-diyl] [17] (Fig. 4)
(v) copolymer isomorphism	poly(acetaldehyde-co-propionaldehyde) poly[oxy(methylmethylene)/oxy(ethylmethylene)] [19] isotactic poly(but-1-ene-co-3-methylbut-1-ene) isotactic poly(1-ethylethane-1,2-diyl/1-isopropylethane-1,2-diyl) [20] isotactic poly(styrene-co-4-fluorostyrene) isotactic poly[1-phenylethane-1,2-diyl/1-(4-fluoro-

TERMINOLOGY

homopolymer isomorphism	phenyl)ethane-1,2-diyl] [21] mixtures of isotactic poly(4-methylpent-1-ene) and isotactic poly(4-methylhex-1-ene) mixtures of isotactic poly(1-isobutylethane-1,2-diyl) and isotactic poly[1-(2-methylbutyl)ethane-1,2-diyl] [22]

2.10.2 chain-orientational disorder

Structural disorder resulting from the statistical coexistence within the *crystals* of identical chains with opposite orientations.

Note: A typical example is provided by the up-down statistical coexistence of anticlined chains in the same *crystal* structure.

2.10.3 configurational disorder

Structural disorder resulting from the statistical co-crystallization of different configurational repeating units (see Fig. 2).

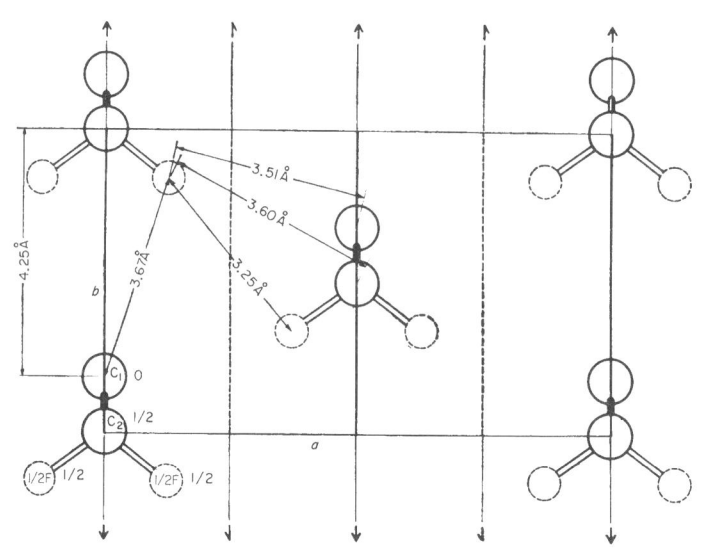

Fig. 2. End projection of atactic poly(vinyl fluoride) [poly(1-fluoroethane-1,2-diyl)] chains in the crystalline state. Broken circles show fluorine atoms with 50% probability [16].

2.10.4 conformational disorder

Structural disorder resulting from the statistical co-existence within the *crystals* of identical configurational units with different conformations (see Figs. 3 and 4).

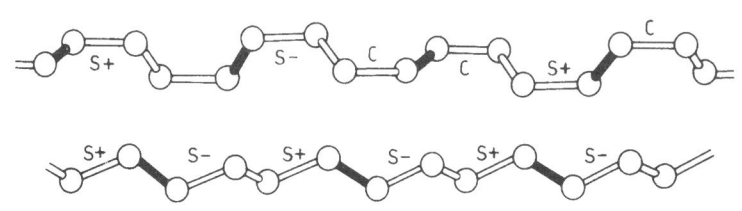

Fig. 3. Chain conformation of the disordered (above) and ordered (below) polymorphs of *trans*-1,4-polybutadiene (poly[(E)-but-1-ene-1,4-diyl]) in the crystalline state. The heavy black lines designate the double bonds and the symbols S+, S- and C the conformation [17].

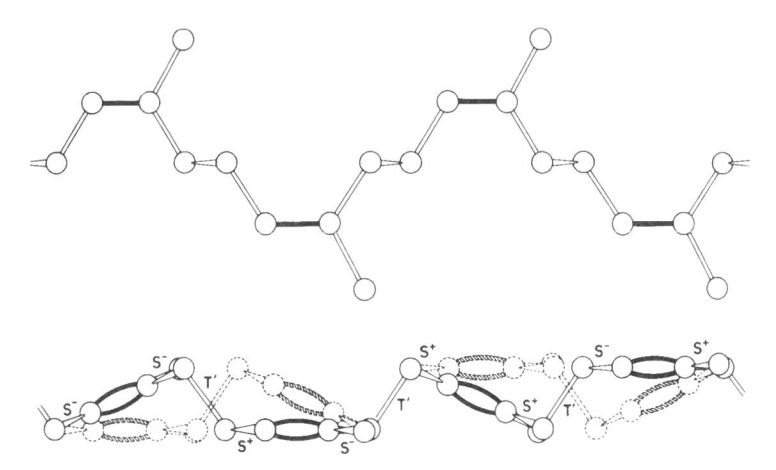

Fig. 4. Different possible conformations of *cis*-1,4-polyisoprene (poly[(Z)-2-methylbut-2-ene-1,4-diyl]) in the crystalline state, as viewed sideways along two orthogonal axes [17].

2.10.5 macromolecular isomorphism
Statistical co-crystallization of different constitutional repeating units, which may either belong to the same copolymer chains (copolymer isomorphism) or originate from different homopolymer chains (homopolymer isomorphism).
Note: Isomorphism is a general term: in the strict sense, the *crystal* structure is essentially the same throughout the range of compositions; in isodimorphism or isopolymorphism, there are two or more *crystal* structures, respectively, depending on composition.

3 TERMINOLOGY RELATING TO MORPHOLOGICAL ASPECTS [9, 10]

3.1 lamellar crystal
Type of *crystal* with a large extension in two dimensions and a uniform thickness.
Note: A lamellar crystal is usually of a thickness in the 5-50 nm range, and it may be found individually or in aggregates. The parallel-chain *stems* intersect the lamellar plane at an angle between 45° and 90°. The lamellae often have pyramidal shape owing to differences in the *fold domains*; as a result, one can deduce different *fold planes* and *fold surfaces* from the lamellar morphology.

3.2 lath crystal
Lamellar crystal prevailingly extended along one lateral dimension.

3.3 multilayer aggregate
Stack of *lamellar crystals* generated by spiral growth at one or more screw dislocations.
Note: The axial displacement over a full turn of the screw (Burgers vector) is usually equal to one lamellar thickness.

3.4 long spacing
Average separation between stacked *lamellar crystals*.

Note: The long spacing is usually measured by small-angle x-ray or neutron diffraction.

TERMINOLOGY

3.5 axialite
Multilayer aggregate, consisting of *lamellar crystals* splaying out from a common edge.

3.6 dendrite
Crystalline morphology produced by skeletal growth, leading to a 'tree-like' appearance.

3.7 fibrous crystal
Type of *crystal* significantly longer in one dimension than in either of the other two.
Note: Fibrous crystals may comprise essentially extended chains parallel to the fibre axis; however, macroscopic polymer fibres containing chain-folded crystals are also known.

3.8 shish-kebab structure
Polycrystalline morphology of double habit consisting of *fibrous crystals* overgrown epitaxially by *lamellar crystals*, the *stems* of which are parallel to the fibre axis.

3.9 spherulite
Polycrystalline, roughly spherical morphology consisting of lath, fibrous or *lamellar crystals* emanating from a common centre.
Note: Space filling is achieved by branching, bending or both, of the constituent fibres or lamellae.

4 TERMINOLOGY RELATING TO MOLECULAR CONFORMATION WITHIN POLYMER CRYSTALS [9, 10]

4.1 tie molecule
Molecule that connects at least two different *crystals*.

4.2 stem
Crystallized, rodlike portion of a polymer chain connected to non-rodlike portions, or chain ends, or both.

4.3 chain fold
Conformational feature in which a loop connects two parallel *stems* belonging to the same *crystal*.

4.4 fold
Loop connecting two different *stems* in a folded chain.

4.5 fold plane
Crystallographic plane defined by a large number of *stems* that are connected by *chain folds*.

4.6 fold surface
Surface approximately tangential to the *folds*.

4.7 fold domain
Portion of a *polymer crystal* wherein the *fold planes* have the same orientation.
Note: The sectors of *lamellar crystals* frequently represent *fold domains*.

CRYSTALLINE POLYMERS

4.8 adjacent re-entry
Model of *crystallinity* in which *chain folds* regularly connect model adjacent *stems*.

4.9 switchboard model
Model of *crystallinity* in which the crystallized segments of a macromolecule belong to the same *crystal*, although the *stems* are connected randomly.

4.10 fringed-micelle model
Model of *crystallinity* in which the crystallized segments of a macromolecule belong predominantly to different *crystals*.

4.11 folded-chain crystal
Polymer crystal consisting predominantly of chains that traverse the *crystal* repeatedly by folding as they emerge at its external surfaces.
Note: The re-entry of the chain into the *crystal* is assumed to be adjacent or near-adjacent within the lattice.

4.12 parallel-chain crystal
Type of *crystal* resulting from parallel packing of *stems*, irrespective of the *stems*' directional sense.

4.13 extended-chain crystal
Polymer crystal in which the chains are in an essentially fully extended conformation.

4.14 globular-chain crystal
Type of *crystal* comprised of macromolecules having globular conformations.
Note: Globular-chain crystals usually occur with globular proteins.

5 TERMINOLOGY RELATING TO CRYSTALLIZATION KINETICS [10, 11]

5.1 nucleation
Formation of the smallest crystalline entity, the further growth of which is thermodynamically favoured.
Note: Nucleation may be classified as primary or secondary. Primary nucleation can be homogeneous or heterogeneous; if heterogeneous nucleation is initiated by entities having the same composition as the crystallizing polymer, it is called self-nucleation. Secondary nucleation is also known as surface nucleation.

5.2 molecular nucleation
Initial crystallization of a small portion of a macromolecule, after which further crystallization is thermodynamically favoured.
Note: Molecular nucleation may give rise to a new *crystal* or increase the size of a pre-existing one.

5.3 Avrami equation
Equation, describing crystallization kinetics, of the form:

$$1-\phi_c = \exp(-Kt^n)$$

where ϕ_c is the crystalline volume fraction developed at time t and constant temperature, and K and n are suitable parameters.

Note 1: K is temperature-dependent.

Note 2: According to the original theory, n should be an integer from 1 to 4, the value of which should depend only on the type of the statistical model; however, it has become customary to regard it as an adjustable parameter that may be non-integral.

Note 3: The Avrami equation addresses the problem that *crystals* growing from different nuclei can overlap. Accordingly, the equation is sometimes called the 'overlap equation'.

5.4 primary crystallization

First stage of crystallization, considered to be ended when most of the *spherulite* surfaces impinge on each other.

Note: In isothermal crystallization, primary crystallization is often described by the *Avrami equation*.

5.5 secondary crystallization

Crystallization occurring after *primary crystallization*, usually proceeding at a lower rate.

5.6 reorganization

Molecular process by which (i) amorphous or poorly ordered regions of a polymer specimen become incorporated into *crystals*, or (ii) a change to a more stable *crystal* structure takes place, or (iii) defects within the *crystals* decrease.

Note 1: *Secondary crystallization* may be involved in the reorganization process.//
Note 2: Reorganization may result from annealing.//
Note 3: (i) and (iii) may also be called crystal perfection.

5.7 recrystallization

Reorganization proceeding through partial melting.

Note: Recrystallization is likely to result in an increase in the *degree of crystallinity*, or crystal perfection, or both.

5.8 segregation

Rejection of a fraction of macromolecules, or of impurities, or both, from growing *crystals*.

Note: The rejected macromolecules are usually those of insufficient relative molecular mass, or differing in constitution or configuration (e.g., branching, tacticity, etc.).

REFERENCES

1. IUPAC. 'Glossary of basic terms in polymer science (IUPAC Recommendations 1996)', *Pure Appl. Chem.* **68**, 2287-2311 (1996). Reprinted as Chapter 1 this volume.
2. IUPAC. 'Definitions of terms relating to individual macromolecules, their assemblies, and dilute polymer solutions (Recommendations 1988)', *Pure Appl. Chem.* **61**, 211-241 (1989). Reprinted as Chapter 3 this volume.
3. The International Union of Crystallography. *International Tables for X-ray Crystallography.* Kynoch Press, Birmingham, UK (1969), Vol. I.
4. H. Tadokoro. *Structure of Crystalline Polymers.* Wiley-Interscience, New York (1979).
5. IUPAC. 'Stereochemical definitions and notations relating to polymers (Recommendations 1980)', *Pure Appl. Chem.* **53**, 733-752 (1981). Reprinted as Chapter 2 this volume.
6a. C. W. Bunn. 'The stereochemistry of chain polymers', *Proc. R. Soc. London, A* **180**, 67 (1942).
6b. M. L. Huggins. 'Comparison of the structures of stretched linear polymers', *J. Chem. Phys.* **13**, 37 (1945).
6c. L. Pauling, R. B. Corey, H. R. Branson. 'The structure of proteins: two hydrogen-bonded helical configurations of the polypeptide chain', *Proc. Natl. Acad. Sci. U.S.A.* **37**, 205 (1951).
6d. G. Natta and P. Corradini. 'Structure of crystalline polyhydrocarbons', *Nuovo Cimento,* Suppl. **15**, 9 (1960).
7. P. Corradini. 'Chain conformation and crystallinity' in *Stereochemistry of Macromolecules* (Ed. A. Ketley). Marcel Dekker, New York (1968), Vol. III, pp. 1-60.
8. R. L. Miller. 'Crystallographic data for various polymers' in *Polymer Handbook,* 2nd ed., J. Brandrup and E. H. Immergut (Eds.). Wiley-Interscience, New York (1975), **III**-1.
9. B. Wunderlich. *Macromolecular Physics.* Academic Press, New York (1973), Vol. 1.
10. P. H. Geil. *Polymer Single Crystals.* Interscience, New York (1963).
11. B. Wunderlich. *Macromolecular Physics.* Academic Press, New York (1976), Vol. 2.
12. G. Natta and P. Corradini. 'Structure and properties of isotactic polypropylene', *Nuovo Cimento,* Suppl. **15**, 40 (1960).
13. G. Natta, P. Corradini, I. W. Bassi. 'Crystal structure of isotactic polystyrene', *Nuovo Cimento,* Suppl. **15**, 68 (1960).
14. Y. Takahashi and H. Tadokoro. 'Short-range order in form II of poly(vinylidene fluoride): antiphase domain structures', *Macromolecules* **16**, 1880 (1983).
15. C. W. Bunn and H. S. Peiser. 'Mixed crystal formation in high polymers', *Nature* **159**, 161 (1947).
16. G. Natta, I. W. Bassi, G. Allegra. 'Struttura crystallina del polivinilfloruro atattico', *Atti Accad. Naz. Lincei Rend., Cl. Sci. Fis., Mat. Nat.* **31**, 350 (1961).
17. P. Corradini. 'Observation of different conformations of a macromolecule in the crystalline state', *J. Polym. Sci., Polym. Symp.* **51**, 1 (1975).
18. G. Allegra and I. W. Bassi. 'Isomorphism in synthetic macromolecular systems', *Adv. Polym. Sci.* **6**, 549 (1969).
19. A. Tanaka, Y. Hozumi, K. Hatada, S. Endo, R. Fujishige. 'Isomorphism phenomena

in polyaldehydes', *J. Polym. Sci., Part B: Polym. Lett.* **2**, 181 (1964).
20. A. Turner Jones. 'Crystalline phases in copolymers of butene and 3-methylbutene', *J. Polym. Sci., Polym. Lett. Ed.* **3**, 591 (1965). *J. Polym. Sci., Part B: Polym. Lett.*
21. G. Natta. P. Corradini, D. Sianesi, D. Morero. 'Isomorphism phenomena in macromolecules', *J. Polym. Sci.* **51**, 527 (1961).
22. G. Natta, G. Allegra, I. W. Bassi, C. Carlini, E. Chiellini, G. Montagnoli. 'Isomorphism phenomena in isotactic poly(4-methyl-substituted α-olefins) and in isotactic poly(alkyl vinyl ethers)', *Macromolecules* **2**, 311 (1969).

7: Definitions of Basic Terms Relating to Low-Molar-Mass and Polymer Liquid Crystals

CONTENTS

1. Preamble
2. General definitions
3. Types of mesophase
4. Textures and defects
5. Physical characteristics of liquid crystals
6. Liquid-crystal polymers
7. References
8. Alphabetical index of terms
9. Glossary of recommended abbreviations and symbols

1 PREAMBLE

This document provides definitions of the basic terms that are widely used in the field of liquid crystals and in polymer science (See references 1-39). It is the first publication of the Commission on Macromolecular Nomenclature dealing specifically with liquid crystals.

The recommendations made, resulting from the joint effort of the IUPAC Commission IV.1 Working Party and members of the International Liquid Crystal Society, are concerned with terminology relating to low-molar-mass and liquid-crystal polymers. Since much of the terminology is common to both classes of liquid crystals, this document has not been divided into sections dealing separately with these two classes of substances. After some general definitions (Section 2), there are sections dealing successively with the structures and optical textures of liquid crystals (Sections 3 and 4), their physical characteristics (including electro-optical and magneto-optical properties) (Section 5) and finally liquid-crystal polymers (Section 6). An alphabetical index of terms and a glossary of recommended symbols are provided for the convenience of the reader.

Implied definitions, occurring in Notes to the main definitions, are indicated by using bold type for the terms so defined.

Originally prepared by a Working Group consisting of C. Noël (France), V. P. Shibaev (Russia), M. Barón (Argentina), M. Hess (Germany), A. D. Jenkins (UK), J.-I. Jin (Korea), A. Sirigu (Italy), R. F. T. Stepto (UK) and W. J. Work (USA), with contributions from G. R. Luckhurst (UK), S. Chandrasekhar (India), D. Demus (Germany), J. W. Goodby (UK), G. W. Gray (UK), S. T. Lagerwall (Sweden), O. D. Lavrentovich (USA), and M. Schadt (Switzerland) of the Internacional Liquid Cristal Society. Reprinted from *Pure Appl. Chem.* **73**, 845-895 (2001).

TERMINOLOGY

2 GENERAL DEFINITIONS

2.1 mesomorphic state
mesomorphous state

State of matter in which the degree of molecular order is intermediate between the perfect three-dimensional, long-range positional and orientational order found in solid crystals and the absence of long-range order found in isotropic liquids, gases and amorphous solids.

Note 1: The term mesomorphic state has a more general meaning than *liquid-crystal state*, but the two are often used as synonyms.

Note 2: The term is used to describe orientationally disordered crystals, crystals with molecules in random conformations (i.e., conformationally disordered crystals), plastic crystals and *liquid crystals*.

Note 3: A compound which can exist in a mesomorphic state is usually called a *mesomorphic compound*.

Note 4: A vitrified substance in the mesomorphic state is called a **mesomorphic glass** and is obtained, for example, by rapid quenching or by crosslinking.

2.2 liquid-crystal state (LC state)
liquid-crystalline state

Mesomorphic state having long-range orientational order and either partial positional order or complete positional disorder.

Note 1: In a LC state, a substance combines the properties of a liquid (e.g., flow, ability to form droplets) and a crystalline solid (e.g., anisotropy of some physical properties).

Note 2: A LC state occurs between the crystalline solid and the isotropic liquid states on varying, for example, the temperature.

2.2.1 liquid-crystalline phase (LC phase)

Phase occurring over a definite temperature range within the *LC state*.

2.3 liquid crystal (LC)

Substance in the *LC state*.

Note: A pronounced anisotropy in the shapes and interactions of molecules, or molecular aggregates is necessary for the formation of liquid crystals.

2.4 mesophase

Phase occurring over a definite range of temperature, pressure or concentration within the *mesomorphic state*.

2.4.1 enantiotropic mesophase

Mesophase that is thermodynamically stable over a definite temperature or pressure range.

Note: The range of thermal stability of an enantiotropic mesophase is limited by the melting point and the *clearing point* of an *LC* compound or by any two successive *mesophase* transitions.

2.4.2 thermotropic mesophase

Mesophase formed by heating a solid or cooling an isotropic liquid, or by heating or cooling a thermodynamically stable *mesophase* at constant pressure.

Note 1: The adjective 'thermotropic' describes a change of phase with a change of temperature. 'Thermotropic' may also be used to qualify types of *mesophase* (e.g., thermotropic *nematic*).

Note 2: Analogous changes can also occur on varying the pressure at constant temperature in which case the *mesophase* may be termed **barotropic mesophase**.

2.4.3 lyotropic mesophase

Mesophase formed by dissolving an amphiphilic *mesogen* in a suitable solvent, under appropriate conditions of concentration, temperature and pressure.

Note 1: The essential feature of a lyotropic liquid crystal is the formation of molecular aggregates or micelles as a result of specific interactions involving the molecules of the *amphiphilic mesogen* and those of the solvent.

Note 2: The mesomorphic character of a lyotropic mesophase arises from the extended, ordered arrangement of the solvent-induced micelles. Hence, such mesophases should be regarded as based not on the structural arrangement of individual molecules (as in a non-amphiphilic or a *thermotropic mesophase*), but on the arrangement within multimolecular *domains*.

2.4.4 amphitropic compound

Compound that can exhibit *thermotropic* as well as *lyotropic mesophases*.

Note: Examples are potassium salts of unbranched alkanoic acids, lecithin, certain polyisocyanates, cellulose derivatives with side-chains, such as (2-hydroxypropyl)cellulose, and cyanobiphenyl derivatives of alkyl(triethyl)ammonium bromide.

2.4.5 monotropic mesophase

Metastable *mesophase* that can be formed by supercooling an isotropic liquid or an *enantiotropic mesophase* at a given pressure to a temperature below the melting point of the crystal.

Note: Monotropic *transition temperatures* are indicated by placing parentheses, (), around the values.

2.5 transition temperature, T_{XY}, SI Unit: K

Temperature at which the transition from *mesophase* X to *mesophase* Y occurs.

Note: *Mesophase* X should be stable at lower temperatures than phase Y. For example, the *nematic*-isotropic transition temperature would be denoted as T_{NI}.

2.6 clearing point, T_{cl} or T_i, SI Unit: K
clearing temperature
isotropization temperature

Temperature at which the transition between the *mesophase* with the highest temperature range and the isotropic phase occurs.

Note: The term should only be used when the identity of the *mesophase* preceding the isotropic phase is unknown.

2.7 virtual transition temperature

Transition temperature that cannot be measured directly, determined by extrapolation of transition lines in binary phase diagrams to 100% of that particular component.

TERMINOLOGY

Note 1: A virtual *transition temperature* lies outside the temperature range over which the (meso) phase implied can be observed experimentally.
Note 2: A virtual *transition temperature* is not well defined; it will, for example, depend on the nature of the liquid-crystal components used to construct the phase diagram.
Note 3: A virtual *transition temperature* is indicated by placing square brackets, [ ], around its value.

2.8   transitional entropy, $\Delta_{XY}S$, SI Unit: J K^{-1} mol^{-1}

Change in entropy on transition from phase X to phase Y.
Note 1: The transitional entropy reflects the change in order, both orientational and translational, at the phase transition.
Note 2: Phase X should be stable at lower temperatures than phase Y.
Note 3: Numerical values of the molar entropy of transition should be given as the dimensionless quantity $\Delta_{XY}S/R$ where R is the gas constant.

2.9   divergence temperature, T^*, SI Unit: K
     pre-transitional temperature

Temperature at which the orientational correlations in an isotropic phase diverge.
Note 1: The divergence temperature is the lowest limit of metastable supercooling of the isotropic phase.
Note 2: The divergence occurs at the point where the isotropic phase would be expected to undergo a second-order transition to the liquid-crystal phase, were it not for the intervention of a first-order transition to the liquid-crystal phase.
Note 3: The divergence temperature for nematogens can be measured by using the Kerr effect or Cotton-Mouton effect or by light-scattering experiments.
Note 4: T^* occurs below the clearing temperature, usually by about 1 K in isotropic-to-nematic transitions and increases to at least 10 K for isotropic-to-smectic transitions.

2.10   mesogenic group
      mesogenic unit
      mesogenic moiety

Part of a molecule or macromolecule endowed with sufficient anisotropy in both attractive and repulsive forces to contribute strongly to *LC mesophase*, or, in particular, to *LC mesophase* formation in low-molar-mass and polymeric substances.
Note 1: '*Mesogen*ic' is an adjective that in the present document applies to molecular moieties that are structurally compatible with the formation of *LC phases* in the molecular system in which they exist.
Note 2: *Mesogen*ic groups occur in both low-molar-mass and polymeric compounds.
Note 3: The majority of *mesogen*ic groups consist of rigid rod-like or disc-like molecular moieties.

Examples

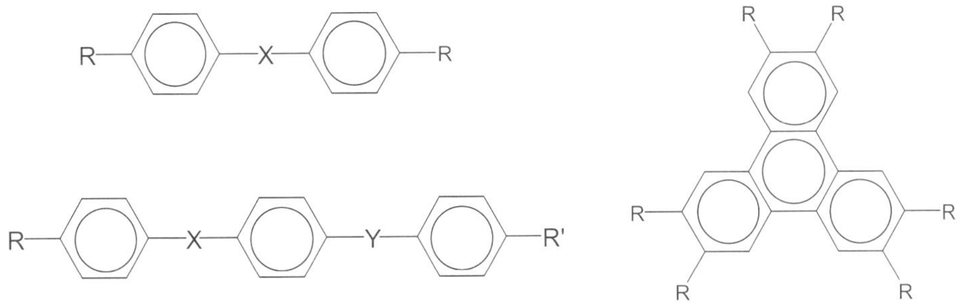

Where R and R' are alkyl groups and X and Y are divalent linking units such as:

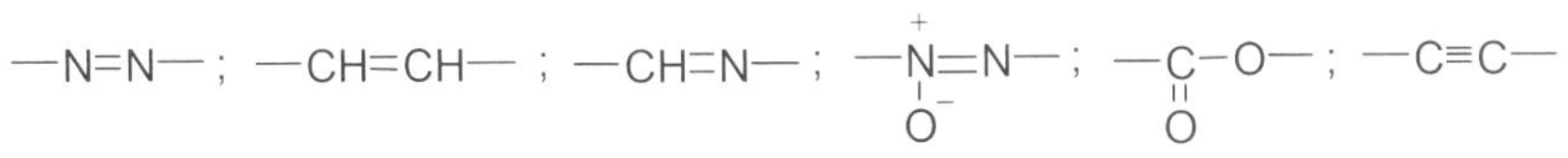

$-N=N-$; $-CH=CH-$; $-CH=N-$; $-\overset{+}{N}=N-$; $-\underset{\parallel}{C}-O-$; $-C\equiv C-$
$\underset{O^-}{}\underset{O}{}$

2.11 mesogen
mesogenic compound
mesomorphic compound

Compound that under suitable conditions of temperature, pressure and concentration can exist as a *mesophase*, or, in particular as a *LC phase*.

Note 1: When the type of *mesophase* formed is known, more precisely qualifying terminology can be used, e.g., **nematogen**, **smectogen** and **chiral nematogen**.

Note 2: When more than one type of *mesophase* can be formed, more than one qualification could apply to the same compound and then the general term mesogen should be used.

2.11.1 amphiphilic mesogen

Mesogen composed of molecules consisting of two parts of contrasting character that are hydrophilic and hydrophobic or lipophobic and lipophilic.

Note 1: Examples of amphiphilic mesogens are soaps, detergents and some block copolymers.

Note 2: Under suitable conditions of temperature and concentration, the similar parts of amphiphilic molecules cluster together to form aggregates or micelles.

2.11.2 non-amphiphilic mesogen

Mesogen that is not of the amphiphilic type.

Note 1: At one time it was thought that a non-amphiphilic molecule had to be long and rod-like for *mesophase* formation, but it has now been established that molecules of other types and shapes, for example, disc-like and banana-shaped molecules, may also form *mesophases*. (See ref. 6).

Note 2: A selection of the types of non-amphiphilic *mesogens* is given in Definitions 2.11.2.1-2.11.2.8.

2.11.2.1 calamitic mesogen

Mesogen composed of rod-like or lath-like molecules.
Note: Examples are:

- 4-butyl-*N*-(4-methoxybenzylidene)aniline (BMBA) (a)
- 4,4′-dimethoxyazoxybenzene (b)
- 4′-pentylbiphenyl-4-carbonitrile (c)
- 4-(*trans*-4-pentylcyclohexyl)benzonitrile (d)
- cholesterol esters and esters of cholest-5-ene-3-carboxylic acid (e).

(a)

(b)

(c)

(d)

(e)

R = $CH_3(CH_2)_nCOO$-, $CH_3(CH_2)_nOCO$-

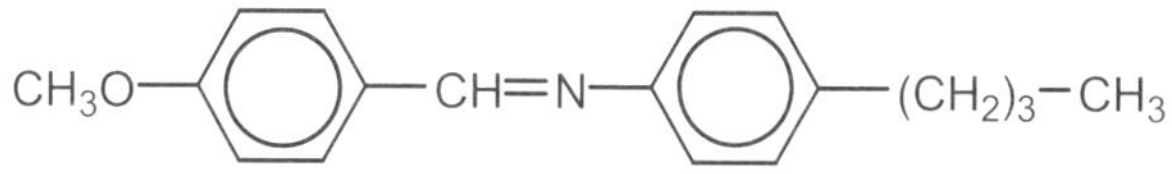

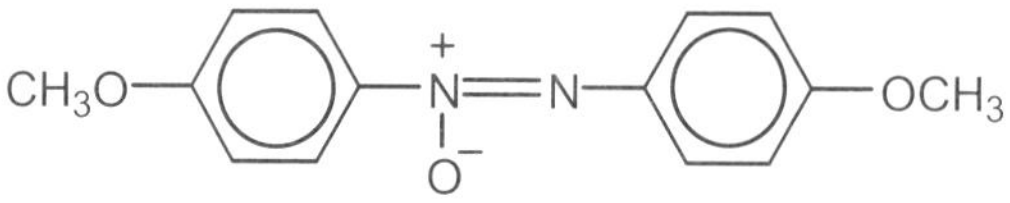

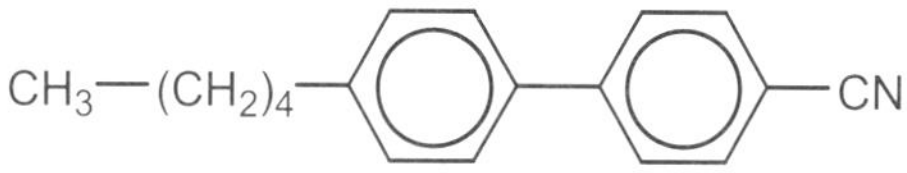

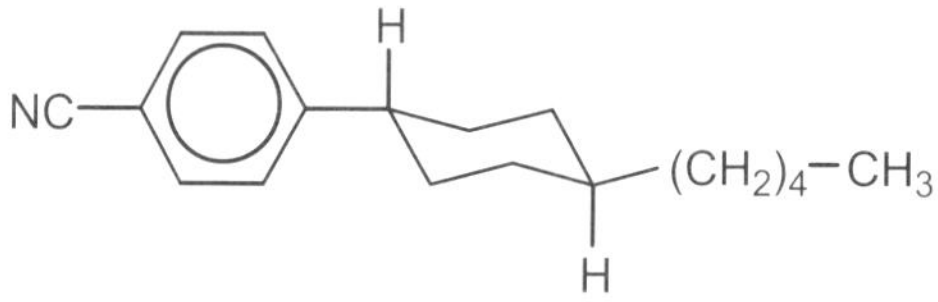

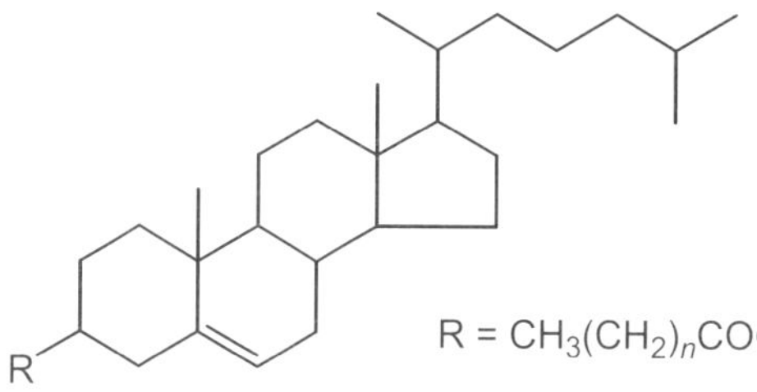

2.11.2.2 discotic mesogen
discoid mesogen

Mesogen composed of relatively flat, disc- or sheet-shaped molecules.

Note 1: Examples are: hexakis(acyloxy)benzenes (a), hexakis(acyloxy)- and hexakis(alkoxy)-triphenylenes (b), 5*H*,10*H*,15*H*-diindeno[1,2-*a*:1′,2′-*c*]fluorene derivatives (c).

(a)

(b)

(c)

R = C_nH_{2n+1}-, $C_nH_{2n+1}O$-, C_nH_{2n+1} COO-, C_nH_{2n+1} COO-

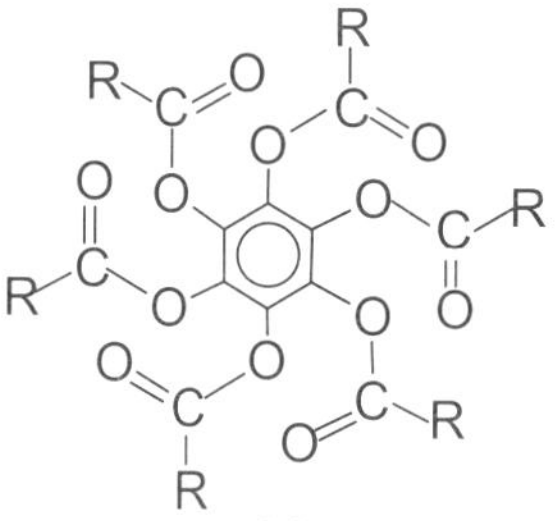

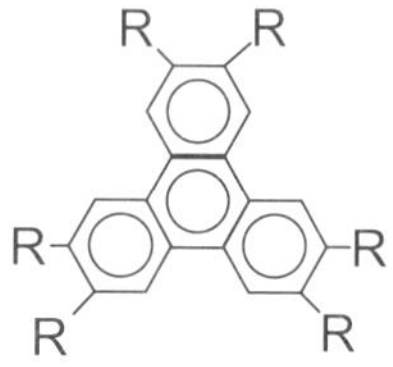

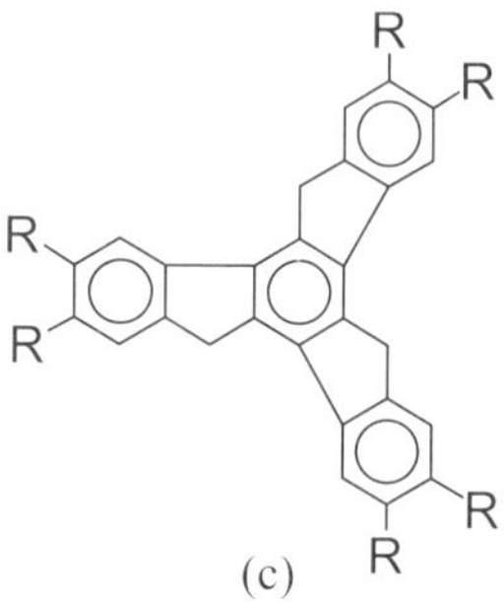

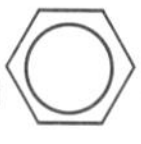

Note 2: The adjective 'discotic' is also employed to describe the *nematic mesophases* formed by discotic *mesogens*. The *mesophases* formed by a columnar stacking of disc-like molecules are described as *columnar mesophases*.

2.11.2.3 pyramidic mesogen
conical or cone-shaped mesogen
bowlic mesogen

Mesogen composed of molecules containing a semi-rigid conical core.
Note: Examples are hexasubstituted 5*H*,10*H*,15*H*-tribenzo[*a,d,g*][9]annulenes.

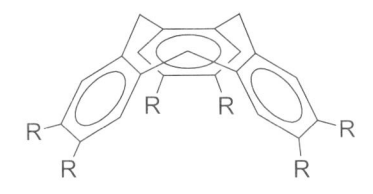

2.11.2.4 sanidic mesogen

Mesogen composed of board-like molecules with the long-range orientational order of the phase reflecting the symmetry of the constituent molecules.
Note: See also Definition 3.4.

2.11.2.5 polycatenary mesogen

Mesogen composed of molecules each having an elongated rigid core with several flexible chains attached to the end(s).
Note 1: The flexible chains are usually aliphatic.
Note 2: The numbers of flexible chains at the ends of the core can be indicated by using the term *m,n*-polycatenary *mesogen*.
Note 3: There exist several descriptive names for these *mesogens*. Examples are: (a) biforked *mesogen* (2,2-polycatenary *mesogen*), (b) hemiphasmidic *mesogen* (3,1-polycatenary *mesogen*), (c) forked hemiphasmidic *mesogen* (3,2-polycatenary *mesogen*) and (d) phasmidic *mesogen* (3,3-polycatenary *mesogen*). Examples of each type with the core represented by ⬬ are given together with a specific example of a forked hemiphasmidic *mesogen* (c).

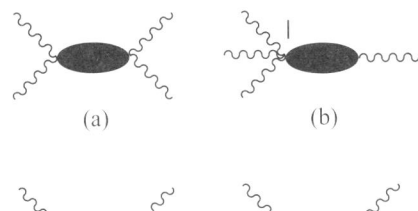

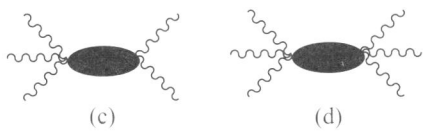

a specific example is:

TERMINOLOGY

2.11.2.6   swallow-tailed mesogen
Mesogen composed of molecules each with an elongated rigid core with, at one end, a branched flexible chain, having branches of about the same length.
Note: A sketch of the structure of a swallow-tailed *mesogen* is

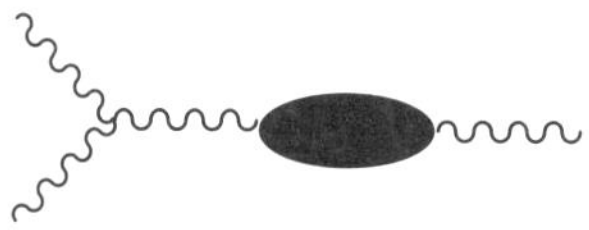

and an example is the fluorene derivative

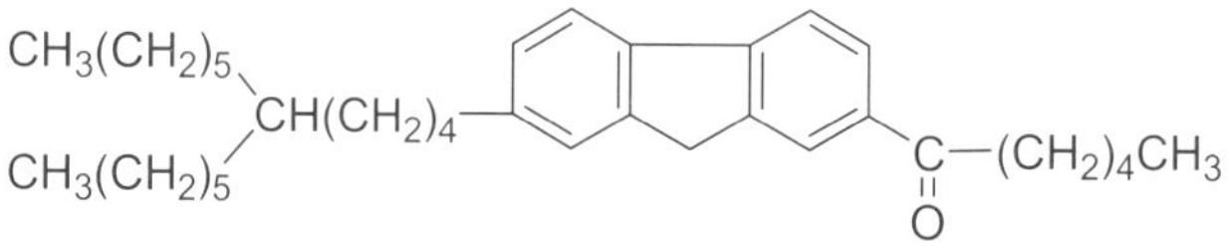

2.11.2.7   bis-swallow-tailed mesogen
Mesogen composed of molecules each with an elongated rigid core and a branched flexible chain, with branches of about the same length, attached at each end.

Example:

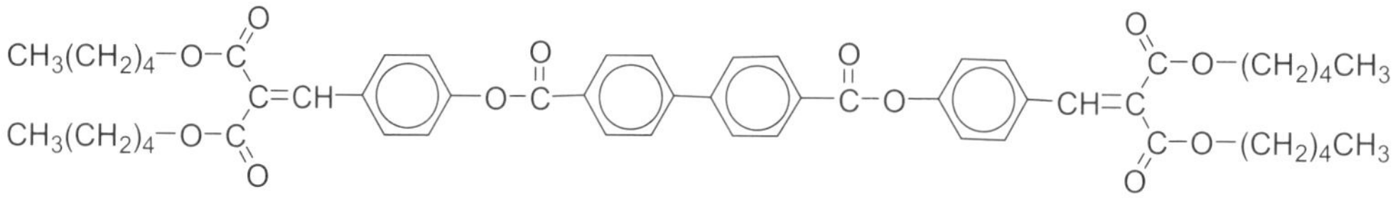

2.11.2.8   laterally branched mesogen
Mesogen composed of rod-like molecules with large lateral branches such as alkyl, alkoxy or ring-containing moieties.

Example:

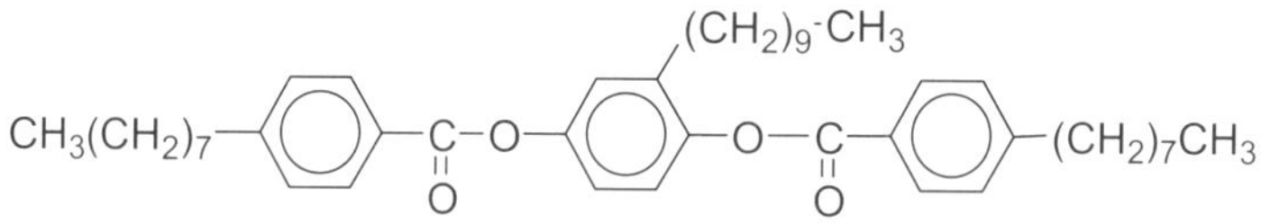

2.11.2.9   liquid-crystal oligomer
          mesogenic oligomer
Mesogen constituted of molecules, each with more than one *mesogenic group*.
Note 1: The *mesogenic groups* usually have identical structures.
Note 2: A liquid-crystal dimer or *mesogenic* dimer is sometimes known as a twin *mesogen*. Use of the terms 'di*mesogenic* compounds' and 'Siamese-twin *mesogen*' for **liquid-crystal dimer** or **mesogenic dimer** is not recommended.
Note 3: Examples of *mesogenic* dimers are: (a) fused twin *mesogen*, where the *mesogenic groups* are linked rigidly by a (usually fused) ring system, (b) ligated twin *mesogen*, in

which the *mesogenic groups* are connected by a *spacer* at a central position, (c) tail-to-tail twin *mesogen*, which has a flexible *spacer* linking the two groups and (d) side-to-tail twin *mesogen*. The structures of these different types of liquid-crystal dimers are illustrated with the *mesogenic groups* represented by ⬬.

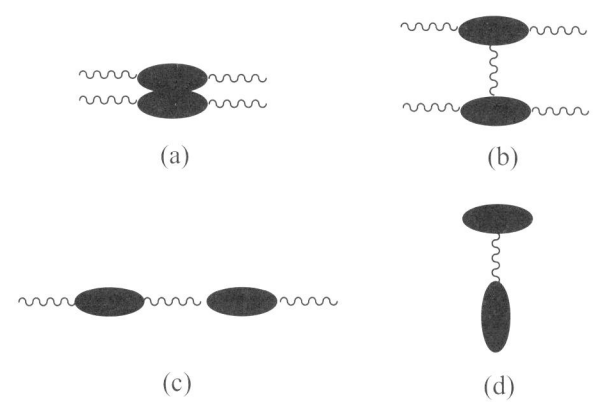

(a) (b)

(c) (d)

A specific example of type (c), a tail-to-tail liquid-crystal dimer, is

$$CH_3-CH_2-O-\text{Ph}-N=N-\text{Ph}-O-\underset{O}{C}-(CH_2)_8-\underset{O}{C}-O-\text{Ph}-N=N-\text{Ph}-O-CH_2-CH_3$$

wherein $-(CH_2)_8-$ is the flexible *spacer* linking the two *mesogenic groups*.

Note 4: A liquid-crystal dimer with different *mesogenic groups* linked by a *spacer* is known as an **asymmetric liquid-crystal dimer**.

Note 5: A liquid-crystal dimer with flexible hydrocarbon chains having an odd number of carbon atoms is called an **odd-membered liquid-crystal dimer**, whilst one with hydrocarbon chains having an even number of carbon atoms is called an **even-membered liquid-crystal dimer**.

2.11.2.10 banana mesogen

Mesogen constituted of bent or so-called banana-shaped molecules in which two *mesogenic groups* are linked through a semi-rigid group in such a way as not to be co-linear.

Note: Examples of such structures are:

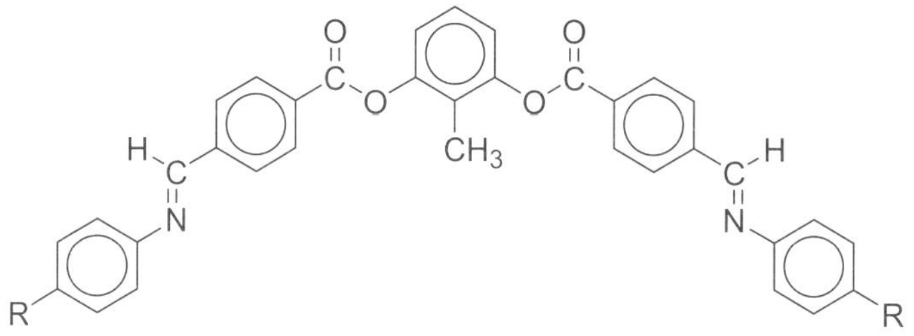

TERMINOLOGY

with the substituent R being an unbranched alkoxy group, ($-O-C_nH_{2n+1}$).

2.11.3 metallomesogen
Mesogen composed of molecules incorporating one or more metal atoms.
Note: Metallo*mesogens* may be either *calamitic* or *discotic mesogens*.
Examples

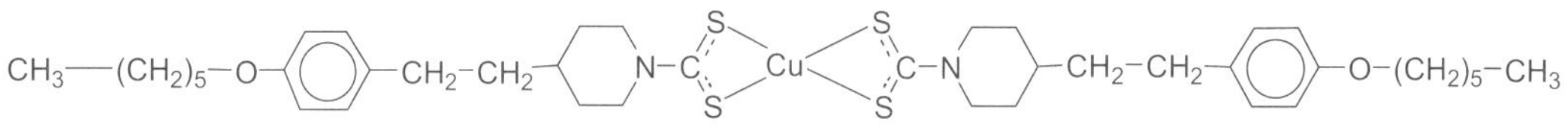

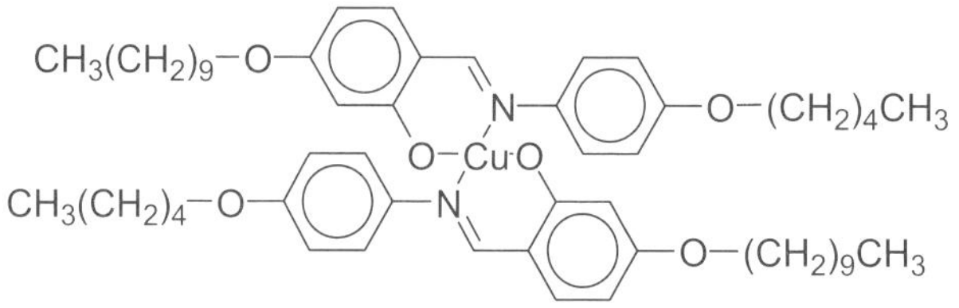

3 TYPES OF MESOPHASE

3.1 mesophases of calamitic mesogens

3.1.1 uniaxial nematic mesophase, N or N_u
nematic

Mesophase formed by a non-chiral compound or by the racemate of a chiral compound in which the spatial distribution of the molecular centers of mass is devoid of long-range positional order and the molecules are, on average, orientationally ordered about a common axis defined as the *director* and represented by the unit vector ***n***.

Note 1: See Fig. 1 for an illustration of the molecular organization in a uniaxial nematic mesophase.
Note 2: The unit vector, ***n***, is defined in 3.1.1.1. (See also Fig. 1.)
Note 3: The direction of ***n*** is usually arbitrary in space.
Note 4: The extent of the positional correlations for the molecules in a nematic phase is comparable to that of an isotropic phase although the distribution function is necessarily anisotropic.
Note 5: From a crystallographic point of view, the uniaxial nematic structure is characterised by the symbol $D_{\infty h}$ in the Schoenflies notation (∞/mm in the International System).
Note 6: Since the majority of nematic phases are uniaxial, if no indication is given, a nematic phase is assumed to be uniaxial but, when there is the possibility of a biaxial as well a uniaxial nematic, a uniaxial phase should be denoted as N_u.

3.1.1.1 director, *n*

Local symmetry axis for the singlet, orientational distribution of the molecules of a *mesophase*.

Note 1: The *director* is defined as a unit vector, but directions *+n* and *-n* are arbitrary.
Note 2: In *uniaxial nematics*, formed by compounds consisting of either rod-like or disc-like molecules, the mean direction of the effective molecular symmetry axis coincides with the *director*.
Note 3: The *director* also coincides with a local symmetry axis of any directional property of the *mesophase*, such as the refractive index or magnetic susceptibility.

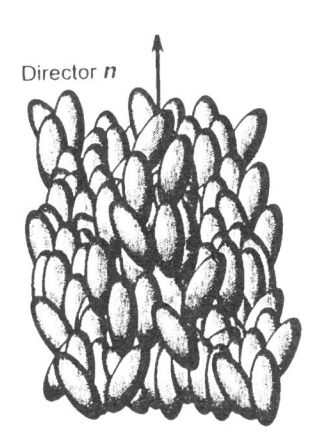

Fig.1. A representation of the molecular organization in a uniaxial nematic mesophase.

3.1.2 cybotactic groups

Assembly of molecules in a *nematic mesophase* having a short-range smectic-like array of the constituent molecules.

Note: Two types of short-range smectic-like structures are possible. One is analogous to a *smectic A mesophase* where the molecules tend to lie along a layer normal and the other is like a *smectic C mesophase* where the molecules tend to be oblique with respect to a layer normal. See Fig. 2 for illustrations of the molecular arrangements in the *smectic A* type structure and the *smectic C* type structure.

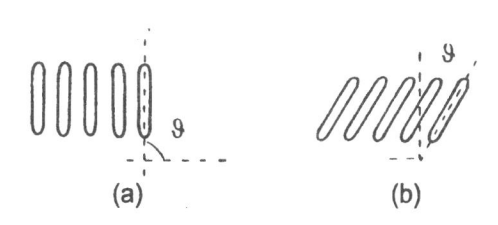

Fig. 2. Schematic representation of the molecules in (a) a smectic A-like local structure and (b) a smectic C-like local structure, making angle ϑ with the layer normal.

TERMINOLOGY

3.1.3 chiral nematic mesophase, N*
chiral nematic
cholesteric mesophase
cholesteric

Mesophase with a helicoidal superstructure of the *director*, formed by *chiral*, *calamitic* or *discotic* molecules or by doping a *uniaxial nematic* host with chiral guest molecules in which the local *director* ***n*** precesses around a single axis.

See Fig. 3 for an illustration of the helicoidal molecular distribution in a chiral nematic mesophase.

Note 1: Locally, a chiral nematic mesophase is similar to a uniaxial nematic, except for the precession of the *director* ***n*** about the axis, Z.

Note 2: The *director* is periodic along Z with the pitch P of the helical structure equal to a turn of the local *director* ***n*** by 2π.

Note 3: Chiral nematic mesophases exhibit Bragg scattering of circularly polarised light at a wavelength λ_R proportional to the pitch P ($\lambda_R = <n>P$, where $<n>$ is the mean refractive index).

Note 4: The *director* precession in a chiral nematic mesophase is spontaneous and should be distinguished from an induced twisted structure produced by a mechanical twist of a *nematic mesophase* between confining surfaces.

Note 5: The term chiral nematic mesophase or chiral nematic is preferred to cholesteric mesophase or cholesteric.

3.1.4 blue phase, (BP)

Mesophase with a three-dimensional spatial distribution of helical *director* axes leading to frustrated structures with defects arranged on a lattice with cubic symmetry and lattice constants of the order of the wavelength of visible light.

Note 1: See Fig. 4 for a possible model for a BP.

Note 2: The name 'blue phase' derives historically from the optical Bragg reflection of blue light but, because of larger lattice constants, BPs can reflect visible light of longer wavelengths.

Note 3: With *chiral nematic* substances forming *chiral nematic mesophases* of short pitch (<700 nm), up to three blue phases occur in a narrow temperature range between the *chiral nematic* phase and the isotropic phase.

Note 4: A BP is optically isotropic and exhibits a Bragg reflection of circularly polarised light.

Note 5: Two BPs of different cubic symmetry (space group I $4_1$32 for BP I and P $4_2$32 for BP II) are presently known, together with a third (BPIII) of amorphous structure. Several other BPs of different cubic symmetry exist but only in the presence of external electric fields.

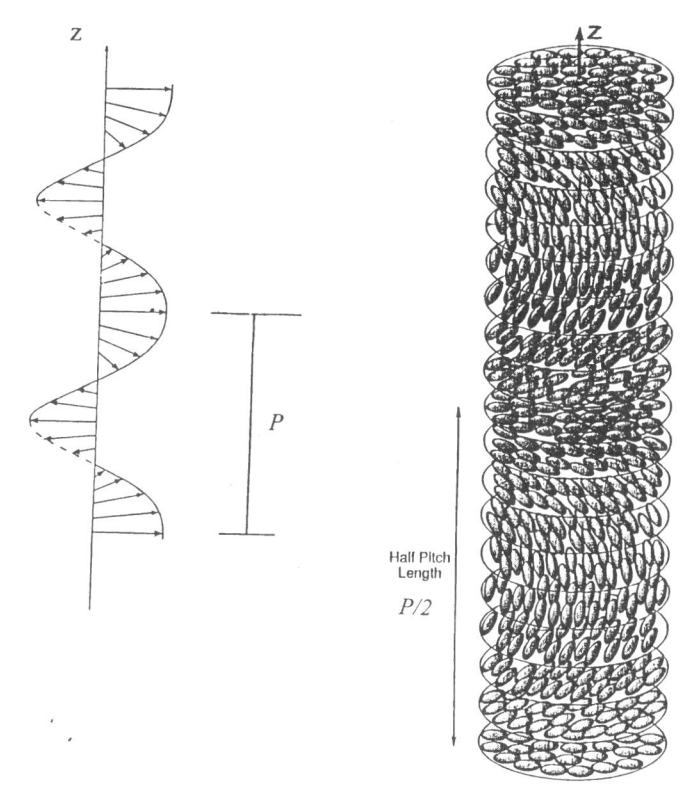

Fig. 3. Illustrating the structure of a chiral nematic mesophase.

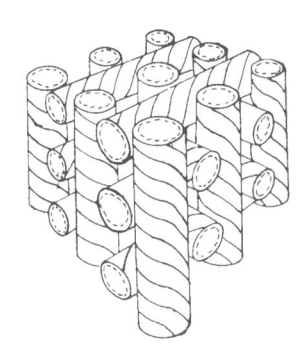

Fig. 4. Illustrating a cubic lattice formed by double-twist cylinders as a possible model of a BP.

3.1.5 smectic mesophase, (Sm)
Mesophase that has the molecules arranged in layers with a well-defined layer spacing or periodicity.
Note 1: There are several types of smectic mesophases, characterised by a variety of molecular arrangements within the layers.
Note 2: Although the total number of smectic mesophases cannot be specified, the following types have been defined: SmA, SmB, SmC, SmF and SmI. The alphabetical order of suffixes merely indicates an order of discovery.
Note 3: The classification of SmD as smectic is largely a consequence of history, and should be discontinued (see Definition 3.1.9).

Note 4: At one time, a number of *mesophases* were identified as smectic on the basis of their *optical textures*, but they are in fact soft crystals characterised by very low yield stresses. Hence, these three-dimensionally ordered phases should no longer be called smectic mesophases. They are akin to plastic crystals with some elementary long-range order and are referred to by the letters E, J, G, H, and K.

Note 5: Tilted smectic mesophases formed by chiral compounds or containing chiral mixtures are designated by the superindex * (SmC*, SmF*, etc.). (See, for example, Definition 3.1.5.1.3.)

3.1.5.1 smectic mesophases with unstructured layers

3.1.5.1.1 smectic A mesophase, (SmA)

Smectic mesophase involving a parallel arrangement of the molecules within layers in which the long axes of the molecules tend to be perpendicular to the layer planes and the molecular centers of mass have no long-range positional order parallel to the layer planes.

Note 1: See Fig. 5 for the molecular organization in a smectic A mesophase

Note 2: Each layer approximates to a true two-dimensional liquid. The system is optically uniaxial and the optic axis, Z, is normal to the layer planes.

Note 3: The directions $+Z$ and $-Z$ are interchangeable.

Note 4: The structure of a smectic A mesophase is characterised by the symbol $D_{\infty h}$ in the Schoenflies notation (∞, 2 in the International System).

Note 5: The *lyotropic* equivalent of a smectic A mesophase is known as a **lamellar mesophase**; where layers of amphiphilic molecules are separated by layers of solvent, normally water, or by oil in an **inverse lamellar mesophase**.

Note 6: A smectic A-phase containing a chiral molecule or dopant, can be called a chiral smectic A-phase. The recommended symbol is SmA* wherein the (*) indicates that the macroscopic structure of the *mesophase* is chiral.

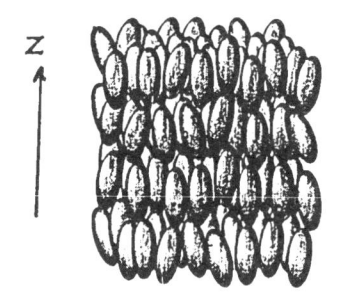

Fig. 5. Illustrating the structure of a smectic A mesophase.

3.1.5.1.2 smectic C mesophase, (SmC)

Analogue of a *smectic A mesophase* involving an approximately parallel arrangement of the molecules within layers in which the *director* is tilted with respect to the layer normal and the molecular centers-of-mass have no long-range positional order parallel to the layer planes (See Fig. 6).

Note 1: See Fig. 6 for an illustration of the molecular organization in a smectic C mesophase.

Note 2: The physical properties of a smectic C mesophase are those of a biaxial crystal.

Note 3: The smectic C structure corresponds to monoclinic symmetry characterised by the symbol C_{2h}, in the Schoenflies notation and the space group $t\ 2/m$ in the International System.

Note 4: The tilt direction varies in a random manner from layer to layer in conventional smectic C *mesophases*. However it can alternate from layer to layer, as in *an antiferroelectric chiral smectic C mesophase* (see Definition 5.9, Note 7) and in the smectic C mesophase formed by certain liquid crystal dimers with an odd-number of carbon atoms in the *spacers*. The recommended symbol for this type of *mesophase* is SmCa.

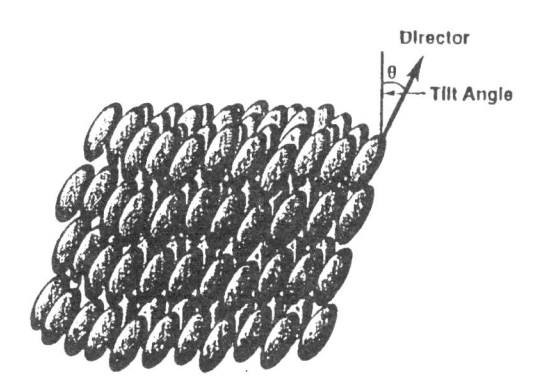

Fig. 6. Illustrating the structure of the smectic C mesophase

3.1.5.1.3 chiral smectic C mesophase, (SmC*)

Smectic C mesophase in which the tilt direction of the *director* in each successive layer is rotated through a certain angle relative to the preceding one so that a helical structure of a constant pitch is formed.

Note 1: See Fig. 7 for an illustration of the molecular organization in a chiral smectic C mesophase.

Note 2: The (*) in SmC* and analogous notations indicate, as in 3.1.5.1.2 (Note 6), that the macroscopic structure of the *mesophase* is chiral. However, it is also used simply to indicate that some of the constituent molecules are chiral even though the microscopic structure may not be.

Note 3: A SmC* *mesophase* is formed by chiral compounds or mixtures containing chiral compounds.

Note 4: Locally, the structure of the chiral smectic C mesophase is essentially the same as that of the achiral smectic C mesophase except that there is a precession of the tilt direction about a single axis. It has the symmetry C_2 in the Schoenflies notation.

Note 5: This chiral smectic C phase is also known as the *ferroelectric* chiral smectic C phase.

Note 6: The helix can be unwound by surface forces to give a surface-stabilised SmC* which has a macroscopic polarization.

3.1.5.2 hexatic smectic mesophase

Smectic mesophase with in-plane short-range positional molecular order, weakly coupled two-dimensional layers and long-range bond orientational molecular order.

Note: There are three types of hexatic smectic mesophases: *smectic B* (SmB), *smectic F* (SmF) and *smectic I* (SmI). Here, the term 'hexatic' may be omitted because it is implicit for this group of *smectic mesophases*.

TERMINOLOGY

3.1.5.2.1 smectic B mesophase, (SmB)
Hexatic smectic mesophase in which the *director* is perpendicular to the layers with the long-range hexagonal bond-orientational order.
Note 1: See Fig. 8 for an illustration of the molecular organization in a smectic B mesophase.
Note 2: Positional molecular order does not propagate over distances larger than a few tens of nanometers but bond orientational molecular order extends over macroscopic distances within and across layers.
Note 3: By contrast with a smectic B mesophase, a crystal B mesophase has correlations of positional order (hexagonal) in three dimensions, i.e., correlations of position occur within and between layers.
Note 4: The structure of a smectic B mesophase is characterised by a D_{6h} point group symmetry, in the Schoenflies notation, by virtue of the bond orientational order.
Note 5: A smectic B mesophase is optically uniaxial.
Note 6: A smectic B mesophase is sometimes denoted SmB_{hex}. The subscript 'hex' denotes the hexagonal structure of the *mesophase*.

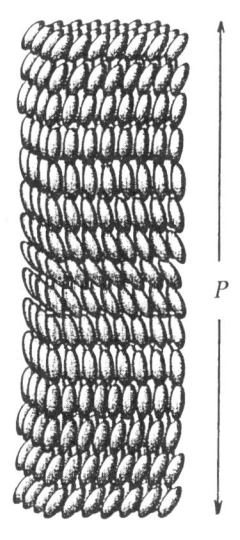

Fig. 7. Illustrating the structure of a chiral smectic C mesophase (P = helical pitch)

3.1.5.2.2 smectic F mesophase, (SmF)
Hexatic smectic mesophase the structure of which may be regarded as a C-centered monoclinic cell with a hexagonal packing of the molecules with the *director* tilted, with respect to the layer normals, towards the sides of the hexagons.
Note 1: See Fig. 9a for an illustration of the molecular organization in a smectic F mesophase, a tilted analogue of the *smectic B mesophase*.
Note 2: A SmF mesophase is characterised by in-plane short-range positional correlations and weak or no interlayer positional correlations.

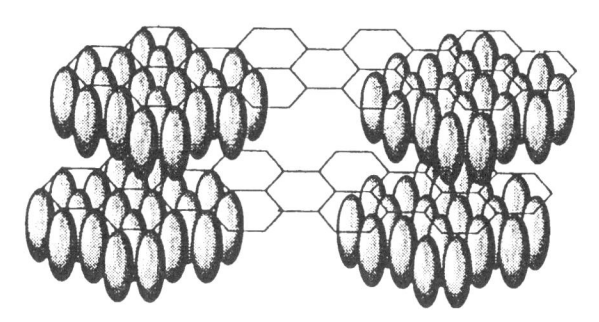

Fig. 8. Illustrating the structure of a smectic B mesophase.

Note 3: Positional molecular order extends over a few tens of nanometers but the bond orientational molecular order is long-range within a layer.
Note 4: The point-group symmetry is C_{2h} ($2/m$) in the Schoenflies notation, and the space group, $t\ 2/m$ in the International System.
Note 5: The smectic F mesophase is optically biaxial.
Note 6: Chiral materials form **chiral smectic F mesophases** denoted by SmF*.

3.1.5.2.3 smectic I mesophase, (SmI)
Hexatic smectic mesophase the structure of which may be regarded as a C-centered monoclinic cell with hexagonal packing of the molecules with the *director* tilted, with respect to the layer normals, towards the apices of the hexagons.
Note 1: See Figs. 9(a) and (b) for illustrations of the molecular organizations of *smectic* F and I *mesophases*. They are tilted analogues of the *smectic B mesophase*.
Note 2: The smectic I mesophase is optically biaxial.
Note 3: The in-plane positional correlations in a smectic I mesophase are slightly greater than in a *smectic F mesophase*.
Note 4: Chiral materials form **chiral smectic I mesophases** denoted by SmI*.

3.1.5.3 crystal B, E, G, H, J and K mesophases
Soft crystals that exhibit long-range positional molecular order, with three-dimensional stacks of layers correlated with each other.
Note 1: Originally, these *mesophases* were designated as smectic, but further investigations have demonstrated their three-dimensional character.
Note 2: In the crystal B and E *mesophases*, the molecular long axes are essentially parallel to the normals to the layer planes while in the G, H, J and K *mesophases* they are tilted with respect to the layer normals.
Note 3: The E, J and K phases have herringbone organizations of the molecular short axes and so the *mesophases* are optically biaxial.

3.1.6   polymorphic modifications of strongly polar compounds

3.1.6.1 re-entrant mesophase
Lowest temperature *mesophase* of certain compounds that exhibit two or more *mesophases* of the same type, over different temperature ranges.
Note 1: The recommended subscript for the designation of re-entrant *mesophases* is: re.
Note 2: Re-entrant *mesophases* are most commonly observed when the molecules have strong longitudinal dipole moments (see example).

TERMINOLOGY

Note 3: Sequences of re-entrant *mesophases* have also been found in binary mixtures of non-polar liquid-crystalline compounds.

Example: The following compound exhibits, as temperature decreases, an isotropic (I) phase, *nematic* (N), *smectic* A (SmA) re-entrant *nematic* (N_{re}), re-entrant *smectic* A (SmA_{re}) *mesophases* and a crystalline (Cr) phase, with transitions at the specified temperatures.

$$CH_3(CH_2)_7-O-\bigcirc-\underset{O}{\overset{\|}{C}}-O-\bigcirc-CH=CH-\bigcirc-CN$$

Cr 349 K SmA_{re} 369 K N_{re} 412 K SmA 520 K N 556 K I

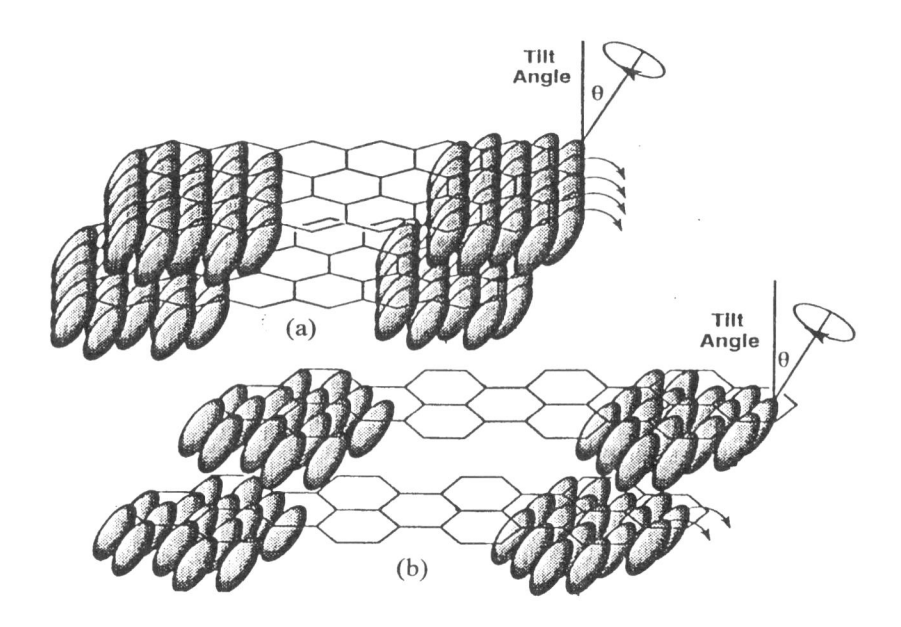

Fig. 9. Illustrating the tilt directions of the *director* in (a) SmF and (b) SmI *mesophases* indicating, respectively, the tilt of the *director* towards the sides of the hexagons (a) and the apices of the hexagons (b).

3.1.6.2 smectic A_1 (SmA_1), A_2 (SmA_2), A_d (SmA_d), C_1 (SmC_1), C_2 (SmC_2), C_d (SmC_d) mesophases

Smectic A and *smectic C mesophases* characterised by anti-parallel (SmA_2, SmA_d, and SmC_2, SmC_d) and random (SmA_1 and SmC_1) alignments of the molecular dipoles within the layer thickness in Fig. 10.

Note 1: See Figs. 10 and 11 for illustrations of the molecular arrangements in the *mesophases*.

Note 2: The subscripts 1, d and 2 indicate that the layer thickness is one, d and two times the fully-extended molecular length, with $1<d<2$.

Note 3: SmA_d and SmC_d *mesophases* form bilayers with partial overlapping of the molecules of adjacent layers.

Note 4: SmA_2 and SmC_2 phases form bilayers with anti-parallel ordering of the molecules.

Note 5: Bilayer structures are also known for *SmB* and *crystal E mesophases*.

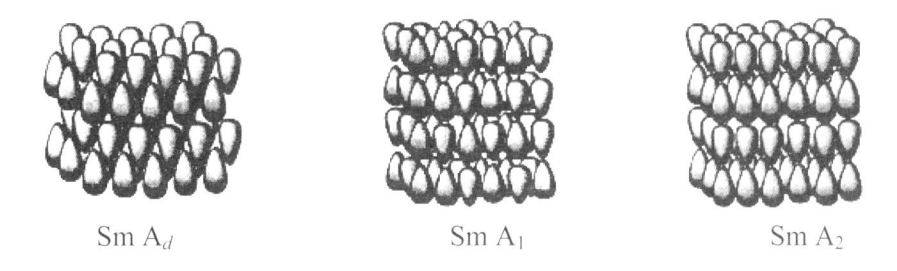

Sm A$_d$ Sm A$_1$ Sm A$_2$

Fig. 10. Illustrating the molecular structures of SmA$_d$, SmA$_1$, and SmA$_2$ *mesophases*.

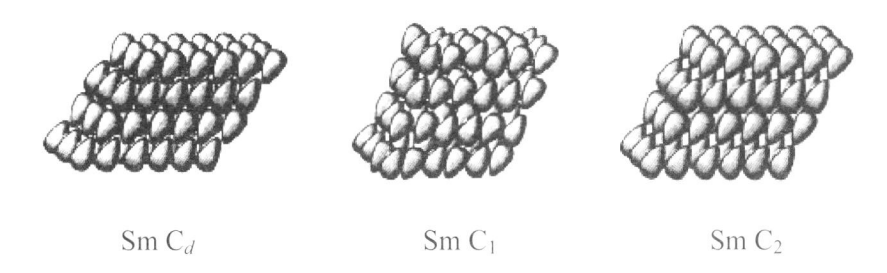

Sm C$_d$ Sm C$_1$ Sm C$_2$

Fig. 11. Illustrating the molecular structures of SmC$_d$, SmC$_1$, and SmC$_2$ *mesophases*.

3.1.6.3 modulated smectic mesophase
Smectic mesophase that has a periodic in-plane density variation.
Note 1: The recommended mark to designate a modulated *smectic mesophase* is a superior tilde (~).
Note 2: See Figs 12a and 12b for illustrations of the molecular arrangements in $\text{Sm}\widetilde{A}$ and $\text{Sm}\widetilde{C}$ *mesophases*.
Note 3: The $\text{Sm}\widetilde{A}$ *mesophase* is also known as a centered rectangular *mesophase* or antimesophase. The dimensional space group is *cmm* in the International System.
Note 4: The $\text{Sm}\widetilde{C}$ *mesophase* is also known as an oblique or ribbon *mesophase*. The dimensional space group is *pmg* in the International System.

3.1.7 intercalated smectic mesophase
Smectic mesophase that has a spacing between layers (smectic periodicity) of approximately one half of the molecular length.
Note 1: The recommended subscript to designate an intercalated *smectic mesophase* is: c
Note 2: Intercalated *smectic mesophases* are commonly observed for liquid-crystal dimers.
Note 3: At present intercalated *smectic* A (SmA$_c$) and *smectic* C (SmC$_c$) as well as intercalated *crystal B* (B$_c$), *G* (G$_c$) and *J* (J$_c$) *mesophases* have been observed.
Note 4: The local structure in the *nematic mesophase* of certain dimers exhibit an intercalated *smectic mesophase*.

TERMINOLOGY

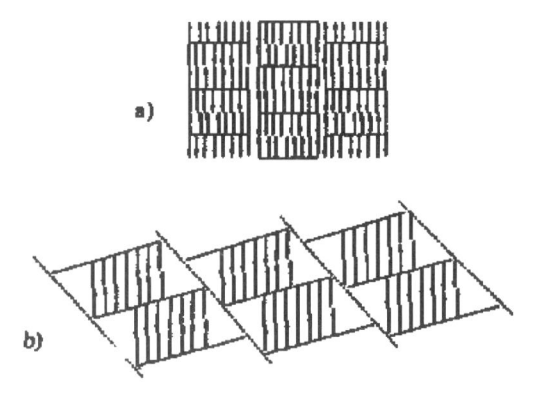

Fig. 12. Schematic drawing of the modulated smectic mesophases (a) Sm$\tilde{\text{A}}$ and (b) Sm$\tilde{\text{C}}$.

3.1.8 induced mesophase
Particular *mesophase* formed by a binary mixture, the components of which do not separately form *mesophases*, with the particular *mesophase* existing above the melting points of both components.
Note 1: The formation of an induced mesophase usually results from an attractive interaction between unlike species, the strength of which exceeds the mean of the strengths of the interactions between like species.
Note 2: Examples of such interactions that have been noted are dipolar/non-polar, charge-transfer and quadrupolar.
Note 3: *Mesophases* can also be induced when the free-volume between the large, irregular molecules of one component is filled by the smaller molecules of the second component. Such *mesophases* have been called filled *smectic mesophases* although the term 'induced' is recommended.
Note 4: A *monotropic mesophase* can be stabilised in a mixture when, as a result of melting-point depression, a metastable *mesophase* becomes stabilised. Such a *mesophase* is distinct from an induced mesophase.

3.1.9 cubic mesophase, (Cub)
Mesophase with an overall three-dimensional order of cubic symmetry in which each micellar unit cell contains several hundred molecules in random configurations, as in a liquid.
Note 1: The *mesophase* formerly designated as *smectic* D (see Definition 3.1.5, Note 3) belongs to this class.
Note 2: A cubic mesophase is optically isotropic; it may be distinguished from an isotropic liquid or a homeotropic phase by the fact that the optically-black isotropic phase or homeotropic phase nucleates in the birefringent SmC phase in straight-edged squares, rhombi, hexagons and rectangles.
Note 3: A cubic mesophase may be formed by rod-like molecules with strong, specific intermolecular interactions, such as hydrogen bonding, between them. However, they are also found in *polycatenary* compounds where there are no specific, strong interactions.
Note 4: Cubic *mesophases* have long been known in thermotropic salt-like compounds and in *lyotropic* liquid-crystals.
Note 5: There are several types of *thermotropic* and *lyotropic* cubic *mesophases*, with different symmetry and miscibility properties; when the space groups of these are known, they should be included in parentheses after the term 'Cub'.

Example: The following compound exhibits a crystalline phase (Cr), smectic SmC, cubic (Cub), *smectic* SmA *mesophases*, and an isotropic (I) phase, with transitions at the specified temperatures:

$$CH_3(CH_2)_{15}-O-\underset{NO_2}{\underset{|}{C_6H_3}}-C_6H_4-COOH$$

Cr 399.95 K   SmC 444.15 K   Cub 471.65 K   SmA 472.15 K   I

3.2 mesophases of disc-like mesogens
discotic mesophases
discotics

3.2.1 discotic nematic mesophase, N
discotic nematic

Nematic mesophase in which disc-shaped molecules, or the disc-shaped portions of macromolecules, tend to align with their symmetry axes parallel to each other and have a random spatial distribution of their centers of mass.

Note 1: See Fig. 13 for an illustration of the molecular arrangement in a *discotic mesophase*.

Note 2: The symmetry and structure of a *nematic mesophase* formed from disc-like molecules is identical to that formed from rod-like molecules. It is recommended therefore, that the subscript 'D' is removed from the symbol 'N_D', often used to denote a *nematic* formed from disc-like molecules.

Note 3: In some cases the discotic nematic mesophase is formed by compounds that do not have molecules of discotic shape (for example, phasmidic compounds, salt-like materials and oligosaccharides).

Note 4: *Chiral discotic nematic mesophases*, N^*, also exist.

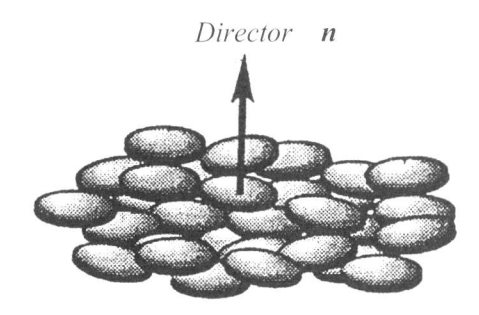

Fig. 13. Illustrating the organization of molecules in a discotic nematic mesophase.

3.2.2 columnar mesophase, (Col)
columnar discotic mesophase
columnar discotic

Mesophase in which disc-shaped molecules, the disc-shaped moieties of macromolecules or wedge-shaped molecules assemble themselves in columns packed parallel on a two-dimensional lattice, but without long-range positional correlations along the columns.

TERMINOLOGY

Note: Depending on the order in the molecular stacking in the columns and the two-dimensional lattice symmetry of the columnar packing, the columnar *mesophases* may be classified into three major classes: hexagonal, rectangular and oblique (see Definitions 3.2.2.1. to 3.2.2.3).

3.2.2.1 columnar hexagonal mesophase, (Col$_h$)
Columnar mesophase characterised by a hexagonal packing of the molecular columns.
Note 1: See Fig. 14 for an illustration of the molecular arrangement in a Col$_h$ mesophase.
Note 2: Hexagonal *mesophases* are often denoted Col$_{ho}$ or Col$_{hd}$ where h stands for hexagonal and o and d refer to the range of positional correlations along the column axes: o stands for ordered and d for disordered. The use of the subscripts o and d should be discontinued. In both cases, the ordering is liquid-like, only the correlation lengths are different.

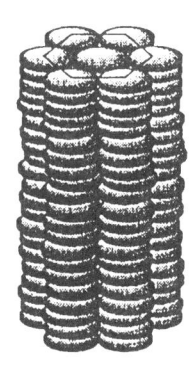

Fig. 14. Illustrating the molecular organization of a columnar hexagonal mesophase.

Note 3: The relevant space group of a Col$_h$ mesophase is *P* 6/*mmm* (equivalent to *P* 6/*m* 2/*m* in the International System and point group D$_{6h}$ in the Schoenflies notation).
Note 4: The *lyotropic* equivalent of a columnar hexagonal mesophase is known as a **hexagonal mesophase**; in it, columns of amphiphilic molecules are surrounded by the solvent, normally water, or an oil in an **inverse hexagonal mesophase**.

3.2.2.2 columnar rectangular mesophase, (Col$_r$)
Columnar mesophase characterised by a liquid-like molecular order along the columns, in which the columns are arranged in a rectangular packing.
Note 1: See Figs 15a–c for illustrations of molecular arrangements in columnar rectangular mesophases.
Note 2: The average orientation of the planes of the molecular discs is not necessarily normal to the column axes.
Note 3: Depending on the plane space-group symmetries, three rectangular *mesophases* are distinguished (See Fig. 15a–c).
Note 4: There also exist chiral columnar rectangular mesophases, with the molecular discs tilted periodically in the columns and with the tilt directions changing regularly down the columns.

3.2.2.3 columnar oblique mesophase, (Col$_{ob}$)
Columnar mesophase characterised by a liquid-like molecular order along the column, in which the columns are arranged with an oblique packing.

Note 1: See Fig. 15d for an illustration of the molecular arrangement in a columnar oblique mesophase.
Note 2: The average of the planes of the molecular discs is not necessarily normal to the columnar axes.
Note 3: The plane space-group symmetry of a Col_{ob} mesophase is P_1 (see Fig. 15d).
Note 4: There also exist chiral columnar oblique mesophases, with the tilt directions of the columnar discs varying regularly along the columns.

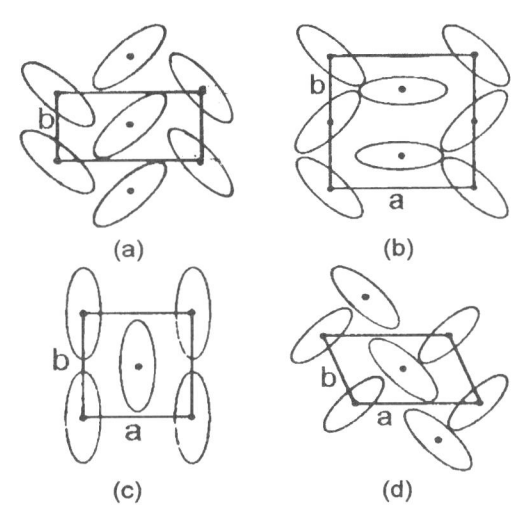

Fig. 15. Plan views of the two-dimensional lattice of the columns in columnar rectangular (a) to (c) and oblique (d) mesophases. Ovals indicate the planes of the molecular discs. Plane space group symmetries in the projection of the International System are:
(a) $P2_1/a$; (b) $P2/a$; (c) $C2/m$; (d) P_1.

3.3 biaxial mesophase

Mesophase composed of board-like molecules in which there are long-range orderings of both the longer and the shorter molecular axes.
Note 1: The recommended subscript to designate a biaxial mesophase is: b
Note 2: A biaxial mesophase has three orthogonal *directors* denoted by the unit vectors l, m and n.
Note 3: The tensorial properties of a biaxial mesophase have biaxial symmetry unlike the uniaxial symmetries of, for example, the *nematic* and *smectic A mesophases*.
Note 4: The biaxiality of the phase does not result from tilted structures as, for example, in a *smectic C mesophase*.
Note 5: Distinct biaxial *mesophases* are created when the molecular centers of mass are correlated within the layers. Such *mesophases* have been proposed for board-like polymers and have been called *sanidic mesophases*.
Note 6: *Sanidic* structures are analogous to the *columnar mesophases* formed by disc-like molecules.

TERMINOLOGY

3.3.1 biaxial nematic mesophase, N_b
biaxial nematic

Mesophase in which the long axes of the molecules are, on average, orientationally ordered about a common *director* and one of the shorter molecular axes is ordered, on average, about a second, orthogonal *director*.

Note 1: See Fig. 16 for an illustration of the molecular arrangement in a N_b mesophase.

Note 2: From a crystallographic point of view, the biaxial *nematic* structure is characterised by the symbol D_{2h} in the Schoenflies notation ($2/m$, m in International System).

Note 3: In *lyotropic* systems, biaxial nematic mesophases have been identified from the biaxial symmetry of their tensorial properties.

Note 4: The situation for *thermotropic calamitic* systems is less clear and for some compounds claimed to form a N_b, detailed investigations have found the *mesophase* to be of type N_u.

Note 5: A biaxial nematic has the same structure as a disordered *sanidic mesophase* (see Definition 3.4, Note 2); it is recommended that the latter name be discontinued and the name biaxial nematic be used.

Director ***n***

Fig. 16. Schematic representation of a biaxial nematic mesophase.

3.3.2 biaxial smectic A mesophase, SmA_b

Smectic A mesophase composed of board-like molecules with the longer and the shorter molecular axes orientationally ordered.

Note: For a SmA_b mesophase, the molecular centers-of-mass have only short-range positional order within a layer.

3.4 sanidic mesophase, Σ

Mesophase in which *board-shaped* molecules assemble in stacks packed parallel to one another on a one- or two-dimensional lattice (see Figs. 17 and 18).

Note 1: See Figs. 17 and 18 for examples of *sanidic mesophases*.

Note 2: Short board-like shaped molecules usually form *biaxial nematic mesophases*. It is recommended that the use of the term disordered *sanidic mesophases* for such *mesophases* be discontinued (see Definition 3.3.1, Note 5).

Note 3: Rotation of the molecules around their long axes is considerably hindered.

3.4.1 rectangular sanidic mesophase, Σ_r

Sanidic mesophase in which the molecular stacks are packed regularly side-by-side with long-range order along a stack normal as well as along the long stack-axis.

Note: See Fig. 17 for an illustration of the molecular arrangement in a Σ_r mesophase.

3.4.2 ordered sanidic mesophase, Σ_o

Mesophase in which the molecular stacks are packed regularly side-by-side with long-range order along a stack normal and no registration along the long stack-axis.
Note: See Fig. 18 for an illustration of the molecular arrangement in a Σ_o mesophase.

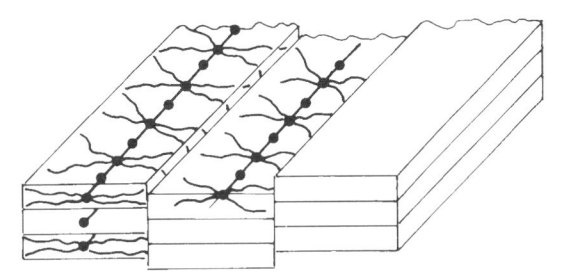

Fig. 17. Illustrating a rectangular sanidic mesophase.

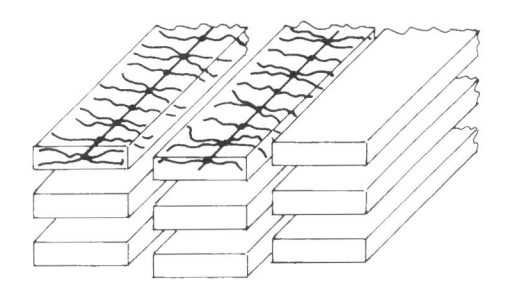

Fig. 18. Illustrating an ordered sanidic mesophase.

3.5 glassy mesophase

Mesophase in which non-vibratory molecular motion is frozen by supercooling a *mesophase* stable at a higher temperature.
Note: The recommended subscript to designate a glassy mesophase is: g.

3.6 twist grain-boundary mesophase (TGB)

Defect-stabilised *mesophase* created when a *smectic A mesophase* is subjected to a twist or bend distortion.
Note 1: The twist and bend distortions can be stabilised by an array of screw or edge dislocations.
Note 2: A TGB mesophase is analogous to the Abrikosov flux-phase of certain superconductors.

3.6.1 twist grain-boundary A* mesophase (TGBA*)

Mesophase with a helicoidal supramolecular structure in which blocks of molecules, with a local structure of the *smectic A* type, have their layer normals rotated with respect to each other and are separated by screw dislocations.

Note 1: See Fig. 19 for an illustration of the molecular arrangement of a TGBA* mesophase.
Note 2: The TGBA* mesophase is formed by a chiral compound or a mixture of chiral compounds.
Note 3: Two TGBA* structures are possible; in one, the number of blocks corresponding to a rotation of the layer normal by 2π is an integer, while in the other, it is a non-integer.
Note 4: A TGBA* is found in a phase diagram between *smectic A* and *chiral nematic mesophases* or between a *smectic A mesophase* and an isotropic phase.
Note 5: The temperature range of existence of a TGBA* mesophase is typically several K.

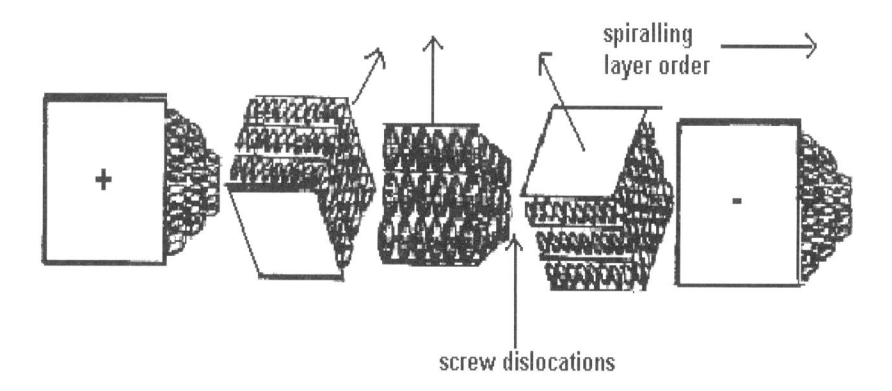

Fig. 19. Illustrating the structure of a TGBA* mesophase corresponding to half of the full helical twist.

3.6.2 twist grain-boundary C* mesophase (TGBC*)

Mesophase of helicoidal supermolecular structure in which blocks of molecules with a local structure of the *smectic C* type, have their layer normals rotated with respect to each other and are separated by screw dislocations.

Note 1: Two forms of TGBC* mesophase are possible: in one form the *director* within the layer is tilted and rotates coherently through the layers in a block as in a *chiral smectic C mesophase*, while in the other form the *director* within a block is simply tilted with respect to the layer normal as in a *smectic C mesophase*.
Note 2: In the case of a short pitch, when P is less than the wavelength λ, the macroscopic extraordinary axis for the refractive index is orthogonal to the *director*.

3.6.3 melted-grain-boundary mesophase (MGBC*)

Mesophase with a helicoidal supramolecular structure of blocks of molecules with a local *smectic C structure*. The layer normal to the blocks rotates on a cone to create a helix-like *director* in the *smectic C**. The blocks are separated by plane boundaries perpendicular to the helical axis. At the boundary, the *smectic* order disappears but the *nematic* order is maintained. In the blocks the *director* rotates from one boundary to the other to allow the rotation of the blocks without any discontinuity in the thermomolecular orientation.
Note: This phase appears between the TGBA and SmC* or N* and SmC* *mesophases*.

4 TEXTURES AND DEFECTS

4.1 domain
Region of a *mesophase* having a single *director*.
Note: See 3.1.1.1 for the definition of a *director*.

4.2 monodomain
Region of a uniaxial *mesophase* or a whole uniaxial *mesophase* having a single *director* or a region of a *biaxial mesophase* or a whole *biaxial mesophase* having two *directors*.
Note 1: See 3.1.1 for the definition of a *uniaxial nematic mesophase*, 5.8.1 for the definition of uniaxial *mesophase* anisotropy, and Definitions 3.3 and 5.8.2 relating to *biaxial mesophases*.
Note 2: For a *smectic mesophase*, the term monodomain also implies a uniform arrangement of the *smectic* layers.

4.3 homeotropic alignment
Molecular alignment of which the *director* is perpendicular to a substrate surface.
Note 1: See Fig. 20a.
Note 2: When the alignment of the *director* in a homeotropic alignment deviates from the perpendicular, the alignment is said to be a **pre-tilted homeotropic alignment**; the pre-tilt angle is the deviation from 90°.
Note 3: **Surface pretilt** is the deviation angle of the *director* away from the surface. It is used to control the threshold voltage and affects viewing angles.

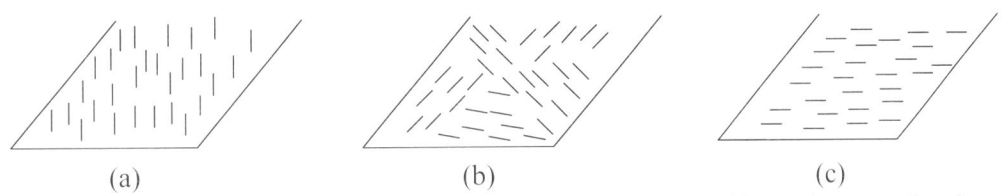

Fig. 20. Representing (a) homeotropic, (b) planar, and (c) uniform planar molecular alignments.

4.4 planar alignment
homogeneous alignment
Molecular alignment in which *directors* lie parallel to a substrate surface.
Note: See Fig. 20b.

4.5 uniform planar alignment
Molecular alignment in which the *director* is parallel to a substrate surface.
Note 1: See Fig. 20c.
Note 2: Sometimes a uniform planar alignment is called a 'uniform homogeneous alignment'. The latter term is not recommended.

4.6 twist alignment
Molecular alignment for which the *director* rotates in a helical fashion when passing between two substrate surfaces having molecules in *uniform planar alignments*.

TERMINOLOGY

Note 1: See Fig. 21. The length of a line in Fig. 21 indicates the length of a *director* projected onto the plane of the page.
Note 2: The orientation of the *directors* on the upper and lower substrate surfaces are usually mutually orthogonal and hence the *directors* undergo a 90° twist over the thickness of the liquid-crystal layer.

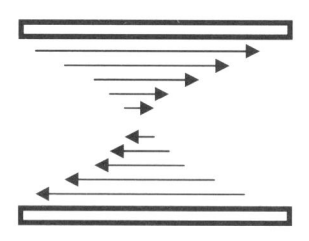

Fig. 21. Illustrating a twist alignment.

4.7 defect
Non-uniform molecular alignment that cannot be transformed into a uniform alignment without creating other defects.
Note 1: Dislocations and disclinations are major types of defects in liquid crystals.
Note 2: Three elementary types of defects may be distinguished in liquid crystals. They are point, line and wall defects,.
Note 3: A discontinuity in the structure (or in the mathematical function describing the structure) is considered as a singularity; in many cases a defect can be regarded as a singularity.

4.7.1 dislocation
Discontinuity in a regular molecular positional arrangement.
Note: Dislocations are found mainly in solid crystals.

4.7.2 disclination
Defect along a line in the regular orientation of *directors*.
Note 1: Disclinations are responsible for some *optical textures* seen with a polarizing microscope, such as the *schlieren texture* formed by disclination lines in nearly vertical orientations, whose projections are seen as dark points with two or four emerging dark stripes or brushes (see Definition 4.9.2).
Note 2: Disclinations are defects in molecular orientational order in contrast to dislocations that are defects in molecular positional order.

4.8 optical texture
Image of a liquid-crystal sample seen with a microscope, usually with crossed polarizers.
Note: An optical texture results from surface orientation of the *directors* at the boundaries of the sample and by defects formed in the sample.

4.9 nematic textures

4.9.1 nematic droplet
Spherical droplet that forms during a transition from an isotropic phase to a *nematic mesophase*. It has characteristic textures that depend on the droplet size and the *director* orientation at the nematic-isotropic interface.
Note: *Nematic* droplets display a texture characteristic of a *nematic mesophase* since they occur nowhere else.

4.9.1.1 bipolar droplet texture
Texture with two point defects at the poles of a *nematic droplet*.
Note 1: A pole is the position of the extreme of a *director* in a droplet.
Note 2: The point defects are called boojums.
Note 3: A bipolar droplet texture occurs when the *director* lies in the plane of a nematic-isotropic interface.

4.9.1.2 radial droplet texture
Texture with one point *defect* at the center of a *nematic droplet*.
Note 1: The point *defect* usually forms when the *director* is normal to the nematic-isotropic interface.
Note 2: The radial droplet texture shows four dark brushes located in the regions where the *director* is in the polarization plane of either the polarizer or the analyser.

4.9.2 schlieren texture
Texture observed for a flat sample between crossed polarizers, showing a network of black brushes connecting centers of point and line defects.
Note 1: The black brushes are also called black stripes or schlieren brushes.
Note 2: Black brushes are located in regions where the *director* lies in the plane of polarization of either the polarizer or the analyser.
Note 3: Schlieren textures observed in *nematic* samples with *planar alignment* show *defect* centers with two or four emerging brushes. Schlieren textures in *nematic* samples with tilted alignments show centers with four brushes; centers with two brushes are caused by *defect* walls.
Note 4: A thin sample of a *smectic C* phase with the layers parallel to the sample surfaces gives schlieren textures with centers that have four brushes. However, a *smectic C* phase formed by odd-membered liquid-crystal dimers (see Definition 2.11.2.9, Note 5) has schlieren textures with two or four brushes.

4.9.2.1 nucleus
Center of a point or line *defect* from which black brushes originate when a liquid crystal is observed between crossed polarizers.
Note: A nucleus can indicate either the end of a *disclination* line terminating at the surface of a sample or an isolated *defect*.

4.9.2.2 disclination strength, s
Number of rotations by 2π of the *director* around the center of the *defect*.

$$|s| = \frac{\text{number of brushes}}{4}$$

TERMINOLOGY

Note 1: s is positive when the brushes turn in the same direction as the polarizer and analyser when they are rotated together, and negative when they turn in the opposite direction.

Note 2: s can be an integer or half-integer since in *nematics* the *directors* +**n** and -**n** are not distinguishable.

Note 3: The angular distribution ϕ of the *director* around a *defect* in a *nematic* planar *texture*, in the X-Y projection, can be expressed in terms of the polar angle θ_r

$$\phi = s\theta_r + \phi_0$$

where θ_r represents the angular polar co-ordinate of a given point with respect to the *disclination* center, ϕ is the angle that the local *director* axis at that point makes with the $\theta_r = 0$ axis and ϕ_0 is a constant ($0 < \phi_0 < 2\pi$) (See Fig. 22).

The product $s\theta_r$ yields the angle by which the *director* turns on a closed curve around the *disclination* center. If a complete circuit is made around the center of an $s = \pm\frac{1}{2}$ *disclination*, the *director* rotates by π. For $s = \pm 1$ a similar circuit yields a total *director* rotation of 2π. So, $s = \pm\frac{1}{2}$ defines a **π-line disclination** and $s = \pm 1$ defines a **2π-line disclination**.

Note 4: *Director* alignments for point defects with different values of s are illustrated in Fig. 23.

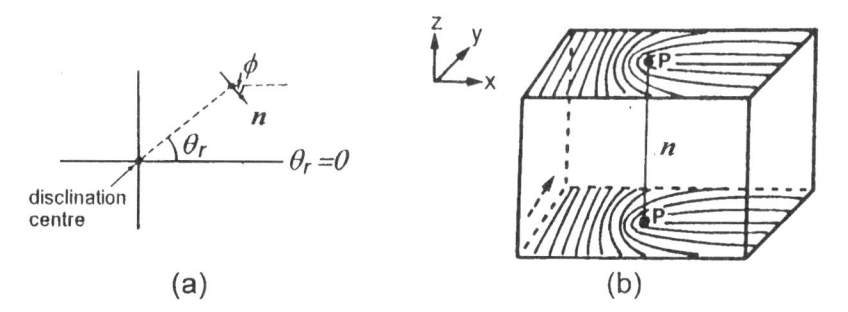

Fig. 22. (a) Identification of the angles ϕ and θ_r used to describe a disclination. (b) *Director* arrangement of an $s = \pm\frac{1}{2}$ singularity line. The end of the line attached to the sample surface appears as the point $s = +\frac{1}{2}$ (points P). The *director* alignment or field does not change along the z direction. The *director* field has been drawn in the upper and the lower surfaces only.

$s = -1$
$\phi_0 = 0$

$s = -\frac{1}{2}$
$\phi_0 = 0$

$s = +\frac{1}{2}$
$\phi_0 = \pi/2$

cont.

POLYMER LIQUID CRYSTALS

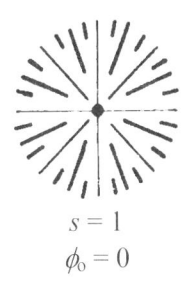

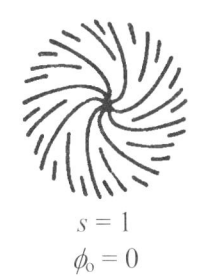

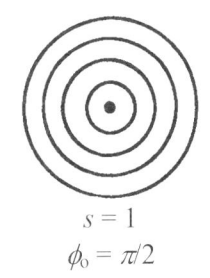

$s = 1$
$\phi_0 = 0$

$s = 1$
$\phi_0 = 0$

$s = 1$
$\phi_0 = \pi/2$

Fig. 23. Schematic representation of the *director* alignments at disclinations with different values of s and ϕ_0; $s = \pm\,\tfrac{1}{2}$ correspond to two-brush defects and $s = \pm\,1$ to four-brush defects.

4.9.3 threaded texture
Texture with π-line disclinations which lie essentially parallel to the surfaces between which a sample is placed, with the ends of the lines attached to the surfaces and the other parts of the lines moving freely in the liquid crystal, appearing as thin thread-like lines.

4.9.3.1 surface disclination line
adhering thread
Thick, thread-like *disclination* line anchored along its length to the upper or the lower of the surfaces between which a sample is placed.

4.9.4 marbled texture
Texture consisting of several areas with different *director* orientations.
Note: On observing a sample with a marbled texture between crossed polarizers, the interference colour is essentially uniform within each individual area, indicating an essentially homogeneous region.

4.10 smectic textures

4.10.1 bâtonnet
Droplet, usually non-spherical, of a *smectic* phase nucleating from an isotropic phase.

4.10.2 focal-conic domain
Domain formed by deformed *smectic* layers that fold around two confocal line defects preserving equidistance of structural layers everywhere except in the vicinity of the *defect* lines.
Note 1: See Fig. 24. The confocal line defects may be an ellipse and a hyperbola or two parabolae.
Note 2: The *smectic* layers within a focal-conic domain adopt the arrangement of Dupin cyclides, since as in these figures there appear concentric circles resulting from the intersection of ellipses and hyperbolae. They also have the distinctive property of preserving an equal distance between them.
Note 3: A focal-conic domain built around an ellipse and an hyperbola is the most common type of *defect* in thermotropic *smectic A* phases. The hyperbola passes through a focus of the ellipse and the ellipse passes through the focus of the hyperbola (see Fig. 24).
Note 4: In a particular limiting case of an ellipse-hyperbola focal-conic domain, the ellipse becomes a straight line passing through the center of a circle.

Note 5: A focal-conic domain built around two confocal parabolae is called a **parabolic focal-conic domain**.

Note 6: At any point inside a focal-conic domain, the *director* is oriented along the straight line drawn through the point and the two *defect* lines (ellipse and hyperbola or two parabolae or circle and straight line). See for example BD, BC and BO in Fig. 24.

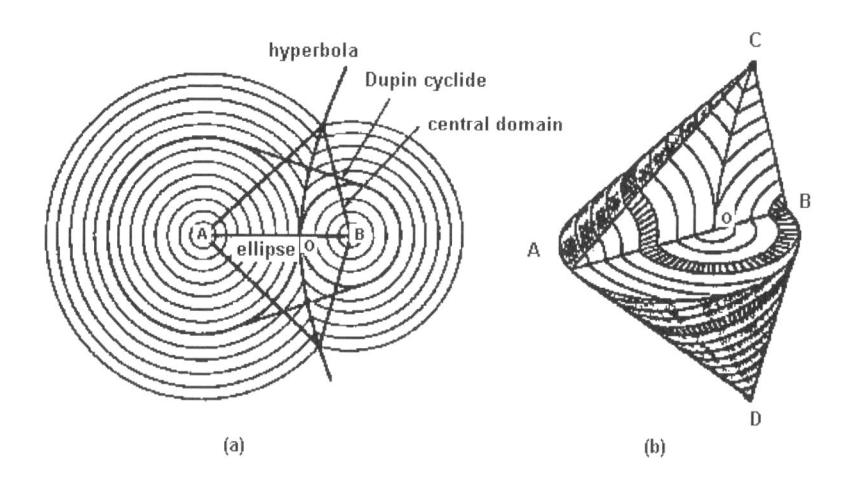

Fig. 24. Dupin cyclide and perfect focal-conic domain construction. (a) Vertical section showing layers of the structure; thick lines indicate the ellipse, hyperbola, Dupin cyclide, and central domain. (b) Focal-conic domain showing structural layers with a representation of the arrangement of the molecules within one of them.

4.10.3 polygonal texture

Texture composed of *focal-conic domains* of the ellipse-hyperbola type with visible ellipses, or parts of ellipses, located at the boundary surfaces.

Note 1: See Figs. 25a and 25b.

Note 2: One branch of the hyperbola (either above or below the plane of the ellipse) is usually missing in the polygonal texture.

Note 3: Neighbouring *domains* form a family with a common apex where the hyperbolae of these *domains* join each other. This common point is located at the surface that is opposite to the surface containing the ellipses (see Fig. 26). Each family is bounded by a polygon formed by hyperbolic and elliptical axes; these are parts of *focal-conic domains* that provide a smooth variation of *smectic* layers between the *domains* of different families. These *domains* are the tetrahedra in Fig. 26.

Note 4: The *smectic* layers pass continuously from one *focal-conic* to the next.

4.10.4 focal-conic, fan-shaped texture

Texture formed partly by *focal-conic domains* with their hyperbolae lying in the plane of observation.

Note 1: See Fig. 27.

Note 2: The layers are aligned almost normal to the sample surfaces. The regular sets of hyperbolae are called '**boundaries of Grandjean**'; they serve as limiting surfaces between *domains* with different *director* orientations.

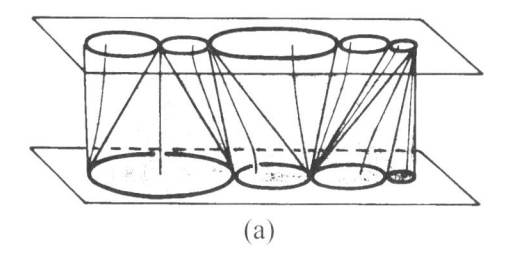

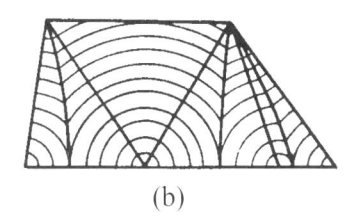

Fig. 25. Arrangement of a smectic A polygonal texture: (a) general view of the focal-conic domains filling space efficiently; (b) cross-section of the *domains* showing arrangement of the smectic layers.

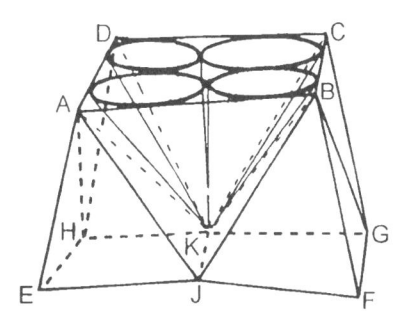

Fig. 26. Elements of a smectic A complex polygonal texture. Upper surface: one polygon with four ellipses. Lower surface: two polygons. The whole space may be divided into three pyramids (ABCDK, AEHKJ, BFGKJ) and three tetrahedra (ABJK, ADHK, BCGK).

5   PHYSICAL CHARACTERISTICS OF LIQUID CRYSTALS

General Note: In this section the *director* ***n*** is treated mathematically as a unit vector, with components n_1, n_2, n_3 along space-fixed axes X_1, X_2, X_3.

5.1   order parameter, $\langle P_2 \rangle$

Parameter characterizing the long-range orientational order with reference to the *director*, with

$$\langle P_2 \rangle = (3\langle \cos^2 \beta \rangle - 1)/2$$

where β is the angle between the molecular symmetry axis and the *director* and $\langle \ \rangle$ denotes an ensemble average.

Note 1: $\langle P_2 \rangle$ characterises long-range molecular order.

Note 2:   For rod-like molecules, the order parameter of the effective molecular symmetry axis at the nematic-isotropic transition is about 0.3 and can increase to about 0.7 in the *nematic mesophase*.

Note 3: Molecules which constitute nematogens are not strictly cylindrically symmetric and have their orientational order given by the Saupe ordering matrix which has elements $S\alpha\beta = (3<|\alpha|\beta> - \delta_{\alpha\beta})/2$, where 1α and 1β are the direction cosines between the *director* and the molecular axes α and β, $\delta_{\alpha\beta}$ is the Kronecker delta and α, β denote the molecular axes X, Y, Z.

Note 4: The constituent molecules of a *nematogen* are rarely rigid and their orientational order is strictly defined, at the second-rank level, by a Saupe ordering matrix for each rigid sub-unit.

Note 5: Even for molecules with cylindrical symmetry, $<P_2>$ does not provide a complete description of the orientational order. Such a description requires the singlet orientational distribution function which can be represented as an expansion in a basis of Legendre polynomials $P_L\cos\beta$, with L an even integer. The expansion coefficients are proportional to the order parameters $<P_L>$ of the same rank. It is these order parameters which provide a complete description of the long-range orientational order.

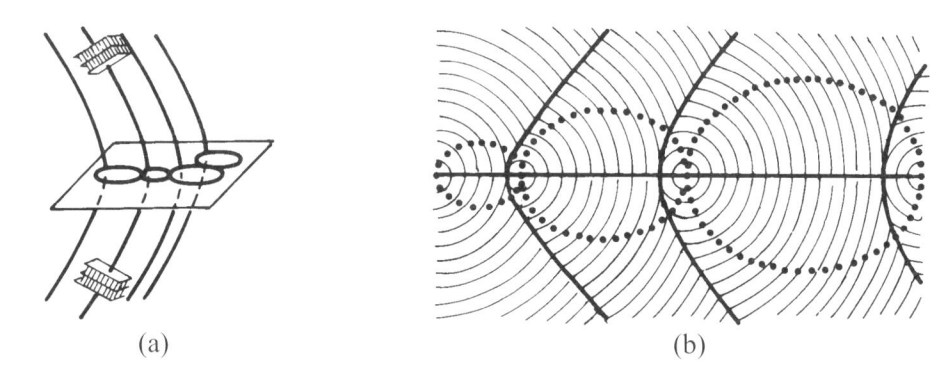

Fig. 27. (a) Illustrating an arrangement of confocal ellipses and hyperbolae. The *directors* become parallel near the extremes of the hyperbolae. (b) Section showing the layer structure. The dotted ellipses in the plane of the drawing are sections perpendicular to the focal-conics.

5.2   distortion in liquid crystals, *a*

Deformation leading to a change in the *director*, where the distortion is described by a tensor of third rank

$$a_{3ij} = n_3 \left(\partial n_i / \partial x_j \right)$$

in which the initial orientation of the *director* n is chosen as the 3-axis; $I = 1, 2$; $j = 1, 2, 3$; $n_3 = 1$, n_i is the *i*th component of the *director* n and x_j is a co-ordinate on axis X_j.

5.2.1 splay deformation, (S-deformation)

Deformation in a direction normal to the initial *director*, **n**, characterised by div **n** ≠ 0.
Note 1: See Fig. 28 and Definition 5.3.
Note 2: A splay deformation is described by the non-zero derivatives $n_3(\partial n_1/\partial x_1)$ and $n_3(\partial n_1/\partial x_1)$, where the symbols are defined in Definition 5.2.

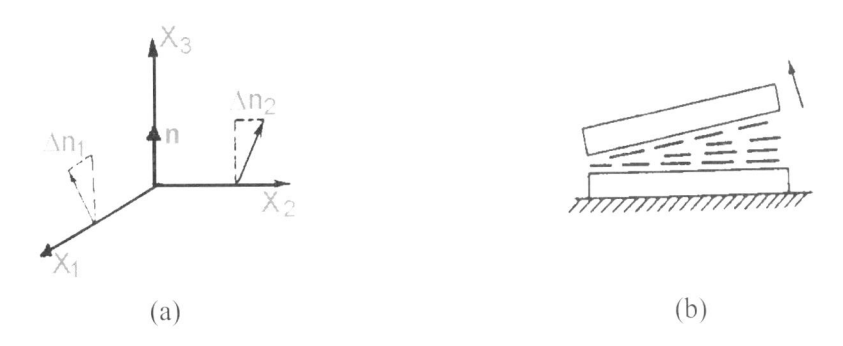

Fig. 28. Schematic representation of a splay deformation: (a) changes in the components of the director **n**, defining the orientational change; (b) splay deformation of an oriented layer of a nematic liquid crystal.

5.2.2 bend deformation, (B-deformation)

Deformation in the direction of the initial *director*, **n**, characterised by **n** × rot **n** ≠ 0.
Note 1: See Fig. 29 and Definition 5.3.
Note 2: The degree of bending is given by the component of rot **n** perpendicular to **n**.
Note 3: A bend deformation is described by the non-zero derivatives $n_3(\partial n_1/\partial x_1)$ and $n_3(\partial n_1/\partial x_1)$, where the symbols are defined in Definition 5.2.

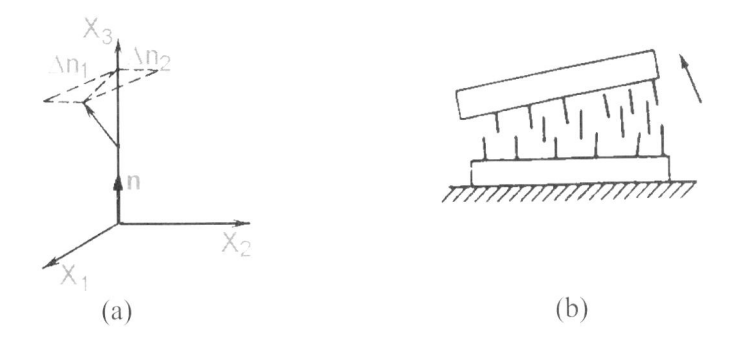

Fig. 29. Schematic representation of a bend deformation: (a) changes in the components of the director, **n** defining the orientation change; (b) bend deformation of an oriented layer of a nematic liquid crystal.

5.2.3 twist deformation, (T-deformation)

Torsional deformation of a planar-oriented layer in the direction of the initial *director*, **n**, characterised by **n**·rot **n** ≠ 0.
Note 1: See Fig. 30 and Definition 5.3.
Note 2: The degree of twisting is given by the component of rot **n** parallel to **n**.

TERMINOLOGY

Note 3: A twist deformation is described by the non-zero derivatives $n_3(\partial n_1/\partial x_1)$ and $n_3(\partial n_1/\partial x_1)$, where the symbols are defined in Definition 5.2.

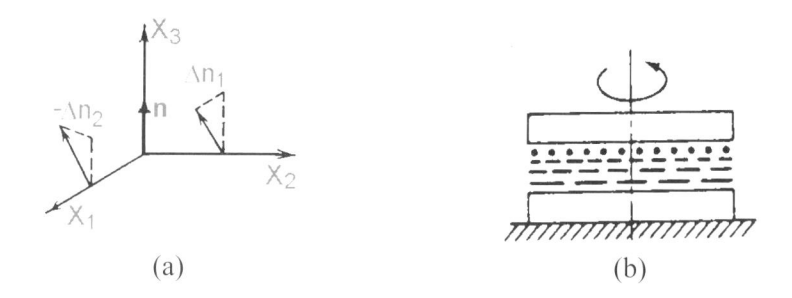

Fig. 30. Schematic representation of the twist deformation: (a) changes in the components of the director $\boldsymbol{n}$, defining the orientation change; (b) twist deformation of an oriented layer of a nematic liquid crystal.

5.3   elastic constants, K_i where $i = 1, 2$ and 3, SI unit: Nm^{-2}
     elasticity moduli

Coefficients K_1, K_2 and K_3, in the expression for the distortion-Gibbs energy density, g, of a bulk *nematic liquid crystal*; with

$$g = g_0 + 1/2[K_1 (\text{div } \boldsymbol{n})^2 + K_2 (\boldsymbol{n} \times \text{rot } \boldsymbol{n})^2 + K_3 (\boldsymbol{n} \times \text{rot } \boldsymbol{n})^2],$$

where g_0 is the Gibbs-energy density of the undistorted liquid crystal, $\boldsymbol{n}$ is the *director*, and K_1, K_2, and K_3 are the elastic constants for splay, twist, and bend deformations, respectively.

Note 1: In the equation for g, the term g_0 is usually equal to zero because the undistorted state of *nematics* is the state of uniform alignment. However, for *chiral nematics*, a non-zero value of g_0 allows for the intrinsic twist in the structure. In order to describe g for *smectic* phases, an additional term must be added, due to the partially solid-like character of the *smectic* state and arising from positional molecular deformations.

Note 2: In low molar mass *nematics* composed of rod-like molecules, the bend constant K_3 is the largest while the twist constant K_2 is the smallest. Typical values of K_1 are 10^{-11} - 10^{-12} N m^{-2}.

Note 3: The names of Oseen, Zocher, and Frank are associated with the development of the theory for the elastic behaviour of *nematics* and so the elastic constants may also be described as the **Oseen-Zocher-Frank constants**, although the term **Frank constants** is frequently used.

5.4   Leslie-Ericksen coefficients, α_i, $i = 1, 2, 3, 4, 5$ and 6, SI unit: Pa s

Six viscosity coefficients required for a description of the dynamics of an incompressible, *nematic liquid crystal*.

Note 1: Assuming Onsager's reciprocal relations for irreversible processes,

$$\alpha_2 + \alpha_3 = \alpha_6 - \alpha_5$$

and the number of independent coefficients reduces to five.

Note 2: For *nematics* formed by low-molar-mass compounds, the Leslie coefficients are typically in the range 10^{-2} to 10^{-1} Pa s.

5.5 Miesowicz coefficient, η_i $i = 1, 2$ and 3, SI unit: Pa s

Ratio of the shear stress, σ, to the shear velocity gradient, γ, for a *nematic* liquid crystal with a particular *director* orientation, denoted by i, under the action of an external field:

$$\eta_i = \sigma/\gamma$$

Note 1: The three Miesowicz coefficients (η_1, η_2, and η_3) describe the shear flow of a *nematic* phase with three different *director* orientations, (see Fig. 31) namely: η_1 for the *director* parallel to the shear-flow axis; η_2 for the *director* parallel to the velocity gradient; and η_3 for the *director* perpendicular to the shear flow and to the velocity gradient.

Note 2: Usually $\eta_1 < \eta_2 < \eta_3$.

Note 3: The Miesowicz coefficients are related to the *Leslie-Ericksen coefficients* by the following equations:

$$\eta_1 = 0.5(\alpha_3 + \alpha_4 + \alpha_6), \quad \eta_2 = 0.5(\alpha_4 + \alpha_5 + \alpha_2), \quad \eta_3 = 0.5(\alpha_4)$$

Note 4: The external field used to align the *director* must be sufficiently large to ensure that it remains aligned during flow.

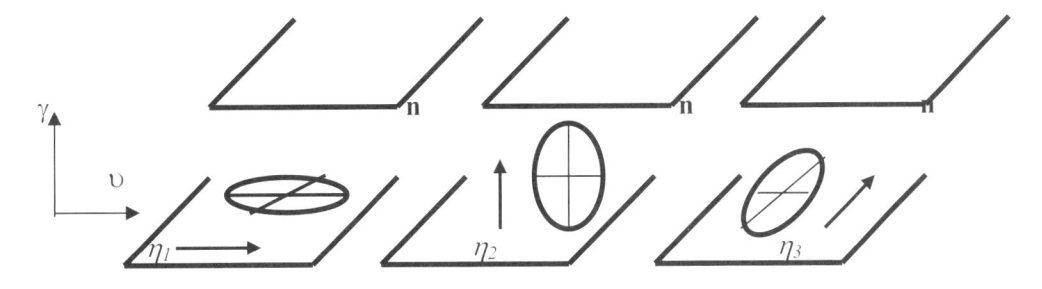

Fig. 31. Scheme of director alignment in the shear flow of velocity υ of a nematic phase and the corresponding Miesowicz coefficients.

5.6 friction coefficients, γ_i $i = 1$ and 2, SI unit: Pa s
   rotational viscosity coefficients

Coefficients that define the energy dissipation associated with a rotation of the *director* in an incompressible, *nematic* liquid crystal.

Note 1: The rotational viscosity coefficients are of the order of 10^{-2}–10^{-1} Pa s for low-molar-mass liquid crystals; for *polymeric liquid-crystals* their values depend strongly on the molar mass of the polymer.

Note 2: The friction coefficients can be expressed in terms of the *Leslie-Ericksen coefficients* as follows:

$$\gamma_1 = \alpha_3 - \alpha_2 \qquad \gamma_2 = \alpha_6 - \alpha_5$$

Note 3: γ_1 is often called the rotational viscosity or the twist viscosity, i.e., the viscosity associated with the rotation of the *director* without material flow.

5.7 backflow
Motion of a liquid crystal associated with the rate of change of the *director* in the direction opposite to that of the action of an external field.

5.8 anisotropy of physical properties
Dependence of certain physical properties, like the electric permittivity, refractive index and magnetic susceptibility on direction. It is created by long-range orientational order in a *mesophase*, provided the corresponding molecular property is anisotropic.
Note: The symmetry of the tensor representing the average anisotropic property cannot be lower than the symmetry of the phase.

5.8.1 uniaxial mesophase anisotropy, $\Delta\widetilde{\chi}$
Value of a property $\widetilde{\chi}$ parallel to the *director* $\widetilde{\chi}_\parallel$ minus that perpendicular to it $\widetilde{\chi}_\perp$:

$$\Delta\widetilde{\chi} = \widetilde{\chi}_\parallel - \widetilde{\chi}_\perp$$

Note 1: The tilde is used to indicate a property of a liquid-crystal *mesophase*.
Note 2: $\Delta\widetilde{\chi}$ provides a practical measure of the orientational order of a *mesophase* and necessarily vanishes in an isotropic phase.
Note 3: For *mesophases* composed of cylindrically symmetric molecules there is a precise relationship between the magnetic anisotropy, $\Delta\widetilde{\chi}$, and the second-rank orientational parameter $\langle P_2 \rangle$.

$$\Delta\widetilde{\chi} = (2/3)(\widetilde{\chi}_{zz} - \widetilde{\chi}_{xx})\langle P_2 \rangle$$

5.8.2 biaxial mesophase anisotropies, $\Delta\widetilde{\chi}$ and $\delta\widetilde{\chi}$
With the principal phase axes, corresponding to the three *directors* *l*, *m*, and *n* (see Definition 3.3, Note 1) labelled X, Y, and Z such that $\widetilde{\chi}_{zz} > \widetilde{\chi}_{xx} > \widetilde{\chi}_{yy}$, the major biaxial mesophase anisotropy, $\Delta\widetilde{\chi}$, is defined by

$$\delta\widetilde{\chi} = (\widetilde{\chi}_{zz} - \widetilde{\chi}_{yy})$$

and the biaxial mesophase anisotropy is

$$\Delta\widetilde{\chi} = \widetilde{\chi}_{zz} - (1/2)(\widetilde{\chi}_{xx} + \widetilde{\chi}_{yy})$$

Note 1: The long-range biaxial ordering of the *mesophase* means that the three principal components of a second-rank tensorial property will not normally be the same, hence the two measures of the anisotropy $\Delta\widetilde{\chi}$ and $\delta\widetilde{\chi}$.
Note 2: At a transition to a *uniaxial mesophase* $\delta\widetilde{\chi}$ vanishes. The **relative biaxiality**, η, is defined as the ratio of $\delta\widetilde{\chi}$ to $\Delta\widetilde{\chi}$.

5.9 ferroelectric effects

Ferroelectric *mesophase* that appears through the breaking of symmetry in a tilted *smectic mesophase* by the introduction of molecular chirality and, hence, *mesophase* chirality.

Note 1: When the numbers of layers with opposite tilt directions are not the same, the *smectic mesophase* has ferroelectric properties.

Note 2: The appearance of a spontaneous polarization, $|P_s|$, in *chiral tilted smectic mesophases* is caused by a long-range ordering of molecular transverse electric dipoles.

Note 3: The polarization $|P_s|$ can be switched between two stable states with an external electric field (E); these states are stable in zero electric field.

Note 4: The switching time, τ, is given by

$$\tau = \gamma_1 \sin\theta / P_s E$$

where γ_1 is the twist viscosity (see Definition 5.6, Note 3) and θ is the tilt angle.

Note 5: The spontaneous polarization depends on the transverse component, μ_t, of the molecular dipole, the number density, ρ, and the polar or first-rank *order parameter*, $<\cos\varphi>$, the ensemble average of $\cos\varphi$, where φ is the angle between the transverse axis and the minor *director*, see Definition 3.3, by

$$|P_s| = \rho\mu_t <\cos\varphi>$$

Note 6: Typical values of the spontaneous polarization, $|P_s|$, in *chiral smectic C mesophases* are between 10^{-3} and 10^{-4} C m^{-2}.

Note 7: When the tilt direction alternates from layer to layer, the *smectic mesophase* is antiferroelectric; such *mesophases* do not possess spontaneous polarization. They can be turned into ferroelectric structures through the application of an electric field.

5.10 Fréedericksz transition

Elastic deformation of the *director*, induced by a magnetic induction or electric field, in a uniformly aligned, thin sample of a *nematic* confined between two surfaces.

Note 1: The Fréedericksz transition occurs when the strength of the applied field exceeds a certain threshold value (see Definition 5.12).

Note 2: For the magnitude of the magnetic induction this threshold has the form

$$B_{th} = (\pi/d)(\mu_0 K_i / \Delta\tilde{\chi})^{1/2}$$

where, B_{th} is the threshold magnetic induction (flux density), d is the thickness of the *nematic* film, μ_0 is the permeability of a vacuum, and $\Delta\tilde{\chi}$ is the *magnetic anisotropy*. The particular *elastic constant* K_i depends on the geometry of the experiment.

Note 3: For an electric field, this threshold has the form

$$E_{th} = (\pi/d)(\varepsilon_0 K_i / \Delta\tilde{\varepsilon})^{1/2}$$

where ε_0 is the permittivity of vacuum and $\Delta\tilde{\varepsilon}$ is the dielectric *anisotropy*.

5.11 electroclinic effect

Tilt in an A *mesophase* is called the electroclinic effect.

Note: In high polarization materials induced tilt angles as high as $10°$ have been observed.

5.12 threshold fields
   threshold electric field, E_{th}, SI unit: V m^{-1}
   threshold magnetic induction, B_{th}, SI unit: T

Critical electric field strength or magnetic induction necessary to change the equilibrium *director* alignment imposed by constraining surfaces.
Note: See Definition 5.10, Notes 1–3.

5.13 electrohydrodynamic instabilities, (EHD instabilities)
Instabilities caused by the anisotropy of conductivity and corresponding to a periodic deformation of the alignment of the *director* in a *nematic monodomain* under the action of a direct current or low-frequency alternating current.
Note 1: See Definition 4.2 for the definition of a *monodomain*.
Note 2: The basic electric parameters determining EHD instabilities are the dielectric anisotropy, $\Delta\tilde{\varepsilon}$, and the anisotropy of the (ionic) conductivity, $\Delta\tilde{\sigma}$.
Distortion of the *director*, space charges and the motion of the fluid are coupled through the applied electric field. Above a given threshold, fluctuations of these quantities are amplified and EHD instabilities develop.

5.14 Williams domains
   Kapustin domains
Regions in a liquid crystal having a specific cellular periodic flow-pattern in the form of long rolls induced by the application of an electric field perpendicular to a *nematic* layer with an initial *planar alignment* of the *director*.
Note 1: The *nematic* liquid crystal must have a negative dielectric anisotropy ($\Delta\tilde{\varepsilon} < 0$), and a positive anisotropy ($\Delta\tilde{\sigma} > 0$). The *optical texture* corresponding to the flow pattern consists of a set of regularly spaced, black and white stripes perpendicular to the initial direction of the *director*. These stripes are caused by the periodicity of the change in the refractive index for the extraordinary ray due to variations in the *director* orientation.
Note 2: The *domains* exist over only a small range of electric-field strength.

5.15 dynamic-scattering mode, (DSM)
State of a liquid crystal that shows a strong scattering of light due to a turbulent flow resulting from an applied voltage greater than a particular critical value.
Note 1: In DSM the *Williams* (*Kapustin*) *domains* become distorted and mobile, and macroscopic *director* alignment is completely disturbed.
Note 2: A liquid crystal in DSM has a complicated *optical texture*.

5.16 flexo-electric effect
Electric polarization resulting from a splay or bend deformation of the *director* of a *nematic* liquid crystal.
Note 1: See Fig. 32.
Note 2: The molecular origins of dipolar flexo-electricity are the particular shape anisotropy (e.g., resembling a pear or banana) of the molecules, each of which must also possess a permanent dipole moment.
Note 3: The net polarization, P, is proportional to the distortion:

$$P = e_1 \, n(\text{div } n) + e_3 \, (\text{rot } n) \times n$$

where e_1 and e_3 are the flexo-electric coefficients. They have the units of an electric potential, namely J C^{-1}, of arbitrary sign.

Note 4: The flexo-electric effect is the analogue of the piezo-electric effect in solids, where the polarization is induced by a strain that produces a translational deformation of the crystal. The flexo-electric effect in a liquid crystal is caused by a purely orientational deformation.

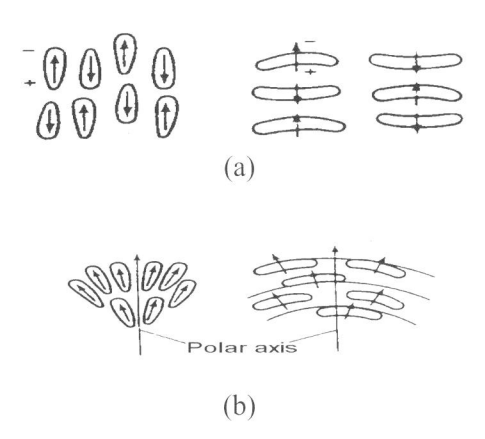

Fig. 32. Schematic representation of the flexo-electric effect. (a) The structure of an undeformed nematic liquid crystal with pear- and banana-shaped molecules; (b) the same liquid crystal subjected to splay and bend deformations, respectively.

5.17 flexo-electric domain
Domain corresponding to a periodic deformation caused by the inverse flexo-electric effect in a *nematic* liquid crystal.

Note: A flexo-electric domain occurs when $\Delta\widetilde{\varepsilon} < 4\pi el/K$ where e is the flexo-electric coefficient and K is the *elastic constant*, assuming $K_1 = K_3 = K$ and $e_1 = -e_3 = e$ (see Definitions 5.3 and 5.16).

5.18 twisted-nematic cell
Twisted *nematic* liquid crystal sandwiched between two glass plates, with the *director* aligned parallel to the plates, with one of the plates turned in its own plane about an axis normal to it.

5.19 'time-on' of the electro-optical effect, τ_{on}, SI unit: s
turn-on time
Time required for the light intensity viewed through crossed polarizers to increase to 90% of the final value from the off-state to the on-state of an electro-optical twisted-nematic cell.

Note: In the off-state the electro-optical cell contains a thin film of a *nematic liquid-crystal* with mutually perpendicular *directors* at the upper and lower glass plates; hence to reach the on-state the *director* performs a 90° twist over the thickness of the liquid crystal film.

TERMINOLOGY

5.20 **'time-off' of the electro-optical effect**, τ_{off}, SI unit: s
turn-off time
The time required for the light intensity viewed through crossed polarizers to decrease by 90% from the on-state to the off-state of an electro-optical twisted-nematic cell.
Note: See the note to Definition 5.19.

5.21 **rise time**, SI unit: s
Time required by an electro-optical *nematic* cell for a light-intensity change from 10% to 90 % of the maximum intensity on going from the off-state to the on-state.
Note: See the note to Definition 5.19.

5.22 **fall time**, SI unit: s
decay time
Time required by an electro-optical twisted-nematic cell for a light-intensity change from 90% to 10 % of the maximum intensity on going from the on-state to the off-state.
Note: See the note to Definition 5.19.

5.23 **guest-host effect**
Field-induced change in the orientation of either dichroic dye molecules (the guest) dissolved in a *mesophase* (the host) or dichroic dye moieties (the guest) of polymers (the host) resulting in changes in the absorption spectrum of a mesomorphic mixture.

6   LIQUID CRYSTAL POLYMERS

6.1 **liquid-crystal polymer**, **(LCP)**
polymer liquid-crystal, **(PLC)**
liquid-crystalline polymer
Polymer material that, under suitable conditions of temperature, pressure and concentration, exists as a *LC mesophase*.

6.2 **main-chain polymer liquid-crystal**, **(MCPLC)**
main-chain liquid-crystalline polymer, **(MCLCP)**
Polymer containing *mesogen*ic units in their main chains but not in side-chains.
Note 1: A MCPLC is formed by linking together suitable relatively rigid units directly or through appropriate functional groups (see Fig. 33).
Note 2: The linkage between the rigid units (I) may be (a) direct or (b-g) via flexible *spacers* (II).
Note 3: A MCPLC with cross-shaped *mesogenic groups* (b or g) is known as a **cruciform (or star) polymer liquid crystal**.
Note 4: The rigid units may, but often do not, possess intrinsic *mesogen*ic character.

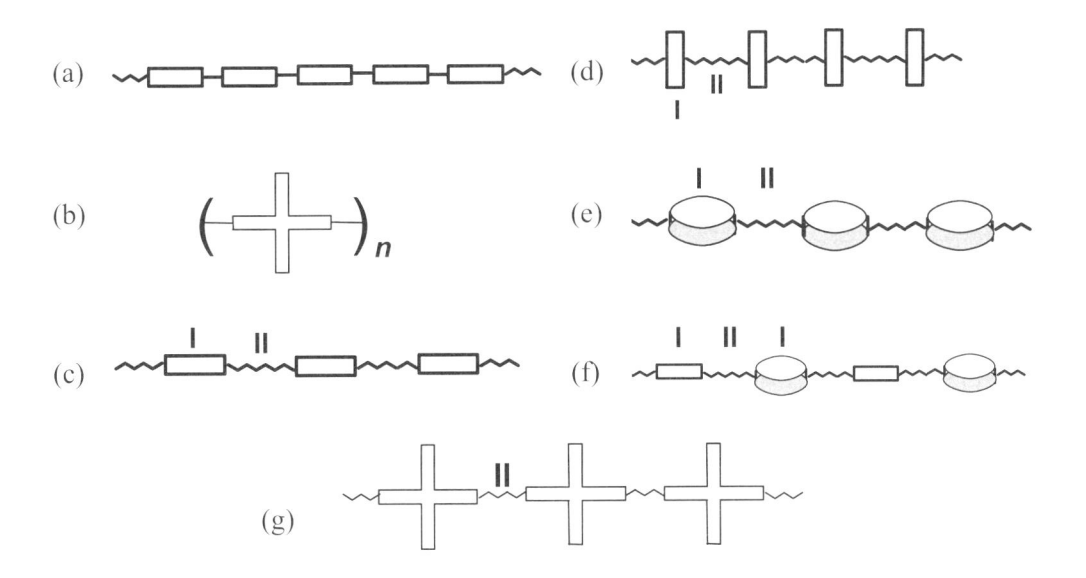

Fig. 33. Examples of main-chain polymer liquid crystals: I - *mesogenic group*; II - *spacer*.

**6.3 side-group or side-chain polymer liquid-crystal, (SGPLC, SCPLC)
side-group or side-chain liquid-crystalline polymer, (SGLCP, SCLCP)
polymer with mesogenic side-groups or side-chains
comb-shaped (comb-like) polymer liquid crystal**

Polymer, the molecules of which have *mesogen*ic units only in the side-groups or side-chains.

Note 1: The *mesogenic groups* (I) in a SGPLC can be connected to the backbone (III) either (a) directly or (b, c) via flexible *spacers* (II) (see Fig. 34).

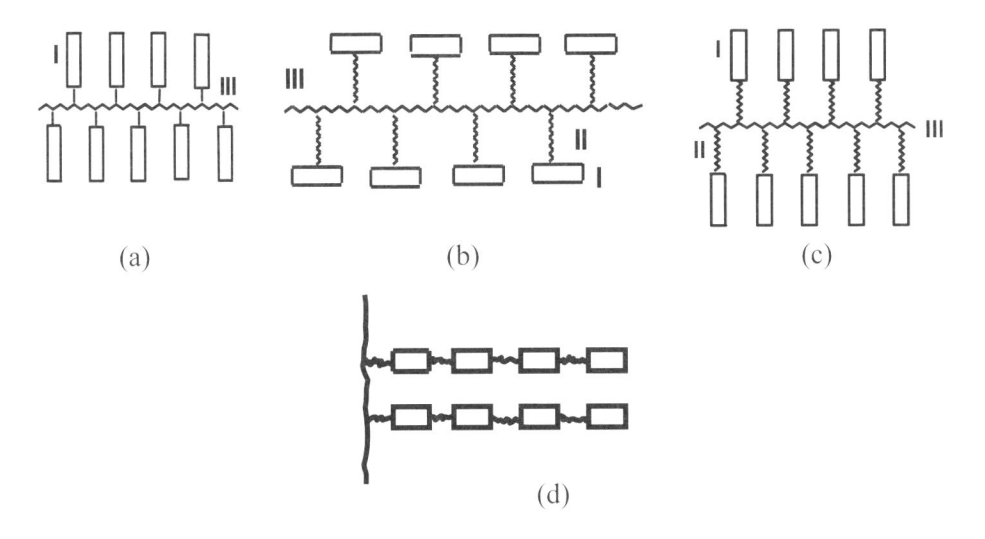

Fig. 34. Examples of side-group polymer liquid crystals: I - *mesogenic group*; II - *spacer*; III - backbone. The terminology 'side-group' is used for (a), 'side-on fixed' is used for (b), 'end-on fixed' for (c) and 'side-chain' for (d).

Note 2: The structures as in Fig. 34 can also be used with the proviso that the side-group units are replaced by chains containing *mesogens*.

Note 3: Examples of polymer backbones are polyacrylates, polymethacrylates, and polysiloxanes; the *spacers* are usually oligomethylene, oligo(oxyethylene), or oligosiloxane.

Note 4: The pendant groups in these polymers have structures compatible with liquid-crystal formation, that is, they are *mesogenic* but not intrinsically mesomorphic. See the examples given in Definitions 2.10; 2.11.2.1.

Note 5: If the *mesogenic* side-groups are rod-like (*calamitic*) in nature, the resulting polymer may, depending upon its detailed structure, exhibit any of the common types of *calamitic mesophases*: *nematic*, *chiral nematic* or *smectic*. Side-on fixed SGPLC, however, are predominantly *nematic* or *chiral nematic* in character. Similarly, disc-shaped side-groups tend to promote *discotic nematic* or *columnar mesophases* while amphiphilic side-chains tend to promote amphiphilic or *lyotropic mesophases*.

Note 6: A plethora of types of copolymers can be produced. For example, non-*mesogenic* side-groups may be used in conjunction with *mesogenic* side-groups and the polymer backbone may be substituted, to various degrees, with side-groups or chains.

6.4 spacer

Flexible segment used to link successive *mesogenic* units in the molecules of MCPLCs or to attach *mesogenic* units as side-groups onto the polymer backbone of SGPLCs.

Note 1: Examples of *spacers* are: oligomethylene, oligooxyethylene, or oligosiloxane chains.

Note 2: The term is also used for the group linking two or more *mesogenic* units in *liquid-crystal oligomers*.

6.5 disruptor

Chemical group used to disrupt the linearity of the backbone of molecules of MCLCPs.

Note: Examples are (a, b) rigid-kink or (c) crankshaft units.

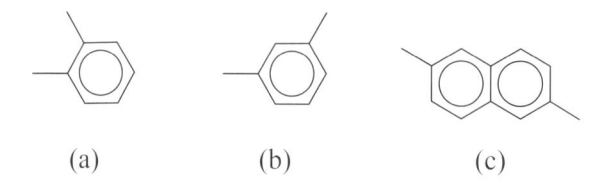

(a)   (b)   (c)

6.6 combined liquid-crystalline polymer

Liquid-crystalline polymer consisting of macromolecules in which *mesogenic groups* are incorporated both in the main-chain and in the side-groups.

Note: See Fig. 35. The *mesogenic* side-groups can be attached either as lateral substituents to the backbone *mesogenic* moieties that are connected to each other either (a) directly or (b) by *spacers* or (c) they can be attached to the *spacer* incorporated into the main-chain.

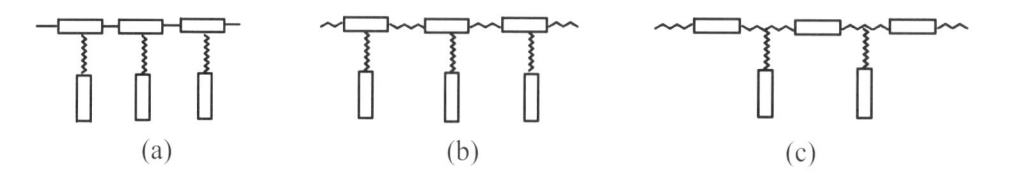

(a)   (b)   (c)

Fig. 35. Examples of combined liquid-crystalline polymers.

6.7 rigid chain
Rod-like chain of a MCPLC with direct links between the *mesogenic groups* for which the persistence length is at least comparable with the contour length and much greater than the diameter.

Note 1: The persistence length is a characteristic of the stiffness of a chain in the limit of infinite chain length.

Note 2: A polymer composed of molecules that have rigid rod-like groups or chains usually does not show thermotropic mesomorphic behaviour because decomposition occurs below its melting point.

Note 3: A polymer composed of molecules that have rigid rod-like groups or chains may form *LC mesophases* in solution under suitable conditions. These are sometimes described as *lyotropic* but, as the solvent does not induce the formation of aggregates or micelles, this term is not appropriate.

6.8 semi-rigid chain
Chain for which the contour length is greater than the persistence length but for which their ratio is still below the Gaussian limit.

Note: Some polymers composed of semi-rigid chains form *amphiphilic mesogens*.

Examples: polyisocyanates and (2-hydroxyethyl)cellulose.

6.9 board-shaped polymer
Polymer chain composed of a rigid backbone to which many lateral side-groups are attached, giving the repeating unit a board-like shape.

Note 1: The rigid backbone often has a polyester, polyamide or poly(ester-amide) type of structure. Examples are:

$R_1 = CH_3(CH_2)_m-O-\overset{O}{\underset{\|}{C}}-$ $m = 11$

$R_2 = CH_3(CH_2)_m-O-$ $m = 7, 11$

$R = CH_3(CH_2)_m-$ $m = 7, 11$

Note 2: A polymer *LC* consisting of macromolecules of board-like shape can be called a board-shaped polymer *LC*. Such polymers can form *sanidic mesophases*.

6.10 liquid-crystal dendrimer
dendrimeric liquid crystal
dendritic liquid crystal
Highly branched oligomer or polymer of dendritic structure containing *mesogenic groups* that can display *mesophase* behaviour.

Note 1: See Fig. 36. The *mesogenic groups* can be located along the chains of the molecule (a) or can occur as terminal groups (b).

TERMINOLOGY

Note 2: The *mesogenic groups* can be, e.g., rod-like or disc-like, and can be attached laterally or longitudinally to the flexible *spacers*.

6.11 hyperbranched polymer liquid-crystal
Polymer composed of highly branched macromolecules containing *mesogenic groups* of which any linear subchain generally may lead in either direction, to at least two other subchains.

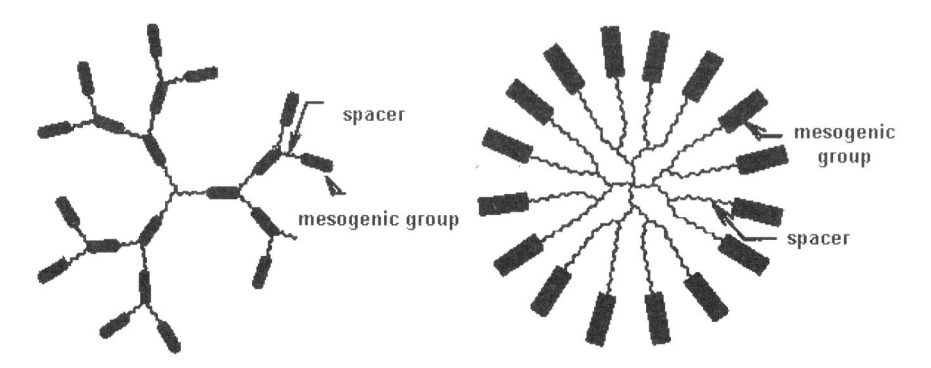

Fig. 36. Liquid-crystal dendrimers: (a) with *mesogenic groups* in the whole volume of a macromolecule; (b) with terminal *mesogenic groups*.

6.12 banded texture
    band texture
Alternating dark and bright bands observed, following shear, in a wide range of main-chain *nematic* and *chiral nematic liquid-crystalline polymers*.
Note 1: The bands always lie perpendicular to the prior shear direction.
Note 2: In general, bands form after the cessation of shear, but, under some circumstances, they may appear during the flow process
Note 3: The bands are associated with a periodic variation in the *director* orientation about the flow axis.

7   REFERENCES

1. D. Demus, G. W. Gray, H. W. Spiess, V. Vill (Eds.). *Handbook of Liquid Crystals*. Wiley-VCH, New York (1998), Vols. Q, 2a, 2b, and 3, J.
2. W. Brostow (Ed.). *Polymer Liquid Crystals Mechanical and Thermophysical Properties*. Chapman & Hall, London (1996).
3. V. Vill. LIQCRYST 2.1, Database of Liquid Crystalline Compounds for Personal Computers, LCI Publisher GmbH, Hamburg (1996).
4. L. M. Blinov and V. G. Chigrinov. *Electrooptic Effects in Liquid Crystal Materials*, Springer-Verlag, New York (1994).
5. V. P. Shibaev and Lui Lam (Eds.). *Liquid Crystalline and Mesomophic Polymers*, Springer-Verlag, New York (1994).
6. H. Stegmeyer (Ed.). *Topics in Physical Chemistry*, Vol. 3, *Liquid Crystals*, Springer-Verlag, New York (1994).
7. G. R. Luckhurst and C. A. Veracini (Eds.). *The Molecular Dynamics of Liquid Crystals*, Kluwer, Dordrecht (1994).

8. P. G. de Gennes and J. Prost. *The Physics of Liquid Crystals*, 2nd ed., Clarendon Press, Oxford (1993).
9. N. A. Platé (Ed.). *Liquid-Crystal Polymers*, Plenum Press, New York (1993).
10. S. Chandrasekhar. *Liquid Crystals*, 2nd ed., University Press, Cambridge (1992).
11. M. Donald and A. H. Windle. *Liquid Crystalline Polymers*, Cambridge University Press, Cambridge (1992).
12. A. A. Collyer (Ed.). *Liquid Crystal Polymers: From Structures to Applications*, Elsevier Applied Science, London, New York (1992).
13. V. Vill. 'Liquid Crystals' in *Landolt-Börnstein. Numerical Data and Functional Relationships in Science and Technology. New Series. Group IV: Macroscopic and Technical Properties of Matter*, Springer Verlag, Berlin (1992), Vol. **7**, Sub-volumes a-f.
14. A. Ciferri (Ed.). *Liquid Crystallinity in Polymers. Principles and Fundamental Properties*, VCH Publishers, New York (1991).
15. C. Noel and P. Navard. *Prog. Polym. Sci.* **16**, 55-110 (1991).
16. J. W. Goodby (Ed.). *Ferroelectric Liquid Crystals*, Gordon and Breach, Philadelphia (1991).
17. B. Bahadur (Ed.). *Liquid Crystals. Applications and Uses*, World Scientific, Singapore (1990/1992), Vols. 1–3.
18. W. Brostow. *Polymer* **31**, 979 (1990).
19. D. Demus. *Liq. Cryst.* **5**, 75 (1989).
20. C. B. McArdle (Ed.). *Side-Chain Liquid Crystal Polymers*, Blackie, Glasgow (1989).
21. N. A. Platé and V. P. Shibaev. *Comb-Shaped Polymers and Liquid Crystals*, Plenum, New York (1988).
22. S. Hsiao, M. T. Shaw, E. T. Samulski. *Macromolecules* **21**, 543 (1988).
23. G. Vertogen and W. H. De Jeu. *Thermotropic Liquid Crystals, Fundamentals* Springer Series in Classical Physics, Springer-Verlag, Berlin (1988), Vol. 45.
24. S. Chandrasekhar and N. V. Madhusudana. *Proc. Indian Acad. Sci. (Chem. Sci.)* **94**, 139-179 (1985).
25. A. Blumstein (Ed.). *Polymeric Liquid Crystals*, Plenum, New York (1985).
26. L. Chapoy (Ed.). *Recent Advances in Liquid-Crystalline Polymers*, Elsevier Applied Science, London (1985).
27. E. T. Samulski. *Faraday Discuss. Chem. Soc.* **79**, 7-20 (1985).
28. G. W. Gray and J. W. Goodby. *Sanidic Liquid Crystals*, Lenard Hill, Glasgow (1984).
29. B. Wunderlich and J. Grebowicz. *Adv. Polym. Sci.* **60/61**, 2-59 (1984).
30. D. Demus, H. Demus, H. Zaschke. *Flüssige Kristalle in Tabellen II*, Verlag für Grundstoffindustrien, Leipzig (1984).
31. W. H. De Jeu. *Philos. Trans. Roy. Soc., London, Ser. A* **304**, 217-29 (1983).
32. A. Ciferri, W. Krigbaum, R. Meyer (Eds.). *Polymer Liquid Crystals*, Academic Press, New-York (1982).
33. H. Kelker and R. Hatz. *Handbook of Liquid Crystals*, Verlag Chemie, Weinheim (1980).
34. G. R. Luckhurst and G. W. Gray (Eds.). *The Molecular Physics of Liquid Crystals*, Academic Press, London (1979).
35. D. Demus and L. Richter. *Textures of Liquid Crystals*, Verlag Chemie, Weinheim (1978).

TERMINOLOGY

36. G. W. Gray and P. A. Winsor (Eds.). *Liquid Crystals and Plastic Crystals*, Ellis Horwood, Chichester (1974), Vols. 1 and 2.
37. D. Demus, H. Demus, H. Zaschke. *Flüssige Krystalle in Tabellen*, Verlag für Grundstoffindustrien, Leipzig (1974).
38. P. H. Hermans. *Contribution to the Physics of Cellulosic Fibres*, Elsevier Publ. Corp. Amsterdam (1946).
39. M. Miesowicz. *Nature* **158**, 27 (1946).

8 ALPHABETICAL INDEX OF TERMS

adhering thread	4.9.3.1	chiral smectic C mesophase	3.1.5.1.3
amphiphilic mesogen	2.11.1	chiral smectic F mesophase	3.1.5.2.2
amphitropic compound	2.4.4	chiral smectic I mesophase	3.1.5.2.3
anisotropy of physical properties	5.8	cholesteric	3.1.3
		cholesteric mesophase	3.1.3
antiferroelectric chiral smectic C mesophase	3.1.5.1.2	clearing point	2.6
		clearing temperature	2.6
antimesophase	3.1.6.3	columnar discotic	3.2.2
asymmetric liquid-crystal dimmer	2.11.2.9	columnar discotic mesophase	3.2.2
		columnar hexagonal mesophase	3.2.2.1
backflow	5.7		
banana mesogen	2.11.2.10	columnar mesophase	3.2.2
band texture	6.12	columnar oblique mesophase	3.2.2.2
banded texture	6.12	columnar rectangular mesophase	3.2.2.2
barotropic mesophase	2.4.2		
bâtonnet	4.10.1	combined liquid-crystalline polymer	6.6
bend deformation	5.2.2		
biaxial mesophase	3.3	comb-shaped (comb-like) polymer liquid-crystal	6.3
biaxial mesophase anisotropies	5.8.2		
biaxial nematic	3.3.1	comb-shaped mesogen	2.11.2.3
biaxial nematic mesophase	3.3.1	conical mesogen	2.11.2.3
biaxial smectic A mesophase	3.3.2	cruciform polymer liquid-crystal	6.2
biforked mesogen	2.11.2.5		
bipolar droplet texture	4.9.1.1	crystal B, E, G, H, J and K mesophases	3.1.5.3
bis-swallow-tailed mesogen	2.11.2.7		
blue phase	3.1.4	cubic mesophase	3.1.9
board-shaped polymer	6.9	cybotactic groups	3.1.2
boojums	4.9.1.1	decay time	5.21
boundaries of Grandjean	4.10.4	defect	4.7
bowlic mesogen	2.11.2.3	dendrimeric liquid-crystal	6.10
calamitic mesogen	2.11.2.1	dendritic liquid-crystal	6.10
centered rectangular mesophase	3.1.6.3	director	3.1.1.1
		disclination	4.7.2
chiral columnar oblique mesophase	3.2.2.3	disclination strength	4.9.2.2
		discoid mesogen	2.11.2.2
chiral nematic	3.1.3	discotic mesogen	2.11.2.2
chiral nematic mesophase	3.1.3	discotic mesophases	3.2
chiral nematogen	2.11	discotic nematic	3.2.1

140

discotic nematic mesophase	3.2.1	lamellar mesophase	3.1.5.1.1
discotics	3.2	laterally branched mesogen	2.11.2.8
dislocation	4.7.1	Leslie-Ericksen coefficients	5.4
disruptor	6.5	ligated twin mesogen	2.11.2.9
distortion in liquid crystals	5.2	liquid crystal	2.3
divergence temperature	2.9	liquid-crystal dendrimer	6.10
domain	4.1	liquid-crystal dimer	2.11.2.9
dynamic-scattering mode	5.15	liquid-crystal oligomer	2.11.2.9
elastic constants	5.3	liquid-crystal polymer	6.1
elasticity moduli	5.3	LIQUID-CRYSTAL POLYMERS	6
electroclinic effect	5.11	liquid-crystal state	2.2
electrohydrodynamic instabilities	5.13	liquid-crystalline phase	2.2.1
enantiotropic mesophase	2.4.1	liquid-crystalline polymer	6.1
end-on fixed side-group polymer liquid-crystal	6.3	liquid-crystalline state	2.2
		lyotropic mesophase	2.4.3
even-membered liquid-crystal dimer	2.11.2.9	magnetic mesophase anisotropy	5.8.1
fall time	5.22	main-chain liquid-crystalline polymer	6.2
ferroelectric effects	5.9		
flexo-electric domain	5.17	main-chain polymer liquid-crystal	6.2
flexo-electric effect	5.16		
focal-conic domain	4.10.2	major biaxial mesophase anisotropy	5.8.2
focal-conic, fan-shaped texture	4.10.4	marbled texture	4.9.4
forked hemiphasmidic mesogen	2.11.2.5	melted-grain boundary mesophase	3.6.3
Frank constants	5.3	mesogen	2.11
Fréedericksz transition	5.10	mesogenic compound	2.11
friction coefficients	5.6	mesogenic dimer	2.11.2.9
fused twin mesogen	2.11.2.9	mesogenic group	2.10
GENERAL DEFINITIONS	2	mesogenic moiety	2.10
glassy mesophase	3.5	mesogenic oligomer	2.11.2.9
guest-host effect	5.23	mesogenic unit	2.10
hemiphasmidic mesogen	2.11.2.5	mesomorphic compound	2.1, 2.11
hexagonal mesophase	3.2.2.1	mesomorphic glass	2.1
hexatic smectic mesophase	3.1.5.2	mesomorphic state	2.1
homeotropic alignment	4.3	mesomorphous state	2.1
homogeneous alignment	4.4	mesophase	2.4
hyperbranched polymer liquid-crystal	6.11	mesophases of calamitic mesogens	3.1
induced mesophase	3.1.8	mesophases of disc-like mesogens	3.2
intercalated smectic mesophase	3.1.7	metallomesogen	2.11.3
inverse hexagonal mesophase	3.2.2.1	Miesowicz coefficient	5.5
inverse lamellar mesophase	3.1.5.1.1	m,n-polycatenary mesogen	2.11.2.5
isotropization temperature	2.6.	modulated smectic mesophase	3.1.6.3
Kapustin domains	5.14	monodomain	4.2

TERMINOLOGY

monotropic mesophase	2.4.5	polymer	6.3
nematic	3.1.1	side-chain polymer liquid-crystal	6.3
nematic droplet	4.9.1		
nematic textures	4.9	side-group liquid-crystalline polymer	6.3
nematogen	2.11		
non-amphiphilic mesogen	2.11.2	side-group polymer liquid-crystal	6.3
nucleus	4.9.2.1		
oblique mesophase	3.1.6.3	side-on fixed side-group polymer liquid crystal	6.3
odd-membered liquid-crystal dimer	2.11.2.9		
		side-to-tail twin mesogen	2.11.2.9
optical texture	4.8	smectic A_1, A_d, A_2, C_1, C_d, C_2 mesophases	3.1.6.2
order parameter	5.1		
ordered sanidic phase	3.4.2	smectic A mesophase	3.1.5.1.1
Oseen-Zocher-Frank constants	5.3	smectic B mesophase	3.1.5.2.1
		smectic C mesophase	3.1.5.1.2
parabolic focal conic domain	4.10.2	smectic F mesophase	3.1.5.2.2
phasmidic mesogen	2.11.2.5	smectic I mesophase	3.1.5.2.3
PHYSICAL CHARACTERISTICS OF LIQUID CRYSTALS	5	smectic mesophase	3.1.5
		smectic mesophases with unstructured layers: SmA and SmC	3.1.5.1
planar alignment	4.4		
polycatenary mesogen	2.11.2.5	smectic textures	4.10
polygonal texture	4.10.3	smectogen	2.11
polymer liquid-crystal	6.1	spacer	6.4
polymer with mesogenic side-groups or side-chains	6.3	splay deformation	5.2.1
		star polymer liquid-crystal	6.2
polymorphic modifications of strongly polar compounds	3.1.6	surface disclination line	4.9.3.1
		surface pre-tilt	4.3
pre-tilted homeotropic alignment	4.3	swallow-tailed mesogen	2.11.2.6
		tail-to-tail twin mesogen	2.11.2.9
pre-transitional temperature	2.9	TEXTURES AND DEFECTS	4
pyramidic mesogen	2.11.2.3	thermotropic mesophase	2.4.1
radial droplet texture	4.9.1.2	threaded texture	4.9.3
rectangular sanidic mesophase	3.4.1	threshold field	5.12
re-entrant mesophase	3.1.6.1	threshold electric field	5.12
relative biaxiality (of a biaxial mesophase)	5.8.2	threshold magnetic field	5.12
		'time-off' of the electro-optical effect	5.20
ribbon mesophase	3.1.6.3		
rigid chain	6.7	'time-on' of the electro-optical effect	5.19
rise time	5.21		
rotational viscosity	5.6	transitional entropy	2.8
rotational viscosity coefficients	5.6	transition temperature	2.5
		turn-off time	5.19
sanidic mesogen	2.11.2.4	turn-on time	5.18
sanidic mesophase	3.4	twin mesogen	2.11.2.9
schlieren texture	4.9.2	twist alignment	4.6
semi-rigid chain	6.8	twist deformation	5.2.3
side-chain liquid-crystallline		twisted-nematic cell	5.18

142

twist grain-boundary mesophase	3.6	uniaxial mesophase anisotropy	5.8.1
		uniaxial nematic mesophase	3.1.1
twist grain-boundary A* mesophase	3.6.1	uniform planar alignment	4.5
		virtual transition temperature	2.7
twist grain-boundary C* mesophase	3.6.2	Williams domains	5.14
		π-line disclination	4.9.2.2
twist viscosity	5.6	2π-line disclination	4.9.2.2
TYPES OF MESOPHASE	3		

9 GLOSSARY OF RECOMMENDED ABBREVIATIONS AND SYMBOLS

9.1 ABBREVIATIONS

B-	bend-
BP	blue phase
Col	columnar discotic mesophase, columnar mesophase
Col_h	columnar hexagonal mesophase
Col_{ob}	columnar oblique mesophase
Col_r	columnar rectangular mesophase
Cub	cubic mesophase
Cr	crystalline phase
DSM	dynamic-scattering mode
EHD	electrohydrodynamic
I	isotropic phase
LC	liquid-crystal,
LCPL	liquid-crystalline phase, liquid-crystalline polymer
MCLCP	main-chain liquid-crystalline polymer
MCPLC	main-chain polymer liquid-crystal
MGBC*	melted-grain-boundary mesophase
PLC	polymer liquid-crystal
re (subscript)	re-entrant mesophase
SCLCP	side-chain liquid-crystal polymer
SCPLC	side-chain polymer liquid-crystal
S-	splay-
SGLCP	side-group liquid-crystal polymer
SGPLC	side-group polymer liquid-crystal
Sm	smectic mesophase
SmA	smectic A mesophases
SmA_1	" " " "
SmA_2	" " " "
SmA_d	" " " "
SmA_b	biaxial smectic A mesophase
SmB, SmB_{hex}	smectic B mesophases
SmC*	chiral smectic C mesophase
SmC	smectic C mesophases
SmC_1	" " " "
SmC_2	" " " "

TERMINOLOGY

SmC$_d$	" " " "
SmF	smectic F mesophase
SmF*	chiral smectic F mesophase
SmI*	smectic I mesophase
SmI*	chiral smectic I mesophase
T-	twist-
TGB	twist grain-boundary mesophases
TGBA*	twist grain-boundary A* mesophase
TGBC*	twist grain-boundary C* mesophase

9.2 SYMBOLS

α_i	Leslie coefficients, Leslie-Ericksen coefficients
$\Delta\widetilde{\varepsilon}$	dielectric anisotropy
γ_i	friction coefficient, rotational viscosity coefficient
$\delta\widetilde{\chi}$	biaxial mesophase anisotropy
η	relative biaxiality of a biaxial mesophase
η_i	Miesowicz coefficient
$\Delta\widetilde{\sigma}$	anisotropy of (ionic) conductivity
τ_{off}	'time-off' of the electro-optical effect
τ_{on}	'time-on' of the electro-optical effect
$\Delta_{XY}S$	transitional entropy
$\Delta\widetilde{\chi}$	magnetic mesophase anisotropy, uniaxial mesophase anisotropy
Σ	sanidic mesophase
Σ_o	ordered sanidic phase
Σ_r	rectangular sanidic mesophase
~	(tilde) modulated smectic mesophases
a	distortion in liquid crystals
b	(subscript) biaxial mesophase
B_{th}	magnitude of threshold magnetic induction
c	(subscript) intercalated smectic mesophase
e_1, e_3	flexo-electric coefficient
E_{th}	magnitude of threshold electric field
g	(subscript) glassy mesophase
K_i	elastic constants, elasticity moduli
l	director in a biaxial mesophase
m	director in a biaxial mesophase
N	nematic, discotic nematic mesophase
n	director
n$_i$	director component
N*	chiral nematic, chiral nematic mesophase, cholesteric mesophase
N$_b$	biaxial nematic mesophase
N$_u$	uniaxial nematic mesophase, nematic
$\langle P_2 \rangle$	order parameter
P	net polarization

P_s	spontaneous polarization
s	disclination strength
T^*	divergence temperature, pretransitional temperature
T_{cl}	clearing point, clearing temperature
T_i	isotropization temperature
T_{XY}	transition temperature, with X and Y being abbreviations for mesophases or a phase and a mesophase
*	tilted smectic mesophase

8: Definition of Terms Relating to the Non-Ultimate Mechanical Properties of Polymers

CONTENTS

Preamble
1. Basic definitions
2. Deformations used experimentally
3. *Stresses* observed experimentally
4. Quantities relating stress and deformation
5. Linear viscoelastic behaviour
6. Oscillatory deformations and *stresses* used experimentally for solids
7. References
8. Alphabetical index of terms
9. Glossary of symbols

PREAMBLE

This document gives definitions of terms related to the non-ultimate mechanical behaviour or mechanical behaviour prior to failure of polymeric materials, in particular of bulk polymers and concentrated solutions and their elastic and viscoelastic properties.

The terms are arranged into sections dealing with basic definitions of stress and strain, deformations used experimentally, *stresses* observed experimentally, quantities relating stress and deformation, linear viscoelastic behaviour, and oscillatory deformations and *stresses* used experimentally for solids. The terms which have been selected are those met in the conventional mechanical characterization of polymeric materials.

To compile the definitions, a number of sources have been used. A number of the definitions were adapted from an International Organization for Standardization (ISO) manuscript on Plastics Vocabulary [1]. Where possible, the names for properties, their definitions and the symbols for linear viscoelastic properties were checked against past compilations of terminology [2-6]. Other documents consulted include ASTM publications [7-13].

The document does not deal with the properties of anisotropic materials. This is an extensive subject in its own right and the reader is referred to specialized texts [14,15] for information.

In the list of contents, main terms separated by / are alternative names, and terms in parentheses give those which are defined in the context of main terms, usually as notes to the definitions of main terms, with their names printed in bold type in the main text. Multicomponent quantities (vectors, tensors, matrices) are printed in bold type. Names

---

Originally prepared by a working group consisting of A. Kaye (UK), R. F. T. Stepto (UK), W. J. Work (USA), J. V. Alemán (Spain) and A. Ya. Malkin (Russia). Reprinted from *Pure Appl. Chem.* **70**, 701-754 (1998).

printed in italics are defined elsewhere in the document and their definitions can be found by reference to the alphabetical list of terms.

NOTE on tensor terminology and symbols*

The tensor quantities given in this chapter are all second rank, and are sometimes referred to as *matrices*, according to common usage, so that the two terms, tensor and matrix, are used interchangeably. In many cases, the components (or coefficients) of second-rank tensors are represented by 3 × 3 matrices. Symbols for tensors (matrices) are printed in bold italic type, while symbols for the components are printed in italic type. In general, the base tensors are those for a rectangular Cartesian coordinate system.

1 BASIC DEFINITIONS

In this section, quantities are expressed with respect to rectangular Cartesian co-ordinate axes, Ox_1, Ox_2, Ox_3, except where otherwise stated. The components of a vector V are denoted V_1, V_2 and V_3 with respect to these axes.

1.1 traction, t, SI unit: Pa
stress vector

Vector force per unit area on an infinitesimal element of area that has a given normal and is at a given point in a body.

Note 1: The components of t are written as t_1, t_2, t_3.

Note 2: t is sometimes called true stress. The term traction (or stress vector) is preferred to avoid confusion with *stress tensor*.

1.2 stress tensor, σ, SI unit: Pa
stress

Tensor with components σ_{ij} which are the components of the *traction* in the Ox_i direction on an element of area whose normal is in the Ox_j direction.

Note 1: A unit vector area with normal n can be resolved into three smaller areas equal to n_1, n_2 and n_3 with normals in the directions of the respective co-ordinate axes. Accordingly, each component of the *traction* on the original area can be considered as the sum of components in the same direction on the smaller areas to give

$$t_i = \sum_{j=1}^{3} \sigma_{ij} n_j \quad , i = 1, 2, 3$$

Note 2: In usual circumstances, in the absence of body couples, $\sigma_{ij} = \sigma_{ji}$.

Note 3: For a homogeneous stress σ is the same at all points in a body.

Note 4: For an inhomogeneous stress $\sigma_{ij} = \sigma_{ij}(x_1, x_2, x_3)$.

Note 5: σ is a **true stress** because its components are forces per unit current area (c.f. Notes 3 and 4).

Note 6: If σ_{13} (= σ_{31}) = σ_{23} (= σ_{32}) = σ_{33} = 0 then the stress is called a **plane stress**. Plane *stresses* are associated with the deformation of a sheet of material in the plane of the sheet.

1.3 deformation of an elastic solid

Deformation of an elastic solid through which a mass point of the solid with co-ordinates X_1, X_2, X_3 in the undeformed state moves to a point with co-ordinates x_1, x_2, x_3 in the deformed state and the deformation is defined by

---
* This note has been added to this edition of the 1997 recommendations.

TERMINOLOGY

$$x_i = x_i(X_1, X_2, X_3), \; i = 1, 2, 3$$

Note 1: A **homogeneous deformation** is one in which the relationships between the co-ordinates in the undeformed and deformed states reduce to

$$x_i = \sum_{j=1}^{3} f_{ij} X_j \quad i = 1, 2, 3$$

where the f_{ij} are constants.

Note 2: An **inhomogeneous deformation** is one in which the incremental changes in the undeformed and deformed co-ordinates are related by

$$dx_i = \sum_{j=1}^{3} f_{ij} dX_j \quad i = 1, 2, 3$$

where $f_{ij} = \partial x_i / \partial X_j$ $i, j = 1, 2, 3$, and where the f_{ij} are the functions of the coordinates x_j.

Note 3: The f_{ij} in notes 1 and 2 are **deformation gradients**.

1.4 deformation gradient tensor for an elastic solid, *F*

Tensor whose components are deformation gradients in an elastic solid.
Note 1: The components of ***F*** are denoted f_{ij}.
Note 2: See Definition 1.3 for the definitions of f_{ij}.

1.5 deformation of a viscoelastic liquid or solid

Deformation of a viscoelastic liquid or solid through which a mass point of the viscoelastic liquid or solid with co-ordinates x'_1, x'_2, x'_3 at time t' moves to a point with co-ordinates x_1, x_2, x_3 at time t such that there are functions g_i, $i = 1, 2, 3$, where

$$g_i(x'_1, x'_2, x'_3, t') = g_i(x_1, x_2, x_3, t).$$

Note 1: t' often refers to some past time and t to the present time.
Note 2: The relationships between the total differentials of the functions g_i define how particles of the material move relative to each other. Thus, if two particles are at small distances dx'_1, dx'_2, dx'_3 apart at time t' and dx_1, dx_2, dx_3 at time t then

$$\sum_{j=1}^{3} g'_{ij} dx'_j = \sum_{j=1}^{3} g_{ij} dx_j$$

where $g'_{ij}(x'_1, x'_2, x'_3, t') = \dfrac{\partial g_i(x'_1, x'_2, x'_3, t')}{\partial x'_j}$

and $g_{ij}(x_1, x_2, x_3, t) = \dfrac{\partial g_i(x_1, x_2, x_3, t)}{\partial x_j}$ where $i, j = 1, 2, 3$.

Note 3: The matrix with elements g_{ij} is denoted ***G*** and the matrix with elements g'_{ij} is denoted ***G'***.
Note 4: A **homogeneous deformation** is one in which the functions g_i are linear functions of the x_j, $i, j = 1, 2, 3$. As a result, the g_{ij} and ***G*** are functions of t only and the equations which define the deformation become

NON-ULTIMATE MECHANICAL PROPERTIES

$$\sum_{j=1}^{3} g'_{ij}(t') x'_j = \sum_{j=1}^{3} g_{ij}(t) x_j$$

Note 5: *Homogeneous deformations* are commonly used or assumed in the methods employed for characterizing the mechanical properties of viscoelastic polymeric liquids and solids.

1.6 deformation gradients in a viscoelastic liquid or solid, f_{ij}

If two mass points of a liquid are at a small distance dx'_1, dx'_2, dx'_3 apart at time t' then the deformation gradients are the rates of change of dx'_i with respect to dx_j, $i, j = 1, 2, 3$.

Note: $f_{ij} = \partial x'_i / \partial x_j$, $i, j = 1, 2, 3$

1.7 deformation gradient tensor for a viscoelastic liquid or solid, F

Tensor whose components are deformation gradients in a viscoelastic liquid or solid.
Note 1: The components of F are denoted f_{ij}.
Note 2: See Definition 1.6 for the definition of f_{ij}.
Note 3: By matrix multiplication, $F = (G')^{-1}G$, where the matrices G and G' are those defined in Definition 1.5.

1.8 strain tensor

Symmetric tensor that results when a *deformation gradient tensor* is factorized into a rotation tensor followed or preceded by a symmetric tensor.
Note 1: A strain tensor is a measure of the relative displacement of the mass points of a body.
Note 2: The *deformation gradient tensor* F may be factorized as

$$F = R U = V R,$$

where R is an orthogonal matrix representing a rotation and U and V are strain tensors which are symmetric.
Note 3: Alternative strain tensors are often more useful, for example:

the **Cauchy** tensor, $C = U^2 = F^T F$
the **Green** tensor, $B = V^2 = F F^T$
the **Finger** tensor, C^{-1}
the **Piola** tensor, B^{-1}

where 'T' denotes transpose and '-1' denotes inverse. B is most useful for solids and C and C^{-1} for viscoelastic liquids and solids.
Note 4: If the 1,3; 3,1; 2,3; 3,2; 3,3 elements of a strain tensor are equal to zero then the strain is termed **plane strain**.

1.9 Cauchy tensor, C

Strain tensor for a viscoelastic liquid or solid, whose elements are

$$c_{ij} = \sum_{k=1}^{3} \frac{\partial x'_k}{\partial x_i} \cdot \frac{\partial x'_k}{\partial x_j},$$

where x'_i and x_i are co-ordinates of a particle at times t' and t, respectively.
Note 1: See Definition 1.5 for the definition of x'_i and x_i.
Note 2: See Definition 1.8 for the definition of a *strain tensor*.

1.10 Green tensor, *B*
Strain tensor for an elastic solid, whose elements are

$$b_{ij} = \sum_{k=1}^{3} \frac{\partial x_i}{\partial X_k} \cdot \frac{\partial x_j}{\partial X_k},$$

where X_i and x_i are co-ordinates in the undeformed and deformed states, respectively.
Note 1: See Definition 1.3 for the definition of X_i and x_i.
Note 2: See Definition 1.8 for the definition of a *strain tensor*.
Note 3: For small strains, *B* may be expressed by the equation

$$\mathbf{B} = \mathbf{I} + 2\boldsymbol{\varepsilon},$$

where *I* is the unit matrix of order three and ε is the **small-strain tensor**. The components of ε are

$$\varepsilon_{ij} = \frac{1}{2}\left(\frac{\partial u_i}{\partial x_j} + \frac{\partial u_j}{\partial x_i}\right)$$

with $u_k = x_k - X_k$, $k = 1, 2, 3$, the displacements due to the deformation.

1.11 Finger tensor, C^{-1}
Strain tensor, for a viscoelastic liquid or solid, whose elements are

$$c_{ij} = \sum_{k=1}^{3} \frac{\partial x_i}{\partial x'_k} \cdot \frac{\partial x_j}{\partial x'_k},$$

where x'_i and x_i are co-ordinates of a particle at times t' and t, respectively.
Note: See Definition 1.5 for the definition of x'_i and x_i.

1.12 rate-of-strain tensor, *D*, SI unit: s^{-1}
Time derivative of a *strain tensor* for a viscoelastic liquid or solid in *homogeneous deformation* at reference time, t.
Note 1: For an *inhomogeneous deformation*, the material derivative has to be used to find time derivatives of strain.
Note 2: $D = \lim_{t' \to t}\left(\frac{\partial U}{\partial t'}\right) = \lim_{t' \to t}\left(\frac{\partial V}{\partial t'}\right)$, where *U* and *V* are defined in Definition 1.8, note 2.
Note 3: The elements of *D* are

$$d_{ij} = \frac{1}{2}\left(\frac{\partial v_i}{\partial x_j} + \frac{\partial v_j}{\partial x_i}\right),$$

where the v_k are the components of the velocity *v* at *x* and time, t.

1.13 vorticity tensor, W, SI unit: s^{-1}

Derivative, for a viscoelastic liquid or solid in *homogeneous deformation*, of the rotational part of the deformation-gradient tensor at reference time, t.

Note 1: For an *inhomogeneous deformation* the material derivative has to be used.

Note 2: $W = \lim\limits_{t' \to t}\left(\dfrac{\partial R}{\partial t'}\right)$, where R is defined in Definition 1.8, note 2.

Note 3: The elements of W are

$$w_{ij} = \frac{1}{2}\left(\frac{\partial v_i}{\partial x_j} - \frac{\partial v_j}{\partial x_i}\right),$$

where the v_k are the components of the velocity v at x and time t.

1.14 Rivlin-Ericksen tensors, A_n, SI unit: s^{-n}

Rivlin-Ericksen tensor of order n, for a viscoelastic liquid or solid in *homogeneous deformation*, is the nth time derivative of the *Cauchy strain tensor* at reference time, t.

Note 1: For an *inhomogeneous deformation* the material derivatives have to be used.

Note 2: $A_n = \lim\limits_{t' \to t}\left(\dfrac{\partial^n C}{\partial t'^n}\right)$, where C is defined in Definition 1.9.

Note 3: $A_0 = I$, where I is the unit matrix of order three.

Note 4: $A_1 = \dot{F}^T + \dot{F} = 2D$ where F is the *deformation gradient tensor*, $\dot{F} = \lim\limits_{t' \to t}\left(\dfrac{\partial F}{\partial t'}\right)$, where 'T' denotes transpose and D is the *rate-of-strain tensor*.

Note 5: In general, $A_{n+1} = \dot{A}_n + \dot{F}^T A_n + A_n \dot{F}$, $n = 0, 1, 2, ...$

2 DEFORMATIONS USED EXPERIMENTALLY

All deformations used in conventional measurements of mechanical properties are interpreted in terms of *homogeneous deformations*.

2.1 general orthogonal homogeneous deformation of an elastic solid

Deformation, such that a mass point of the solid with co-ordinates X_1, X_2, X_3 in the undeformed state moves to a point with co-ordinates x_1, x_2, x_3 in the deformed state, with

$$x_i = \lambda_i X_i, \; i = 1, 2, 3,$$

where the λ_i are constants.

Note 1: The relationships between the x_i and X_i for orthogonal *homogeneous deformations* are a particular case of the general relationships given in Definition 1.3, provided the deformation does not include a rotation and the co-ordinate axes are chosen as the principal directions of the deformation.

Note 2: The λ_i are effectively **deformation gradients**, or, for finite deformations, the **deformation ratios** characterizing the deformation.

Note 3: For an incompressible material, $\lambda_1 \lambda_2 \lambda_3 = 1$.

TERMINOLOGY

Note 4: The λ_i are elements of the *deformation gradient tensor* **F** and the resulting *Cauchy* and *Green tensors* **C** and **B** are

$$C = B = \begin{bmatrix} \lambda_1^2 & 0 & 0 \\ 0 & \lambda_2^2 & 0 \\ 0 & 0 & \lambda_3^2 \end{bmatrix}$$

2.2 uniaxial deformation of an elastic solid
Orthogonal, *homogeneous deformation* in which, say,

$$\lambda_1 = \lambda \quad \text{and} \quad \lambda_2 = \lambda_3.$$

Note 1: See Definition 2.1 for the definition of λ_i, $i = 1, 2, 3$.
Note 2: For an incompressible material, $\lambda_2 = \lambda_3 = 1/\lambda^{1/2}$.

2.3 uniaxial deformation ratio, λ
    deformation ratio
Quotient of the length (l) of a sample under uniaxial tension or compression and its original length (l_0)

$$\lambda = l / l_0$$

Note 1: In tension λ (>1) may be termed the **extension ratio**.
Note 2: In compression λ (<1) may be termed the **compression ratio**.
Note 3: λ is equivalent to λ_1 in Definitions 2.1 and 2.2.

2.4 uniaxial strain, ε
    engineering strain
Change in length of a sample in uniaxial tensile or compressive deformation divided by its initial length

$$\varepsilon = (l_1 - l_0) / l_0$$

where l_0 and l_1 are, respectively, the initial and final lengths.
Note 1: $\varepsilon = \lambda - 1$, where λ is the *uniaxial deformation ratio*.
Note 2: $\varepsilon > 0$ is referred to as **(uniaxial) tensile strain**.
Note 3: $\varepsilon < 0$ is referred to as **(uniaxial) compressive strain**.

2.5 Hencky strain, ε_H
Integral over the total change in length of a sample of the incremental strain in uniaxial tensile deformation

$$\varepsilon_H = \int_{l_0}^{l_1} dl/l = \ln(l_1/l_0)$$

l_0, l_1 and l are, respectively, the initial, final and instantaneous lengths.
Note 1: See *uniaxial strain* (Definition 2.4).

Note 2: The same equation can be used to define a quantity ε_H (< 0) in compression.

2.6 Poisson's ratio, μ

In a sample under small *uniaxial deformation*, the negative quotient of the lateral strain (ε_{lat}) and the longitudinal strain (ε_{long}) in the direction of the uniaxial force

$$\mu = -\left(\frac{\varepsilon_{lat}}{\varepsilon_{long}}\right)$$

Note 1: **Lateral strain** ε_{lat} is the strain normal to the *uniaxial deformation*.
Note 2: $\varepsilon_{lat} = \lambda_2 - 1 = \lambda_3 - 1$ (see Definitions 2.2 and 2.4).
Note 3: For an isotropic, incompressible material, $\mu = 0.5$. It should be noted that, in materials referred to as incompressible, volume changes do in fact occur in deformation, but they may be neglected.
Note 4: For an anisotropic material, μ varies with the direction of the *uniaxial deformation*.
Note 5: Poisson's ratio is also sometimes called the **lateral contraction ratio** and is sometimes used in cases of non-linear deformation. The present definition will not apply in such cases.

2.7 pure shear of an elastic solid

Orthogonal, *homogeneous deformation* in which

$$\lambda_1 = \lambda$$
$$\lambda_2 = 1/\lambda$$
$$\lambda_3 = 1$$

Note: See Definition 2.1 for the definition of λ_i, $i = 1, 2, 3$.

2.8 simple shear of an elastic solid

Homogeneous deformation, such that a mass point of the solid with co-ordinates X_1, X_2, X_3 in the undeformed state moves to a point with co-ordinate x_1, x_2, x_3 in the deformed state, with

$$x_1 = X_1 + \gamma X_2$$
$$x_2 = X_2$$
$$x_3 = X_3$$

where γ is constant.
Note 1: The relationships between the x_i and X_i, $i = 1, 2, 3$, in simple shear are a particular case of the general relationships given in Definition 1.3.
Note 2: γ is known as the **shear** or **shear strain**.
Note 3: The *deformation gradient tensor* for the simple shear of an *elastic solid* is

$$F = \begin{pmatrix} 1 & \gamma & 0 \\ 0 & 1 & 0 \\ 0 & 0 & 1 \end{pmatrix}$$

TERMINOLOGY

and the *Cauchy* (C) and *Green* (B) *strain tensors* are

$$C = \begin{pmatrix} 1 & \gamma & 0 \\ \gamma & 1+\gamma^2 & 0 \\ 0 & 0 & 1 \end{pmatrix} \quad \text{and} \quad B = \begin{pmatrix} 1+\gamma^2 & \gamma & 0 \\ \gamma & 1 & 0 \\ 0 & 0 & 1 \end{pmatrix}$$

2.9 bulk compression, χ

Fractional decrease in volume (V) caused by a hydrostatic pressure

$$\chi = -\Delta V/V.$$

Note: Also referred to as **volume compression**, **isotropic compression** and **bulk compressive strain**.

2.10 general homogeneous deformation or flow of a viscoelastic liquid or solid

Flow or deformation such that a particle of the viscoelastic liquid or solid with co-ordinate vector X' at time t' moves to a point with co-ordinate vector X at time t with

$$G'X' = GX$$

where G' and G are tensors defining the type of deformation or flow and are functions of time only.

Note 1: The definition is equivalent to that given in Definition 1.5, note 4. Accordingly, the elements of G' and G are denoted $g'_{ij}(t')$ and $g_{ij}(t)$ and those of X' and X, (x'_1, x'_2, x'_3) and (x_1, x_2, x_3).

Note 2: For an incompressible material

$$\det G = 1, \text{ where } \det G \text{ is the determinant of } G.$$

Note 3: Deformations and flows used in conventional measurements of properties of viscoelastic liquids and solids are usually interpreted assuming incompressibility.

2.11 homogeneous orthogonal deformation or flow of an incompressible viscoelastic liquid or solid

Deformation or flow, as defined in Definition 2.10, such that

$$G = \begin{pmatrix} g_{11}(t) & 0 & 0 \\ 0 & g_{22}(t) & 0 \\ 0 & 0 & g_{33}(t) \end{pmatrix}.$$

Note 1: The g_{ii} are defined in Definition 1.5, notes 2 to 4.
Note 2: If $g_{22} = g_{33} = 1/g_{11}^{1/2}$ the elongational deformation or flow is **uniaxial**.
Note 3: The *Finger strain tensor* for a homogeneous orthogonal deformation or flow of incompressible, viscoelastic liquid or solid is

NON-ULTIMATE MECHANICAL PROPERTIES

$$C^{-1} = \begin{pmatrix} \left(\dfrac{g'_{11}(t')}{g_{11}(t)}\right)^2 & 0 & 0 \\ 0 & \left(\dfrac{g'_{22}(t')}{g_{22}(t)}\right)^2 & 0 \\ 0 & 0 & \left(\dfrac{g'_{33}(t')}{g_{33}(t)}\right)^2 \end{pmatrix}.$$

2.12 steady uniaxial homogeneous elongational deformation or flow of an incompressible viscoelastic liquid or solid

Uniaxial homogeneous elongational flow in which

$$g_{11}(t) = \exp(-\dot{\gamma}_E t)$$

where $\dot{\gamma}_E$ is a constant, E denotes elongational and $g_{22} = g_{33} = 1/g_{11}^{1/2}$.

Note 1: $g_{11}(t)$, $g_{22}(t)$ and $g_{33}(t)$ are elements of the tensor G defined in Definition 1.5.
Note 2: From the definition of *general homogeneous flow* (Definition 1.5) ($G'X' = GX$ = constant) in the particular case of *steady uniaxial elongation flow*

$$x_1 g_{11}(t) = x_1 \exp(-\dot{\gamma}_E t) = \text{constant}$$

and differentiation with respect to time gives

$$\dot{\gamma}_E = (1/x_1)(dx_1/dt).$$

Hence, $\dot{\gamma}_E$ is the **elongational** or **extensional strain rate**.

Note 3: The *Finger strain tensor* for a steady uniaxial homogeneous elongation deformation or flow of an incompressible viscoelastic liquid or solid is

$$C^{-1} = \begin{pmatrix} \exp(2\dot{\gamma}_E(t-t')) & 0 & 0 \\ 0 & \exp(-\dot{\gamma}_E(t-t')) & 0 \\ 0 & 0 & \exp(-\dot{\gamma}_E(t-t')) \end{pmatrix}.$$

2.13 homogeneous simple shear deformation or flow of an incompressible viscoelastic liquid or solid

Flow or deformation such that

$$G = \begin{pmatrix} 1 & -\gamma(t) & 0 \\ 0 & 1 & 0 \\ 0 & 0 & 1 \end{pmatrix}$$

where $\gamma(t)$ is the **shear**.

Note 1: The general tensor G is defined in Definition 1.5.
Note 2: $\dot{\gamma} = d\gamma(t)/dt$ is the **shear rate**. The unit of $\dot{\gamma}$ is s^{-1}.

TERMINOLOGY

Note 3: If $\gamma(t) = \dot{\gamma} t$, where $\dot{\gamma}$ is a constant, then the flow has a constant shear rate and is known as **steady (simple) shear flow**.
Note 4: If $\gamma(t) = \gamma_0 \sin 2\pi \nu t$ then the flow is **oscillatory (simple) shear flow** of amplitude γ_0 and frequency ν expressed in Hz.
Note 5: The *Finger strain tensor* for *simple shear flow* is

$$\mathbf{C}^{-1} = \begin{pmatrix} 1+(\gamma(t)-\gamma(t'))^2 & \gamma(t)-\gamma(t') & 0 \\ \gamma(t)-\gamma(t') & 1 & 0 \\ 0 & 0 & 1 \end{pmatrix}$$

where $\gamma(t) - \gamma(t')$ is the amount of *shear* given to the liquid between the times t' and t. For *steady simple shear* flow $\mathbf{C}^{-1}$ becomes

$$\mathbf{C}^{-1} = \begin{pmatrix} 1+\dot{\gamma}^2(t-t')^2 & \dot{\gamma}(t-t') & 0 \\ \dot{\gamma}(t-t') & 1 & 0 \\ 0 & 0 & 1 \end{pmatrix}.$$

3 STRESSES OBSERVED EXPERIMENTALLY

For a given deformation or flow, the resulting *stress* depends on the material. However, the *stress tensor* does take particular general forms for experimentally used deformations (see **section 2**). The definitions apply to elastic solids, and viscoelastic liquids and solids.

3.1 stress tensor resulting from an orthogonal deformation or flow, σ, SI unit: Pa

For an orthogonal deformation or flow the *stress tensor* is diagonal with

$$\boldsymbol{\sigma} = \begin{pmatrix} \sigma_{11} & 0 & 0 \\ 0 & \sigma_{22} & 0 \\ 0 & 0 & \sigma_{33} \end{pmatrix}.$$

Note 1: See Definition 1.2 for the general definition of σ.
Note 2: If the *strain tensor* is diagonal for all time then the *stress tensor* is diagonal for all time for isotropic materials.
Note 3: For a **uniaxial (orthogonal) deformation or flow** $\sigma_{22} = \sigma_{33}$.
Note 4: For a **pure shear deformation or flow** the *stresses* (σ_{11}, σ_{22}, σ_{33}) are usually all different from each other.
Note 5: The *stress tensor* resulting from a *pure shear deformation or flow* is called a **pure shear stress**.

3.2 tensile stress, σ, SI unit: Pa

Component σ_{11} of the *stress tensor* resulting from a tensile *uniaxial deformation*.
Note 1: The *stress tensor* for a *uniaxial deformation* is given in Definition 3.1.
Note 2: The Ox_1 direction is chosen as the direction of the *uniaxial deformation*.

NON-ULTIMATE MECHANICAL PROPERTIES

3.3 compressive stress, σ, SI unit: Pa
Component σ_{11} of the *stress tensor* resulting from a compressive *uniaxial deformation*.
Note: See notes 1 and 2 of Definition 3.2.

3.4 nominal stress, σ, SI unit: Pa
   engineering stress
Force resulting from an applied tensile or compressive *uniaxial deformation* divided by the initial cross-sectional area of the sample normal to the applied deformation.
Note: The term *engineering* or *nominal stress* is often used in circumstances when the deformation of the body is not infinitesimal and its cross-sectional area changes.

3.5 stress tensor resulting from a simple shear deformation or flow, σ, SI unit: Pa
For a simple shear deformation or flow the *stress tensor* takes the form

$$\sigma = \begin{pmatrix} \sigma_{11} & \sigma_{12} & 0 \\ \sigma_{21} & \sigma_{22} & 0 \\ 0 & 0 & \sigma_{33} \end{pmatrix}$$

where σ_{21} is numerically equal to σ_{12}.
Note 1: See Definition 1.2 for the general definition of σ.
Note 2: σ_{ii}, $i = 1, 2, 3$ are denoted **normal stresses**.
Note 3: σ_{12} is called the **shear stress**.

3.6 first normal-stress difference, N_1, SI unit: Pa
   first normal-stress function
Difference between the first two normal *stresses* σ_{11} and σ_{22} in simple shear flow

$$N_1 = \sigma_{11} - \sigma_{22}.$$

Note 1: See Definition 3.5 for the definition of σ_{11} and σ_{22}.
Note 2: For *Newtonian liquids* $N_1 = 0$.

3.7 second normal-stress difference, N_2, SI unit: Pa
   second normal-stress function
Difference between the second and third normal-*stresses* ($\sigma_{22} - \sigma_{33}$) in simple shear flow

$$N_2 = \sigma_{22} - \sigma_{33}.$$

Note 1: See Definition 3.5 for the definition of σ_{22} and σ_{33}
Note 2: For *Newtonian liquids*, $N_2 = 0$.

4 QUANTITIES RELATING STRESS AND DEFORMATION

4.1 constitutive equation for an elastic solid
Equation relating *stress* and strain in an elastic solid.
Note 1: For an elastic solid, the constitutive equation may be written

TERMINOLOGY

$$\sigma = \frac{2}{I_3^{1/2}} \left(\frac{\partial W}{\partial I_1} \mathbf{B} + \frac{\partial W}{\partial I_2} (I_1 \mathbf{B} - \mathbf{B}^2) + I_3 \frac{\partial W}{\partial I_3} \mathbf{I} \right),$$

where $\mathbf{B}$ is the *Green strain tensor*.
I_1, I_2, I_3 are invariants of $\mathbf{B}$, with
  $I_1 = \text{Tr}(\mathbf{B})$
  $I_2 = 1/2 \{(\text{Tr}(\mathbf{B}))^2 - \text{Tr}(\mathbf{B}^2)\}$
  $I_3 = \det \mathbf{B}$,
where 'Tr' denotes trace and 'det' denotes determinant. The invariants are independent of the co-ordinate axes used and for symmetric tensors there are three independent invariants. W is a function of I_1, I_2 and I_3 and is known as the **stored energy function** and is the increase in energy (stored energy) per unit initial volume due to the deformation.
Note 2: For small deformations, the constitutive equation may be written

$$\sigma = 2G\varepsilon + l\mathbf{I}\text{Tr}(\varepsilon),$$

where G is the *shear modulus*, ε is the *small-strain tensor* and l is a **Lamé constant**.
Note 3: The Lamé constant, (l), is related to the *shear modulus* (G) and *Young's modulus* (E) by the equation

$$l = G(2G - E)/(E - 3G).$$

Note 4: For an incompressible elastic solid, the constitutive equation may be written

$$\sigma + P\mathbf{I} = 2\frac{\partial W}{\partial I_1} \mathbf{B} - 2\frac{\partial W}{\partial I_2} \mathbf{B}^{-1},$$

where P is the hydrostatic (or isotropic) pressure, $I_3 = 1$ and W is a function of I_1 and I_2, only.
Note 5: For small deformations of an incompressible, inelastic solid, the constitutive equation may be written

$$\sigma + P\mathbf{I} = 2G\varepsilon.$$

4.2 constitutive equation for an incompressible viscoelastic liquid or solid
Equation relating *stress* and deformation in an incompressible viscoelastic liquid or solid.
Note 1: A possible general form of constitutive equation when there is no dependence of *stress* on amount of strain is

$$\sigma + P\mathbf{I} = f(A_1, A_2, \ldots, A_n),$$

where $A_1, A_2, \ldots$ are the *Rivlin-Ericksen tensors*.
Note 2: For a *non-Newtonian liquid* (see note 3), a form of the general constitutive equation which may be used is

$$\sigma + P\mathbf{I} = \eta A_1^2 + \alpha A_1 + \beta A_2,$$

where η is the *viscosity* and α and β are constants.

Note 3: A **Newtonian liquid** is a liquid for which the constitutive equation may be written

$$\sigma + PI = \eta A_1 = 2\eta D,$$

where D is the *rate-of-strain tensor*. Liquids which do not obey this constitutive equation are termed **non-Newtonian liquids**.

Note 4: For cases where there is a dependence of *stress* on strain history the following constitutive equation may be used, namely

$$\sigma + PI = 2\int_{-\infty}^{t}\left(\frac{\partial \Omega}{\partial I_1}C^{-1} - \frac{\partial \Omega}{\partial I_2}C\right)dt',$$

where C is the *Cauchy strain tensor* and Ω is a function of the invariants I_1, I_2 and I_3 of C^{-1} and the time interval $t-t'$. Ω is formally equivalent to the *stored-energy function*, W, of a solid.

4.3 **modulus**, in general M, SI unit: Pa
      in bulk compressive deformation K, SI unit: Pa
      in *uniaxial deformation* E, SI unit: Pa
      in shear deformation G, SI unit: Pa

Quotient of *stress* and strain where the type of *stress* and strain is defined by the type of deformation employed.

Note 1: The detailed definitions of K, E and G are given in Definitions 4.5, 4.7 and 4.10.
Note 2: An **elastic modulus** or **modulus of elasticity** is a modulus of a body which obeys Hooke's law (*stress* $\propto$ strain).

4.4 **compliance**, in general C, SI unit: Pa
      in bulk compressive deformation B, SI unit: Pa
      in *uniaxial deformation* D, SI unit: Pa
      in shear deformation J, SI unit: Pa

Quotient of strain and *stress* where the type of strain and *stress* is defined by the type of deformation employed.

Note 1: $C = 1/M$, where M is *modulus*.
Note 2: The detailed definitions of B, D and J are given in Definitions 4.6, 4.8 and 4.11.

4.5 **bulk modulus**, K, SI unit: Pa

Quotient of hydrostatic pressure (P) and *bulk compression* (χ)

$$K = P/\chi.$$

Note 1: Also known as **bulk compressive modulus**.
Note 2: For the definition of χ, see Definition 2.9.
Note 3: At small deformations, the *bulk modulus* is related to Young's modulus (E) by

$$K = \frac{E}{3(1-2\mu)}$$

where μ is *Poisson's ratio*.

TERMINOLOGY

4.6 bulk compliance, B, SI unit: Pa^{-1}
Quotient of *bulk compression* (χ) and hydrostatic pressure (P)

$$B = \chi/P.$$

Note 1: Also known as **bulk compressive compliance**.
Note 2: For the definition of χ, see Definition 2.9.
Note 3: $B = 1/K$, where K is the *bulk modulus*.

4.7 Young's modulus, E, SI unit: Pa
Quotient of uniaxial stress (σ) and strain (ε) in the limit of zero strain

$$E = \lim_{\varepsilon \to 0} (\sigma/\varepsilon).$$

Note 1: The *stress* is a *true stress*, as in Definitions 3.2 and 3.3, and not a *nominal stress*, as in Definition 3.4.
Note 2: ε is defined in Definition 2.4.
Note 3: *Young's modulus* may be evaluated using *tensile* or *compressive uniaxial deformation*. If determined using tensile deformation it may be termed **tensile modulus**.
Note 4: For non-Hookean materials the *Young's modulus* is sometimes evaluated as:
  (i) the **secant modulus** – the quotient of *stress* (σ) and strain at some nominal *strain* (ε) in which case

$$E = \sigma/\varepsilon$$

  (ii) the **tangent modulus** – the slope of the *stress*-strain curve at some nominal *strain* (ε'), in which case

$$E = (d\sigma/d\varepsilon)_{\varepsilon=\varepsilon'}$$

4.8 uniaxial compliance, D, SI unit: Pa^{-1}
Quotient of *uniaxial strain* (ε) and uniaxial stress (σ) in the limit of zero strain

$$D = \lim_{\varepsilon \to 0} (\varepsilon/\sigma).$$

Note 1: The *stress* is a *true stress* as in Definitions 3.2 and 3.3, and not a *nominal stress*, as in Definition 3.4.
Note 2: ε is defined in Definition 2.4.
Note 3: Uniaxial compliance may be evaluated using *tensile or compressive uniaxial deformation*. If determined using tensile deformation it may be termed **tensile compliance**.
Note 4: $D = 1/E$, where E is *Young's modulus*.

4.9 extensional viscosity, η_E, SI unit: Pa s
   elongational viscosity
Quotient of the difference between the longitudinal *stress* (σ_{11}) and the lateral *stress* (σ_{22}) and the elongational strain rate ($\dot{\gamma}_E$) in steady uniaxial flow

$$\eta_E = (\sigma_{11} - \sigma_{22})/\dot{\gamma}_E.$$

NON-ULTIMATE MECHANICAL PROPERTIES

Note: See Definitions 3.1 and 2.12 for the definitions of σ_{11}, σ_{22} and $\dot{\gamma}_E$.

4.10   shear modulus, G, SI unit: Pa
Quotient of shear stress (σ_{12}) and shear strain (γ)

$$G = \sigma_{12}/\gamma.$$

Note 1: See Definition 2.8 for the definitions of γ for an elastic solid and Definition 3.5 for the definition of σ_{12}.
Note 2: The *shear modulus* is related to *Young's modulus* (E) by the equation

$$G = \frac{E}{2(1 + \mu)}$$

where μ is *Poisson's ratio*.
Note 3: For elastomers, which are assumed incompressible, the *modulus* is often evaluated in *uniaxial tensile* or *compressive deformation* using $\lambda - \lambda^{-2}$ as the strain function (where λ is the *uniaxial deformation ratio*). In the limit of zero deformation the *shear modulus* is evaluated as

$$\frac{d\sigma}{d(\lambda - \lambda^{-2})} = \frac{E}{3} = G \text{ (for } \mu = 0.5\text{)},$$

where σ is the *tensile* or *compressive stress*.

4.11   shear compliance, J, SI unit: Pa^{-1}
Quotient of shear strain (γ) and shear stress (σ_{12})

$$J = \gamma/\sigma_{12}.$$

Note 1: See Definition 2.8 for the definition of γ for an elastic solid and 3.5 for the definition of σ_{12}.
Note 2: $J = 1/G$, where G is the *shear modulus*.

4.12   shear viscosity, η, SI unit: Pa s
       coefficient of viscosity
       viscosity
Quotient of shear stress (σ_{12}) and shear rate ($\dot{\gamma}$) in steady, simple shear flow

$$\eta = \sigma_{12}/\dot{\gamma}.$$

Note 1: See Definitions 3.5 and 2.13 for the definitions of σ_{12} and $\dot{\gamma}$.
Note 2: For *Newtonian liquids*, σ_{12} is directly proportional to $\dot{\gamma}$ and η is constant.
Note 3: For *non-Newtonian liquids*, when σ_{12} is not directly proportional to $\dot{\gamma}$, η varies with $\dot{\gamma}$. The value of η evaluated at a given value of $\dot{\gamma}$ is termed the **non-Newtonian viscosity**.

TERMINOLOGY

Note 4: Some experimental methods, such as capillary flow and flow between parallel plates, employ a range of shear rates. The value of η evaluated at some nominal average value of $\dot{\gamma}$ is termed the **apparent viscosity** and given the symbol η_{app}. It should be noted that this is an imprecisely defined quantity.

Note 5: Extrapolation of η or η_{app} for *non-Newtonian liquids* to zero $\dot{\gamma}$ gives the **zero-shear viscosity**, which is given the symbol η_0.

Note 6: Extrapolation of η and η_{app} for *non-Newtonian liquids* to infinite $\dot{\gamma}$ gives the **infinite-shear viscosity**, which is given the symbol η_∞.

4.13   first normal-stress coefficient, ψ_1, SI unit: Pa s^2

Quotient of the first normal *stress* difference (N_1) and the square of the shear rate ($\dot{\gamma}$) in the limit of zero shear rate

$$\psi_1 = \lim_{\dot{\gamma} \to 0} (N_1/\dot{\gamma}^2).$$

Note: See Definitions 3.6 and 2.13 for the definitions of N_1 and $\dot{\gamma}$.

4.14   second normal-stress coefficient, ψ_2, SI unit: Pa s^2

Quotient of the second normal *stress* difference (N_2) and the square of the shear rate ($\dot{\gamma}$) in the limit of zero shear rate

$$\psi_2 = \lim_{\dot{\gamma} \to 0} (N_2/\dot{\gamma}^2).$$

Note: See Definitions 3.7 and 2.13 for the definitions of N_2 and $\dot{\gamma}$.

5   LINEAR VISCOELASTIC BEHAVIOUR

5.1   viscoelasticity

Time-dependent response of a liquid or solid subjected to *stress* or strain.

Note 1: Both viscous and elastic responses to *stress* or strain are required for the description of viscoelastic behaviour.

Note 2: Viscoelastic properties are usually measured as responses to an instantaneously applied or removed constant *stress* or strain or a **dynamic stress or strain**. The latter is defined as a sinusoidal *stress* or strain of small amplitude, which may or may not decrease with time.

5.2   linear viscoelastic behaviour

Interpretation of the viscoelastic behaviour of a liquid or solid in *simple shear* or *uniaxial deformation* such that

$$P(D)\sigma = Q(D)\gamma,$$

where σ is the shear stress or uniaxial stress, γ is the shear strain or *uniaxial strain*, and $P(D)$ and $Q(D)$ are polynomials in D, where D is the differential coefficient operator d/dt.

Note 1: In *linear viscoelastic behaviour*, *stress* and strain are assumed to be small so that the squares and higher powers of σ and γ may be neglected.
Note 2: See Definitions 3.5 and 2.13 for the definitions of σ and γ in *simple shear*.
Note 3: See Definitions 3.2 and 2.12 for definitions of σ and γ ($\equiv \gamma_E$) in *uniaxial deformations*.
Note 4: The polynomials $Q(D)$ and $P(D)$ have the forms:

$Q(D) = a(D + q_0) \ldots (D + q_n)$
(a polynomial of degree $n + 1$)
$P(D) = (D + p_0)(D + p_1) \ldots (D + p_n)$
(a polynomial of degree $n + 1$)
$P(D) = (D + p_0)(D + p_1) \ldots (D + p_{n-1})$
(a polynomial of degree n)

where

(i) a is a constant
(ii) $q_0 \geq 0$, $p_0 > 0$ and p_s, $q_s > 0$, $s = 1, \ldots, n$.
(iii) $q_i < p_i < q_{i+1}$ and $q_n < p_n$ (if p_n exists) with p_i and q_i related to *relaxation* and *retardation times*, respectively.

Note 5: If $q_0 = 0$, the material is a liquid, otherwise it is a solid.
Note 6: Given that $Q(D)$ is a polynomial of degree $n + 1$; if $P(D)$ is also of degree $n + 1$ the material shows instantaneous elasticity; if $P(D)$ is of degree n, the material does not show instantaneous elasticity (i.e. elasticity immediately the deformation is applied.)
Note 7: There are definitions of linear viscoelasticity which use integral equations instead of the differential equation in Definition 5.2. (See, for example, [11].) Such definitions have certain advantages regarding their mathematical generality. However, the approach in the present document, in terms of differential equations, has the advantage that the definitions and descriptions of various viscoelastic properties can be made in terms of commonly used mechano-mathematical models (e.g. the *Maxwell* and *Voigt-Kelvin models*).

5.3 Maxwell model
Maxwell element
Model of the linear viscoelastic behaviour of a liquid in which

$(\alpha D + \beta)\sigma = D\gamma$

where α and β are positive constants, D the differential coefficient operator d/dt, and σ and γ are the *stress* and strain in *simple shear* or *uniaxial deformation*.
Note 1: See Definition 5.2 for a discussion of σ and γ.
Note 2: The relationship defining the Maxwell model may be written

$d\sigma/dt + (\beta/\alpha)\sigma = (1/\alpha)d\gamma/dt.$

Note 3: Comparison with the general definition of *linear viscoelastic behaviour* shows that the polynomials $P(D)$ and $Q(D)$ are of order one, $q_0 = 0$, $p_0 = \beta/\alpha$ and $a = 1/\alpha$. Hence, a

material described by a Maxwell model is a *liquid* ($q_0 = 0$) having instantaneous elasticity (P(D) and Q(D) are of the same order).
Note 4: The Maxwell model may be represented by a spring and a dashpot filled with a *Newtonian liquid* in series, in which case $1/\alpha$ is the **spring constant** (force = $1/\alpha$·extension) and $1/\beta$ is the **dashpot constant** (force = $(1/\beta)$·rate of extension).

5.4 Voigt-Kelvin model
Voigt-Kelvin element

Model of the *linear viscoelastic behaviour* of a solid in which

$$\sigma = (\alpha + \beta D)\gamma$$

where α and β are positive constants, D is the differential coefficient operator d/dt, and σ and γ are the *stresses* and strain in *simple shear* or *uniaxial deformation*.
Note 1: The Voigt-Kelvin model is also known as the **Voigt model** or **Voigt element**.
Note 2: See Definition 5.2 for a discussion of σ and γ.
Note 3: The relationship defining the Voigt-Kelvin model may be written

$$\sigma = \alpha\gamma + \alpha\beta\, d\gamma/dt.$$

Note 4: Comparison with the general definition *of linear viscoelastic behaviour* shows that the polynomial P(D) is of order zero, Q(D) is of order one, $aq_0 = \alpha$ and $a = \beta$. Hence, a material described by the Voigt-Kelvin model is a *solid* ($q_0 > 0$) without instantaneous elasticity (P(D) is a polynomial of order one less than Q(D)).
Note 5: The Voigt-Kelvin model may be represented by a spring and a dashpot filled with a Newtonian liquid in parallel, in which case α is the **spring constant** (force = α·extension) and β is the **dashpot constant** (force = β·rate of extension).

5.5 standard linear viscoelastic solid
Model of the *linear viscoelastic behaviour* of a solid in which

$$(\alpha_1 + \beta_1 D)\sigma = (\alpha_2 + \beta_2 D)\gamma$$

where α_1, β_1, α_2 and β_2 are positive constants, D is the differential coefficient operator d/dt, and σ and γ are the *stress* and strain in *simple shear* or *uniaxial deformation*.
Note 1: See Definition 5.2 for a discussion of σ and γ.
Note 2: The relationship defining *the standard linear viscoelastic solid* may be written

$$\alpha_1 \sigma + \beta_1\, d\sigma/dt = \alpha_2 \gamma + \beta_2\, d\gamma/dt$$

Note 3: Comparison with the general definition of a *linear viscoelastic behaviour* shows that the polynomial P(D) and Q(D) are of order one, $q_0 = \alpha_2/\beta_2$, $a = \beta_2/\beta_1$ and $p_0 = \alpha_1/\alpha_2$. Hence, the standard linear viscoelastic solid is a solid ($aq_0 > 0$) having instantaneous elasticity (P(D) and Q(D) are of the same order).
Note 4: The standard linear viscoelastic solid may be represented by:
  (i) a *Maxwell model* (of spring constant h_2 and dashpot constant k_2) in parallel with a spring (of spring constant h_1) in which case $\alpha_1 = h_2$, $\beta_1 = k_2$, $\alpha_2 = h_1 h_2$ and $\beta_2 = h_1 k_2 + h_2 k_2$.

(ii) a *Voigt-Kelvin model* (of spring constant h_2 and dashpot constant k_2) in series with a spring (of spring constant h_1) in which case $\alpha_1 = h_1 + h_2$, $\beta_1 = k_2$, $\alpha_2 = h_1 h_2$ and $\beta_2 = h_1 k_2$.

Note 5: The standard linear viscoelastic solid can be used to represent both *creep* and *stress relaxation* in materials in terms of single *retardation* and *relaxation times*, respectively.

5.6 relaxation time, τ, SI unit: s

Time characterizing the response of a viscoelastic liquid or solid to the instantaneous application of a constant strain.

Note 1: The response of a material to the instantaneous application of a constant strain is termed *stress relaxation*.

Note 2: The relaxation time of a *Maxwell element* is $\tau = 1/p_0 = \alpha/\beta$.

Note 3: The relaxation time of a *standard linear viscoelastic solid* is $\tau = 1/p_0 = \beta_1/\alpha_1$.

Note 4: Generally, a *linear viscoelastic material* has a spectrum of relaxation times, which are the reciprocals of p_i, $i = 0, 1, \ldots, n$ in the polynomial $P(D)$ (see Definition 5.2).

Note 5: The **relaxation spectrum (spectrum of relaxation times)** describing *stress relaxation* in polymers may be considered as arising from a group of *Maxwell elements* in parallel.

5.7 stress relaxation

Change in *stress* with time after the instantaneous application of a constant strain.

Note 1: The applied strain is of the form $\gamma = 0$ for $t < 0$ and $\gamma = \gamma_0$ for $t > 0$ and is usually a *uniaxial extension* or a *simple shear*.

Note 2: For *linear viscoelastic behaviour*, the *stress* takes the form

$$\sigma(t) = (c + \overline{\psi}(t))\gamma_0$$

c is a constant that is non-zero if the material has instantaneous elasticity and $\overline{\psi}(t)$ is the relaxation function.

Note 3: $\overline{\psi}(t)$ has the form

$$\overline{\psi}(t) = \sum_{i=0}^{n} \beta_i e^{-p_i t}$$

where the β_i are functions of the p_i and q_i of the polynomials $P(D)$ and $Q(D)$ defining the *linear viscoelastic material*.

Note 4: The *relaxation times* of the material are $1/p_i$.

5.8 retardation time, τ, SI unit: s

Time characterizing the response of a viscoelastic material to the instantaneous application of a constant *stress*.

Note 1: The response of a material to the instantaneous application of a constant *stress* is termed *creep*.

Note 2: The *retardation time* of a *Voigt-Kelvin element* is $\tau = 1/q_0 = \beta/\alpha =$ (dashpot constant)/(spring constant).

Note 3: The *retardation time* of a *standard linear viscoelastic solid* is $\tau = 1/q_0 = \beta_2/\alpha_2$.

TERMINOLOGY

Note 4: Generally, a *linear viscoelastic material* has a spectrum of *retardation times*, which are reciprocals of q_i, $i = 0, 1, \ldots, n$ in the polynomial $Q(D)$.
Note 5: The **retardation spectrum** (spectrum of retardation times) describing *creep* in polymers may be considered as arising from a group of *Voigt-Kelvin elements* in series.

5.9 creep

Change in strain with time after the instantaneous application of a constant *stress*.
or
Time-dependent change of the dimensions of a material under a constant load.
Note 1: The applied *stress* is of the form $\sigma = 0$ for $t < 0$ and $\sigma = \sigma_0$ for $t > 0$ and is usually a *uniaxial stress* or a *simple shear stress*.
Note 2: For *linear viscoelastic behaviour*, the *strain* usually takes the form

$$\gamma(t) = (a + bt + \psi(t))\sigma_0$$

a is a constant that is non-zero if the material has instantaneous elasticity and b is a constant that is non-zero if the material is a liquid. $\psi(t)$ is the **creep function**. In addition,

$$J(t) = \gamma(t)/\sigma_0$$

is sometimes called the **creep compliance**.
Note 3: The creep function has the form

$$\psi(t) = \sum_i A_i e^{-q_i t}$$

where the summation runs from $i = 0$ to n for a *solid* and 1 to n for a *liquid*. The A_i are functions of the p_i and q_i of the polynomials $P(D)$ and $Q(D)$ defining the *linear viscoelastic material* and the q_i are the q_i of the polynomial $Q(D)$ (see Definition 5.2).
Note 4: The *retardation times* of the material are $1/q_i$.
Note 5: Creep is sometimes described in terms of non-linear viscoelastic behaviour, leading, for example, to evaluation of recoverable shear and steady-state recoverable *shear compliance*. The definitions of such terms are outside the scope of this document.

5.10 forced oscillation

Deformation of a material by the application of a small sinusoidal strain (γ) such that

$$\gamma = \gamma_0 \cos \omega t$$

where γ_0 and ω are positive constants.
Note 1: γ may be in *simple shear* or *uniaxial deformation*.
Note 2: γ_0 is the **strain amplitude**.
Note 3: ω is the **angular velocity** of the circular motion equivalent to a sinusoidal frequency v, with $\omega = 2\pi v$. The unit of ω is rad s^{-1}.
Note 4: For *linear viscoelastic behaviour*, a sinusoidal *stress* (σ) results from the sinusoidal strain with

$$\sigma = \sigma_0 \cos(\omega t + \delta) = \sigma_0 \cos\delta \cos\omega t - \sigma_0 \sin\delta \sin\omega t.$$

σ_0 is the **stress amplitude**. δ is the **phase angle** or **loss angle** between *stress* and strain.
Note 5: Alternative descriptions of the sinusoidal *stress* and strain in a viscoelastic material under forced oscillations are:

(i) $\gamma = \gamma_0 \sin \omega t$ $\sigma = \sigma_0 \sin(\omega t + \delta) = \sigma_0 \sin \delta \cos \omega t + \sigma_0 \cos \delta \sin \omega t$

(ii) $\gamma = \gamma_0 \cos(\omega t - \delta) = \gamma_0 \cos \delta \cos \omega t + \gamma_0 \sin \delta \sin \omega t$ $\sigma = \sigma_0 \cos \omega t$

5.11   **loss factor**, tan δ
            loss tangent

Tangent of the *phase angle* difference (δ) between *stress* and strain during forced oscillations.
Note 1: tan δ is calculated using

$\gamma = \gamma_0 \cos \omega t$ and $\sigma = \sigma_0 \cos(\omega t + \delta)$.

Note 2: tan δ is also equal to the ratio of *loss* to *storage modulus*.
Note 3: A plot of tan δ vs. temperature or frequency is known as a **loss curve**.

5.12   **storage modulus**   in general M', SI unit: Pa
                                  in simple shear deformation G', SI unit: Pa
                                  in *uniaxial deformation* E', SI unit: Pa

Ratio of the amplitude of the *stress* in phase with the strain ($\sigma_0 \cos \delta$) to the amplitude of the strain (γ_0) in the *forced oscillation* of a material

$$M' = \sigma_0 \cos \delta / \gamma_0.$$

Note: See Definition 5.10 for the definition of a *forced oscillation* in which $\gamma = \gamma_0 \cos \omega t$ and $\sigma = \sigma_0 \cos(\omega t + \delta)$.

5.13   **loss modulus**   in general M'', SI unit: Pa
                               in simple shear deformation G'', SI unit: Pa
                               in *uniaxial deformation* E'', SI unit: Pa

Ratio of the amplitude of the *stress* 90° out of phase with the strain ($\sigma_0 \sin \delta$) to the amplitude of the strain (γ_0) in the *forced oscillation* of a material

$$M'' = \sigma_0 \sin \delta / \gamma_0.$$

Note: See Definition 5.10 for the definition of a *forced oscillation* in which $\gamma = \gamma_0 \cos \omega t$ and $\sigma = \sigma_0 \cos(\omega t + \delta)$.

5.14   **absolute modulus**   in general $|M^*|$, SI unit: Pa
                                   in simple shear deformation $|G^*|$, SI unit: Pa
                                   in *uniaxial deformation* $|E^*|$, SI unit: Pa

Ratio of the amplitude of the *stress* (σ_0) to the amplitude of the strain (γ_0) in the *forced oscillation* of a material

TERMINOLOGY

$|M^*| = \sigma_0/\gamma_0.$

Note 1: See Definition 5.10 for the definition of a *forced oscillation* in which $\gamma = \gamma_0 \cos \omega t$ and $\sigma = \sigma_0 \cos(\omega t + \delta)$.

Note 2: The absolute modulus is related to the *storage modulus* and the *loss modulus* by the relationship

$$|M^*| = \left(\frac{\sigma_0^2 \cos^2 \delta}{\gamma_0^2} + \frac{\sigma_0^2 \sin^2 \delta}{\gamma_0^2}\right) = (M'^2 + M''^2)^{1/2}.$$

5.15 **complex modulus** in general M^*, SI unit: Pa
in simple shear deformation G^*, SI unit: Pa
in *uniaxial deformation* E^*, SI unit: Pa

Ratio of complex *stress* (σ^*) to complex strain (γ^*) in the *forced oscillation* of material

$M^* = \sigma^*/\gamma^*.$

Note 1: See Definition 5.10 for the definition of a *forced oscillation* in which $\gamma = \gamma_0 \cos \omega t$ and $\sigma = \sigma_0 \cos(\omega t + \delta)$.

Note 2: The **complex strain** $\gamma^* = \gamma_0 e^{i\omega t} = \gamma_0 (\cos \omega t + i \sin \omega t)$, where $i = \sqrt{-1}$, so that the real part of the complex strain is that actually applied to the material.

Note 3: SI unit: Pa The **complex stress** $\sigma^* = \sigma_0 e^{i(\omega t + \delta)} = \sigma_0 (\cos(\omega t + \delta) + i \sin(\omega t + \delta))$, so that the real part of the complex strain is that actually experienced by the material.

Note 4: The *complex modulus* is related to the *storage* and *loss moduli* through the relationships

$M^* = \sigma^*/\gamma^* = \sigma_0 e^{i\delta}/\gamma_0 = (\sigma_0/\gamma_0)(\cos \delta + i \sin \delta) = M' + iM''.$

Note 5: For linear viscoelastic behaviour interpreted in terms of *complex stress* and *strain* (see notes 2 and 3)

$P(D)\sigma^* = Q(D)\gamma^*$

(see Definition 5.2). Further as $D\sigma^* = d\sigma^*/dt = i\omega\sigma^*$ and $D\gamma^* = i\omega\gamma^*$,

$M^* = \sigma^*/\gamma^* = Q(i\omega)/P(i\omega).$

5.16 **storage compliance** in general C', SI unit: Pa^{-1}
in simple shear deformation J', SI unit: Pa^{-1}
in *uniaxial deformation* D', SI unit: Pa^{-1}

Ratio of the amplitude of the strain in phase with the *stress* ($\gamma_0 \cos \delta$) to the amplitude of the *stress* (σ_0) in the *forced oscillation* of a material

$C' = \gamma_0 \cos \delta/\sigma_0.$

Note: See Definition 5.10, note 5 for the definition of a *forced oscillation* in which $\gamma = \gamma_0 \cos(\omega t - \delta)$ and $\sigma = \sigma_0 \cos \omega t$.

5.17 **loss compliance** in general C", SI unit: Pa^{-1}
in simple shear deformation J'', SI unit: Pa^{-1}
in *uniaxial deformation* D'', SI unit: Pa^{-1}

Ratio of the amplitude of the strain 90° out of phase with the *stress* ($\gamma_0 \sin \delta$) to the amplitude of the *stress* (σ_0) in the *forced oscillation* of a material

$$C'' = \gamma_0 \sin \delta / \sigma_0.$$

Note: See Definition 5.10 for the definition of a *forced oscillation* in which $\gamma = \gamma_0 \cos(\omega t - \delta)$ and $\sigma = \sigma_0 \cos \omega t$.

5.18 **absolute compliance** in general $|C^*|$, SI unit: Pa^{-1}
in simple shear deformation $|J^*|$, SI unit: Pa^{-1}
in *uniaxial deformation* $|D^*|$, SI unit: Pa^{-1}

Ratio of the amplitude of the strain (γ_0) to the amplitude of the *stress* (σ_0) in the *forced oscillation* of a material

$$|C^*| = \gamma_0 / \sigma_0.$$

Note 1: See Definition 5.10, note 5 for the definition of a *forced oscillation* in which

$$\gamma = \gamma_0 \cos(\omega t - \delta) \text{ and } \sigma = \sigma_0 \cos \omega t.$$

Note 2: The absolute compliance is related to the *storage compliance* (5.16) and the *loss compliance* (Definition 5.17) by the relationship

$$|C^*| = \left(\frac{\gamma_0^2 \cos^2 \delta}{\sigma_0^2} + \frac{\gamma_0^2 \sin^2 \delta}{\sigma_0^2} \right)^{1/2} = (C'^2 + C''^2)^{1/2}.$$

Note 3: The absolute compliance is the reciprocal of the *absolute modulus*.

$$|C^*| = 1/|M^*|$$

5.19 **complex compliance** in general C^*, SI unit: Pa^{-1}
in simple shear deformation J^*, SI unit: Pa^{-1}
in *uniaxial deformation* D^*, SI unit: Pa^{-1}

Ratio of complex strain (γ^*) to complex *stress* (σ^*) in the *forced oscillation* of a material

$$C^* = \gamma^*/\sigma^*.$$

Note 1: See Definition 5.10 for the definition of a *forced oscillation* in which $\gamma = \gamma_0 \cos(\omega t - \delta)$ and $\sigma = \sigma_0 \cos \omega t$.

Note 2: The **complex strain** $\gamma^* = \gamma_0 e^{i(\omega t - \delta)} = \gamma_0 (\cos(\omega t - \delta) + i \sin(\omega t - \delta))$, where $i = \sqrt{-1}$, so that the real part of the complex strain is that actually experienced by the material.

Note 3: The **complex stress** $\sigma^* = \sigma_0 e^{i\omega t} = \sigma_0 (\cos \omega t + i \sin \omega t)$, so that the real part of the complex *stress* is that actually applied to the material.

TERMINOLOGY

Note 4: The complex compliance is related to the *storage* and *loss compliances* through the relationships

$$C^* = \gamma^*/\sigma^* = \gamma_0 e^{-i\delta}/\sigma_0 = (\gamma_0/\sigma_0)(\cos\delta - i\sin\delta) = C' - iC''$$

Note 5: The complex compliance is the reciprocal of the *complex modulus*.

$$C^* = 1/M^*$$

5.20 dynamic viscosity, η', SI unit: Pa s
Ratio of the *stress* in phase with the rate of strain ($\sigma_0 \sin\delta$) to the amplitude of the rate of strain ($\omega\gamma_0$) in the *forced oscillation* of a material

$$\eta' = \sigma_0 \sin\delta / \omega\gamma_0.$$

Note 1: See Definition 5.10, note 5 for the definition of a *forced oscillation* in which $\gamma = \gamma_0 \sin\omega t$ and $\sigma = \sigma_0 \sin(\omega t + \delta)$, so that $\dot{\gamma} = \omega\gamma_0 \cos\omega t$ and $\sigma = \sigma_0 \sin\delta \cos\omega t + \sigma_0 \cos\delta \sin\omega t$.
Note 2: See Definition 5.2, note 6: $\eta' = M''/\omega$ may be used for evaluating the *dynamic viscosity*. The same expression is often used to evaluate the *shear viscosity*. The latter use of this expression is not recommended.

5.21 out-of-phase viscosity, η'', SI unit: Pa s
Ratio of the *stress* 90° out of phase with the rate of strain ($\sigma_0 \cos\delta$) to the amplitude of the rate of strain ($\omega\gamma_0$) in the *forced oscillation* of a material

$$\eta'' = \sigma_0 \cos\delta / \omega\gamma_0.$$

Note 1: See Definition 5.10, note 5 for the definition of a *forced oscillation* in which $\gamma = \gamma_0 \sin\omega t$ and $\sigma = \sigma_0 \sin(\omega t + \delta)$, so that $\dot{\gamma} = \omega\gamma_0 \cos\omega t$ and $\sigma = \sigma_0 \sin\delta \cos\omega t + \sigma_0 \cos\delta \sin\omega t$.
Note 2: See Definition 5.22, note 6: $\eta'' = M'/\omega$ may be used to evaluate the *out-of-phase viscosity*.

5.22 complex viscosity, η^*, SI unit: Pa s
Ratio of complex *stress* (σ^*) to complex rate of strain ($\dot{\gamma}^*$) in the *forced oscillation* of a material

$$\eta^* = \sigma^*/\dot{\gamma}^*.$$

Note 1: See Definition 5.10, note 5 for the definition of a *forced oscillation* in which $\gamma = \gamma_0 \sin\omega t$ and $\sigma = \sigma_0 \cos(\omega t + \delta)$ and the rate of strain $\dot{\gamma} = \omega_0 \cos\omega t$.
Note 2: The **complex rate of strain** $\dot{\gamma}^* = i\omega\gamma_0 e^{i\omega t} = i\omega\gamma_0(\cos\omega t + i\sin\omega t)$, where $i = \sqrt{-1}$.
Note 3: The **complex stress** $\sigma^* = \sigma_0 e^{i(\omega t + \delta)} = \sigma_0(\cos(\omega t + \delta) + i\sin(\omega t + \delta))$.
Note 4: The *complex viscosity* may alternatively be expressed as

NON-ULTIMATE MECHANICAL PROPERTIES

$$\eta^* = \sigma^*/\dot{\gamma}^* = (\sigma_0 e^{i\delta})/(i\,\omega\gamma_0) = M^*/i\,\omega$$

where M^* is the *complex modulus*.

Note 5: The complex viscosity is related to the *dynamic* and *out-of-phase viscosities* through the relationships

$$\eta^* = \sigma^*/\dot{\gamma}^* = \sigma_0(\cos\delta + i\sin\delta)/(i\,\omega\gamma_0) = \eta' - i\,\eta''.$$

Note 6: The *dynamic* and *out-of-phase viscosities* are related to the *storage* and *loss moduli* by the relationships $\eta^* = \eta' - i\,\eta'' = M^*/i\,\omega = (M' + i\,M'')/i\,\omega$, so that $\eta' = M''/\omega$ and $\eta'' = M'/\omega$.

6 OSCILLATORY DEFORMATIONS AND STRESSES USED EXPERIMENTALLY

There are three modes of **free** and **forced** oscillatory deformations which are commonly used experimentally, **torsional oscillations**, **uniaxial extensional oscillations** and **flexural oscillations**.

The oscillatory deformations and *stresses* can be used for solids and liquids. However, the apparatuses employed to measure them are usually designed for solid materials. In principle, they can be modified for use with liquids.

Analyses of the results obtained depend on the shape of the specimen, whether or not the distribution of mass in the specimen is accounted for and the assumed model used to represent the linear viscoelastic properties of the material. The following terms relate to analyses which generally assume small deformations, specimens of uniform cross-section, non-distributed mass and a *Voigt-Kelvin* solid. These are the conventional assumptions.

6.1 free oscillation
Oscillatory deformation of a material specimen with the motion generated without the continuous application of an external force.
Note: For any real sample of material the resulting oscillatory deformation is one of decaying amplitude.

6.2 damping curve
Decreased deformation of a material specimen vs. time when the specimen is subjected to a *free oscillation*.
Note 1: See Definition 6.1 for the definition of a *free oscillation*.
Note 2: The term 'damping curve' is sometimes used to describe a *loss curve*.
Note 3: A damping curve is usually obtained using a **torsion pendulum**, involving the measurement of decrease in the axial, torsional displacement of a specimen of uniform cross-section of known shape, with the torsional displacement initiated using a torsion bar of known moment of inertia.
Note 4: Damping curves are conventionally analysed in terms of the *Voigt-Kelvin* solid giving a decaying amplitude and a single frequency.
Note 5: Given the properties of a *Voigt-Kelvin* solid, a damping curve is described by the equation

$$X = A\exp(-\beta t)\sin(\omega t - \phi),$$

where X is the displacement from equilibrium (for torsion $X = \theta$, the angular displacement), t is time, A is the amplitude, β is the *decay constant*, ω is the *angular velocity* corresponding to the *decay frequency* and ϕ is the *phase angle*.

6.3 decay constant, β, SI unit: s^{-1}

Exponential coefficient of the time-dependent decay of a *damping curve*, assuming Voigt-Kelvin behaviour.
Note 1: See *damping curve* and the equation therein (Definition 6.2, note 5).
Note 2: See *Voigt-Kelvin* solid.
Note 3: For small damping, β is related to the *loss modulus* (M''), through the equation

$$M'' = 2\beta\omega/H,$$

where ω is the *angular velocity* corresponding to the *decay frequency*. H is a parameter that depends on the cross-sectional shape of the specimen and the type of deformation. (For example, for the axial torsion of a circular rod of radius a and length l using a *torsion pendulum* with a torsion bar of moment of inertia I

$$H = \pi a^4/(2Il)$$

and $M'' \cong G''$, the *loss modulus* in simple shear).

6.4 decay frequency, ν, SI unit: Hz

Frequency of a *damping curve* assuming *Voigt-Kelvin* behaviour.
Note 1: See *damping curve* and the equation therein.
Note 2: See *Voigt-Kelvin* solid.
Note 3: $\nu = \omega/2\pi$, where ω is the *angular velocity* corresponding to ν.
Note 4: For small damping, the *storage modulus* (M') may be evaluated from ω through the equation

$$M' = \omega^2/H,$$

where H is as defined in Definition 6.3, note 3. Again, for torsion, $M' \cong G'$, the *storage modulus* in simple shear.

6.5 logarithmic decrement, Λ

Natural logarithm of the ratio of the displacement of a *damping curve* separated by one period of the displacement.
Note 1: *Voigt-Kelvin* behaviour is assumed so that the displacement decays with a single period T, where

$$T = \frac{1}{\nu} = \frac{2\pi}{\omega},$$

with ν the frequency and ω is the *angular velocity* corresponding to ν.

Note 2: The *logarithmic decrement* can be used to evaluate the *decay constant*, β. From the equation for the *damping curve* of a *Voigt-Kelvin* solid.

$$\Lambda = \ln(X_n/X_{n+1}) = \beta(t_{n+1} - t_n) = \beta T,$$

where X_n and t_n are the displacement and time at a chosen point (usually near a maximum) in the *n*-th period of the decay, and X_{n+1} and t_{n+1} are the corresponding displacement and time one period later.

Note 3: Λ can also be defined using displacements k periods apart, with

$$\Lambda = (1/k)\ln(X_n/X_{n+k}).$$

Note 4: For small damping, Λ is related to the *loss tangent*, $\tan\delta$ by

$$\tan\delta = M''/M' = 2\beta/\omega = 2\Lambda/T\omega = \Lambda/\pi$$

(See Definitions 6.3 and 6.4 for expressions for M' and M'').

6.6 forced uniaxial extensional oscillations

Uniaxial extensional deformation of a material specimen of uniform cross-sectional area along its long axis by the continuous application of a sinusoidal force of constant amplitude.

Note 1: For a specimen of negligible mass, the linear-viscoelastic interpretation of the resulting deformation gives

$$(A/L)Q(D)l = P(D)f_0 \cos\omega t$$

where $P(D)$ and $Q(D)$ are the polynomials in D $(=d/dt)$ characterizing the *linear-viscoelastic behaviour*, A is the cross-sectional area of the specimen, L its original length, l is here the change in length, f_0 the amplitude of the applied force of *angular velocity* ω and t the time.

Note 2: For a *Voigt-Kelvin* solid, with $P(D)=1$ and $Q(D)=\alpha+\beta D$, where α is the spring constant and β the dashpot constant, the equation describing the deformation becomes

$$(A/L)\beta(dl/dt) + (A/L)\alpha l = f_0 \cos\omega t$$

or, in terms of *stress* and *strain*,

$$\alpha\varepsilon + \beta\frac{d\varepsilon}{dt} = \sigma_0 \cos\omega t$$

where $\varepsilon = l/L$ is the *uniaxial strain* and $\sigma_0 = f_0/A$ is the amplitude of the *stress*. The solution of the equation is

$$\varepsilon = \frac{\sigma_0}{\left(\alpha^2 + \beta^2\omega^2\right)^{1/2}} \cos(\omega t - \delta) = \varepsilon_0 \cos(\omega t - \delta)$$

where δ is the *phase angle* with $\tan \delta = \beta\omega/\alpha$.
Note 3: From Definition 5.14, the *absolute modulus* in *uniaxial deformation*

$$|E^*| = \sigma_0/\varepsilon_0 = (\alpha^2 + \beta^2\omega^2)^{1/2}$$

where $\alpha = E'$, $\beta\omega = E''$ and $\tan \delta = E''/E'$ equal to the *loss tangent*.
Note 4: If one end of the specimen is fixed in position and a mass m is attached to the moving end, the *linear-viscoelastic* interpretation of the resulting deformation gives

$$mP(D)(d^2l/dt^2) + (A/L)Q(D)l = P(D)f_0 \cos \omega t$$

where the symbols have the same meaning as in note 1.
Note 5: For a *Voigt-Kelvin* solid (cf. note 2), the equation in note 4 describing the deformation becomes

$$m(d^2l/dt^2) + (A/L)\beta(dl/dt) + (A/L)\alpha l = f_0 \cos \omega t$$

with the solution

$$\varepsilon = \frac{\sigma_0 \cdot (A/(Lm))}{\left(\left(\frac{A\alpha}{Lm} - \omega^2\right)^2 + \omega^2\left(\frac{A\beta}{Lm}\right)^2\right)^{1/2}} \cos(\omega t - \theta) = \varepsilon_0 \cos(\omega t - \theta)$$

where $\tan \theta = \dfrac{(A\beta/Lm)\omega}{(A\alpha/Lm) - \omega^2}$.

Note 6: The amplitude of the strain ε_0 is maximal when

$$\omega^2 = A\alpha/(Lm) = \omega_R^2$$

giving the value of the *angular velocity* (ω_R) of the *resonance frequency* of the specimen in forced uniaxial extensional oscillation.
Note 7: Notes 2 and 5 show that application of a sinusoidal uniaxial force to a *Voigt-Kelvin* solid of negligible mass, with or without added mass, results in an out-of-phase sinusoidal uniaxial extensional oscillation of the same frequency.

6.7 forced flexural oscillation

Flexural deformation (bending) of a material specimen of uniform cross-sectional area perpendicular to its long axis by the continuous application of a sinusoidal force of constant amplitude.
Note 1: There are three modes of flexure in common use.
(i) Application of the flexural force at one end of the specimen with the other end clamped.
(ii) Application of the flexural force at the centre of the specimen with the two ends clamped (**three-point bending or flexure**).
(iii) Application of the flexural force at the centre of the specimen with the two ends resting freely on supports (also known as **three-point bending** or **flexure**).

Note 2: For specimens *without mass*, the linear-viscoelastic interpretation of the resulting deformations follows a differential equation of the same form as that for *a uniaxial extensional forced oscillation*, namely

$$(HJ/L^3)Q(D)y = P(D)f_0 \cos \omega t$$

where $P(D)$, $Q(D)$, f_0, ω and t have the same meaning as for a *forced uniaxial extensional oscillation* and H is a constant. The length of the specimen is $2L$. For mode of flexure (i) $H=3$, for (ii) $H=24$ and for (iii) $H=6$. J is the **second moment of area** of the specimen, defined by

$$J = \int_A q^2 dA$$

where dA is an element of the cross-sectional area (A) of the specimen and q is the distance of that element from the **neutral axis or plane** of the specimen, lying centrally in the specimen and defined by points which experience neither compression nor extension during the flexure. For a specimen of circular cross-section $J=\pi r^2/4$, where r is the radius, and for one of rectangular cross-section $J=4ab^3/3$, where $2a$ and $2b$ are the lateral dimensions with flexure along the b dimension. Finally, y is the *flexural deflection* of the specimen at the point of application of the force, of either the end (mode of flexure (i)) or the middle (modes of flexure (ii) and (iii)).

Note 3: For a *Voigt-Kelvin* solid, the equation describing the deformation becomes

$$(HJ/L^3)\alpha y + (HJ/L^3)\beta(dy/dt) = f_0 \cos \omega t$$

with solution

$$y = \frac{f_0 L^3}{HJ(\alpha^2 + \beta^2\omega^2)^{1/2}} \cos(\omega t - \delta)$$

where δ is the *phase angle* with

$$\tan \delta = \beta\omega/\alpha$$

equal to the *loss tangent*.

Note 4: Unlike the strain in *forced uniaxial extensional oscillations*, those in *forced flexural deformations* are not homogeneous. In the latter modes of deformation, the strains vary from point-to-point in the specimen. Hence, the equation defining the displacement y in terms of the amplitude of applied force (f_0) cannot be converted into one defining strain in terms of amplitude of *stress*.

Note 5: If a mass m is attached to the specimen at the point of application of the force, the linear-viscoelastic interpretation of the resulting deformation gives

$$m \cdot P(D)(d^2y/dt^2) + (HJ/L^3)Q(D)y = P(D)f_0 \cos \omega t$$

(cf. Definition 6.6, note 4).

TERMINOLOGY

Note 6: For a *Voigt-Kelvin solid* (cf. note 3 and Definition 6.6, note 5), the equation describing the deformation becomes

$$m(d^2y/dt^2) + (HJ/L^3)\beta(dy/dt) + (HJ/L^3)\alpha y = f_0 \cos \omega t$$

with the solution

$$y = \frac{f_0/m}{\left[\left(\frac{HJ\alpha}{L^3 m} - \omega^2\right)^2 + \omega^2\left(\frac{HJ\beta}{L^3 m}\right)^2\right]^{1/2}} \cos(\omega t - \delta)$$

where $\tan \delta = \dfrac{(HJ\beta/L^3 m)\omega}{(HJ\alpha/L^3 m) - \omega^2}$.

Note 7: The *flexural deflection y* is maximal when

$$\omega^2 = HJ\alpha/(L^3 m) = \omega_R^2$$

giving the value of the *angular velocity* (ω_R) of the *resonance frequency* of the specimen in forced flexural oscillations.

Note 8: Notes 3 and 6 show that the application of the defined sinusoidal flexural forces (i), (ii) and (iii) (note 1) to a *Voigt-Kelvin* solid of negligible mass, with or without added mass at the points of application of the forces, results in out-of-plane sinusoidal flexural oscillations of the same frequency.

6.8   flexural force, f_0, SI unit: N
Amplitude of the force applied to a material specimen to cause a *forced flexural oscillation*.
Note: A related quantity is the **flexural stress** which is somewhat arbitrarily defined as the amplitude of the *stress* in the convex, outer surface of a material specimen in *forced flexural oscillation*.

6.9   flexural deflection, y, SI unit: m
Deflection of a specimen subject to a *forced flexural oscillation* at the point of application of the *flexural force*.

6.10   flexural modulus, $|E^*|$, SI unit: Pa
Modulus measured using *forced flexural oscillations*.
Note 1: For a *Voigt-Kelvin solid* of negligible mass, the *absolute modulus* can be evaluated from the ratio of the *flexural force* (f_0) and the amplitude of the *flexural deflection* (y) with
$$f_0/Y_0 = (HJ/L^3)(\alpha^2 + \beta^2\omega^2)^{1/2}$$

where Y_0 is the amplitude of the *flexural deflection*,

$$|E^*| = (\alpha^2 + \beta^2\omega^2)^{1/2}$$

(see Definitions 5.14 and 6.6, note 3) and the remaining symbols are as defined in Definition 6.7, note 2.

Note 2: The ratio of the loss to the storage flexural modulus (E''/E') is derived from the *loss tangent* ($\tan \delta$) of the *forced flexural oscillation* with

$$\tan \delta = \beta \omega / \alpha = E''/E'$$

(see Definitions 5.11 and 6.7, note 3).

Note 3: The flexural modulus has been given the same symbol as the *absolute modulus* in *uniaxial deformation* as it becomes equal to that quantity in the limit of zero amplitudes of applied force and deformation. Under real experimental conditions it is often used as an approximation to $|E^*|$.

6.11 resonance curve, $A(\nu)$, SI unit: that of the amplitude A
Curve of the frequency dependence of the amplitude of the displacement of a material specimen subject to *forced oscillation*s in the region of a *resonance frequency*.
Note: See Definitions 6.6 and 6.7 for the description of modes of *forced oscillation* commonly used.

6.12 resonance frequency, ν_R, SI unit: Hz
Frequency at a maximum of a *resonance curve*.
Note 1: Material specimens subject to a *forced oscillation* in general have a spectrum of resonance frequencies.
Note 2: In cases of a single *resonance frequency*, the *resonance frequency* is proportional to the square root of the *storage modulus* (M') of the material.
Note 3: A material specimen which behaves as a *Voigt-Kelvin solid* under *forced oscillation*s with a mass added at the point of application of the applied oscillatory force has a single resonance frequency.
Note 4: Under a *forced uniaxial extensional oscillation* the resonance frequency

$$\nu_R = \omega_R / 2\pi = \left(\frac{A\alpha}{Lm}\right)^{1/2} \bigg/ 2\pi = \left(\frac{AE'}{Lm}\right)^{1/2} \bigg/ 2\pi$$

(see Definition 6.6 for the origin of the equation and definitions of symbols). E' is the *storage modulus in uniaxial extension*.
Note 5: Under a *forced flexural oscillation* the resonance frequency

$$\nu_R = \omega_R / 2\pi = \left(\frac{HJ\alpha}{L^3 m}\right)^{1/2} \bigg/ 2\pi = \left(\frac{HJE'}{L^3 m}\right)^{1/2} \bigg/ 2\pi$$

(see Definition 6.7 for the origin of the equation and the definition of symbols).

6.13 width of the resonance curve, $\Delta \nu$, SI unit: Hz
Magnitude of the difference in frequency between two points on a resonance curve on either side of ν_R which have amplitudes equal to $(1/\sqrt{2})A(\nu_R)$.

TERMINOLOGY

Note 1: For a material specimen which behaves as a *Voigt-Kelvin solid* under *forced uniaxial extensional oscillation* with mass added at the point of application of the applied oscillatory force, Δv is proportional to the *loss modulus* (E'').

$$2\pi\Delta v = \frac{A\beta}{Lm} = \frac{AE''}{Lm\omega_R}$$

In addition the *storage modulus* (E') may be evaluated from

$$\omega_R^2 = \frac{A\alpha}{Lm} = \frac{AE'}{Lm}$$

(see Definition 6.6 for the definition of symbols).

Note 2: For a material specimen which behaves as *Voigt-Kelvin solid* under *forced flexural oscillations* with added mass at the point of application of the applied oscillatory force, Δv is proportional to the *loss modulus* (E'')

$$2\pi\Delta v = \frac{HJ\beta}{L^3 m} = \frac{HJE''}{L^3 m\omega_R}.$$

In addition, the *storage modulus* (E') may be evaluated from

$$\omega_R^2 = \frac{HJ\alpha}{L^3 m} = \frac{HJE'}{L^3 m}$$

(see Definition 6.7 for the definition of symbols).

Note 3: For the *Voigt-Kelvin* behaviours specified in notes 1 and 2, the ratio of the *width of the resonance curve* (Δv_R) to the *resonance frequency* (v_R) is equal to the *loss tangent* ($\tan \delta$).

Under *forced uniaxial extensional oscillation*

$$\frac{\Delta v_R}{v_R} = \left(\frac{A}{Lm}\right)\beta\omega_R \cdot \frac{Lm}{A\alpha} = \frac{\beta}{\alpha}\omega_R = \frac{E''}{E'} = \tan \delta.$$

Under *forced flexural oscillation*

$$\frac{\Delta v_R}{v_R} = \left(\frac{HJ}{L^3 m}\right)\beta\omega_R \cdot \frac{L^3 m}{HJ\alpha} = \frac{\beta}{\alpha}\omega_R = \frac{E''}{E'} = \tan \delta.$$

NON-ULTIMATE MECHANICAL PROPERTIES

7 REFERENCES

1. ISO 472, ISO TC-61, Dec. 1991 Draft, Plastics Vocabulary.
2. C. L. Sieglaff. *Trans. Soc. Rheol.* **20**, 311 (1976).
3. D. W. Hadley and J. D. Weber. *Rheol. Acta* **14**, 1098 (1975).
4. D. W. Hadley and J. D. Weber. *Bull. Br. Soc. Rheol.* **22**, 4 (1979).
5. M. Reiner and G. W. Scott Blair. *Rheology*, F. R. Eirich (Ed.), Academic Press, New York (1967), Vol. 4.
6. J. M. Dealy. *J. Rheol.* **28**, 181 (1984); **39**, 253 (1995).
7. ASTM E 6-88, *Standard Definitions of Terms Relating to Methods of Mechanical Testing*, Annual Book of ASTM Standards (1989), Vol. 8.03, p. 672.
8. ASTM D4092-83a, *Standard Definitions and Descriptions of Terms Relating to Dynamical Mechanical Measurements on Plastics*, Annual Book ASTM Standards (1989), Vol. 8.03, p. 334.
9. ASTM D 883-88, *Standard Definitions of Terms Relating to Plastics*, Annual Book of ASTM Standards (1989), Vol. 8.01, p. 333.
10. G. Astarita and G. Marucci. *Principles of Non-Newtonian Fluid Mechanics*, McGraw-Hill Book Company (UK) Ltd., Maidenhead (1974).
11. J. D. Ferry. *Viscoelastic Properties of Polymers*, 3rd ed., John Wiley and Sons, New York (1980).
12. R. I. Tanner. *Engineering Rheology*, Clarendon Press, Oxford (1985).
13. H. A. Barnes, J. F. Hutton, K. Walters. *An Introduction to Rheology*, Elsevier, Amsterdam (1989).
14. J. F. Nye. *Physical Properties of Crystals*, Oxford University Press (1969).
15. I. M. Ward and D. W. Hadley. *An Introduction to the Mechanical Properties of Solid Polymers*, John Wiley and Sons, Chichester (1993).

8 ALPHABETICAL INDEX OF TERMS

absolute compliance	5.18
absolute modulus	5.14
angular velocity (of a forced oscillation)	5.10
angular velocity of resonance frequency	6.7
apparent viscosity	4.12
bulk compression	2.9
bulk compressive compliance	4.6
bulk compressive modulus	4.5
bulk compressive strain	2.9
bulk modulus	4.5
Cauchy tensor	1.8, 1.9
coefficient of viscosity	4.12
complex compliance	5.19
complex modulus	5.15
complex rate of strain	5.22
complex strain	5.15, 5.19
complex stress	5.15, 5.19, 5.22
complex viscosity	5.22

TERMINOLOGY

compliance	4.4
compressive strain	2.4
compressive stress	3.3
constitutive equation for an elastic solid	4.1
constitutive equation for an incompressible viscoelastic liquid or solid	4.2
creep	5.9
creep compliance	5.9
creep function	5.9
damping curve	6.2
dashpot constant	5.3, 5.4
decay constant	6.3
decay frequency	6.4
deformation gradients in an elastic solid	1.3
deformation gradients in a viscoelastic liquid or solid	1.6
deformation gradient in the orthogonal deformation of an elastic solid	2.1
deformation gradient tensor for an elastic solid	1.4
deformation gradient tensor for a viscoelastic liquid or solid	1.7
deformation of an elastic solid	1.3
deformation of a viscoelastic liquid or a solid	1.5
deformation ratio	2.3
deformation ratio in the orthogonal deformation of an elastic solid	2.1
dynamic stress	5.1
dynamic viscosity	5.20
elastic modulus	4.3
elongational strain rate	2.12
elongational viscosity	4.9
engineering strain	2.4
engineering stress	3.4
extensional strain rate	2.12
extensional viscosity	4.9
extension ratio	2.3
Finger tensor	1.8, 1.11
first normal-stress coefficient	4.13
first normal-stress difference	3.6
first normal-stress function	3.6
flexural deflection	6.9
flexural force	6.8
flexural modulus	6.10
flexural stress	6.8
forced flexural oscillation	6.7
forced oscillation	5.10
forced uniaxial extensional oscillation	6.6
free oscillation	6.1
general homogenous deformation or flow of a viscoelastic liquid or solid	2.10
general orthogonal homogeneous deformation of an elastic solid	2.1
Green tensor	1.8, 1.10
Hencky strain	2.5
homogeneous deformation of elastic solids	1.3

homogeneous deformation of viscoelastic liquids and solids	1.5
homogeneous orthogonal deformation or flow of an incompressible viscoelastic liquid or solid	2.11
homogeneous simple shear deformation or flow of an incompressible viscoelastic liquid or solid	2.13
infinite-shear viscosity	4.12
inhomogeneous deformation of elastic solids	1.3
isotropic compression	2.9
lateral contraction ratio	2.6
lateral strain	2.6
linear viscoelastic behaviour of a liquid	5.2
linear viscoelastic behaviour of a solid	5.2
logarithmic decrement	6.5
loss angle of a forced oscillation	5.10
loss compliance	5.17
loss curve	5.11
loss factor	5.11
loss modulus	5.13
loss tangent	5.11
Maxwell element	5.3
Maxwell model	5.3
modulus	4.3
modulus of elasticity	4.3
neutral axis (in forced flexural oscillation)	6.7
neutral plane (in forced flexural oscillation)	6.7
Newtonian liquid	4.2
nominal stress	3.4
non-Newtonian liquid	4.2
normal stresses	3.5
out-of-phase viscosity	5.21
phase angle (of a forced oscillation)	5.10
Piola tensor	1.8
plane strain	1.8
plane stress	1.2
Poisson's ratio	2.6
pure shear deformation or flow	3.1
pure shear of an elastic solid	2.7
pure shear stress	3.1
relaxation function	5.7
relaxation spectrum	5.6
relaxation time	5.6
resonance curve	6.11
resonance frequency	6.12
resonance frequency (in forced flexural oscillation)	6.7
resonance frequency (in forced uniaxial extensional oscillation)	6.7
retardation spectrum	5.8
retardation time	5.8
Rivlin-Ericksen tensors	1.14
secant modulus	4.7

TERMINOLOGY

second moment of area (in forced flexural oscillation)	6.7
second normal-stress coefficient	4.14
second normal-stress difference	3.7
second normal-stress function	3.7
shear	2.8, 2.13
shear compliance	4.11
shear modulus	4.10
shear rate	2.13
shear strain	2.8
shear stress	3.5
shear viscosity	4.12
simple shear of an elastic solid	2.8
small-strain tensor	1.10
spectrum of relaxation times	5.6
spring constant	5.3, 5.4
standard linear viscoelastic solid	5.5
steady (simple) shear flow	2.13
steady uniaxial homogeneous elongational deformation or flow of an incompressible viscoelastic liquid or solid	2.12
storage compliance	5.16
storage modulus	5.12
stored energy function	4.1
strain amplitude (of a forced oscillation)	5.10
strain tensor	1.8
stress	1.2
stress amplitude (of a forced oscillation)	5.10
stress relaxation	5.7
stress tensor	1.2
stress tensor resulting from an orthogonal deformation or flow	3.1
stress tensor resulting from a simple shear deformation or flow	3.5
stress vector	1.1
tensile compliance	4.8
tensile modulus	4.7
tensile strain	2.4
tensile stress	3.2
three-point bending	6.7
three-point flexure	6.7
torsion pendulum	6.2
traction	1.1
true stress	1.2
uniaxial compliance	4.8
uniaxial deformation of an elastic solid	2.2
uniaxial deformation or flow of an incompressible viscoelastic liquid or solid	2.11
uniaxial deformation ratio	2.3
uniaxial orthogonal deformation or flow	3.1
uniaxial strain	2.4
viscoelasticity	5.1
viscosity	4.12

Voigt-Kelvin element		5.4
Voigt-Kelvin model		5.4
Voigt element		5.4
Voigt model		5.4
volume compression		2.9
vorticity tensor		1.13
width of the resonance curve		6.13
Young's modulus		4.7
zero-shear viscosity		4.12

9   GLOSSARY OF SYMBOLS

Symbol	Description	Section
$A(\nu)$	resonance curve	6.11
A_n	Rivlin-Ericksen tensors	1.14
B	compliance in bulk compressive deformation	4.4
B	bulk compliance/bulk compressive compliance	4.6
$\mathbf{B}$	Green tensor	1.8, 1.10
$\mathbf{B}^{-1}$	Piola tensor	1.8
C	compliance (general symbol)	4.4
C'	storage compliance (general symbol)	5.16
C''	loss compliance (general symbol)	5.17
C^*	complex compliance (general symbol)	5.19
$|C^*|$	absolute compliance (general symbol)	5.18
$\mathbf{C}$	Cauchy tensor	1.8, 1.9
$\mathbf{C}^{-1}$	Finger tensor	1.8, 1.11
D	differential coefficient operator d/dt	5.2
D	compliance in uniaxial deformation	4.4
D	uniaxial compliance/tensile compliance	4.8
D'	storage compliance in uniaxial deformation	5.16
D''	loss compliance in uniaxial deformation	5.17
D^*	complex compliance in uniaxial deformation	5.19
$|D^*|$	absolute compliance in uniaxial deformation	5.18
$\mathbf{D}$	rate-of-strain tensor	1.12
E	modulus in uniaxial deformation	4.3
E	Young's modulus/tensile modulus/secant modulus/tangent modulus	4.7
E'	storage modulus in uniaxial deformation	5.12
E''	loss modulus in uniaxial deformation	5.13
E^*	complex modulus in uniaxial deformation	5.15
$|E^*|$	absolute modulus in uniaxial deformation	5.14
$|E^*|$	flexural modulus	6.10
$\mathbf{F}$	deformation gradient tensor for an elastic solid	1.4
$\mathbf{F}$	deformation gradient tensor for a viscoelastic liquid or solid	1.7
f_{ij}	deformation gradients in a viscoelastic liquid or solid	1.6
f_0	flexural force	6.8
G	modulus in shear deformation	4.3
G	shear modulus	4.10

TERMINOLOGY

G'	storage modulus in simple shear deformation	5.12
G''	loss modulus in simple shear deformation	5.13
G^*	complex modulus in simple shear deformation	5.15
$|G^*|$	absolute modulus in simple shear deformation	5.14
J	compliance in shear deformation	4.7
J	shear compliance	4.11
J	creep compliance	5.9
J	second moment of area (in a forced flexural oscillation)	6.7
J'	storage compliance in simple shear deformation	5.16
J''	loss compliance in simple shear deformation	5.17
J^*	complex compliance in simple shear deformation	5.19
$|J^*|$	absolute compliance in simple shear deformation	5.18
K	modulus in bulk compressive deformation	4.3
K	bulk modulus/bulk compressive modulus	4.5
M	modulus (general symbol)	4.3
M'	storage modulus (general symbol)	5.12
M''	loss modulus (general symbol)	5.13
M^*	complex modulus (general symbol)	5.15
$|M^*|$	absolute modulus (general symbol)	5.14
N_1	first normal-stress difference/first normal-stress function	3.6
N_2	second normal-stress difference/second normal-stress function	3.7
t	traction	1.1
$\tan \delta$	loss factor/loss tangent	5.11
W	stored energy function	4.1
W	vorticity tensor	1.13
y	flexural deflection	6.9
β	decay constant (of a damping curve)	6.2, 6.3
Δv	width of the resonance curve	6.13
γ	shear/shear strain	2.8
$\dot{\gamma}$	shear rate	2.13
γ_0	strain amplitude (of a forced oscillation)	5.10
$\dot{\gamma}_E$	elongational strain rate/extension strain rate	2.12
γ^*	complex strain (of a forced oscillation)	5.15, 5.19
$\dot{\gamma}^*$	complex rate of strain (of a forced oscillation)	5.22
δ	phase angle (of a forced oscillation)/loss angle of a forced oscillation	5.10
ε	uniaxial strain/engineering strain/(uniaxial) tensile strain/(uniaxial) compressive strain	2.4
ε	small-strain tensor	1.10
ε_H	Hencky strain	2.5
ε_{lat}	lateral strain	2.6
η	shear viscosity/coefficient of viscosity/viscosity	4.12
η'	dynamic viscosity	5.20
η''	out-of-phase viscosity	5.21
η_{app}	apparent viscosity	4.12

η_E	extensional viscosity/elongational viscosity	4.9
η_0	zero shear viscosity	4.12
η^*	complex viscosity	5.22
η_∞	infinite-shear viscosity	4.12
λ	uniaxial deformation ratio/deformation ratio/extension ratio/compression ratio	2.3
λ_i	deformation gradients/deformation ratios; $i = 1, 2, 3$	2.1
Λ	logarithmic decrement (of a decay curve)	6.5
μ	Poisson's ratio	2.6
ν	decay frequency (of a damping curve)	6.4
ν_R	resonance frequency	6.12
σ	tensile stress	3.2
σ	compressive stress	3.3
σ	engineering stress	3.4
σ_{ii}	normal *stresses*; $i = 1, 2, 3$	3.5
σ_0	stress amplitude (of a forced oscillation)	5.10
σ_{12}	shear stress	3.5
σ	stress/stress tensor	1.2, 3.1, 3.5
σ^*	complex stress (in a forced oscillation)	5.15, 5.19, 5.22
τ	relaxation time	5.6
τ	retardation time	5.8
χ	bulk compression/volume compression/isotropic compression/bulk compressive strain	2.9
Ψ_1	first normal-stress coefficient	4.13
Ψ_2	second normal-stress coefficient	4.14
$\Psi(t)$	creep function	5.9
$\overline{\psi}(t)$	relaxation function	5.7
ω	angular velocity (of a forced oscillation)	5.10
ω	angular velocity (of a decay frequency)	6.2
ω_R	angular velocity of the resonance frequency (of a forced flexural oscillation)	6.7

9: Definitions of terms related to polymer blends, composites, and multiphase polymeric materials

CONTENTS

Introduction
1. Basic Terms in Polymer Mixtures
2. Phase Domain Behaviour
3. Domain Morphologies
4. References
5. Bibliography
6. Alphabetical Index of Terms

INTRODUCTION

It is the intent of this document to define the terms most commonly encountered in the field of polymer blends and composites. The scope has been limited to mixtures in which the components differ in chemical composition or molar mass or both and in which the continuous phase is polymeric. Many of the materials described by the term 'multiphase' are two-phase systems that may show a multitude of finely dispersed phase domains. Hence, incidental thermodynamic descriptions are mainly limited to binary mixtures, although they can be and, in the scientific literature, have been generalized to multicomponent mixtures. Crystalline polymers and liquid-crystal polymers have been considered in other documents [1,2] and are not discussed here.

This document is organized into three sections. The first defines terms basic to the description of polymer mixtures. The second defines terms commonly encountered in descriptions of phase-domain behaviour of polymer mixtures. The third defines terms commonly encountered in the descriptions of the morphologies of phase-separated polymer mixtures.

General terms describing the composition of a system as defined in ref. [3] are used without further definition throughout the document. Implicit definitions are identified in boldface type throughout the document.

1 BASIC TERMS IN POLYMER MIXTURES

1.1 polymer blend
Macroscopically homogeneous mixture of two or more different species of polymer [3,4].
Note 1: See the Gold Book, p. 312 [3].

Originally prepared by a working group consisting of W. J. Work (USA), K. Horie (Japan), M. Hess (Germany) and R. F. T. Stepto (UK). Reprinted from *Pure Appl. Chem.*, 76, 1985–2007 (2004)

BLENDS, COMPOSITES, AND MULTIPHASE MATERIALS

Note 2: In most cases, blends are homogeneous on scales larger than several times the wavelengths of visible light.
Note 3: In principle, the constituents of a blend are separable by physical means.
Note 4: No account is taken of the *miscibility* or immiscibility of the constituent macromolecules, i.e., no assumption is made regarding the number of *phase domains* present.
Note 5: The use of the term '*polymer alloy*' for 'polymer blend' is discouraged, as the former term includes *multiphase copolymers* but excludes incompatible polymer blends (see Definition 1.3).
Note 6: The number of polymeric components which comprises a blend is often designated by an adjective, viz., binary, ternary, quaternary.

1.2 miscibility

Capability of a mixture to form a single phase over certain ranges of temperature, pressure, and composition.
Note 1: Whether or not a single phase exists depends on the chemical structure, molar-mass distribution, and molecular architecture of the components present.
Note 2: The single phase in a mixture may be confirmed by light scattering, X-ray scattering, and neutron scattering.
Note 3: For a two-component mixture, a necessary and sufficient condition for stable or metastable equilibrium of a homogeneous single phase is

$$\left(\frac{\partial^2 \Delta_{mix} G}{\partial \phi^2}\right)_{T,p} > 0,$$

where $\Delta_{mix}G$ is the Gibbs energy of mixing and ϕ the composition, where ϕ is usually taken as the volume fraction of one of the components. The system is unstable if the above second derivative is negative. The borderline (*spinodal*) between (meta)stable and unstable states is defined by the above second derivative equalling zero. If the compositions of two conjugate (coexisting) phases become identical upon a change of temperature or pressure, the third derivative also equals zero (defining a critical state).
Note 4: If a mixture is thermodynamically metastable, it will demix if suitably nucleated (see Definition 2.5). If a mixture is thermodynamically unstable, it will demix by *spinodal decomposition* or by nucleation and growth if suitably nucleated, provided there is minimal kinetic hindrance.

1.3 miscible polymer blend
homogeneous polymer blend

Polymer blend that exhibits *miscibility*.
Note 1: For a *polymer blend* to be miscible, it must satisfy the criteria of *miscibility*.
Note 2: *Miscibility* is sometimes erroneously assigned on the basis that a blend exhibits a single T_g or optical clarity.
Note 3: A miscible system can be thermodynamically stable or metastable (see Note 4 in Definition 1.2).
Note 4: For components of chain structures that would be expected to be miscible, *miscibility* may not occur if molecular architecture is changed, e.g., by crosslinking.

TERMINOLOGY

1.4 homologous polymer blend
Mixture of two or more fractions of the same polymer, each of which has a different molar-mass distribution.

1.5 isomorphic polymer blend
Polymer blend of two or more different semi-crystalline polymers that are miscible in the crystalline state as well as in the molten state.
Note 1: Such a blend exhibits a single, composition-dependent glass-transition temperature, T_g, and a single, composition-dependent melting point, T_m, subscript
Note 2: This *behaviour* is extremely rare; very few cases are known.

1.6 polymer–polymer complex
Complex, at least two components of which are different polymers [4].
Note 1: A **complex** is a molecular entity formed from two or more components that can be ionic or uncharged [3].
Note 2: Although the intrinsic binding energy between the individual interacting sites giving rise to the complex is weaker than a covalent bond, the total binding energy for any single molecule may exceed the energy of a single covalent bond.
Note 3: The properties of a complex defined here differ from those given in ref. [3] because, owing to the repeating nature of a polymer molecule, many interacting sites may be present, which together will provide stronger bonding than a single covalent bond.

1.7 metastable miscibility
Capability of a mixture to exist for an indefinite period of time as a single phase that is separated by a small or zero energy barrier from a thermodynamically more stable multiphase system [3].
Note: Mixtures exhibiting metastable miscibility may remain unchanged or they may undergo phase separation, usually by nucleation or *spinodal decomposition*.

1.8 metastable miscible polymer blend
Polymer blend that exhibits *metastable miscibility*.
Note: In polymers, because of the low mobility of polymer chains, particularly in a glassy state, metastable mixtures may exist for indefinite periods of time without phase separation. This has frequently led to confusion when metastable *miscible polymer blends* are erroneously claimed to be miscible.

1.9 interpenetrating polymer network, (IPN)
Polymer comprising two or more polymer networks which are at least partially interlaced on a molecular scale, but not covalently bonded to each other and cannot be separated unless chemical bonds are broken [3,4].
Note 1: A mixture of two or more preformed polymer networks is not an interpenetrating polymer network.
Note 2: An IPN may be further described by the process by which it is synthesized. When an IPN is prepared by a process in which the second component network is formed following the completion of formation of the first component network, the IPN may be referred to as a **sequential IPN**. When an IPN is prepared by a process in which both component networks are formed concurrently, the IPN may be referred to as a **simultaneous IPN.**

BLENDS, COMPOSITES, AND MULTIPHASE MATERIALS

1.10 semi-interpenetrating polymer network, (SIPN)
Polymer comprising one or more polymer network(s) and one or more linear or branched polymer(s) characterized by the penetration on a molecular scale of at least one of the networks by at least some of the linear or branched chains [3,4].
Note 1: Semi-interpenetrating polymer networks are different from *interpenetrating polymer networks* because the constituent linear-chain or branched-chain macromolecule(s) can, in principle, be separated from the constituent polymer network(s) without breaking chemical bonds, and, hence, they are *polymer blends*.
Note 2: Semi-interpenetrating polymer networks may be further described by the process by which they are synthesized. When an SIPN is prepared by a process in which the second component polymer is formed or incorporated following the completion of formation of the first component polymer, the SIPN may be referred to as a **sequential SIPN**. When an SIPN is prepared by a process in which both component polymers are formed concurrently, the SIPN may be referred to as a **simultaneous SIPN**. (This note has been changed from that which appears in ref. [4] to allow for the possibility that a linear or branched polymer may be incorporated into a network by means other than polymerization, e.g., by swelling of the network and subsequent diffusion of the linear or branched chain into the network.).

1.11 immiscibility
Inability of a mixture to form a single phase.
Note 1: Immiscibility may be limited to certain ranges of temperature, pressure, and composition.
Note 2: Immiscibility depends on the chemical structures, molar-mass distributions, and molecular architectures of the components.

1.12 immiscible polymer blend
      heterogeneous polymer blend
Polymer blend that exhibits *immiscibility*.

1.13 composite
Multicomponent material comprising multiple, different (non-gaseous) *phase domains* in which at least one type of *phase domain* is a *continuous phase*.
Note: Foamed substances, which are multiphase materials that consist of a gas dispersed in a liquid or solid, are not normally considered to be *composites*.

1.14 polymer composite
Composite in which at least one component is a polymer.

1.15 nanocomposite
Composite in which at least one of the phases has at least one dimension of the order of nanometers.

1.16 laminate
Material consisting of more than one layer, the layers being distinct in composition, composition profile, or anisotropy of properties.
Note 1: Laminates may be formed by two or more layers of different polymers.
Note 2: *Composite* laminates generally consist of one or more layers of a substrate, often fibrous, impregnated with a curable polymer, curable polymers, or liquid reactants.

TERMINOLOGY

Note 3: The substrate is usually a sheet-like woven or non-woven material (e.g., glass fabric, paper, copper foil).
Note 4: A single layer of a laminate is termed a **lamina.**

1.17 lamination
Process of forming a *laminate*.

1.18 delamination
Process that separates the layers of a *laminate* by breaking their structure in planes parallel to those layers.

1.19 impregnation
Penetration of monomeric, oligomeric, or polymeric liquids into an assembly of fibers.
Note 1: The term as defined here is specific to polymer science. An alternative definition of 'impregnation' applies in some other fields of chemistry [3].
Note 2: Impregnation is usually carried out on a woven fabric or a yarn.

1.20 prepreg
Sheets of a substrate that have been impregnated with a curable polymer, curable polymers, or liquid reactants, or a thermoplastic, and are ready for fabrication of *laminates*.
Note 1: See Definition 1.16 notes 2 and 3.
Note 2: During the *impregnation* the curable polymer, curable polymers, or liquid reactants may be allowed to react to a certain extent (sometimes termed **degree of ripening).**

1.21 intercalation
Process by which a substance becomes transferred into pre-existing spaces of molecular dimensions in a second substance.
Note: The term as defined here is specific to polymer science. An alternative definition of 'intercalation' applies in some other fields of chemistry [3].

1.22 exfoliation
Process by which the layers of a multi-layered structure separate.
Note: In the context of a *nanocomposite* material, the individual layers are of the order of at most a few nanometers in thickness.

1.23 wetting
Process by which an interface between a solid and a gas is replaced by an interface between the same solid and a liquid.

1.24 adhesion
Holding together of two bodies by interfacial forces or mechanical interlocking on a scale of micrometers or less.

1.25 chemical adhesion
Adhesion in which two bodies are held together at an interface by ionic or covalent bonding between molecules on either side of the interface.

1.26 interfacial adhesion
   tack

Adhesion in which interfaces between phases or components are maintained by intermolecular forces, chain entanglements, or both, across the interfaces.

Note 1: **Adhesive strength**, F_a, SI unit: Nm^{-2} is the force required to separate one condensed *phase domain* from another at the interface between the two *phase domains* divided by the area of the interface.

Note 2: **Interfacial tension**, γ, unit: Nm^{-1}, Jm^{-2} is the change in Gibbs energy per unit change in interfacial area for substances in physical contact.

Note 3: Use of the term **interfacial energy** for *interfacial tension* is not recommended.

1.27 interfacial bonding

Bonding in which the surfaces of two bodies in contact with one another are held together by intermolecular forces.

Note: Examples of intermolecular forces include covalent, ionic, van der Waals, and hydrogen bonds.

1.28 interfacial fracture

Brittle fracture that takes place at an interface.

1.29 craze

Crack-like cavity formed when a polymer is stressed in tension that contains load-bearing fibrils spanning the gap between the surfaces of the cavity.

Note: Deformation of continua occurs with only minor changes in volume; hence, a craze consists of both fibrils and voids.

1.30 additive

Substance added to a polymer.

Note 1: The term as defined here is specific to polymer science. An alternative definition of 'additive' applies in some other fields of chemistry [3].

Note 2: An additive is usually a minor component of the mixture formed and usually modifies the properties of the polymer.

Note 3: Examples of additives are antioxidants, plasticizers, flame retardants, processing aids, other polymers, colorants, UV absorbers, and *extenders* [7].

1.31 interfacial agent

Additive that reduces the *interfacial energy* between *phase domains*.

1.32 compatibility

Capability of the individual component substances in either an *immiscible polymer blend* or a *polymer composite* to exhibit *interfacial adhesion*.

Note 1: Use of the term 'compatibility' to describe miscible systems is discouraged.

Note 2: Compatibility is often established by the observation of mechanical integrity under the intended conditions of use of a *composite* or an *immiscible polymer blend*.

1.33 compatibilization

Process of modification of the interfacial properties in an *immiscible polymer blend* that results in formation of the *interphases* and stabilization of the morphology, leading to the creation of a *compatible polymer blend*.

TERMINOLOGY

Note: Compatibilization may be achieved by addition of suitable copolymers or by chemical modification of interfaces through physical treatment (i.e., irradiation or thermal) or reactive processing.

1.34 degree of compatibility
Measure of the strength of the *interfacial bonding* between the component substances of a *composite* or *immiscible polymer blend*.
Note 1: Estimates of the degree of compatibility are often based upon the mechanical performance of the *composite*, the *interphase thickness*, or the sizes of the *phase domains* present in the *composite*, relative to the corresponding properties of *composites* lacking *compatibility*.
Note 2: The term **degree of incompatibility** is sometimes used instead of degree of compatibility. Such use is discouraged as incompatibility is related to the weakness of *interfacial bonding*.

1.35 compatible polymer blend
Immiscible polymer blend that exhibits macroscopically uniform physical properties.
Note: The macroscopically uniform physical properties are usually caused by sufficiently strong interactions between the component polymers.

1.36 compatibilizer
Polymer or copolymer that, when added to an *immiscible polymer* blend, modifies its interfacial character and stabilizes its morphology.
Note: Compatibilizers usually stabilize *morphologies* over distances of the order of micrometers or less.

1.37 coupling agent
adhesion promoter
Interfacial agent comprised of molecules possessing two or more functional groups, each of which exhibits preferential interactions with the various types of *phase domains* in a *composite*.
Note 1: A coupling agent increases *adhesion* between *phase domains*.
Note 2: An example of the use of a coupling agent is in a mineral-filled polymer material where one part of the coupling agent molecule can chemically bond to the inorganic mineral while the other part can chemically bond to the polymer.

1.38 polymer alloy
Polymeric material, exhibiting macroscopically uniform physical properties throughout its whole volume, that comprises a *compatible polymer blend*, a *miscible polymer blend*, or a *multiphase copolymer*.
Note 1: See note Note 5 in Definition 1.1.
Note 2: The use of the term polymer alloy for a *polymer blend* is discouraged.

1.39 dispersion
Material comprising more than one phase where at least one of the phases consists of finely divided *phase domains*, often in the colloidal size range, distributed throughout a *continuous phase domain*.
Note 1: The term as defined here is specific to polymer science. An alternative definition of 'dispersion' applies in some other fields of chemistry [3].

BLENDS, COMPOSITES, AND MULTIPHASE MATERIALS

Note 2: Particles in the colloidal size range have linear dimensions [3] between 1 nm and 1 μm.
Note 3: The finely divided domains are called the dispersed or *discontinuous phase domains*.
Note 4: For a definition of *continuous phase domain* see 3.12.
Note 5: A dispersion is often further characterized on the basis of the size of the *continuous phase domains* as a **macrodispersion** or a **microdispersion**. To avoid ambiguity when using these terms, the size of the domain should also be defined.

1.40 dispersing agent
dispersing aid
dispersant

Additive, exhibiting surface activity, that is added to a suspending medium to promote uniform and maximum separation of extremely fine solid particles, often of colloidal size (see Note 2 in Definition 1.39)

Note: Although dispersing agents achieve results similar to *compatibilizers*, they function differently in that they reduce the attractive forces between fine particles, which allows them to be more easily separated and dispersed.

1.41 agglomeration
aggregation

Process in which dispersed molecules or particles assemble rather than remain as isolated single molecules or particles [3].

1.42 agglomerate
aggregate

Cluster of molecules or particles that results from *agglomeration*.
Note: The term as defined here is specific to polymer science. An alternative definition of 'aggregate' is used in some other fields of chemistry [3] (see also Chapter 11).

1.43 extender

Substance, especially a diluent or modifier, added to a polymer to increase its volume without substantially altering the desirable properties of the polymer.
Note: An extender may be a liquid or a solid.

1.44 filler

Solid *extender*.
Note 1: The term as defined here is specific to polymer science. An alternative definition of 'filler' applies in some other fields of chemistry [3].
Note 2: Fillers may be added to modify mechanical, optical, electrical, thermal, flammability properties, or simply to serve as *extenders*.

1.45 fill factor, ϕ_{fill}

Maximum volume fraction of a particulate *filler* that can be added to a polymer while maintaining the polymer as the continuous *phase domains*.

TERMINOLOGY

1.46 thermoplastic elastomer
Either
Melt-processable *polymer blend* or copolymer in which a continuous elastomeric *phase domain* is reinforced by dispersed hard (glassy or crystalline) *phase domains* that act as junction points over a limited range of temperature.
or
Elastomer comprising a thermoreversible network.
Note 1: The *behaviour* of the hard *phase domains* as junction points is thermally reversible.
Note 2: The interfacial interaction between hard and soft *phase domains* in a thermoplastic elastomer is often the result of covalent bonds between the phases and is sufficient to prevent the flow of the elastomeric *phase domains* under conditions of use.
Note 3: Examples of thermoplastic elastomers include block copolymers and blends of plastics and rubbers.

2 PHASE DOMAIN BEHAVIOR

2.1 miscibility window
Range of copolymer compositions in a polymer mixture, at least one component substance of which is a copolymer, that gives *miscibility* over a range of temperatures and pressures.
Note 1: Outside the miscibility window immiscible mixtures are formed.
Note 2: The compositions of the copolymers within the miscibility window usually exclude the homopolymer compositions of the monomers from which the copolymers are prepared.
Note 3: The miscibility window is affected by the molecular weights of the component substances.
Note 4: The existence of miscibility windows has been attributed to an average force between the monomer units of the copolymer that leads to those units associating preferentially with the monomer units of the other polymers.

2.2 miscibility gap
Area within the coexistence curve of an isobaric phase diagram (temperature vs. composition) or an isothermal phase diagram (pressure vs. composition).
Note: A miscibility gap is observed at temperatures below an *upper critical solution temperature* (UCST) or above the *lower critical solution temperature* (LCST). Its location depends on pressure. In the miscibility gap, there are at least two phases coexisting.

2.3 Flory–Huggins theory
    Flory–Huggins–Staverman theory
Statistical thermodynamic mean-field theory of polymer solutions, formulated independently by Flory, Huggins, and Staverman, in which the thermodynamic quantities of the solution are derived from a simple concept of combinatorial entropy of mixing and a reduced Gibbs-energy parameter, the χ *interaction parameter* [3].
Note 1: The Flory–Huggins theory has often been found to have utility for *polymer blends*; however, there are many equation-of-state theories that provide more accurate descriptions of polymer–polymer interactions.
Note 2: The present definition has been modified from that which appears in ref. [5] to acknowledge the contributions of Staverman and to further clarify the statistical basis of the theory.

2.4 χ interaction parameter, χ

Interaction parameter, employed in the *Flory–Huggins theory*, to account for the contribution of the noncombinatorial entropy of mixing and the enthalpy of mixing to the Gibbs energy of mixing.

Note 1: The definition and the name of the term have been modified from that which appears in ref. [5] to reflect its broader use in the context of *polymer blends*. In its simplest form, the χ parameter is defined according to the Flory–Huggins equation for binary mixtures

$$\frac{\Delta_{mix}}{RT} = n_1 \ln \phi_1 + n_2 \ln \phi_2 + \chi x_1 n_1 \ln \phi_2$$

for a mixture of amounts of substance n_1 and n_2 of components denoted 1 and 2, giving volume fractions ϕ_1 and ϕ_2, with the molecules of component 1 each conceptually consisting of x_1 segments whose Gibbs energy of interaction with segments of equal volume in the molecules of component 2 is characterized by the interaction parameter χ.

Note 2: The χ interaction parameters characterizing a given system vary with composition, molar mass, and temperature.

Note 3: B is an alternative parameter to χ, where $B = \chi RT/V_m$, in which V_m is the molar volume of one of the components of the mixture.

2.5 nucleation of phase separation

Initiation of *phase domains* formation through the presence of heterogeneities [3].

Note: In a metastable region of a phase diagram (see Definition 1.2), phase separation is initiated only by nucleation.

2.6 binodal
binodal curve
coexistence curve

Curve defining the region of composition and temperature in a phase diagram for a binary mixture across which a transition occurs from *miscibility* of the components to conditions where single-phase mixtures are metastable or unstable (see Note 4 in Definition 1.2).

Note: Binodal compositions are defined by pairs of points on the curve of Gibbs energy of mixing vs. composition that have common tangents, corresponding to compositions of equal chemical potentials of each of the two components in two phases.

2.7 spinodal
spinodal curve

Curve defining the region of composition and temperature for a binary mixture across which a transition occurs from conditions where single-phase mixtures are metastable to conditions where single-phase mixtures are unstable and undergo phase separation by *spinodal decomposition*.

Note 1: The spinodal curve for a binary mixture is defined as the geometrical locus of all states with

$$\left(\frac{\partial^2 \Delta_{mix} G}{\partial \phi^2} \right)_{T,p} = 0, \text{ (see Definition 1.2, note 4)}.$$

TERMINOLOGY

Note 2: In the unstable region bounded by the spinodal curve, *phase domains* separation is spontaneous, i.e., no nucleation step is required to initiate the separation process.

2.8 spinodal decomposition
spinodal phase-demixing

Long-range, diffusion-limited, spontaneous *phase domains* separation initiated by delocalized concentration fluctuations occurring in an unstable region of a mixture bounded by a *spinodal curve*.

Note: Spinodal decomposition occurs when the magnitude of Gibbs energy fluctuations with respect to composition are zero.

2.9 cloud point

Experimentally measured point in the phase diagram of a mixture at which a loss in transparency is observed due to light scattering caused by a transition from a single- to a two-phase state.

Note 1: The phenomenon is characterized by the first appearance of turbidity or cloudiness.

Note 2: A cloud point is heating rate- or cooling rate-dependent.

2.10 cloud-point curve

Curve of temperature vs. composition defined by the *cloud points* over range of compositions of two substances.

Note: Mixtures are observed to undergo a transition from a single- to a two-phase state upon heating or cooling.

2.11 cloud-point temperature

Temperature at a *cloud point*.

2.12 critical point

Point in the isobaric temperature-composition plane for a binary mixture where the compositions of all coexisting phases become identical.

Note 1: An alternative definition of 'critical solution point' refers strictly to liquid-vapor equilibria [3].

Note 2: Unless specified atmospheric pressure is assumed.

Note 3: In a phase diagram, the slope of the tangent to the *spinodal* is zero at this point.

Note 4: At a critical point, *binodals* and *spinodals* coincide.

Note 5: Although the definition holds strictly for binary mixtures, it is often erroneously applied to multicomponent mixtures.

Note 6: See note 3 in Definition 1.2.

2.13 lower critical solution temperature, (LCST)

Critical temperature below which a mixture is miscible [3].

Note 1: Below the LCST and above the *upper critical solution temperature* (UCST), if it exists, a single phase exists for all compositions.

Note 2: The LCST depends upon pressure and the molar-mass distributions of the constituent polymer(s).

Note 3: For a mixture containing or consisting of polymeric components, these may be different polymers or species of different molar mass of the same polymer.

BLENDS, COMPOSITES, AND MULTIPHASE MATERIALS

2.14 upper critical solution temperature, (UCST)
Critical temperature above which a mixture is miscible [3].
Note 1: Above the UCST and below the *lower critical solution temperature* (LCST), if it exists, a single phase exists for all compositions
Note 2: The UCST depends upon the pressure and molar-mass distributions of the constituent polymer(s).
Note 3: For a mixture containing or consisting of polymeric components, these may be different polymers or species of different molar mass of the same polymer.

2.15 phase inversion
Process by which an initially *continuous phase domain* becomes the *dispersed phase domain* and the initially *dispersed phase domain* becomes the *continuous phase domains* [3].
Note 1: Phase inversion may be observed during the polymerization or melt processing of *polymer blend* systems.
Note 2: The phenomenon is usually observed during polymerization of a monomer containing a dissolved polymer.

2.16 interdiffusion
Process by which homogeneity in a mixture is approached by means of spontaneous mutual molecular diffusion.

2.17 blooming
Process in which one component of a polymer mixture, usually not a polymer, undergoes phase separation and migration to an external surface of the mixture.

2.18 coalescence
Process in which two *phase domains* of essentially identical composition in contact with one another form a larger *phase domain* [3].
Note 1: Coalescence reduces the total interfacial area.
Note 2: The flocculation of a polymer colloid, through the formation of *aggregates*, may be followed by coalescence.

2.19 morphology coarsening
      phase ripening
Process by which *phase domains* increase in size during the aging of a multiphase material.
Note 1: In the coarsening at the late stage of phase separation, volumes and compositions of phase domains are conserved.
Note 2: Representative mechanisms for coarsening at the late stage of phase separation are: (1) material flow in domains driven by *interfacial tension* (observed in a co-continuous morphology), (2) the growth of domain size by evaporation from smaller droplets and condensation into larger droplets, and (3) *coalescence* (fusion) of more than two droplets. The mechanisms are usually called (1) Siggia's mechanism, (2) Ostwald ripening (or the Lifshitz-Slyozov mechanism), and (3) *coalescence*.
Note 3: Morphology coarsening can be substantially stopped by, for example, vitrification, crosslinking, and **pinning**, the slowing down of molecular diffusion across *domain interfaces*.

TERMINOLOGY

3 DOMAINS AND MORPHOLOGIES

Many types of morphologies have been reported in the literature of multiphase polymeric materials. It is the intent of this document to define only the most commonly used terms. In addition, some morphologies have historically been described by very imprecise terms that may not have universal meanings. However, if such terms are widely used they are defined here.

3.1 morphology

Shape, optical appearance, or form of *phase domains* in substances, such as high polymers, *polymer blends*, *composites*, and crystals.

Note: For a *polymer blend* or *composite*, the morphology describes the structures and shapes observed, often by microscopy or scattering techniques, of the different *phase domains* present within the mixture.

3.2 phase domain

Region of a material that is uniform in chemical composition and physical state.

Note 1: A phase in a multiphase material can form domains differing in size.

Note 2: The term 'domain' may be qualified by the adjective microscopic or nanoscopic or the prefix micro- or nano- according to the size of the linear dimensions of the domain.

Note 3: The prefixes micro-, and nano- are frequently incorrectly used to qualify the term 'phase' instead of the term 'domain'; hence, 'microphase domain', and 'nanophase domain' are often used. The correct terminology that should be used is **phase microdomain** and **phase nanodomain**.

3.3 multiphase copolymer

Copolymer comprising phase-separated domains.

3.4 domain interface
domain boundary

Surface forming a boundary between two *phase domains*.

Note: A representation of the domain interface as a two-dimensional surface oversimplifies the actual structure. All interfaces have a third dimension, namely, the *interphase* or *interfacial region*.

3.5 domain structure

Morphology of individual *phase domain*s in a multiphase system.

Note: Domain structures may be described for *phase domains* or domains that are themselves multiphase structures.

3.6 interfacial region
interphase

Region between *phase domains* in an *immiscible polymer blend* in which a gradient in composition exists [3].

3.7 phase interaction

Molecular interaction between the components present in the interphases of a multiphase mixture.

Note: The **interphase elasticity** is the capability of a deformed interphase to return to its original dimensions after the force causing the deformation has been removed.

BLENDS, COMPOSITES, AND MULTIPHASE MATERIALS

3.8 interfacial-region thickness
interphase thickness
interfacial width

Linear extent of the composition gradient in an *interfacial region* [3].

Note: The width at half the maximum of the composition profile across the *interfacial region* or the distance between locations where $d\phi/dr$ (with ϕ the composition of a component and r the distance through the *interfacial region*) has decreased to $1/e$ are used as measures of the *interfacial-region thickness*.

3.9 hard-segment phase domain

Phase domain of microscopic or smaller size, usually in a block, graft, or *segmented copolymer*, comprising essentially those segments of the polymer that are rigid and capable of forming strong intermolecular interactions.

Note: Hard-segment *phase domains* are typically of 2–15 nm linear size.

3.10 soft-segment phase domain

Phase domain of microscopic or smaller size, usually in a block, graft, or *segmented copolymer*, comprising essentially those segments of the polymer that have glass transition temperatures lower than the temperature of use.

Note: Soft-segment *phase domains* are often larger than *hard-segment phase domains* and are often continuous.

3.11 segmented copolymer

Copolymer containing *phase domains* of microscopic or smaller size, with the domains constituted principally of single types of constitutional unit.

Note: The types of domain in a segmented copolymer usually comprise *hard-* and *soft-segment phase domains*.

3.12 continuous phase domain
matrix phase domain

Phase domain consisting of a single phase in a heterogeneous mixture through which a continuous path to all *phase domain* boundaries may be drawn without crossing a *phase domain* boundary.

Note: In a *polymer blend*, the continuous phase domain is sometimes referred to as the **host polymer**, **bulk substance**, or **matrix**.

3.13 discontinuous phase domain
discrete phase domain
dispersed phase domain

Phase domain in a phase-separated mixture that is surrounded by a continuous phase but isolated from all other similar *phase domains* within the mixture.

Note: The discontinuous phase domain is sometimes referred to as the **guest polymer.**

3.14 dual phase domain continuity
co-continuous phase domains

Topological condition, in a phase-separated, two-component mixture, in which a continuous path through either *phase domain* may be drawn to all *phase domain* boundaries without crossing any *phase domain* boundary.

TERMINOLOGY

3.15 core-shell morphology
Two-phase domain *morphology*, of approximately spherical shape, comprising two polymers, each in separate *phase domains*, in which *phase domains* of one polymer completely encapsulate the *phase domains* of the other polymer.
Note: This *morphology* is most commonly observed in copolymers or blends prepared in emulsion polymerization by the sequential addition and polymerization of two different monomer compositions.

3.16 cylindrical morphology
Phase domain morphology, usually comprising two polymers, each in separate *phase domains*, in which the *phase domains* of one polymer are of cylindrical shape.
Note 1: *Phase domains* of the constituent polymers may alternate, which results in many cylindrical layers surrounding a central core domain.
Note 2: Cylindrical *morphologies* can be observed, for example, in triblock copolymers.

3.17 fibrillar morphology
Morphology in which *phase domains* have shapes with one dimension much larger than the other two dimensions.

3.18 lamellar domain morphology
Morphology in which *phase domains* have shapes with two dimensions much larger than the third dimension.
Note: Plate-like *phase domains* have the appearance of extended planes that are often oriented essentially parallel to one another.

3.19 microdomain morphology
Morphology consisting of phase microdomains.
Note 1: See Defintion 3.2.
Note 2: Microdomain *morphologies* are usually observed in block, graft, and *segmented copolymers*.
Note 3: The type of *morphology* observed depends upon the relative abundance of the different types of constitutional units and the conditions for the generation of the *morphology*. The most commonly observed *morphologies* are spheres, cylinders, and lamellae.

3.20 nanodomain morphology
Morphology consisting of phase nanodomains.
Note: See Defintion 3.2.

3.21 onion morphology
Multiphase *morphology* of roughly spherical shape that comprises alternating layers of different polymers arranged concentrically, all layers being of similar thickness.

3.22 ordered co-continuous double gyroid morphology
Co-continuous *morphology* in which a set of two gyroid-based *phase domains* exhibits a highly regular, three-dimensional lattice-like *morphology* with Ia3d space group symmetry.
Note 1: The domains are composed of tripoidal units as the fundamental building structures.
Note 2: The two domains are interlaced.

3.23 multicoat morphology

Morphology observed in a blend of a block copolymer with the homopolymer of one of the blocks and characterized by alternating concentric shells of the copolymer and the homopolymer.

Note: The *morphology* is identical to *onion morphology* within a matrix of homopolymer [6].

3.24 rod-like morphology

Morphology characterized by cylindrical *phase domains*.

3.25 multiple inclusion morphology
   salami-like morphology

Multiphase *morphology* in which dispersed *phase domains* of one polymer contain and completely encapsulate many *phase domains* of a second polymer that may have the same composition as the continuous *phase domain*.

4 REFERENCES

1. IUPAC. 'Definitions of terms relating to crystalline polymers (IUPAC Recommendations 1988)', *Pure Appl. Chem.* **61**, 769–785 (1989). Reprinted as Chapter 6 this volume.
2. IUPAC. 'Definitions of basic terms relating to low-molar-mass and polymer liquid crystals (IUPAC Recommendations 2001)', *Pure Appl. Chem.* **75**, 845–895 (2001). Reprinted as Chapter 7 this volume.
3. IUPAC. Compendium of Chemical Terminology, 2nd ed. (the 'Gold Book'). Compiled by A. D. McNaught and A. Wilkinson. Blackwell Scientific Publications, Oxford (1997). XML on-line corrected version: http://goldbook.iupac.org (2006-) created by M. Nic, J. Jirat, B. Kosata; updates compiled by A. Jenkins.
4. IUPAC. 'Glossary of basic terms in polymer science (IUPAC Recommendations 1996)', *Pure Appl. Chem.* **68**, 2287–2311 (1996). Reprinted as Chapter 1 this volume.
5. D. K. Carpenter. 'Solution properties', in *Encyclopedia of Polymer Science and Engineering*, Vol. 15, 2nd ed., J. I. Kroschwitz (Ed.), pp. 419–481, Wiley Interscience, New York (1989).
6. J. M. G. Cowie. 'Miscibility', in *Encyclopedia of Polymer Science and Engineering*, 2nd ed., J. I. Kroschwitz (Ed.), Supplement, pp. 455–480, Wiley Interscience, New York (1989).
7. IUPAC. 'Definitions of terms relating to degradation, aging, and related chemical transformations of polymers (IUPAC Recommendations 1996)', *Pure Appl. Chem.*, **68**, 2313-2323 (1996). Reprinted as Chapter 13 this volume.

5 BIBLIOGRAPHY

1. IUPAC. 'Definitions of terms relating to degradation, aging, and related chemical transformations of polymers (IUPAC Recommendations 1996)', *Pure Appl. Chem.* **68**, 2313–2323 (1996). Reprinted as Chapter 13 this volume.
2. *ASTM Glossary of ASTM Definitions*, 2nd ed., American Society for Testing and Materials, Philadelphia, PA (1973).

TERMINOLOGY

3. A. N. Gent and G. R. Hamed. 'Adhesion', in *Encyclopedia of Polymer Science and Engineering*, Vol. 1, 2nd ed., J. I. Kroschwitz (Ed.), pp. 476–517, Wiley Interscience, New York (1985).
4. L. Leibler. 'Phase transformations', in *Encyclopedia of Polymer Science and Engineering*, Vol. 11, 2nd ed., J. I. Kroschwitz, (Ed.), pp. 30–45, Wiley Interscience, New York (1988).
5. J. Koberstein. 'Interfacial properties', in *Encyclopedia of Polymer Science and Engineering*, Vol. 8, 2nd ed., J. I. Kroschwitz (Ed.), pp. 237–279, Wiley Interscience, New York (1987).
6. D. W. Fox and R. B. Allen. 'Compatibility', in *Encyclopedia of Polymer Science and Engineering*, Vol. 3, 2nd ed., J. I. Kroschwitz (Ed.), pp. 758–775, Wiley Interscience, New York (1985).
7. R. A. Orwoll. 'Solubility of polymers', *Encyclopedia of Polymer Science and Engineering*, Vol. 15, 2nd ed., J. I. Kroschwitz (Ed.), pp. 380–402, Wiley Interscience, New York (1989).
8. L. H. Sperling. 'Microphase structure', in *Encyclopedia of Polymer Science and Engineering*, Vol. 9, 2nd ed., J. I. Kroschwitz (Ed.), pp. 760–788, Wiley Interscience, New York (1987).
9. D. R. Paul, J. W. Barlow, H. Keskkula. 'Polymer blends' in *Encyclopedia of Polymer Science and Engineering*, Vol. 12, 2nd ed., J. I. Kroschwitz (Ed.), pp. 399–461, Wiley Interscience, NewYork (1988).
10. D. R. Paul and S. Newman. *Polymer Blends*, Academic Press, New York (1978).
11. D. R. Paul and C. B. Bucknall. *Polymer Blends: Formulation and Performance*, John Wiley, New York (1999).
12. L. A. Utracki. *Polymer Alloys and Blends*, Hanser Publishers, New York (1990).

6   ALPHABETICAL INDEX OF TERMS

additive	1.30
adhesion	1.24
adhesion promoter	1.37
adhesive strength	1.26
agglomerate	1.42
agglomeration	1.41
aggregate	1.42
aggregation	1.41
binodal	2.6
binodal curve	2.6
blooming	2.17
bulk substance	3.12
chemical adhesion	1.25
cloud point	2.9
cloud-point curve	2.10
cloud-point temperature	2.11
co-continuous phase domains	3.14
coalescence	2.18
coexistence curve	2.6
compatibility	1.32
compatibilization	1.33
compatibilizer	1.36
compatible polymer blend	1.35
complex	1.6
composite	1.13
continuous phase domain	3.12
core-shell morphology	3.15
coupling agent	1.37
craze	1.29
critical point	2.12
cylindrical morphology	3.16
degree of compatibility	1.34
degree of incompatibility	1.34
degree of ripening	1.20
delamination	1.18
discontinuous phase domain	3.13
discrete phase domain	3.13
dispersant	1.40
dispersed phase domain	3.13
dispersing agent	1.40
dispersing aid	1.40
dispersion	1.39
domain boundary	3.4
domain interface	3.4
domain structure	3.5
dual phase domain continuity	3.14
exfoliation	1.22
fibrillar morphology	3.17
fill factor	1.45
filler	1.44
Flory–Huggins theory	2.3
Flory–Huggins–Staverman theory	2.3
guest polymer	3.13
hard-segment phase domain	3.9
heterogeneous polymer blend	1.12
homogeneous polymer blend	1.3
homologous polymer blend	1.4
host polymer	3.12
immiscibility	1.11
immiscible polymer blend	1.12
impregnation	1.19
intercalation	1.21
interdiffusion	2.16
interfacial adhesion	1.26
interfacial agent	1.31
interfacial bonding	1.27
interfacial energy	1.26
interfacial fracture	1.28
interfacial region	3.6
interfacial-region thickness	3.8
interfacial tension	1.26
interfacial width	3.8
interpenetrating polymer network (IPN)	1.9
interphase	3.6
interphase elasticity	3.7
interphase thickness	3.8
isomorphic polymer blend	1.5
lamellar domain morphology	3.17
lamina	1.16
laminate	1.16
lamination	1.17
lower critical solution temperature (LCST)	2.13
macrodispersion	1.39
matrix	3.12
matrix phase domain	3.12
metastable miscibility	1.7
metastable miscible polymer blend	1.8
microdispersion	1.39
microdomain morphology	3.19
miscibility	1.2
miscibility gap	2.2
miscibility window	2.1

TERMINOLOGY

miscible polymer blend	1.3	polymer–polymer complex	1.6
morphology	3.1	prepreg	1.20
morphology coarsening	2.19	rod-like morphology	3.24
multicoat morphology	3.23	salami-like morphology	3.25
multiphase copolymer	3.3	segmented copolymer	3.11
multiple inclusion morphology	3.25	sequential IPN	1.9
nanocomposite	1.15	sequential SIPN	1.10
nanodomain morphology	3.20	semi-interpenetrating polymer network (SIPN)	1.10
nucleation of phase separation	2.5	simultaneous IPN	1.9
onion morphology	3.21	simultaneous SIPN	1.10
ordered co-continuous double gyroid morphology	3.22	soft-segment phase domain	3.10
phase domain	3.2	spinodal	2.7
phase interaction	3.7	spinodal curve	2.7
phase inversion	2.15	spinodal decomposition	2.8
phase microdomain	3.2	spinodal phase-demixing	2.8
phase nanodomain	3.2	tack	1.26
phase ripening	2.19	thermoplastic elastomer	1.46
pinning	2.19	upper critical solution temperature (UCST)	2.14
polymer alloy	1.38	wetting	1.23
polymer blend	1.1	χ interaction parameter	2.4
polymer composite	1.14		

10: Terminology of polymers containing ionizable or ionic groups and of polymers containing ions

CONTENTS

Introduction
List of Terms
References

INTRODUCTION

This document defines the most commonly used terms relating to polymers containing ionizable or ionic groups and to polymers containing ions. Inorganic materials, such as certain phosphates, silicates, etc., that also may be considered ionic polymers are excluded from the present document. Only those terms that could be defined without ambiguity are considered. Cross-references to terms defined elsewhere within the document are printed in italic type.

LIST OF TERMS

1 ampholytic polymer
   polyampholyte
Polyelectrolyte composed of macromolecules containing both cationic and anionic groups, or corresponding ionizable groups.
Note: An ampholytic polymer in which ionic groups of opposite sign are incorporated into the same pendant groups is called, depending on the structure of the pendant groups, a *zwitterionic polymer*, *polymeric inner salt*, or *polybetaine*.

2 anion-exchange polymer
See *ion-exchange polymer*.

3 anionic polymer
Polymer composed of negatively charged macromolecules and an equivalent amount of countercations.

Originally prepared by a working group consisting of M. Hess (Germany) R. G. Jones (UK), J. Kahovec (Czech Republic), T. Kitayama (Japan), P. Kratochvíl (Czech Republic), P. Kubisa (Poland), W. Mormann (Germany), R. F. T. Stepto (UK), D. Tabak (Brazil), J. Vohlídal (Czech Republic), E. S. Wilks (USA). Reprinted from *Pure Appl. Chem.*, **78**, 2067–2074 (2006).

TERMINOLOGY

Note 1: If a substantial fraction of constitutional units carries negative charges, then an anionic polymer is a *polyelectrolyte*.
Note 2: The term *anionic polymer* should not be used to denote a polymer prepared by anionic polymerization.

4 cation-exchange polymer
See *ion-exchange polymer*.

5 cationic polymer
Polymer composed of positively charged macromolecules and an equivalent amount of counteranions.
Note 1: If a substantial fraction of constitutional units carries positive charges, then a *cationic polymer* is a *polyelectrolyte*.
Note 2: The positive charges may be fixed on groups located in main chains as in an *ionene* or in pendant groups.
Note 3: The term *cationic polymer* should not be used to denote a polymer prepared by cationic polymerization.

6 conducting polymer composite
Electrically conducting composite comprising a non-conducting polymer matrix and an electrically conducting material.
Note 1: Examples of the electrically conducting materials are carbon black and metal particles.
Note 2: See also *solid polymer electrolyte*.

7 critical ion-concentration in an ionomer
Concentration of ionic groups in an ionomer matrix above which ionic aggregation occurs.

8 dopant
Charge-transfer agent used to generate, by oxidation or reduction, positive or negative charges in an intrinsically conducting polymer.
Note: Examples of dopants include AsF_5 or I_2 as oxidizing agents, generating cation radicals on the chains of an *intrinsically conducting polymer* (so-called holes), or a solution of sodium naphthalenelide in tetrahydrofuran as a reducing agent, generating anion radicals on the chains of an *intrinsically conducting polymer*.

9 doping
Oxidation or reduction process brought about by a *dopant*.

10 electrically conducting polymer
Polymeric material that exhibits bulk electric conductivity.
Note: Definition same as definition 3.2, ref. [1].

11 halatopolymer
See *halato-telechelic polymer*.

12 halato-telechelic polymer
Polymer composed of linear macromolecules having ionic or ionizable end-groups.
Note 1: The term *halato-telechelic polymer* is used to denote a polymer composed of macromolecules having stable (long-lived) ionic or ionizable groups, such as carboxylate

or quaternary ammonium groups, as chain ends. It should not be used to describe a polymer composed of macromolecules having chain ends that are transient intermediates in ionic polymerizations initiated by difunctional initiators.

Note 2: The term *halatopolymer* is used for a linear polymer formed by the coupling of halato-telechelic polymer molecules, for example, for the linear polymer formed by the coupling of carboxylate end-groups with divalent metal cations [2].

13 intrinsically conducting polymer

Electrically conducting polymer composed of macromolecules having fully conjugated sequences of double bonds along the chains.

Note 1: The bulk electrical conductivity of an *intrinsically conducting polymer* is comparable to that of some metals and results from its macromolecules acquiring positive or negative charges through oxidation or reduction by an electron acceptor or donor (charge-transfer agent), termed a *dopant*.

Note 2: Examples of *intrinsically conducting polymers* are polyacetylene, polythiophene, polypyrrole, or polyaniline.

Note 3: Unlike *polymeric electrolytes*, in which charge is transported by dissolved ions, charge in *intrinsically conducting polymers* is transported along and between polymer molecules via generated charge carriers (e.g., holes, electrons).

Note 4: An *intrinsically conducting polymer* should be distinguished from a *conducting polymer composite* and from a *solid polymer electrolyte*.

14 ion-containing polymer

See *ionic polymer*.

15 ionene

Polymer composed of macromolecules in which ionized or ionic groups are parts of main chains.

Note: Most commonly, the ionic groups in ionenes are quaternary ammonium groups.

16 ion-exchange membrane

See *ion-exchange polymer*.

17 ion-exchange polymer

Polymer that is able to exchange ions (cations or anions) with the ionic components in solution.

Note 1: Definition same as definition 2.2, ref. [1]

Note 2: See ref. [3] for ion exchange.

Note 3: Depending on which ion can be exchanged, the polymer is referred to as an *anion exchange polymer* or *cation exchange polymer*.

Note 4: An ion-exchange polymer in ionized form may also be referred to as a *polyanion* or *polycation*.

Note 5: Synthetic ion-exchange organic polymers are often network *polyelectrolytes*.

Note 6: A membrane having ion-exchange groups is called an ion-exchange membrane.

Note 7: Use of the term *ion-exchange resin* for *ion-exchange polymer* is strongly discouraged.

18 ionic aggregates in an ionomer

Domains enriched with ionic groups within an ionomer matrix.

TERMINOLOGY

19 ionic polymer
    ion-containing polymer
Polymer composed of macromolecules containing ionic or ionizable groups, or both, irrespective of their nature, content, and location.

20 ionomer
Polymer composed of macromolecules in which a small but significant proportion of the constitutionalunits has ionic or ionizable groups, or both.
Note 1: Definition same as definition 1.66, ref. [4].
Note 2: Ionic groups are usually present in sufficient amounts (typically less than 10 % of constitutional units) to cause micro-phase separation of ionic domains from the continuous polymer phase. The ionic domains act as physical crosslinks.

21 ionomer cluster
Ionic aggregate, in a polymer matrix of low polarity, formed through interactions of ionomer multiplets.
Note 1: The mobility of the polymer segments surrounding the multiplets is reduced relative to that of bulk material. With increasing ion content, the number density of the ionomer multiplets increases, leading to overlapping of the restricted mobility regions around the multiplets and the formation of clusters.
Note 2: Typically, an ionomer exhibits two glass transition temperatures (T_g), one for the nonpolar matrix and the other for clusters.

22 ionomer multiplet
Ionic aggregate, in a polymer matrix of low polarity, formed through the association of ion pairs in an ionomer.

23 polyacid
In polymer terminology, polyelectrolyte composed of macromolecules containing acid groups on a substantial fraction of the constitutional units.
Note: Most commonly, the acid groups are $-COOH$, $-SO_3H$, or $-PO_3H_2$.

24 polyampholyte
See *ampholytic polymer*.

25 polybase
In polymer terminology, polyelectrolyte composed of macromolecules containing basic groups on a substantial fraction of the constitutional units.
Note: Most commonly, the basic groups are amino groups.

26 polybetaine
Ampholytic polymer in which pendant groups have a betaine-type structure.
Note 1: For the definition of a betaine-type structure, see ref. [3].
Note 2: A *polybetaine* is a type of *zwitterionic polymer*.

27 polyelectrolyte
    polymer electrolyte
    polymeric electrolyte
Polymer composed of macromolecules in which a substantial portion of the constitutional units contains ionic or ionizable groups, or both.

POLYMERS CONTAINING IONIZABLE OR IONIC GROUPS

Note 1: Definition 24 is consistent with definition 1.65, ref. [4] and supersedes the definition given in [3].
Note 2: The terms polyelectrolyte, polymer electrolyte, and *polymeric electrolyte* should not be confused with the term *solid polymer electrolyte*.
Note 3: Polyelectrolytes can be either synthetic or natural. Nucleic acids, proteins, teichoic acids, some polypeptides, and some polysaccharides are examples of natural polyelectrolytes.

28   polyelectrolyte complex
Neutral polymer-polymer complex composed of macromolecules carrying charges of opposite sign causing the macromolecules to be bound together by electrostatic interactions.
Note: A *polyelectrolyte complex* is also called a polysalt. Use of this term is not recommended.

29   polyelectrolyte network
Polymer network containing ionic or ionizable groups in a substantial fraction of its constitutional units.
Note 1: A polyelectrolyte network is sometimes called a cross-linked polyelectrolyte. Use of the latter term is not recommended unless the polyelectrolyte network is formed by the cross-linking of existing polyelectrolyte macromolecules rather than by nonlinear polymerization. (See the definition of a crosslink, definition 1.59, ref. [4].)
Note 2: In contrast to a *polyelectrolyte*, a polyelectrolyte network is always insoluble, although swelling or contraction can occur when it is immersed in a solvent.
Note 3: A polyelectrolyte network in contact with a solution of a salt is able to exchange counterions (cations or anions) with ionic species in the solution and act as an ion exchanger. Therefore, a polelectrolyte network is frequently described as an ion-exchange polymer.

30   polymer electrolyte
See *polyelectrolyte*.

31   polymeric electrolyte
See *polyelectrolyte*.

32   polymeric inner salt
See *zwitterionic polymer*.

33   solid polymer electrolyte
Electrically conducting solution of a salt in a polymer.
Note 1: An example of a *solid polymer electrolyte* is a solution of a lithium salt in a poly(oxyethylene) matrix; the ionic conductivity of such material is due to the mobility of lithium cations and their counterions in an electric field.
Note 2: Although the adjective 'solid' is used, the material may be a liquid.
Note 3: The term *solid polymer electrolyte* should not be confused with the term *polymeric electrolyte*.
Note 4: See also *conducting polymer composite*.

TERMINOLOGY

**34   zwitterionic polymer
      polymeric inner salt**

Ampholytic polymer containing ionic groups of opposite sign, commonly on the same pendant groups.

REFERENCES

1. IUPAC. 'Definitions of terms relating to reactions of polymers and to functional polymeric materials', *Pure Appl. Chem.* **76**, 889 (2004). Reprinted as Chapter 12 this volume.
2. M. S. Alger (Ed.). *Polymer Science Dictionary*, Elsevier Applied Science, London (1990).
3. IUPAC. *Compendium of Chemical Terminology* (the 'Gold Book'), 2nd ed., compiled by A. D. McNaught, A. Wilkinson, Blackwell Science, Oxford (1997 XML on-line corrected version: http://goldbook.iupac.org (2006-) created by M. Nic, J. Jirat, B. Kosata; updates compiled by A. Jenkins.
4. IUPAC. 'Glossary of basic terms in polymer science', *Pure Appl. Chem.* **68**, 2287 (1996). Reprinted as Chapter 1 this volume.

11: Definitions of terms relating to the structure and processing of sols, gels, networks and inorganic-organic hybrid materials

CONTENTS

1. Introduction
2. Precursors
3. Gels
4. Solids
5. Processes
6. References
7. Alphabetical Index of Terms

1 INTRODUCTION

This document provides definitions of the terms most commonly used in relation to sol-gel processing and ceramization. It embraces all categories of materials and their processing. The definitions result from the efforts of a working party drawn from the membership of the IUPAC Polymer and Inorganic Chemistry Divisions.

As depicted in Figure 1, the various terms relating both to materials and processing can be described within a grid that correlates type of material (precursor, gel, or solid) with class of material (inorganic, hybrid, or polymeric). Accordingly, the terms in this document are grouped first by type (the columns in Figure 1) and then by type-to-type conversion, i.e., process (the rows in Figure 1). Where it is necessary to indicate or differentiate between classes of material (inorganic, hybrid and polymer), this has been done within the definition of the relevant term.

For ease of reference, the terms in each section, sub-section, etc. are listed alphabetically and numbered sequentially. Cross references to terms defined elsewhere in the document are denoted in italic typeface. If there are two terms in an entry on successive lines, the second is a synonym.

For the present document, the terms describing the structure and behaviour of networks have been restricted to those that are commonly used, rely on clearly defined theoretical concepts and are unambiguous in their meanings. This area of terminology will be dealt with in more detail in a later document.

Originally prepared by a working group consisting of J. Alemán (Spain), A. V. Chadwick (UK), J. He (China), M. Hess (Germany), K. Horie (Japan), R.G.Jones (UK), P. Kratochvíl (Czech Republic), I. Meisel (Germany), I. Mita (Japan), G. Moad (Australia), S. Penczek (Poland) and R. F. T. Stepto (UK). Reprinted from *Pure Appl. Chem.*, **79**, 1801-1829 (2007)

FIGURE 1. Block diagram depicting the categories of materials and processes addressed by the terms within this document.

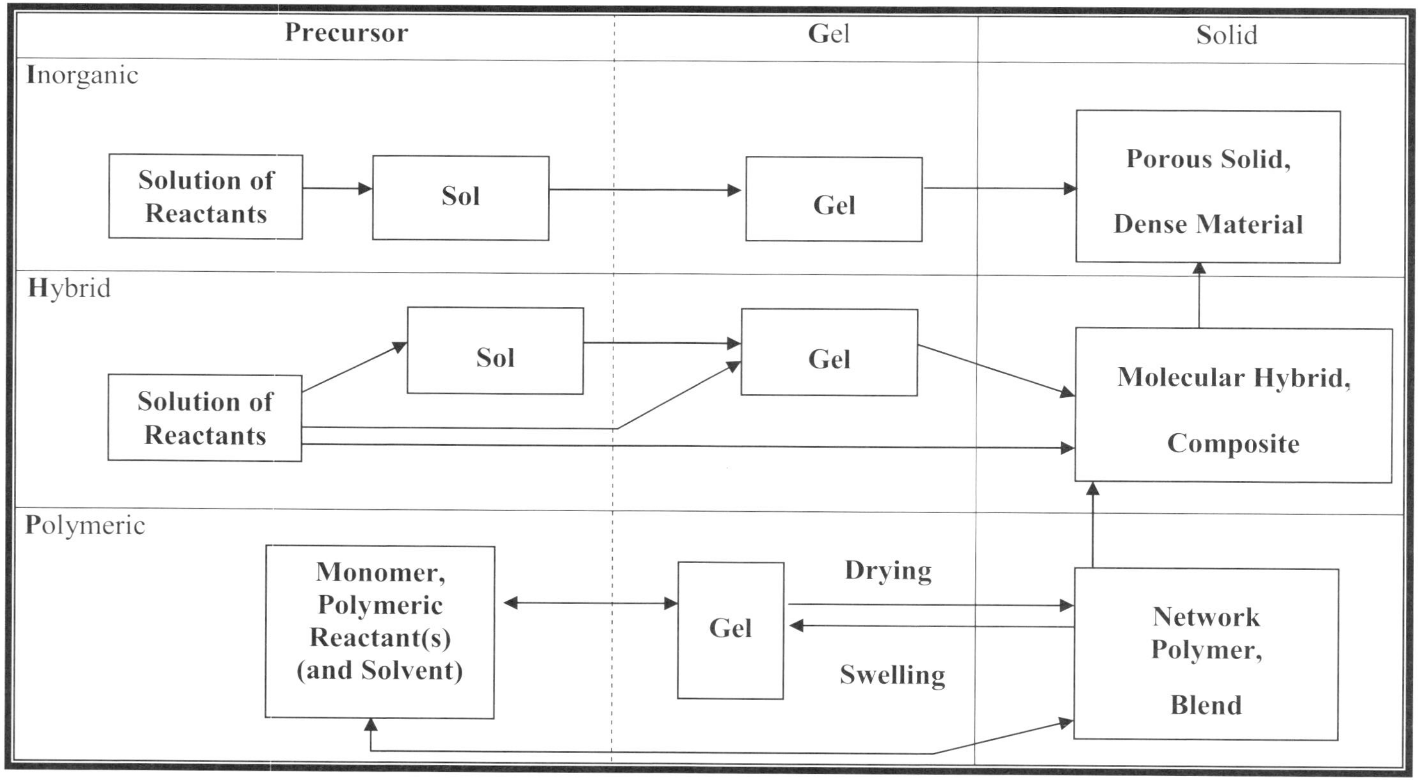

2 PRECURSORS

2.1 agglomerate (except in polymer science)
Cluster of primary particles held together by physical interactions.
Note 1: A primary particle is the smallest discrete identifiable entity observable by a specified identification technique, e.g., transmission electron microscopy, scanning electron microscopy, etc.
Note 2: The particles that comprise agglomerates can usually be readily dispersed.
See the Note of **2.3**.

2.2 agglomerate (in polymer science)
  aggregate (in polymer science)
Cluster of molecules or particles that results from *agglomeration*.
See Definition 1.42 in [3]

2.3 aggregate (except in polymer science)
Cluster of primary particles interconnected by chemical bonds.
Note: Alternative definitions of aggregate and agglomerate are used in catalysis [2]. The distinction offered by these definitions is in conflict with the distinction understood in the wider context and with the concepts of aggregation and agglomeration. To avoid confusion the definitions proposed here are recommended.

2.4 aggregate (in polymer science)
See *agglomerate (in polymer science)*.

2.5 chemical functionality
Ability of functional groups present within a polymer or *polymer network* to participate in chemical reactions.
Note: The chemical functionality of a network formed by a *sol-gel process* from a precursor such as $(RO)_3Si–CH=CH_2$ is that of the vinyl group.

2.6 colloid
A short synonym for a *colloidal* system. (Gold Book online, 1972 entry [2].)

2.7 colloidal
State of subdivision, implying that the molecules or polymolecular particles dispersed in a medium have at least in one direction a dimension roughly between 1 nm and 1 μm, or that in a system discontinuities are found at distances of that order. (Gold Book online, 1972 entry [2].)

2.8 colloidal dispersion
System in which particles of *colloidal* size of any nature (e.g., solid, liquid or gas) are dispersed in a continuous phase of a different composition (or state). (Gold Book online, 1972 entry [2].)
Note: The name dispersed phase for the particles should be used only if they have essentially the properties of a bulk phase of the same composition.

2.9 colloidal sol
See *sol*.

TERMINOLOGY

2.10 colloidal suspension
Suspension in which the size of the particles lies in the *colloidal* range. (Gold Book online, 1972 entry [2].)
Note: A colloidal suspension is *colloidal dispersion* of a solid in a liquid.

2.11 connectivity
Number of covalent bonds emanating from a constitutional unit of an oligomer molecule or a macromolecule.
Note: The definition within [2] is compatible with this definition but is too general to be readily understood in the present context.

2.12 functionality (of a monomer), f
Number of covalent bonds that a monomer molecule or monomeric unit (see Definition 1.8 in [1]) in a macromolecule or oligomer molecule can form with other reactants.
Note 1: There are no monofunctional monomers.
Note 2: If $f = 2$, a linear chain macromolecule or a macrocycle (see Definition 1.57 in [1]) can be formed.
Note 3: If $f > 2$, a branch point (see definition 1.54 in [1]) can be formed leading to a branched macromolecule, a *network* or a *micronetwork*.
Note 4: Ethene and ethylene glycol[*] are examples of difunctional monomers, glycerol[†] is an example of a trifunctional monomer, and divinylbenzene and pentaerythritol[‡] are examples of tetrafunctional monomers.

2.13 pre-gel regime
Stage of a network-forming polymerization or crosslinking process extending up to, but not beyond, the *gel point*.
Note: The pre-gel regime may be expressed as the length of time or the chemical conversion required to reach the *gel point* from the start of a polymerization or *crosslinking* process.

2.14 pre-gel state
State of a network-forming polymerization or crosslinking process in the *pre-gel regime*.
Note: In the pre-gel state, the *sol fraction* is equal to unity. All the molecules formed have finite (statistically definable) relative molecular masses.

2.15 slip
Ceramic precursor dispersed in a liquid.

2.16 sol
    colloidal sol
Fluid *colloidal* system of two or more components. (Gold Book online, 1972 entry [2].)
Note: Examples of *colloidal* sols are protein sols, gold sols, emulsions and surfactant solutions above their critical micelle concentrations.

2.16.1 aerosol
Sol in which the dispersed phase is a solid, a liquid or a mixture of both and the continuous phase is a gas (usually air).

---
[*] ethane-1,2-diol, [†] 2-hydroxypropane-1,3-diol , [‡] 2,2-bis(hydroxymethyl)propan-1,3-diol

Note 1: Owing to their size, the particles of the dispersed phase have a comparatively small settling velocity and hence exhibit some degree of stability in the earth's gravitational field.
Note 2: An aerosol can be characterized by its chemical composition, its radioactivity (if any), the particle size distribution, the electrical charge and the optical properties.
Modified from [2], within which particles with equivalent diameters usually between 0.01 μm and 100 μm are specified. This extends beyond the size range specified for a *colloidal* system. To avoid confusion the definition proposed here is recommended.

2.16.2   particulate sol
Sol in which the dispersed phase consists of solid particles.

2.16.3   polymeric sol
Sol in which the dispersed phase consists of particles having a polymeric structure.

2.16.4   sonosol
Sol produced by the action of ultrasonically induced cavitation.

2.17   sol fraction
Mass fraction of the dissolved or dispersed material resulting from a *network*-forming polymerization or *crosslinking* process that is constituted of molecules of finite (statistically definable) relative molecular masses.
See also *gel fraction*.

3   GELS

3.1   gel
Non-fluid *colloidal network* or *polymer network* that is expanded throughout its whole volume by a fluid.
Note 1: A gel has a finite, usually rather small, yield stress,
Note 2: A gel can contain:
(i)   a covalent *polymer network*, e.g., a network formed by crosslinking polymer chains or by non-linear polymerization;
(ii) a *polymer network* formed through the physical *aggregation* of polymer chains, caused by hydrogen bonds, crystallization, helix formation, complexation, *etc*, that results in regions of local order acting as the network *junction points*. The resulting swollen network may be termed a *thermoreversible gel* if the regions of local order are thermally reversible;
(iii) a polymer network formed through glassy *junction points*, e.g., one based on block copolymers. If the *junction points* are thermally reversible glassy domains, the resulting swollen network may also be termed a *thermoreversible gel*;
(iv) lamellar structures including mesophases {[2] defines *lamellar crystal* and *mesophase*}, e.g., soap gels, phospholipids and clays;
(v)   particulate disordered structures, e.g., a flocculent precipitate usually consisting of particles with large geometrical anisotropy, such as in V_2O_5 gels and globular or fibrillar protein gels.
Corrected from [2], where the definition is *via* the property identified in Note 1 (above) rather than of the structural characteristics that describe a gel.

TERMINOLOGY

3.1.1 aerogel
Gel comprised of a microporous solid in which the dispersed phase is a gas.
Note: Microporous silica, microporous glass and zeolites are common examples of aerogels.
Corrected from [2], where the definition is a repetition of the incorrect definition of a gel followed by an inexplicit reference to the porosity of the structure.

3.1.2 alcogel
Gel in which the swelling agent consists predominantly of an alcohol or a mixture of alcohols.

3.1.3 aquagel
Hydrogel in which the *network* component is a *colloidal network*.

3.1.4 colloidal gel
Gel in which the network component comprises particles of colloidal dimensions.
See also *colloidal network*.

3.1.5 gel microparticle
See *microgel*

3.1.6 gel nanoparticle
See *nanogel*

3.1.7 humming gel
See *ringing gel*

3.1.8 hydrogel
Gel in which the swelling agent is water.
Note 1: The *network* component of a hydrogel is usually a *polymer network*.
Note 2: A hydrogel in which the *network* component is a *colloidal network* may be referred to as an *aquagel*.

3.1.9 microgel
      gel microparticle
Particle of *gel* of any shape with an equivalent diameter of approximately 0.1 to 100 μm.
Modified from [2]. The definition proposed here is recommended for its precision and because it distinguishes between a microgel and a *nanogel*.

3.1.10 nanogel
      gel nanoparticle
Particle of *gel* of any shape with an equivalent diameter of approximately 1 to 100 nm.

3.1.11 neutralized gel
Gel containing acidic or basic groups that have been neutralized.

3.1.12 particulate gel
Gel in which the network component comprises solid particles.

SOLS, GELS, NETWORKS AND HYBRIDS

3.1.13 polyelectrolyte gel
Polymer gel in which the *polymer network* contains ionic or ionizable groups in a significant fraction of its constitutional units.

3.1.14 polymer gel
Gel in which the *network* component is a polymer.

3.1.15 responsive gel
Gel that responds to external electrical, mechanical, thermal, light-induced or chemical stimulation
Note: The use of the term *intelligent gel* is discouraged.

3.1.16 rheopexic gel
rheotropic gel
Gel for which the time of solidification after discontinuation of a relatively high shear rate, is reduced by applying a small shear rate.

3.1.17 rheotropic gel
See *rheopexic gel*.

3.1.18 ringing gel
humming gel
Gel with energy dissipation in the acoustic frequency range.
Note: A ringing gel is often a *hydrogel* with a surfactant as a third component and has a composition within an isotropic, one-phase region of its ternary phase diagram

3.1.19 sonogel
Colloidal gel produced by the action of ultrasonically induced cavitation.

3.1.20 thermoreversible gel
Swollen *network* in which the *junction points* are thermally reversible.
See also *gel*.

3.1.21 thixotropic gel
Gel which has a reduced viscosity on the application of a finite shear but which recovers its original viscosity when the shear is discontinued.

3.1.22 xerogel
Open *network* formed by the removal of all *swelling agents* from a *gel*.
Note: Examples of xerogels include silica gel and dried out, compact macromolecular structures such as gelatin or rubber.
Modified from [2]. The definition proposed here is recommended as being more explicit.
See also *swelling*.

3.2 drying control chemical additive, (DCCA)
Co-solvent included to facilitate the rapid drying of *gels* without cracking.

TERMINOLOGY

3.3 gel point
gelation point
Point of incipient *network* formation in a process forming a chemical or physical *polymer network*.

Note 1: In both *network*-forming polymerization and the *crosslinking* of polymer chains, the gel point is expressed as an extent of chemical reaction (*c.f., gel time*).

Note 2: At the gel point a solid (*network*) material spanning the entire system is formed. See also *gel fraction*.

Note 3: The gel point is often detected using rheological methods. Different methods can give different gel points because viscosity is tending to infinity at the gel point and a unique value cannot be measured directly.

3.4 gel temperature
See *gelation temperature*.

3.5 gel time
gelation time
Time interval from the start of a *network*-forming process to the *gel point*.

3.6 gelation point
See *gel point*.

3.7 gelation temperature
gel temperature
Temperature threshold for the formation of a *thermoreversible gel*.

Note 1: A *thermoreversible gel* is usually formed by cooling a polymer solution. In these cases, the gel temperature is a maximum temperature at which the presence of *network* is observed.

Note 2: Since gel temperatures depend on the method of determination, this should always be indicated.

3.8 gelation time
See *gel time*.

3.9 swelling agent
Fluid used to swell a *gel*, *network* or solid.
See also *swelling*.

4 SOLIDS

4.1 Terms Describing Materials

4.1.1 ceramer
Chemically bonded hybrid material which is a crosslinked *organic-inorganic polymer*.

Note: Ceramers are usually prepared by *sol-gel processing* of oligomers or polymers with reactive silyloxy substituents.

SOLS, GELS, NETWORKS AND HYBRIDS

4.1.2 ceramic
Rigid material that consists of an infinite three-dimensional *network* of sintered crystalline grains comprising metals bonded to carbon, nitrogen or oxygen.
Note: The term ceramic generally applies to any class of inorganic, non-metallic product subjected to high temperature during manufacture or use.

4.1.3 ceramic precursor
  pre-ceramic
  pre-ceramic material
Material that is converted to a ceramic through *pyrolysis*.
Note: Examples include poly(dimethylsilanediyl), poly(carbasilane)s, poly(silazane)s, etc.

4.1.4 ceramic-reinforced polymer
Polymer composite consisting of a polymer continuous phase and disperse phase domains of microscopic ceramic particles.
See also Definition 3.2 in [3].

4.1.5 ceramic yield
Mass of ceramic expressed as a percentage of the mass of the *ceramic precursor* used in the *ceramization* process.

4.1.5.1 theoretical ceramic yield
Ceramic yield based on the stoichiometry of the *ceramization* process.

4.1.6 composite
Multicomponent material comprising multiple, different (non-gaseous) phase domains in which at least one type of phase domain is a continuous phase.[3]
Note: A foamed substance, which is a multiphase material that consists of a gas dispersed in a liquid or solid, is not normally considered to be a composite.

4.1.7 creep
Either
Time-dependent change of the dimensions of a material under a constant load.
or
Change in strain with time after the instantaneous application of a constant stress.

4.1.8 elastomer
Polymer that displays rubber-like elasticity.

4.1.8.1 thermoplastic elastomer
Either
Elastomer comprising a *thermoreversible network*.
or
Melt-processable *polymer blend* or copolymer in which a continuous elastomeric *phase domain* is reinforced by dispersed hard (glassy or crystalline) *phase domains* that act as junction points over a limited range of temperature.

4.1.9 fractal agglomerate
Agglomerate having the same *fractal dimension* as the constituent particles.

4.1.10 fractal dimension, d
mass fractal dimension
Hausdorff dimension

Parameter that provides a mathematical description of the fractal structure of a *polymer network*, an aggregated *particulate sol*, or of the particles that comprise them.

Note 1: $m \propto r^d$ in which m is the mass contained within a radius, r, measured from any site or bond within a fractal structure.

Note 2: For a Euclidean object of constant density, $d = 3$, but for a fractal object, $d < 3$, such that its density decreases as the object gets larger.

Note 3: For the surface area of a fractal object, $s \propto r^{d'}$ in which s is the surface area contained within a radius, r, measured from any site or bond and d' is termed the surface fractal dimension.

4.1.11 gel fraction

Mass fraction of the *network* material resulting from a network-forming polymerization or *crosslinking* process.

Note: The gel fraction comprises a single molecule spanning the entire volume of the material sample.

See also *sol fraction*.

4.1.12 green body

Object formed from a *preceramic* material prior to *pyrolysis*.

4.1.13 Hausdorff dimension

See *fractal dimension*.

4.1.14 hybrid material

Material composed of an intimate mixture of inorganic components, organic components or both types of component.

Note: The components usually interpenetrate on scales of less than 1 µm.

4.1.14.1 chemically bonded hybrid (material)

Hybrid material in which the different components are bonded to each other by covalent or partially covalent bonds.

4.1.14.2 clay hybrid
polymer-clay hybrid
polymer-clay composite

Organic-inorganic composite material in which one of the components is a clay, the particles of which are dispersed in a polymer.

4.1.14.3 hybrid polymer

Polymer or *polymer network* comprised of inorganic and organic components.

Note: Examples include *inorganic-organic polymers* and *organic-inorganic polymers*.

4.1.14.4 polymer-clay composite

See *clay hybrid*

SOLS, GELS, NETWORKS AND HYBRIDS

4.1.14.5 polymer-clay hybrid
See *clay hybrid*

4.1.15 inorganic-organic polymer, (IOP)
Polymer or *polymer network* with a skeletal structure comprising inorganic and organic units. [4]
Note 1: Examples include poly(carbasilane)s, poly(phenylenesilanediyl)s, poly(sulfanylphenylene)s, etc.
Note 2: c.f. 4.1.22.

4.1.15.1 inorganic polymer
Polymer or *polymer network* with a skeletal structure that does not include carbon atoms.
Note: Examples include polyphosphazenes, polysilicates, polysiloxanes, polysilanes, polysilazanes, polygermanes, polysulfides, etc.

4.1.16 mass fractal dimension
See *fractal dimension*

4.1.17 mixed ceramic
Ceramic material consisting of co-continuous interpenetrating networks of two or more metal carbides, nitrides or oxides.

4.1.18 monolith
Shaped, fabricated, intractable article with a homogeneous microstructure which does not exhibit any structural components distinguishable by optical microscopy.
Note: The article is usually fabricated by cold pressing or hot pressing of a polymeric material, or by using a reactive processing technique such as *reaction injection moulding, crosslinking, sol-gel processing, sintering,* etc.

4.1.19 multiphase copolymer
Copolymer comprising phase-separated domains.
See Definition 3.3 in [3].

4.1.20 nanocomposite
Composite in which at least one of the phase domains has at least one dimension of the order of nanometres.
Corrected from Definition 1.15 in [3], which refers to phases instead of phase domains.

4.1.21 network
Highly ramified structure in which essentially each constitutional unit is connected to each other constitutional unit and to the macroscopic phase boundary by many paths through the structure, the number of such paths increasing with the average number of intervening constitutional units; the paths must on average be co-extensive with the structure.
Note: Usually, and in all systems that exhibit rubber elasticity, the number of distinct paths is very high, but, in most cases, some constitutional units exist that are connected by a single path only.
Modified from [2]. The definition proposed here is a generalization to cover both polymeric networks and networks comprised of particles.

TERMINOLOGY

4.1.21.1 colloidal network
Network comprising particles of colloidal dimensions.

4.1.21.2 network polymer
See *polymer network*

4.1.21.3 polymer network
          network polymer
Polymer composed of one or more *networks*. (Gold Book online, 1996 entry [2].)

4.1.21.3.1 bimodal network
              bimodal polymer network
Polymer network comprising polymer chains having two significantly different molar-mass distributions between adjacent *junction points*.

4.1.21.3.2 bimodal polymer network
See *bimodal network*.

4.1.21.3.3 covalent network
              covalent polymer network
Network in which the permanent paths through the structure are all formed by covalent bonds.
Modification of the entry given as a note within the definition of network (in polymer chemistry) in [2].

4.1.21.3.4 covalent polymer network
See *covalent network*.

4.1.21.3.5 entanglement network
Polymer network with junction points or zones formed by physically entangled chains.
See also *physical network* and *chain entanglement*.

4.1.21.3.6 interpenetrating polymer network, (IPN)
Polymer comprising two or more *networks* that are at least partially interlaced on a molecular scale but not covalently bonded to each other and cannot be separated unless chemical bonds are broken.
Note: A mixture of two or more preformed polymer *networks* is not an IPN.

4.1.21.3.6.1 sequential interpenetrating polymer network
Interpenetrating polymer network prepared by a process in which the second component *network* is formed following the formation of the first component *network*.

4.1.21.3.6.2 simultaneous interpenetrating polymer network
Interpenetrating polymer network prepared by a process in which the component *networks* are formed concurrently.

4.1.21.3.7 micronetwork
Polymer network that has dimensions of the order of 1 nm to 1 µm.

Modified from [2]. The definition proposed here is recommended as being more explicit.

4.1.21.3.8 model network
Polymer network synthesized using a reactant or reactants of known molar mass or masses and chemical structure.
Note 1: A model network can be prepared using a non-linear polymerization or by crosslinking of existing polymer chains.
Note 2: A model network is not necessarily a *perfect network*. If a non-linear polymerization is used to prepare the *network*, non-stoichiometric amounts of reactants or incomplete reaction can lead to network containing *loose ends*. If the *crosslinking* of existing polymer chains is used to prepare the *network*, then two *loose ends* per existing polymer chain result. In the absence of chain entanglements, *loose ends* can never be elastically active network chains.
Note 3: In addition to *loose ends*, model networks usually contain ring structures as *network* imperfections.
Note 4: *Loose ends* and ring structures reduce the concentration of *elastically active network chains* and result in the shear modulus and Young's modulus of the rubbery *networks* being less than the values expected for a *perfect network* structure.
Note 5: Physical entanglements between *network* chains can lead to an increase in the concentration of *elastically active network chains* and, hence, increases in the shear modulus and the Young's modulus above the values expected for a *perfect network* structure.

4.1.21.3.9 oxide network
Network comprising only metal-oxygen linkages.

4.1.21.3.10 perfect network
perfect polymer network
Polymer network composed of chains all of which are connected at both of their ends to different junction points.
Note: If a perfect *network* is in the rubbery state then, on macroscopic deformation of the *network*, all of its chains are elastically active and display rubber elasticity.

4.1.21.3.11 perfect polymer network
See *perfect network*

4.1.21.3.12 physical network
Polymer network with *junction points* or zones formed by physically interacting chains which need not be permanent.
Note 1: The *junction points* or zones need not be permanent over the time scale of the observation or measurement.
Note 2: The interaction can be due to hydrogen bonds, π-π interactions, chain entanglements, etc.
Modification of the entry given as a note within the definition of network (in polymer chemistry) in [2].

4.1.21.3.13 reversible network
Polymer network that forms or breaks up as the temperature is changed or under the action of a force.

TERMINOLOGY

Note: The *junction points* in a reversible network are usually small crystallites or glassy domains such as those formed within block copolymers.

4.1.21.3.13.1 thermoreversible network
Reversible network that forms or breaks up as the temperature is changed.

4.1.21.3.14 semi-interpenetrating polymer network, (SIPN)
Polymer comprising one or more *polymer networks* and one or more linear or branched polymers characterized by the penetration on a molecular scale of at least one of the *networks* by at least some of the linear or branched macromolecules.
Note: A SIPN is distinguished from an *IPN* because the constituent linear or branched macromolecules can, in principle, be separated from the constituent *polymer network(s)* without breaking chemical bonds; it is a *polymer blend*.

4.1.21.3.14.1 sequential semi-interpenetrating polymer network
Semi-interpenetrating polymer network prepared by a process in which the linear or branched components are formed following the completion of the reactions that lead to the formation of the *network(s)* or *vice versa*.

4.1.21.3.14.2 simultaneous semi-interpenetrating polymer network
Semi-interpenetrating polymer network prepared by a process in which the *networks* and the linear or branched components are formed concurrently.

4.1.21.3.15 transient network
Network that exists only transiently.
Note: The *network* structure of a transient *polymer network* is based on transient *junction points* or *crosslinks* arising from interactions between polymer chains.

4.1.22 organic-inorganic polymer, (OIP)
Polymer or *polymer network* with a skeletal structure comprised only of carbon but which has side groups that include inorganic components [4].
Note: c.f. 4.1.15.

4.1.23 organically-modified ceramic
organomodified ceramic
Chemically bonded hybrid material which is a crosslinked *inorganic-organic polymer*.
Note 1: Organically modified ceramics are hybrid polymers with inorganic and organic moieties linked by stable covalent bonds and based on organically modified alkoxysilanes, functionalized organic polymers or both.
Note 2: Though it is a commonly used acronym for organically modified ceramic, 'Ormocer' is a registered trademark and as such its terminological use is strongly discouraged.

4.1.24 organically modified silica (silicate)
Silica modified by organic groups.
Note 1: Organically modified silicas can be obtained by *sol-gel processing*.
Note 2: An organically modified silica is of general structure $(RO)_a Si(B)_b (C)_c (D)_d$, where $(a + b + c + d) = 4$, R is any alkyl, aryl or heteroaryl group and B, C and D are generally organic groups.

SOLS, GELS, NETWORKS AND HYBRIDS

Note 3: Though it is a commonly used acronym for organically modified silica, 'Ormosil' is a trademark and as such its terminological use is discouraged.

4.1.25 organomodified ceramic
See *organically modified ceramic*.

4.1.26 polymer alloy
Polymeric material, exhibiting macroscopically uniform physical properties throughout its whole volume, that comprises a compatible *polymer blend*, a miscible *polymer blend* or a multiphase copolymer.
See Definition 1.38 in [3].

4.1.27 polymer blend
Macroscopically homogeneous mixture of two or more different species of polymer. (Gold Book online, 1997 entry [2].)
Note 1: In most cases, blends are homogeneous on scales larger than several times the wavelengths of visible light.
Note 2: In principle, the constituents of a blend are separable by physical means.
Note 3: No account is taken of the miscibility or immiscibility of the constituent macromolecules, i.e., no assumption is made regarding the number of phase domains present.
Note 4: The use of the term *polymer alloy* for a polymer blend is discouraged, as the former term includes multiphase copolymers but excludes incompatible polymer blends.
Note 5: The number of polymeric components which comprise a blend is often designated by an adjective, *viz.* binary, ternary, quaternary, etc.

4.1.27.1 compatible polymer blend
Immiscible *polymer blend* that exhibits macroscopically uniform physical properties.
Note: The macroscopically uniform properties are usually caused by sufficiently strong interactions between the component polymers.

4.1.27.2 homogeneous polymer blend
          miscible polymer blend
Polymer blend that is a single-phase structure.
Note 1: For a *polymer blend* to be miscible it must obey the thermodynamic criteria of miscibility.
Note 2: Miscibility is sometimes assigned erroneously on the basis that a blend exhibits a single T_g or is optically clear.
Note 3: The miscible system can be thermodynamically stable or metastable.
Note 4: For components of chain structures that would be expected to be miscible, miscibility may not occur if molecular architecture is changed, e.g., by *crosslinking*.
Modified from Definition 1.3 in [3]. The definition proposed here is preferred because it emphasizes the requirement for homogeneity over miscibility.

4.1.27.3 miscible polymer blend
See *homogeneous polymer blend*

4.1.28 polymer-derived ceramic, (PDC)
Ceramic derived from a polymeric *ceramic precursor*.

TERMINOLOGY

4.1.29 pre-ceramic
          pre-ceramic material
See *ceramic precursor*

4.1.30 pre-ceramic material
See *ceramic precursor*

4.1.31 sol-gel material
Material formed through a *sol-gel process*.

4.1.31.1 sol-gel coating
Coating formed through a *sol-gel process*.

4.1.31.2 sol-gel metal oxide
Metal oxide formed through a *sol-gel process*.

4.1.31.3 sol-gel silica
Silica formed through a *sol-gel process*.

4.1.32 surface fractal dimension
See *fractal dimension*.

4.2 Terms Describing the Molecular Structure and Behaviour of Networks

4.2.1 affine chain behaviour
Behaviour of a *polymer network* in which the junction points deform uniformly with the macroscopic deformation of the *network*.
Note: In reality, affine chain behaviour can apply only at small deformations.

4.2.2 branch point
Point on a polymer chain at which a branch is attached. (Definition 1.54 in [1] and Gold Book online, 1996 entry [2].)
Note: The Gold Book entry has notes which define an *f-functional branch point* and a *junction point*, both of which are explicitly defined in the present document.

4.2.2.1 *f*-functional branch point
Branch point from which f linear chains emanate.
Note 1: Examples are three-, four- and five-functional, branch points.
Note 2: Alternatively the terms trifunctional, tetrafunctional, pentafunctional, etc. may be used. See also *functionality*.

4.2.3 chain entanglement
Interlocking of polymer chains in a polymer material forming a transient or permanent network junction over the time-scale of the measurement.

4.2.3.1 bowtie entanglement
          butterfly entanglement
Chain entanglement with topology similar to that of a bowtie.

4.2.3.2 butterfly entanglement
See *bowtie entanglement*.

4.2.4 crosslink
Small region in a macromolecule from which at least four chains emanate and which is formed by reactions involving sites or groups on existing macromolecules or by interactions between existing macromolecules. (Definition 1.59 in [1] and Gold Book online, 1996 entry [2].)
Note 1: The small region may be an atom, a group of atoms, a *branch point* or a number *of branch points* connected by bonds, groups of atoms or oligomeric chains.
Note 2: In the majority of cases, a crosslink is a covalent structure but the term is also used to describe a region of weaker chemical interaction, portions of crystallites or even physical interactions and entanglements.
See also *crosslinking*.

4.2.4.1 permanent crosslink
Crosslink formed by covalent bonds, intermolecular or intramolecular interactions that are stable under the conditions of use of the material formed.

4.2.4.2 transient crosslink
Crosslink formed by intermolecular or intramolecular interactions that are unstable under the conditions of use of the material formed.

4.2.5 crosslink density
Number of *crosslinks* per unit volume in a *polymer network*.
See also *junction-point density*.

4.2.6 crosslinking site
Site on a macromolecule or region in a polymer material that takes part in the formation of chemical or physical *crosslinks*.

4.2.7 elastically active network chain
Segment of a chain between two successive *crosslinks* in a *polymer network* that is long enough to show entropic elasticity.

4.2.8 inter-junction molar mass
See *network-chain molar mass*

4.2.9 junction point
Branch point in a *polymer network*.

4.2.9.1 thermoreversible junction point
Junction point in a *polymer network* that can be destroyed and formed reversibly by a change of temperature.

4.2.9.2 transient junction point
Junction point in a *polymer network* that exists only for a finite period of time.
See also *crosslink*, Note 2.

4.2.10 junction-point density
Number of *junction points* per unit volume in a *polymer network*.
See also *crosslink density*.

TERMINOLOGY

4.2.11 loose end
Polymer chain within a *network* which is connected by a *junction-point* at one end only.
Modified from [2].

4.2.12 network-chain molar mass, M_c, SI unit: kg mol^{-1}.
    inter-junction molar mass
Number-average molar mass of polymer chains between two adjacent *crosslinks* or *junction points* in a *polymer network*.

4.2.13 network defect
Elastically ineffective chains in a *polymer network*.
Note: A network defect is caused by a *loose end* or a cyclic structure.

4.2.14 phantom chain behaviour
Hypothetical behaviour in which chains can move freely through one another when a *network* is deformed.

5 PROCESSES

5.1 aerosol hydrolysis
Hydrolysis of the dispersed component of an *aerosol*.

5.2 agglomeration (except in polymer science)
    coagulation
    flocculation
Process of contact and adhesion whereby dispersed particles are held together by weak physical interactions ultimately leading to phase separation by the formation of precipitates of larger than colloidal size.
Note: Agglomeration is a reversible process.
Modified from [2]. The definition proposed here is recommended for distinguishing agglomeration from *aggregation*.

5.3 agglomeration (in polymer science)
    aggregation (in polymer science)
Process in which dispersed molecules or particles assemble rather than remain as isolated single molecules or particles.
See Definition 1.41 in [3].

5.4 aggregation (except in polymer science)
Process whereby dispersed molecules or particles form *aggregates*.

5.5 aggregation (in polymer science)
See *agglomeration (in polymer science)*.

5.6 calcination
Heating to high temperatures in air or oxygen.
Note 1: The term is most likely to be applied to a step in the preparation of a catalyst.
Note 2: In *sol-gel processing* the term applies to the heating of a *polymer network* containing metal compounds to convert it into an *oxide network*.

SOLS, GELS, NETWORKS AND HYBRIDS

Modified from [2]. The definition proposed here is more explicit about the elevated temperatures that are required.

5.7 carbo-reduction
Process in which a metal oxide is reduced in the presence of carbon or a carbon-containing compound.

5.8 ceramization
Process in which a *ceramic precursor* is converted into a *ceramic*.

5.9 coagulation
See *agglomeration*

5.10 colloidal processing
Sol-gel processing in which a *network* of precipitated colloidal particles is treated by a conventional processing technique, such as cold pressing, hot pressing or *sintering*, in order to produce a ceramic article.

5.11 critical concentration
See *sol-gel critical concentration*

5.12 crosslinking
Reaction involving sites or groups on existing macromolecules or an interaction between existing macromolecules that results in the formation of a small region in a macromolecule from which at least four chains emanate. [6]
Note 1: The small region may be an atom, a group of atoms, or a number of *branch points* connected by *bonds*, groups of atoms, or oligomeric chains.
Note 2: A reaction of a reactive chain end of a linear macromolecule with an internal reactive site of another linear macromolecule results in the formation of a *branch point*, but is not regarded as a crosslinking reaction.
See also *crosslink*.

5.13 curing
Chemical process of converting a prepolymer or a polymer into a polymer of higher molar mass and then into a *network*.
Note 1: Curing is achieved by the induction of chemical reactions which might or might not require mixing with a chemical curing agent.
Note 2: Physical aging, crystallization, physical crosslinking and postpolymerization reactions are sometimes referred to as 'curing'. Use of the term 'curing' to describe such processes is deprecated.
See also Definition 1.4 in [6] and *vulcanization*.

5.13.1 EB curing
See *electron beam curing*.

5.13.2 electron-beam curing
EB curing
Curing induced by electron beam irradiation.

5.13.3 photochemical curing
photo-curing
Curing induced by photo-irradiation.

5.13.4 photo-curing
See *photochemical curing.*

5.13.5 thermal curing
Curing induced by heating.

5.14 deflocculation
See *peptization*

5.15 densification
Removal of impurities and the elimination of pores from a *xerogel* to give a material of as near bulk density as possible.

5.16 exfoliation
Process by which the layers of a multi-layered structure separate.

5.17 flocculation
See *agglomeration*

5.18 gel aging
Time-dependent changes in the chemical or the physical structure and the properties of a *gel*.
Note: The aging of a *gel* can involve polymerization, crystallization, *aggregation*, *syneresis*, phase changes, formation of *branch points* and *junction points* as well as scission and chemical changes to constitutional units of network chains.

5.19 gelation
Process of passing through the *gel point* to form a *gel* or *network*.

5.20 hipping
hot isostatic pressing
Isostatic pressing process carried out at elevated temperatures.
Note 1: The pressurizing fluid used in this process is usually a gas.
Note 2: The temperature is usually in excess of 600 °C.

5.21 hot isostatic pressing
See *hipping*

5.22 hydrolysis ratio, r_w
Mole ratio of water to alkoxy groups used in *sol-gel processing* of metal alkoxides.

5.23 insertion reaction
intercalation reaction
Reaction, generally reversible, that involves the penetration of a host material by guest species without causing a major structural modification of the host. (Gold Book online, 1994 entry [2].)

SOLS, GELS, NETWORKS AND HYBRIDS

Note 1: Intercalation can refer to the insertion of a guest species into a one-, two- or three-dimensional host structure.
Note 2: The guest species is not distributed randomly but occupies positions predetermined by structure of the host material.
Note 3: Examples of intercalation reactions are the insertion of lithium into layered TiS_2 (Li_xTiS_2 ($0 \leq x \leq 1$)) and of potassium into the layers of graphite (C_8K)

5.24 *in-situ* composite formation
Process for preparing a polymer *composite* by (a) forming the filler or reinforcement in an existing polymer or (b) polymerizing monomers in the presence of dispersed filler.

5.25 isostatic pressing
Application of a hydrostatic pressure through a liquid to achieve densification prior to the production of a uniform compact *monolith* through *ceramization* of the densified liquid.

5.26 net shaping
Production of an object in, or as close as possible to, its final shape prior to ceramization.

5.27 Ostwald ripening
Dissolution of small crystals or *sol* particles and the redeposition of the dissolved species on the surfaces of larger crystals or *sol* particles.
Note: The process occurs because smaller particles have a higher surface energy, hence higher total Gibbs energy, than larger particles, giving rise to an apparent higher solubility.
Modified from [2]. The definition proposed here is recommended for its inclusion of sol particles.

5.28 peptization
deflocculation
Reversal of *coagulation* or *flocculation*, i.e., the dispersion of *agglomerates* to form a colloidally stable suspension or emulsion. [2]

5.29 precipitation
Sedimentation of a solid material (a precipitate) from a liquid solution in which the material is present in amounts greater than its solubility in the liquid. (Gold Book online, 1990 entry [2].)
Note: When precipitation occurs in *sol-gel processing*, *sol* particles have aggregated to a size where gravitational forces cause them to sink or float. Generally, *aggregation* arises from a change in the *sol* that reduces the interparticle repulsion.

5.30 pyrolysis
Thermolysis, usually associated with exposure to a high temperature.
Note 1: The term generally refers to reaction in an inert environment.
Note 2: Pyrolysis is the commonly used term for a high-temperature treatment that converts a *ceramic precursor* to a *ceramic*.
Modified from [2]. The definition proposed here is more explicit about the elevated temperatures involved.

5.31 reaction injection moulding, (RIM)
Reactive polymer processing that produces polymer *monoliths* by low-pressure injection and mixing of low viscosity precursors into moulds.

TERMINOLOGY

Note: Reaction injection moulding commonly uses two-component precursors that produce *polymer networks* after mixing.
See also reaction blending, Definition 1.19 in [6]

5.31.1 reinforced reaction injection moulding, (RRIM)
Reaction injection moulding within which glass fibres are included to increase the strength of the moulding.

5.32 reactive polymer processing
Process whereby a polymeric *monolith* is produced through an *in situ* polymerization or polymer modification reaction.
Note 1: The polymerization or modification reaction and the transformation of the resulting polymer into a shaped product is accomplished in the same processing equipment.
Note 2: This type of processing is commonly accomplished by extrusion or injection molding.
Note 3: *Reaction injection moulding* and *reinforced reaction injection moulding* are types of reactive polymer processing.

5.33 sedimentation (in chemistry)
Separation of a dispersed system under the action of a gravitational or centrifugal field according to the different densities of the components.

5.34 shrinkage
Decrease in volume of a *network*, *gel* or solid associated with the exudation of a fluid.

5.35 sintering
Temperature induced coalescence and *densification* of porous solid particles below the melting points of their major components.
Note: The term was originally coined for the process by which fly ash produced in combustion of fuels such as coal is baked at a very high temperature. The sintered material is used in the manufacture of cinder blocks and other ceramic products.
Modified from [2]. The definition proposed here is recommended as being more explicit.

5.36 slip casting
Procedure in *ceramic* processing whereby *slip* is contained in a porous plaster mould prior to *pyrolysis*.

5.37 sol-gel critical concentration
critical concentration
Concentration of an added electrolyte above which a *particulate sol* undergoes *coagulation* instead of *gelation*.

5.38 sol-gel process
Process through which a *network* is formed from solution by a progressive change of liquid precursor(s) into a *sol*, to a *gel*, and in most cases finally to a dry *network*.
Note: An *inorganic polymer*, e.g., silica gel, or an organic-inorganic hybrid can be prepared by sol-gel processing.

SOLS, GELS, NETWORKS AND HYBRIDS

5.39 sol-gel transition
Transition of a *sol* to a *gel* at the *gel point*.
Corrected from [2], within which the definition improperly attempts a redefinition of the terms *sol* and *gel*. The definition proposed here is recommended for its precision through cross-reference to the properly defined terms.

5.40 supercritical drying of a gel
Drying of a *gel* using a supercritical fluid.
Note: Since liquid and vapour are indistinguishable in a supercritical fluid, there is no capillary pressure to cause shrinkage and cracking of the pores formed in the *gel*.

5.41 swelling
Increase in volume of a *gel* or solid associated with the uptake of a liquid or gas. (Gold Book online, entry 1972 [2].)

5.42 syneresis
Spontaneous shrinking of a *gel* with exudation of liquid. (Gold Book online, 1972 entry [2].)
Note: Bond formation or attraction between particles or network chains within a *gel* induces the contraction and thereby the exudation of liquid from the *network*.

5.42.1 microsyneresis
Syneresis in which the exudation of the liquid is from microscopic regions within a *network*.

5.43 thermolysis
Uncatalysed cleavage of one or more covalent bonds resulting from exposure of a compound to a raised temperature, or a process in which such cleavage is an essential part. (Gold Book online, 1994 entry [2].) See also *pyrolysis*.

5.44 uniaxial pressing
Application of pressure in one direction during *ceramization* to achieve a uniform densification and the production of a compact *monolith*.

5.45 viscous flow sintering
See *viscous sintering*

5.46 viscous sintering
viscous flow sintering
Sintering process by which it is possible to *densify gels* to glasses and *ceramics* at elevated temperatures.

5.47 vulcanization
Chemical *crosslinking* of high molar-mass linear or branched polymers to give a *polymer network*.
Note 1: The *polymer network* formed often displays rubberlike elasticity. However, a high concentration of *crosslinks* can lead to rigid materials.
Note 2: A classic example of vulcanization is the crosslinking of *cis*-polyisoprene through sulfide bridges in the thermal treatment of natural rubber with sulfur or a sulfur-containing compound.

TERMINOLOGY

6 REFERENCES

1. IUPAC. 'Glossary of Basic Terms in Polymer Science (IUPAC Recommendations 1996)', A. D. Jenkins, P. Kratochvíl, R. F. T. Stepto and U. W. Suter. *Pure Appl. Chem.* **68**, 2287-2311 (1996). Reprinted as Chapter 1 this volume.
2. IUPAC. Compendium of Chemical Terminology, 2nd ed. (the 'Gold Book'). Compiled by A. D. McNaught and A.Wilkinson. Blackwell Scientific Publications, Oxford (1997). XML on-line corrected version: http://goldbook.iupac.org (2006-) created by M. Nic, J. Jirat, B. Kosata; updates compiled by A. Jenkins.
3. IUPAC. 'Definitions of Terms Related to Polymer Blends, Composites and Multiphase Polymeric Materials (IUPAC Recommendations 2004)', W. J. Work, K. Horie, M. Hess and R. F. T. Stepto, *Pure Appl. Chem.* **76**, 1985-2007 (2004). Reprinted as Chapter 9 this volume.
4. 'Hybrid Inorganic/Organic Polymers with Nanoscale Building Blocks: Precursors, Processing, Properties and Applications', K.-H. Haas and K. Rose, *Rev. Adv. Mater. Sci.* **5**, 47-52 (2003).
5. IUPAC. 'Source-Based Nomenclature for Non-Linear Macromolecules and Macromolecular Assemblies (IUPAC Recommendations 1997)', J. Kahovec, P. Kratochvíl, A. D. Jenkins, I. Mita, I. M. Papisov, L. H. Sperling and R. F. T. Stepto, *Pure Appl. Chem.* **69**, 2511-2521 (1997). Reprinted as Chapter 21 this volume.
6. IUPAC. 'Definitions of Terms Relating to Reactions of Polymers and to Functional Polymeric Materials (IUPAC Recommendations 2003)', K. Horie, M. Barón, R. B. Fox, J. He, M. Hess, J. Kahovec, T. Kitayama, P. Kubisa, E. Maréchal, W. Mormann, R. F. T. Stepto, D. Tabak, J. Vohlidal, E.S.Wilks and W. J. Work, *Pure Appl. Chem.* **76**, 889-906 (2004). Reprinted as Chapter 12 this volume.

ALPHABETICAL INDEX OF TERMS

aerogel	3.1.1
aerosol	2.16.1
aerosol hydrolysis	5.1
affine chain behaviour	4.2.1
agglomerate	2.1, 2.2
agglomeration	5.2, 5.3
aggregate	2.3, 2.4
aggregation	5.4, 5.5
alcogel	3.1.2
aquagel	3.1.3
bimodal network	4.1.21.3.1
bimodal polymer network	4.1.21.3.2
bowtie entanglement	4.2.3.1
branch point	4.2.2
butterfly entanglement	4.2.3.2
calcinations (in polymer networks)	5.6
carbo-reduction	5.7
ceramer	4.1.1
ceramic	4.1.2
ceramic precursor	4.1.3
ceramic-reinforced polymer	4.1.4
ceramic yield	4.1.5
ceramization	5.8
chain entanglement	4.2.3
chemical functionality	2.5
chemically bonded hybrid (material)	4.1.14.1
clay hybrid	4.1.14.2
coagulation	5.9
colloid	2,6
colloidal	2,7
colloidal dispersion	2.8
colloidal gel	3.1.4
colloidal network	4.1.21.1
colloidal processing	5.10
colloidal sol	2.9
colloidal suspension	2.10
compatible polymer blend	4.1.27.1
composite	4.1.6
connectivity	2.11
covalent network	4.1.21.3.3
covalent polymer network	4.1.21.3.4

creep	4.1.7	insertion reaction.	5.23
critical concentration	5.11	*in-situ* composite formation	5.24
crosslink	4.2.4	intercalation reaction	5.23
crosslink density	4.2.5	inter-junction molar mass	4.2.8
crosslinking	5.12	interpenetrating polymer network	4.1.21.3.6
crosslinking site	4.2.6	isostatic pressing	5.25
curing	5.13	junction point	4.2.9
deflocculation	5.14	junction point density	4.2.10
densification	5.15	loose end	4.2.11
drying control chemical additive	3.2	mass fractal dimension	4.1.16
EB curing	5.13.1	microgel	3.1.9
elastically active network chain	4.2.7	micronetwork	4.1.21.3.7
		microsyneresis	5.42.1
elastomer	4.1.8	miscible polymer blend	4.1.27.3
electron-beam curing	5.13.2	mixed ceramic	4.1.17
entanglement network	4.1.21.3.5	model network	4.1.21.3.8
exfoliation	5.16	monolith	4.1.18
f-functional branch point	4.2.2.1	multiphase copolymer	4.1.19
flocculation	5.17	nanocomposite	4.1.20
fractal agglomerate	4.1.9	nanogel	3.1.10
fractal dimension	4.1.10	net shaping	5.26
functionality (of a monomer)	2.12	network.	4.1.21
		network-chain molar mass	4.2.12
GELS	**3**	network defect	4.2.13
gel	3.1	network polymer	4.1.21.2
gel aging	5.18	neutralized gel	3.1.11
gel fraction	4.1.11	organic-inorganic polymer	4.1.22
gel microparticle	3.1.5	organically modified ceramic	4.1.23
gel nanoparticle	3.1.6	organically modified silicate	4.1.24
gel point	3.3		
gel temperature	3.4	organomodified ceramic	4.1.25
gel time	3.5	Ormocer	4.1.23
gelation 5.19		Ormosil	4.1.24
gelation point	3.6	Ostwald ripening	5.27
gelation temperature	3.7	oxide network	4.1.21.3.9
gelation time	3.8	particulate gel	3.1.12
green body	4.1.12	particulate sol	2.16.2
Hausdorff dimension	4.1.13	peptization	5.28
hipping	5.20	perfect network	4.1.21.3.10
homogeneous polymer blend	4.1.27.2	perfect polymer network	4.1.21.3.11
hot isostatic pressing	5.21	permanent crosslink	4.2.4.1
humming gel	3.1.7	phantom chain behaviour	4.2.14
hybrid material	4.1.14	photochemical curing	5.13.3
hybrid polymer	4.1.14.3	photo-curing	5.13.4
hydrogel	3.1.8	physical network	4.1.21.3.12
hydrolysis ratio	5.22	polyelectrolyte gel	3.1.13
inorganic-organic polymer	4.1.15	polymer alloy	4.1.26
inorganic polymer	4.1.15.1	polymer blend	4.1.27

TERMINOLOGY

polymer-clay composite	4.1.14.4
polymer-clay hybrid	4.1.14.5
polymer derived ceramic	4.1.28
polymer gel	3.1.14
polymer network	4.1.21.3
polymeric sol	2.16.3
pre-ceramic	4.1.29
pre-ceramic material	4.1.30
precipitation	5.29
PRECURSORS	**2**
pre-gel regime	2.13
pre-gel state	2.14
PROCESSES	**5**
pyrolysis	5.30
reaction injection moulding	5.31
reactive polymer processing	5.32
reinforced reaction injection moulding	5.31.1
reversible network	4.1.21.3.13
responsive gel	3.1.15
rheopexic gel	3.1.16
rheotropic gel	3.1.17
ringing gel	3.1.18
sedimentation	5.33
semi-interpenetrating polymer network	4.1.21.3.14
sequential interpenetrating polymer network	4.1.21.3.6.1
sequential semi-interpenetrating polymer network	4.1.21.3.14.1
shrinkage	5.34
simultaneous inter-penetrating polymer network	4.1.21.3.6.2
simultaneous semi-inter-penetrating polymer network	4.1.21.3.14.2
sintering	5.35
slip	2.15
slip casting	5.36
sol	2.16
sol fraction	2.17
sol-gel coating	4.1.31.1
sol-gel critical concentration	5.37
sol-gel material	4.1.31
sol-gel metal oxide	4.1.31.2
sol-gel process	5.38
sol-gel silica	4.1.31.3
sol-gel transition	5.39
SOLIDS	**4**
sonogel	3.1.19
sonosol	2.16.4
supercritical drying of a gel	5.40
surface fractal dimension	4.1.32
swelling	5.41
swelling agent	3.9
syneresis	5.42
Terms Describing Materials	**4.1**
Terms Describing the Molecular Structure and Behaviour of Networks	**4.2**
theoretical ceramic yield	4.1.5.1
thermal curing	5.13.5
thermolysis	5.43
thermoplastic elastomer	4.1.8.1
thermoreversible gel	3.1.20
thermoreversible junction point	4.2.9.1
thermoreversible network	4.1.21.3.13.1
thixotropic gel	3.1.21
transient crosslink	4.2.4.2
transient junction point	4.2.9.2
transient network	4.1.21.3.15
uniaxial pressing	5.44
viscous flow sintering	5.45
viscous sintering	5.46
vulcanization	5.47
xerogel	3.1.2

12: Definitions of terms relating to reactions of polymers and to functional polymeric materials

CONTENTS

Introduction
1. Reactions involving polymers
2. Polymer reactants and reactive polymeric materials
3. Functional polymeric materials
References
Alphabetical index of terms

INTRODUCTION

Chemical reactions of polymers have received much attention during the last two decades. Many fundamentally and industrially important reactive polymers and functional polymers are prepared by the reactions of linear or cross-linked polymers and by the introduction of reactive, catalytically active, or other groups onto polymer chains. Characteristics of polymer reactions may be appreciably different from both reactions of low-molar-mass compounds and polymerization reactions. Basic definitions of polymerization reactions have been included in the Glossary of Basic Terms in Polymer Science published by the IUPAC Commission on Macromolecular Nomenclature and reproduced as Chapter 1 in this book. The Basic Classification and Definitions of Polymerization Reactions and Definitions of Terms Relating to Degradation, Aging, and Related Chemical Transformations of Polymers are also reproduced herein as Chapters 4 and 13 respectively. However, in spite of the growing importance of the field, a clear and uniform terminology covering the field of reactions and the functionalization of polymers has not been presented until now. For example, combinatorial chemistry using reactive polymer beads has become a new field in recent years. The development of a uniform terminology for such multidisciplinary areas can greatly aid communication and avoid confusion.

This document presents clear concepts and definitions of general and specific terms relating to reactions of polymers and functional polymers. The document is divided into three sections. In Section 1, terms relating to reactions of polymers are defined. Names of individual chemical reactions (e.g., chloromethylation) are omitted from this document, even in cases where the reactions are important in the field of polymer reactions, because such names are usually already in widespread use and are well defined in organic

---

Originally prepared by a working group consisting of K. Horie (Japan), M. Barón (Argentina), R. B. Fox (USA), J. He (China), M. Hess (Germany), J. Kahovec (Czech Republic), T. Kitayama (Japan), P. Kubisa (Poland), E. Maréchal (France), W. Mormann (Germany), R. F. T. Stepto (UK), D. Tabak (Brazil), J. Vohlídal (Czech Republic), E. S. Wilks (USA), and W. J. Work (USA). Reprinted from *Pure Appl. Chem.* **76**, 889–906 (2004)

TERMINOLOGY

chemistry and other areas of chemistry [1]. Sections 2 and 3 deal with the terminology of reactive and functional polymers. The term 'functional polymer' has two meanings: (a) a polymer bearing functional groups (such as hydroxy, carboxy, or amino groups) that make the polymer reactive and (b) a polymer performing a specific function for which it is produced and used. The function in the latter case may be either a chemical function such as a specific reactivity or a physical function like electrical conductance. Polymers bearing reactive functional groups are usually regarded as polymers capable of undergoing chemical reactions. Thus, Section 2 deals with polymers and polymeric materials that undergo various kinds of chemical reactions (i.e., show chemical functions). Section 3 deals with terms relating to polymers and polymeric materials exhibiting some specific physical functions. For definitions of some physical functions, see also *Compendium of Chemical Terminology* ('Gold Book') [2].

A functional polymer according to Definition 3.6 of the present document is a polymer that exhibits specified chemical reactivity or has specified physical, biological, pharmacological, or other uses that depend on specific chemical groups. Thus, several terms concerned with properties or the structure of polymers are included in Section 3 whenever they are closely related to specific functions. Terms that are defined implicitly in the notes and related to the main terms are given in bold type.

1 REACTIONS INVOLVING POLYMERS

1.1 chemical amplification
Process consisting of a chemical reaction that generates a species that catalyzes another reaction and also the succeeding catalyzed reaction.
Note 1: Chemical amplification can lead to a change in structure and by consequence to a change in the physical properties of a polymeric material.
Note 2: The term 'chemical amplification' is commonly used in photoresist lithography employing a **photo-acid generator** or **photo-base generator**.
Note 3: An example of chemical amplification is the transformation of [(*tert*-butoxy-carbonyl)oxy]phenyl groups in polymer chains to hydroxyphenyl groups catalyzed by a photogenerated acid.
Note 4: The term 'amplification reaction' as used in analytical chemistry is defined in [2], p. 21.

1.2 chemical modification
Process by which at least one feature of the chemical constitution of a polymer is changed by chemical reaction(s).
Note: A configurational change (e.g., *cis–trans* isomerization) is not usually referred to as a chemical modification.

1.3 cross-linking
Reaction involving sites or groups on existing macromolecules or an interaction between existing macromolecules that results in the formation of a small region in a macromolecule from which at least four chains emanate.
Note 1: See [2], p. 94 and Definition 1.59 in Chapter 1 for crosslink.
Note 2: The small region may be an atom, a group of atoms, or a number of branch points connected by bonds, groups of atoms, or oligomeric chains.

REACTIONS OF POLYMERS

Note 3: A reaction of a reactive chain end of a linear macromolecule with an internal reactive site of another linear macromolecule results in the formation of a branch point, but is not regarded as a cross-linking reaction.

1.4 curing
Chemical process of converting a prepolymer or a polymer into a polymer of higher molar mass and connectivity and finally into a network.
Note 1: Curing is typically accomplished by chemical reactions induced by heating (**thermal curing**), photo-irradiation (**photo-curing**), or electron-beam irradiation (**EB curing**), or by mixing with a chemical curing agent.
Note 2: Physical aging, crystallization, physical *cross-linking*, and post-polymerization reactions are sometimes referred to as 'curing'. Use of the term 'curing' in these cases is discouraged.
Note 3: See also Definition **1.22**.

1.5 depolymerization
Process of converting a polymer into its monomer or a mixture of monomers (see [2], p. 106 and Definition 3.25 in Chapter 1).

1.6 grafting
Reaction in which one or more species of block are connected to the main chain of a macromolecule as side chains having constitutional or configurational features that differ from those in the main chain.
Note: See [2], p. 175 and Definition 1.28 in Chapter 1 for graft macromolecule.

1.7 interchange reaction
Reaction that results in an exchange of atoms or groups between a polymer and low-molar-mass molecules, between polymer molecules, or between sites within the same macromolecule.
Note: An interchange reaction that occurs with polyesters is called **transesterification**.

1.8 main-chain scission
Chemical reaction that results in the breaking of main-chain bonds of a polymer molecule.
Note 1: See [2], p. 64 and Definition 3.24 in Chapter 1 for chain scission.
Note 2: Some main-chain scissions are classified according to the mechanism of the scission process: **hydrolytic**, **mechanochemical**, **thermal**, **photochemical**, or **oxidative scission**. Others are classified according to their location in the backbone relative to a specific structural feature, for example, α**-scission** (a scission of the C-C bond alpha to the carbon atom of a photo-excited carbonyl group) and β**-scission** (a scission of the C-C bond beta to the carbon atom bearing a radical), etc.

1.9 mechanochemical reaction
Chemical reaction that is induced by the direct absorption of mechanical energy.
Note: Shearing, stretching, and grinding are typical methods for the mechanochemical generation of reactive sites, usually macroradicals, in polymer chains that undergo mechanochemical reactions.

1.10 photochemical reaction
Chemical reaction that is caused by the absorption of ultraviolet, visible, or infrared radiation ([2], p. 302).

TERMINOLOGY

Note 1: Chemical reactions that are induced by a reactive intermediate (e.g., radical, carbene, nitrene, or ionic species) generated from a photo-excited state are sometimes dealt with as a part of photochemistry.
Note 2: An example of a photochemical reaction concerned with polymers is **photopolymerization**.
Note 3: See also Definitions 1.1, 1.18, 3.14, and 3.25.

1.11 polymer complexation
polymer complex formation
Process that results in the formation of a polymer–polymer complex or a complex composed of a polymer and a low-molar-mass substance.

1.12 polymer cyclization
Chemical reaction that leads to the formation of ring structures in or from polymer chains.
Note 1: Examples of cyclization along polymer chains are: (a) cyclization of polyacrylonitrile, (b) acetalization of poly(vinyl alcohol) with an aldehyde, (c) cyclization of polymers of conjugated dienes such as polyisoprene or polybutadiene leading to macrocycles.
Note 2: Examples of cyclization of polymer molecules are: (a) cyclization of poly(dimethylsiloxane), (b) back-biting reaction during ionic polymerizations of heterocyclic monomers.

1.13 polymer degradation
Chemical changes in a polymeric material that usually result in undesirable changes in the in-use properties of the material.
Note 1: In most cases (e.g., in vinyl polymers, polyamides) degradation is accompanied by a decrease in molar mass. In some cases (e.g., in polymers with aromatic rings in the main chain), degradation means changes in chemical structure. It can also be accompanied by *cross-linking*.
Note 2: Usually, degradation results in the loss of, or deterioration in useful properties of the material. However, in the case of **biodegradation** (degradation by biological activity), polymers may change into environmentally acceptable substances with desirable properties (see Definition 3.1
Note 3: See Definition 16 in Chapter 13 for degradation.

1.14 polymer functionalization
Introduction of desired chemical groups into polymer molecules to create specific chemical, physical, biological, pharmacological, or other properties.

1.15 polymer reaction
Chemical reaction in which at least one of the reactants is a high-molar-mass substance.

1.16 polymer-supported reaction
Chemical reaction in which at least one reactant or a catalyst is bound through chemical bonds or weaker interactions such as hydrogen bonds or donor–acceptor interactions to a polymer.
Note 1: The easy separation of low-molar-mass reactants or products from the polymer-supported species is a great advantage of polymer-supported reactions.
Note 2: Typical examples of polymer-supported reactions are: (a) reactions performed by use of polymer-supported catalysts, (b) solid-phase peptide synthesis, in which

REACTIONS OF POLYMERS

intermediate peptide molecules are chemically bonded to beads of a suitable polymer support.

1.17 protection of a reactive group
Temporary chemical transformation of a reactive group into a group that does not react under conditions where the nonprotected group reacts.
Note: For example, **trimethylsilylation** is a typical transformation used to protect reactive functional groups such as hydroxy or amino groups from their reaction with growing anionic species in anionic polymerization.

1.18 radiation reaction
Chemical reaction that is induced by ionizing radiation with γ-ray, X-ray, electron, or other high-energy beams.
Note 1: Radiation reactions involving polymers often lead to chain scission and *cross-linking*.
Note 2: A *photochemical reaction* is sometimes regarded as a type of radiation reaction.

1.19 reactive blending
Mixing process that is accompanied by the chemical reaction(s) of components of a polymer mixture.
Note 1: Examples of reactive blending are: (a) blending accompanied by the formation of a polymer-polymer complex, (b) the formation of block or graft copolymers by a combination of radicals formed by the *mechanochemical* scission of polymers during blending.
Note 2: Reactive blending may also be carried out as reactive extrusion or reaction injection molding (RIM).

1.20 sol-gel process
Process through which a *network* is formed from solution by a progressive change of liquid precursor(s) into a *sol*, to a *gel*, and in most cases finally to a dry *network*.
Note: An *inorganic polymer*, e.g., silica gel, or an organic-inorganic hybrid can be prepared by sol-gel processing.

1.21 surface grafting
Process in which a polymer surface is chemically modified by *grafting* or by the generation of active sites that can lead to the initiation of a graft polymerization.
Note 1: Peroxidation, ozonolysis, high-energy irradiation, and plasma etching are methods of generating active sites on a polymer surface.
Note 2: See also Definition 1.6.

1.22 vulcanization
Chemical *crosslinking* of high-molar-mass linear or branched polymer or polymers to give a polymer network.
Note 1: The polymer network formed often displays rubberlike elasticity. However, a high concentration of cross-links can lead to rigid materials.
Note 2: A classic example of vulcanization is the *cross-linking* of *cis*-polyisoprene through sulfide bridges in the thermal treatment of natural rubber with sulfur or a sulfur-containing compound.

TERMINOLOGY

2 POLYMER REACTANTS AND REACTIVE POLYMERIC MATERIALS

2.1 chelating polymer
Polymer containing ligand groups capable of forming bonds (or other attractive interactions) between two or more separate binding sites within the same ligand group and a single metal atom.
Note 1: Chelating polymers mostly act as ion-exchange polymers specific to metal ions that form chelates with chelating ligands of the polymer.
Note 2: See [2], p. 68 for chelation.

2.2 ion-exchange polymer
Polymer that is able to exchange ions (cations or anions) with ionic components in solution.
Note 1: See [2], p. 208 for ion exchange.
Note 2: An ion-exchange polymer in ionized form may also be referred to as a **polyanion** or a **polycation**.
Note 3: Synthetic ion-exchange organic polymers are often network polyelectrolytes.
Note 4: A membrane having ion-exchange groups is called an **ion-exchange membrane**.
Note 5: Use of the term 'ion-exchange resin' for 'ion-exchange polymer' is strongly discouraged.

2.3 living polymer
Polymer with stable, polymerization-active sites formed by a chain polymerization in which irreversible chain transfer and chain termination are absent.
Note: See [2], p. 236 and Definition 3.21 in Chapter 1 for living polymerization.

2.4 macromonomer
Polymer or oligomer whose molecules each have one end-group that acts as a monomer molecule, so that each polymer or oligomer molecule contributes only a single monomer unit to a chain of the product polymer.
Note 1: The homopolymerization or copolymerization of a macromonomer yields a comb or graft polymer.
Note 2: In the present definition, Definition 2.35 in Chapter 1 has been combined with Definition 1.9 in Chapter 1. See also [2], p. 241.
Note 3: Macromonomers are also sometimes referred to as macromers®. The use of the term 'macromer' is strongly discouraged.

2.5 polymer catalyst
Polymer that exhibits catalytic activity.
Note 1: Certain synthetic polymer catalysts can behave like enzymes.
Note 2: Poly(4-vinylpyridine) in its basic form and sulfonated polystyrene in its acid form are examples of polymers that can act as catalysts in some base- and acid-catalyzed reactions, respectively.

2.6 polymer-metal complex
Complex comprising a metal and one or more polymeric ligands.

2.7 polymer phase-transfer catalyst
Polymer that acts as a phase-transfer catalyst and thereby causes a significant enhancement of the rate of a reaction between two reactants located in neighboring phases owing to its

REACTIONS OF POLYMERS

catalysis of the extraction of one of the reactants across the interface to the other phase where the reaction takes place.
Note 1: Polymer phase-transfer catalysts in the form of beads are often referred to as **triphase catalysts** because such catalysts form the third phase of the reaction system.
Note 2: See [2], p. 299 for phase-transfer catalyst.

2.8 polymer-supported catalyst
Catalyst system comprising a polymer support in which catalytically active species are immobilized through chemical bonds or weaker interactions such as hydrogen bonds or donor–acceptor interactions.
Note 1: Polymer-supported catalysts are often based on network polymers in the form of beads. They are easy to separate from reaction media and can be used repeatedly.
Note 2: Examples of polymer-supported catalysts are: (a) a polymer-metal complex that can coordinate reactants, (b) colloidal palladium dispersed in a swollen network polymer that can act as a hydrogenation catalyst.
Note 3: **Polymer-supported enzymes** are a type of polymer-supported catalysts.

2.9 polymer reactant
   polymer reagent
   polymer-supported reagent
Reactant (reagent) that is or is attached to a high-molar-mass linear polymer or a polymer network.
Note: The attachment may be by chemical bonds, by weaker interactions such as hydrogen bonds, or simply by inclusion.

2.10 prepolymer
Polymer or oligomer whose molecules are capable of entering, through reactive groups, into further polymerization and thereby contributing more than one constitutional unit to at least one type of chain of the final polymer.
Note: Definition 2.37 in Chapter 1 has been combined with Definition 1.11 in Chapter 1. See also [2], p. 318.

2.11 reactive polymer
Polymer having reactive functional groups that can undergo chemical transformation under the conditions required for a given reaction or application.

2.12 redox polymer
   electron-exchange polymer
   oxidation-reduction polymer
Polymer containing groups that can be reversibly reduced or oxidized.
Note 1: Reversible redox reaction can take place in a polymer main-chain, as in the case of polyaniline and quinone/hydroquinone polymers, or on side-groups, as in the case of a polymer carrying ferrocene side-groups.
Note 2: See Chapter 7.
Note 3: Use of the term 'redox resin' is strongly discouraged.

2.13 resin
Soft solid or highly viscous substance, usually containing prepolymers with reactive groups.

TERMINOLOGY

Note 1: This term was used originally because of its analogy with a natural resin (rosin) and designated, in a broad sense, any polymer that is a basic material for plastics, organic coatings, or lacquers. However, the term is now used in a more narrow sense to refer to prepolymers of thermosets (thermosetting polymers).

Note 2: The term is sometimes used not only for prepolymers of thermosets, but also for cured thermosets (e.g., epoxy resins, phenolic resins). Use of the term for cured thermosets is strongly discouraged.

Note 3: Use of the term 'resin' to describe the polymer beads used in solid-phase synthesis and as polymer supports, catalysts, reagents, and scavengers is also discouraged.

2.14 telechelic polymer
telechelic oligomer

Prepolymer capable of entering into further polymerization or other reactions through its reactive endgroups.

Note 1: Reactive end-groups in telechelic polymers come from initiator or termination or chain transfer agents in chain polymerizations, but not from monomer(s) as in polycondensations and polyadditions.

Note 2: See [2], p. 414 and the Note to Definition 1.11 in Chapter 1 for telechelic molecule.

2.15 thermosetting polymer

Prepolymer in a soft solid or viscous state that changes irreversibly into an infusible, insoluble polymer network by *curing*.

Note 1: *Curing* can be induced by the action of heat or suitable radiation, or both.

Note 2: A cured thermosetting polymer is called a **thermoset**.

3 FUNCTIONAL POLYMERIC MATERIALS

3.1 biodegradable polymer

Polymer susceptible to degradation by biological activity, with the degradation accompanied by a lowering of its molar mass.

Note 1: See also Note 2 to Definition 1.13.

Note 2: See [2], p. 43 for biodegradation. In the case of a polymer, its biodegradation proceeds not only by catalytic activity of enzymes, but also by a wide variety of biological activities.

3.2 conducting polymer

Polymeric material that exhibits bulk electric conductance.

Note 1: See [2], p. 84 for conductivity.

Note 2: The electric conductivity of a conjugated polymer is markedly increased by doping it with an electron donor or acceptor, as in the case of polyacetylene doped with iodine.

Note 3: A polymer showing a substantial increase in electric conductivity upon irradiation with ultraviolet or visible light is called a **photoconductive polymer**; an example is poly(N-vinylcarbazole) (see [2], p. 302 for photoconductivity).

Note 4: A polymer that shows electric conductivity due to the transport of ionic species is called an **ion-conducting polymer**; an example is sulfonated polyaniline. When the transported ionic species is a proton as, e.g., in the case of fuel cells, it is called a **proton-conducting polymer**.

REACTIONS OF POLYMERS

Note 5: A polymer that shows electric semiconductivity is called a **semiconducting polymer** (See [2], p. 372 for semiconductor).

Note 6: Electric conductance of a nonconducting polymer can be achieved by dispersing conducting particles (e.g., metal, carbon black) in the polymer. The resulting materials are referred to as conducting **polymer composites** or **solid polymer-electrolyte composites**.

3.3 electroluminescent polymer

Polymeric material that shows luminescence when an electric current passes through it such that charge carriers can combine at luminescent sites to give rise to electronically excited states of luminescent groups or molecules.

Note 1: Electroluminescent polymers are often made by incorporating luminescent groups or dyes into conducting polymers.

Note 2: Electrogenerated chemiluminescence (see [2], p. 130) directly connected with electrode reactions may also be called electroluminescence.

3.4 ferroelectric polymer

Polymer in which spontaneous polarization arises when dipoles become arranged parallel to each other by electric fields.

Note 1: See [2], p. 153 for ferroelectric transition.

Note 2: Poly(vinylidene fluoride) after being subjected to a corona discharge is an example of a ferroelectric polymer.

3.5 ferromagnetic polymer

Polymer that exhibits magnetic properties because it has unpaired electron spins aligned parallel to each other or electron spins that can easily be so aligned.

3.6 functional polymer

(a) Polymer that bears specified chemical groups, or (b) Polymer that has specified physical, chemical, biological, pharmacological, or other uses which depend on specific chemical groups.

Note: Examples of functions of functional polymers under definition (b) are catalytic activity, selective binding of particular species, capture and transport of electric charge carriers or energy, conversion of light to charge carriers and vice versa, and transport of drugs to a particular organ in which the drug is released.

3.7 impact-modified polymer

Polymeric material whose impact resistance and toughness have been increased by the incorporation of phase microdomains of a rubbery material.

Note: An example is the incorporation of soft polybutadiene domains into glassy polystyrene to produce high-impact polystyrene.

3.8 liquid-crystalline polymer

Polymeric material that, under suitable conditions of temperature, pressure, and concentration, exists as a liquid-crystalline mesophase (Definition 6.1 in Chapter 7).

Note 1: See Chapter 13, for liquid-crystal.

Note 2: A liquid-crystalline polymer can exhibit one or more liquid state(s) with one- or two-dimensional, long-range orientational order over certain ranges of temperatures either in the melt (**thermotropic liquid-crystalline polymer**) or in solution (**lyotropic liquid-crystalline polymer**).

TERMINOLOGY

3.9   macroporous polymer
Glass or rubbery polymer that includes a large number of macropores (50 nm–1 μm in diameter) that persist when the polymer is immersed in solvents or in the dry state.
Note 1:   Macroporous polymers are often network polymers produced in bead form. However, linear polymers can also be prepared in the form of macroporous polymer beads.
Note 2:   Macroporous polymers swell only slightly in solvents.
Note 3:   Macroporous polymers are used, for example, as precursors for ion-exchange polymers, as adsorbents, as supports for catalysts or reagents, and as stationary phases in size-exclusion chromatography columns.
Note 4:   Porous polymers with pore diameters from ca. 2 to 50 nm are called **mesoporous polymers**.

3.10   nonlinear optical polymer
Polymer that exhibits an optical effect brought about by electromagnetic radiation such that the magnitude of the effect is not proportional to the irradiance.
Note 1:   See [2], p. 275 for nonlinear optical effect.
Note 2:   An example of nonlinear optical effects is the generation of higher harmonics of the incident light wave.
Note 3:   A polymer that exhibits a nonlinear optical effect due to anisotropic electric susceptibilities when subjected to electric field together with light irradiation is called an **electro-optical polymer**. A polymer that exhibits electro-optical behavior combined with photoconductivity is called a **photorefractive polymer**.

3.11   optically active polymer
Polymer capable of rotating the polarization plane of a transmitted beam of linear-polarized light.
Note 1:   See [2], p. 282 for optical activity.
Note 2:   The optical activity originates from the presence of chiral elements in a polymer such as chiral centers or chiral axes due to long-range conformational order in a polymer (helicity) (see [2], p. 182 for helicity).

3.12   photoelastic polymer
Polymer that under stress exhibits birefringence.

3.13   photoluminescent polymer
Polymer that exhibits luminescence (i.e., fluorescence or phosphorescence arising from photoexcitation).
Note: See [2], p. 304 for photoluminescence.

3.14   photosensitive polymer
Polymer that responds to ultraviolet or visible light by exhibiting a change in its physical properties or its chemical constitution.
Note 1:   Examples of the changes in photosensitive polymers are a change in molecular shape (**photoresponsive polymer**), a change in its constitution (**photoreactive polymer**), and a reversible change in color (**photochromic polymer**).
Note 2:   Photosensitivity in photosensitive polymers means that the polymers are sensitive to the irradiated light leading to some change in their properties or structure. It is different from photosensitization defined in [2], p. 307.
Note 3:   See [2], p. 307 for photoreaction and [2], p. 302 for photochromism.

REACTIONS OF POLYMERS

3.15   piezoelectric polymer
(a) Polymer that exhibits a change in dielectric properties on application of pressure, or (b) Polymer that shows a change in its dimensions when subjected to an electric field.

3.16   polyelectrolyte
Polymer composed of molecules in which a portion of the constitutional units has ionizable or ionic groups, or both.
Note 1:   A polymer bearing both anionic and cationic groups in the same molecule is called an **amphoteric polyelectrolyte**.
Note 2:   A polymer bearing acid or basic groups is called a **polymer acid** or a **polymer base**, respectively.
Note 3:   A polymer acid or a polymer base can be used as a matrix for ion-conducting polymers.
Note 4:   Definition 2.38 in Chapter 1 has been combined with Definition 1.65 in Chapter 1. The present definition replaces the one in [2], p. 312.

3.17   polymer compatibilizer
Polymeric additive that, when added to a blend of immiscible polymers, modifies their interfaces and stabilizes the blend.
Note: Typical polymer compatibilizers are block or graft copolymers.

3.18   polymer drug
Polymer that contains either chemically bound drug molecules or pharmacologically active moieties.
Note: A polymer drug is usually used to provide drug delivery targeted to an organ and controlled release of an active drug at the target organ.

3.19   polymer gel
Gel in which the network component is a polymer network.
Note 1:   A gel is an elastic colloid or polymer network that is expanded throughout its whole volume by a fluid.
Note 2:   The polymer network can be a network formed by covalent bonds or by physical aggregation with region of local order acting as network junctions.
Note 3:   An example of covalent polymer gels is *net*-poly(*N*-isopropylacrylamide) swollen in water, which shows volume phase transition during heating.
Note 4:   Examples of physically aggregated polymer gels are poly(vinyl alcohol) gel and agarose gel, which show reversible sol-gel transitions.
Note 5:   See Definition 1.58 in Chapter 1 for network.
Note 6:   The definition for gel in [2], p. 170 does not include a polymer gel.

3.20   polymer membrane
Thin layer of polymeric material that acts as a barrier permitting mass transport of selected species.
Note: See [2], p. 251 for membrane.

3.21   polymer solvent
Polymer that acts like a solvent for compounds of low molar mass.
Note: An example of a polymer solvent is poly(oxyethane-1,2-diyl); it can dissolve various inorganic salts by complexation.

TERMINOLOGY

3.22 polymer sorbent
Polymer that adsorbs or absorbs a certain substance or certain substances from a liquid or a gas.
Note 1: A polymer sorbent may be a **polymer adsorbent** or a **polymer absorbent**. The former acts by surface sorption and the latter by bulk sorption.
Note 2: See [2], p. 383 for sorption, [2], p. 11 for adsorption, and [2], p. 3 for absorption.

3.23 polymer support
Polymer to or in which a reagent or catalyst is chemically bound, immobilized, dispersed, or associated.
Note 1: A polymer support is usually a network polymer.
Note 2: A polymer support is usually prepared in bead form by suspension polymerization.
Note 3: The location of active sites introduced into a polymer support depends on the type of polymer support. In a **swollen-gel-bead polymer support** the active sites are distributed uniformly throughout the beads, whereas in a **macroporous-bead polymer support** they are predominantly on the internal surfaces of the macropores.

3.24 polymer surfactant
Polymer that lowers the surface tension of the medium in which it is dissolved, or the interfacial tension with another phase, or both.
Note: See [2], p. 409 for surfactant.

3.25 resist polymer
Polymeric material that, when irradiated, undergoes a marked change in solubility in a given solvent or is ablated.
Note 1: A resist polymer under irradiation either forms patterns directly or undergoes chemical reactions leading to pattern formation after subsequent processing.
Note 2: A resist material that is optimized for use with ultraviolet or visible light, an electron beam, an ion beam, or X-rays is called a **photoresist** (see [2], p. 307), **electron-beam resist**, **ion-beam resist**, or **X-ray resist**, respectively.
Note 3: In a **positive-tone resist**, also called a **positive resist**, the material in the irradiated area not covered by a mask is removed, which results in an image with a pattern identical with that on the mask. In a **negative-tone resist**, also called a **negative resist**, the non-irradiated area is subsequently removed, which results in an image with a pattern that is the complement of that on the mask.

3.26 shape-memory polymer
Polymer that, after heating and being subjected to a plastic deformation, resumes its original shape when heated above its glass-transition or melting temperature.
Note: Crystalline *trans*-polyisoprene is an example of a shape-memory polymer.

3.27 superabsorbent polymer
Polymer that can absorb and retain extremely large amounts of a liquid relative to its own mass.
Note 1: The liquid absorbed can be water or an organic liquid.
Note 2: The swelling ratio of a superabsorbent polymer can reach the order of 1000:1.
Note 3: Superabsorbent polymers for water are frequently polyelectrolytes.

REFERENCES

1. R. A. Y. Jones and J. F. Bunnett. 'Nomenclature for organic chemical transformations (IUPAC Recommendations 1989)', *Pure Appl. Chem.* **61**, 725–768 (1989).
2. IUPAC. Compendium of Chemical Terminology, 2nd ed. (the 'Gold Book'). Compiled by A. D. McNaught and A.Wilkinson. Blackwell Scientific Publications, Oxford (1997). XML on-line corrected version: http://goldbook.iupac.org (2006-) created by M. Nic, J. Jirat, B. Kosata; updates compiled by A. Jenkins.

ALPHABETICAL INDEX OF TERMS

α-scission	1.8
β-scission	1.8
amphoteric polyelectrolyte	3.16
biodegradable polymer	3.1
biodegradation	1.13
chelating polymer	2.1
chemical amplification	1.1
chemical modification	1.2
conducting polymer	3.2
conducting polymer composite	3.2
cross-linking	1.3
curing	1.4
depolymerization	1.5
EB curing	1.4
electro-optical polymer	3.10
electroluminescent polymer	3.3
electron-beam resist	3.25
electron-exchange polymer	2.12
ferroelectric polymer	3.4
ferromagnetic polymer	3.5
functional polymer	3.6
grafting	1.6
hydrolytic scission	1.8
impact-modified polymer	3.7
interchange reaction	1.7
ion-beam resist	3.25
ion-conducting polymer	3.2
ion-exchange membrane	2.2
ion-exchange polymer	2.2
liquid-crystalline polymer	3.8
living polymer	2.3
lyotropic liquid-crystalline polymer	3.8
macromonomer	2.4
macroporous-bead polymer support	3.23
macroporous polymer	3.9
main-chain scission	1.8
mechanochemical reaction	1.9
mechanochemical scission	1.8
mesoporous polymer	3.9
negative resist	3.25
negative-tone resist	3.25
nonlinear-optical polymer	3.10
optically active polymer	3.11
oxidation-reduction polymer	2.12
oxidative scission	1.8
photo-acid generator	1.1
photo-base generator	1.1
photo-curing	1.4
photochemical reaction	1.10, 1.18
photochemical scission	1.8
photochromic polymer	3.14
photoconductive polymer	3.2
photoelastic polymer	3.12
photoluminescent polymer	3.13
photopolymerization	1.10
photoreactive polymer	3.14
photorefractive polymer	3.10
photoresist	3.25
photoresponsive polymer	3.14
photosensitive polymer	3.14
piezoelectric polymer	3.15
polyanion	2.2
polycation	2.2
polyelectrolyte	3.16
polymer absorbent	3.22
polymer acid	3.16
polymer adsorbent	3.22
polymer base	3.16
polymer catalyst	2.5
polymer compatibilizer	3.17
polymer complex formation	1.11
polymer complexation	1.11
polymer cyclization	1.12

TERMINOLOGY

polymer degradation	1.13	radiation reaction	1.18
polymer drug	3.18	reactive blending	1.19
polymer functionalization	1.14	resist polymer	3.25
polymer gel	3.19	semiconducting polymer	3.2
polymer membrane	3.20	shape-memory polymer	3.26
polymer phase-transfer catalyst	2.7	sol-gel process	1.20
polymer reactant	2.9	solid polymer-electrolyte composite	3.2
polymer reaction	1.15	superabsorbent polymer	3.27
polymer reagent	2.9	surface grafting	1.21
polymer solvent	3.21	swollen-gel-bead polymer support	3.23
polymer sorbent	3.22	telechelic oligomer	2.14
polymer support	3.23	telechelic polymer	2.14
polymer surfactant	3.24	thermal curing	1.4
polymer-metal complex	2.6	thermal scission	1.8
polymer-supported catalyst	2.8	thermoset	2.15
polymer-supported enzyme	2.8	thermosetting polymer	2.15
polymer-supported reaction	1.16	thermotropic liquid-crystalline polymer	3.8
polymer-supported reagent	2.9	transesterification	1.7
positive resist	3.25	trimethylsilylation	1.17
positive-tone resist	3.25	triphase catalyst	2.7
prepolymer	2.10	vulcanization	1.22
protection of a reactive group	1.17	X-ray resist	3.25
proton-conducting polymer	3.2		

13: Definitions of terms relating to degradation, aging, and related chemical transformations of polymers

INTRODUCTION

Chemical reactions of polymers are of great importance. First, chemical modifications of polymers have been used widely to improve their properties. Second, owing to physical and chemical factors such as abrasion, heat, light, radiation, or the action of chemicals or micro-organisms, all polymers tend to undergo spontaneous changes in structure and physico-chemical properties in the course of synthesis, processing and application. These types of transformations are extremely important from economic and environmental points of view, however, their terminology has developed only on an *ad hoc* basis and the development of a self-consistent terminology for the field will greatly aid communication and alleviate confusion.

This document presents an alphabetical list of terms relating to degradation, aging, and related chemical transformations of polymers; an index is appended. The general definitions in this document are often modified by one or more prefixes or adjectives describing type or cause. Thus, definitions of basic terms (such as aging, degradation, and stability) have notes appended relating to specific cases. For example, artificial, cosmic, ground, and radiation aging are covered by notes under the general term 'aging'. Prefix- or adjective-modified terms that are frequently used by themselves are defined separately; biodegradation, photostabilizer, and chain-terminating antioxidant are typical examples.

1   ablation
Removal of surface layers of polymers through heat generated by external forces, such as by the action of high-speed hot-gas flow or a laser beam.

2   ablator
Substance that forms a protective surface on a spacecraft or a missile, and is consumed in an *ablation* process.

3   abrasion
Removal of surface material from a solid, particularly through the frictional action of solids, liquids, or gases.

Originally prepared by a working group consisting of K. Hatada (Japan), R. B. Fox (USA),
J. Kahovec (Czech Republic), E. Maréchal (France), I. Mita (Japan), V. Shibaev (Russia). Reprinted from *Pure Appl. Chem.*, **68**, 2313-2323 (1996)

TERMINOLOGY

4 aging
Processes that occur in a polymeric material during a specified period of time, and that usually result in changes in physical and or chemical structure and the values of the properties of the material.
Note 1: Thermodynamic processes that produce reversible changes in the physical structure of a polymeric material are termed **physical aging**.
Note 2: If desired, the term aging may be qualified. For example, aging by the action of water or aqueous solutions is termed **aqueous aging**; aging by the direct or indirect effect of living organisms is termed **biologically-induced aging**; aging in extraterrestrial space or under conditions that simulate outer-space is termed **cosmic aging**; aging through contact with ground or soil is termed **underground aging**; aging caused by the action of an oxidizing agent, especially oxygen, is termed **oxidative aging**; aging induced by the combined action of light and oxygen is termed **photo-oxidative aging**, by the action of heat and **oxygen thermo-oxidative aging**, by the action of heat alone **thermal aging**, and by the action of visible or ultraviolet light **photochemical aging**.

5 antagonism
Opposing action of two or more agents that results in an effect smaller than would be expected from the individual action of each.
Note: Used especially for *antioxidants* and *stabilizers*.

6 anti-fatigue agent
Agent used to inhibit the *fatigue* of a polymer.

7 antioxidant
Substance that inhibits or retards oxidation.
Note 1: Antioxidants acting under specified conditions may be referred to as thermal antioxidants, photo-antioxidants, mechano-antioxidants, etc.
Note 2: See also *chain-terminating antioxidant*.

8 antiradiant
Additive that protects a polymer against ionizing radiation.

9 ashing
Process in which a polymer is burned to a powdery residue.

10 autoxidation
Oxidation in which the intermediate products increase the rate of reaction.
Note: Autoxidation is usually the result of chain-reaction with air or oxygen, and the intermediate products are usually peroxidic in nature.

11 biodegradation
Degradation of a polymeric material caused, at least in part, by a biological process.
Note 1: See also *degradation*.
Note 2: Usually biodegradation takes place through enzymatic processes resulting from the action of bacteria or fungi.
Note 3: Biodegradation of a polymer is sometimes desirable.

12 carbonization
Transformation of an organic polymer into a material that consists largely of carbon.

DEGRADATION AND AGING

**13   chain-terminating antioxidant
      chain-breaking antioxidant**
Antioxidant capable of interrupting *autoxidation* by reacting with the propagating free radicals to form inactive products or products of reduced activity.

14   cracking
Formation of cracks in a polymeric material.
Note 1: Cracking by the action of chemicals, ozone, oxidizing agent, solvent (liquid), ultraviolet or other electromagnetic radiation is termed **chemical cracking**, **ozone cracking**, **oxidative cracking**, **solvent cracking**, **UV cracking** or **radiation cracking**, respectively.
Note 2: See also *environmental stress cracking*.

15   crazing
Formation of cavities, when a polymer is stressed, which contain load-bearing fibrils spanning the gap between the surfaces of each cavity.
Note: One of the dimensions of a cavity is usually less than a few micrometers.

16   degradation
Chemical changes in a polymeric material that result in undesirable changes in the values of in-use properties of the material.
Note 1: In some cases, degradation is accompanied by a lowering of molecular weight.
Note 2: Causes of degradation may be specified by prefixes or by adjectives preceding the term degradation. For example, degradation caused by exposure to visible or ultraviolet light is termed **photodegradation**; degradation induced by the action of oxygen or by the combined action of light and oxygen is termed **oxidative degradation** or **photo-oxidative degradation**, respectively; degradation induced by the action of heat or by the combined effect of chemical agents and heat is termed **thermal degradation** or **thermochemical degradation**, respectively; degradation induced by the combined action of heat and oxygen is termed **thermo-oxidative degradation**.

17   denaturation
Change in the native conformation of proteins or nucleic acids resulting in loss of their biological activity.
Note: Denaturation is caused by factors such as heating, change in pH, or treatment with chemicals.

18   durability
Ability of a polymeric material to retain the values of its properties under specified conditions.

**19   environmental stress cracking
      stress cracking**
Cracking caused by the combined actions of (i) mechanical stress and (ii) chemical agents or radiation or both.
Note: See also *cracking*.

TERMINOLOGY

20 environmentally degradable polymer
Polymer that can be degraded by the action of the environment, through, for example, air, light, heat, or micro-organisms.
Note 1: The *degradation* of an environmentally degradable polymer after use is sometimes desirable.
Note 2: A **controlled-degradable polymer** is a polymer designed to degrade into products at a predictable rate. Such products are usually of lower molecular weight than the original polymer.

21 fatigue
Process of progressive, localized, permanent, structural change occurring in a material subjected to fluctuating external stimuli.
Note: Mechanical stimuli such as stresses and strains may produce cracks or fracture in a material. Loss of function of a photochromic material resulting from cyclic irradiation may also be considered fatigue.

22 fire retardant
Additive that increases the fire resistance of a material.
Note 1: A fire retardant is sometimes called a **flame retardant**.
Note 2: The ability of a material to resist fire is called **fire retardancy** or **flame retardancy**.

23 heat endurance
Ability of a polymer to retain its function under the application of heat.

24 inhibitor
Substance that stops a chemical reaction.
Note 1: See also *retarder*.
Note 2: In a general chemical terminology, the term inhibitor is defined as a substance that diminishes the rate of a chemical reaction [1].

25 lifetime
Time during which a polymer keeps a fraction of its original property values to such an extent to be useful in an intended application.

26 metal deactivator
Complexing agent that deactivates or reduces the ability of metal ions to initiate or to catalyze the *degradation* of a polymer.

27 peroxidation
Process of the formation of a hydroperoxide or peroxide by oxidation.

28 peroxide decomposer
Agent that transforms peroxides into stable compounds without the formation of free radicals.
Note: A decomposer for hydroperoxides is termed a hydroperoxide decomposer.

29 photosensitizer
Substance that permits or enhances the initiation of a photochemical reaction.
Note: The process involved is called **photosensitization**.

30 photostabilizer
Additive used to protect a polymer from *photodegradation*.
Note: A photostabilizer is also called **light stabilizer** or **photoprotective agent**.

31 pyrolysis
Thermolysis, usually associated with exposure to a high temperature [1].
Note 1: **thermolysis** is defined as the uncatalyzed cleavage of one or more covalent bonds resulting from exposure of the compound to an elevated temperature, or a process in which such cleavage is an essential part [1].
Note 2: Self-sustained pyrolysis in which the reaction is sufficiently supported, once initiated, by the exothermic heat of reaction is termed **auto-pyrolysis**.

32 retarder
Substance that decreases the rate of a reaction.
Note: See also *inhibitor*.

33 sensitizer
Substance that permits or enhances the initiation of chemical change in a polymeric material.

34 stability
Ability of a polymer to maintain the values of its properties over a specified period of time.
Note: Particular types of stability may be specified by adjectives preceding the term 'stability'. For example, the ability of a polymer to resist *biologically-induced aging* or *biodegradation* is termed **biological stability**; the abilities of a polymer to resist the actions of chemicals, light, ionizing radiation, or heat are termed **chemical stability**, **photostability**, **radiation stability** or **thermal stability** (or **thermostability**), respectively; the ability of a polymer to resist oxidation is termed **oxidative stability**; the abilities of a polymer to resist the combined action of light and oxygen or oxygen and heat are termed **photo-oxidative stability** and **thermo-oxidative stability**, respectively.

35 stabilization
Treatment of a polymer to improve its *stability*.
Note 1: Stabilization by introducing certain additives to a polymer or by modifying the chemical structure of polymer molecules may be termed **chemical stabilization**.
Note 2: Stabilization achieved through physical (e.g. mechanical or thermal) treatment may be termed **physical stabilization**.

36 stabilizer
Additive that increases the *stability* of a polymer.
Note: Additives used for specific purposes are termed UV stabilizers, *photostabilizers*, thermal stabilizers, etc.

37 weak link
Chemical bond in the main chain of a polymer molecule that is most susceptible to scission.

TERMINOLOGY

38 wear
Loss or deterioration of a polymer due to continued use, friction or exposure to other natural destructive agencies.
Note: See also *abrasion*. Wear due to frictional action may be termed **abrasive wear**.

39 weathering
Exposure of a polymeric material to a natural or simulated environment.
Note 1: Weathering results in changes in appearance or mechanical properties.
Note 2: Weathering in which the rate of change has been artificially increased is termed **accelerated weathering**.
Weathering in a simulated environment is termed **artificial weathering**.
Note 3: The ability of a polymer to resist weathering is termed **weatherability**.

REFERENCE

1. IUPAC. Compendium of Chemical Terminology, 2nd ed. (the 'Gold Book'). Compiled by A. D. McNaught and A. Wilkinson. Blackwell Scientific Publications, Oxford (1997). XML on-line version: http://goldbook.iupac.org (2006-) created by M. Nic, J. Jirat, B. Kosata; updates compiled by A. Jenkins.

ALPHABETICAL INDEX OF TERMS

ablation	1
ablator	2
abrasion	3
abrasive wear	38
accelerated weathering	39
aging	4
aqueous	4
biologically-induced	4
cosmic	4
oxidative	4
photochemical	4
photo-oxidative	4
physical	4
thermal	4
thermo-oxidative	4
underground	4
antagonism	5
anti-fatigue agent	6
antioxidant	7
chain-breaking	13
chain-terminating	13
antiradiant	8
aqueous aging	4
artificial weathering	39
ashing	9
auto-pyrolysis	31
autoxidation	10
biodegradation	11
biologically-induced aging	4
biological stability	34
carbonization	12
chain-breaking antioxidant	13
chain-terminating antioxidant	13
chemical cracking	14
chemical stability	34
chemical stabilization	35
controlled-degradable polymer	20
cosmic aging	4
cracking	14
chemical	14
environmental stress	19
oxidative	14
ozone	14
radiation	14
solvent	14
UV	14
crazing	15
degradation	16
bio-	11
oxidative	16
photo-	16
photo-oxidative	16

thermal	16	radiation cracking	14
thermochemical	16	radiation stability	34
thermo-oxidative	16	retarder	32
denaturation	17	sensitizer	33
durability	18	photo-	29
environmental stress cracking	19	solvent cracking	14
environmentally degradable polymer	20	stability	34
		biological	34
fatigue	21	chemical	34
fire retardancy	22	oxidative	34
fire retardant	22	photo-	34
flame retardancy	22	photo-oxidative	34
flame retardant	22	radiation	34
heat endurance	23	thermal	34
inhibitor	24	thermo-	34
lifetime	25	thermo-oxidative	34
light stabilizer	3	stabilization	35
metal deactivator	26	chemical	35
oxidative aging	4	physical	35
photo-	4	stabilizer	36
thermo-	4	light	30
oxidative cracking	14	photo-	30
oxidative degradation	16	stress cracking	19
oxidative stability	34	thermal aging	4
photo-	34	thermal degradation	16
thermo-	34	thermal stability	34
ozone cracking	14	thermochemical degradation	16
peroxidation	27	thermolysis	31
peroxide decomposer	28	thermo-oxidative aging	4
photochemical aging	4	thermo-oxidative degradation	16
photodegradation	16	thermo-oxidative stability	34
photo-oxidative aging	4	thermostability	34
photo-oxidative degradation	16	underground aging	4
photo-oxidative stability	34	UV cracking	14
photoprotective agent	30	weak link	37
photosensitization	29	wear	38
photosensitize	29	abrasive	38
photostability	34	weatherability	39
photostabilizer	30	weathering	39
physical aging	4	accelerated	39
physical stabilization	35	artificial	39
pyrolysis	31		
auto-	31		

NOMENCLATURE

The nomenclature developed and recommended by IUPAC has emphasized the generation of unambiguous names that accord with the historical development of chemistry. However, as a consequence of the explosion in the circulation of information and the globalization of human activities, it is now deemed necessary to have a common language that would be useful in legal situations, manifestations in patents, export-import regulations, environmental and health and safety information, etc. Hence, rather than recommending only a single 'unique name' for each structure, rules have been developed for assigning 'preferred IUPAC names', while continuing to allow alternatives in order to preserve the diversity and adaptability of the nomenclature to daily activities in chemistry and in science in general. Thus, the existence of preferred IUPAC names would not prevent the use of other names to take into account a specific context or to emphasize structural features common to a series of compounds. Any name other than a preferred IUPAC name, as long as it is unambiguous and broadly follows the principles of the IUPAC recommendations, is acceptable as an IUPAC name.

The formation of a systematic name for a polymer requires the identification and naming of a preferred constitutional repeating unit (CRU). This basic name is then modified by prefixes, which convey precisely the structural identity of the polymer in question. Such names are referred to as structure-based names. However, polymers can also be named as being derived from a monomer (or precursors), named according to IUPAC rules. Such names are referred to as source-based names. Over the years, rules for determining polymer nomenclature under these two systems have developed in parallel. An example of the modification of the IUPAC name of an organic molecule to IUPAC structure-based and source-based names of a polymer is illustrated below.

Monomer: $CH_3CH_2CH_2CH=CH_2$ Preferred IUPAC name: pent-1-ene

Polymer: $-(CH-CH_2)_n$ Structure-based name: poly(1-propylethane-1,2-diyl)
           $\quad CH_2CH_2CH_3$ Source-based name: poly(pent-1-ene)

In addition to structure-based and source-based names, there are traditional names (or retained names) for polymers which are widely used, particularly in industry but also in academia. When they meet the requirements of utility and when they fit into the general pattern of systematic nomenclature, these traditional names are retained. The following

NOMENCLATURE

table illustrates the distinction between the traditional, structure-based and source-based names of three common examples: polyethylene (PE), polypropylene (PP) and polystyrene (PS).

traditional name	structure-based name	source-based name
polyethylene (PE)	poly(methylene)	polyethene
polypropylene (PP)	poly(1-methylethane-1,2-diyl)	polypropene
polystyrene (PS)	poly(1-phenylethane-1,2-diyl)	poly(ethenylbenzene) or poly(vinylbenzene)

14: Introduction to Polymer Nomenclature

CONTENTS

1. Preamble
   1.1 The principles of source-based nomenclature
   1.2 The principles of structure-based nomenclature
2. IUPAC nomenclature
   2.1 Source-based nomenclature
   2.2 Structure-based nomenclature
3. Abbreviations
4. References

1 PREAMBLE

The aim of this chapter is to introduce and summarize the work on polymer nomenclature which has emanated, firstly, from the Commission on Macromolecular Nomenclature of the IUPAC Macromolecular Division and, latterly, from the Sub-Committee on Polymer Terminology of the IUPAC Macromolecular (now Polymer) Division, jointly with the IUPAC Chemical Nomenclature and Structure Representation Division. The Commission on Macromolecular Nomenclature is henceforth denoted as 'the Commission'.

Two systems of polymer nomenclature have been introduced – the source-based and the structure-based. The latter cannot be used for all types of macromolecule, e.g., statistical copolymer molecules and polymer networks. IUPAC expresses no strong preference for the use of structure-based nomenclature *versus* source-based nomenclature, but for certain purposes one system of naming may be preferred to the other. .

1.1 The Principles of Source-Based Nomenclature

Traditionally, a polymer has been named by attaching the prefix 'poly' to the name of the real or assumed monomer (the 'source') from which it is derived. Thus, 'polystyrene' is the name of the polymer made from styrene. When the name of the monomer comprises two or more words, parentheses should be used [1], as in poly(vinyl acetate), poly(methyl methacrylate), etc. Failure to use parentheses can lead to ambiguity. For example, 'polychlorostyrene' could be the name of either a polychlorinated (monomeric) styrene or of a polymer derived from chlorostyrene; similarly, 'polyethylene oxide' could refer to the polymer **1**, the polymer **2** or the macrocycle **3**.

Prepared by E. S. Wilks. This chapter is based on the article 'Nomenclature' by J. L. Schultz and E. S. Wilks, which appeared in the *Encyclopedia of Polymer Science and Technology*, 3rd ed., John Wiley & Sons, New York (2003), Vol. 7, pp. 262-292. Permission of the publisher is gratefully acknowledged. The generous assistance of N. Bikales is gratefully acknowledged

NOMENCLATURE

$$\text{H}\!\!-\!\!\!\left(\!\text{O}\!-\!\text{CH}_2\!-\!\text{CH}_2\!\right)_{\!n}\!\!-\!\!\text{OH} \qquad \text{H}\!\!-\!\!\!\left(\!\text{CH}_2\!-\!\text{CH}_2\!\right)_{\!n}\!\!-\!\!\text{O}\!\!-\!\!\!\left(\!\text{CH}_2\!-\!\text{CH}_2\!\right)_{\!n}\!\!-\!\!\text{H}$$

1   2   3

These problems are easily overcome with parentheses: names such as 'poly(chlorostyrene)' and 'poly(ethylene oxide)' clearly indicate the part of the name to which the prefix 'poly' refers. However, the omission of parentheses is, unfortunately, still common in the literature.

The principal deficiency of source-based nomenclature is that the chemical structure of the monomeric unit in a polymer is not identical with that of the monomer, e.g., –CHX–CH$_2$– *versus* CHX=CH$_2$. The structure of the constitutional repeating unit (CRU) may also not be clearly identified in this scheme; for example, the name 'polyacrylaldehyde' does not indicate whether (i) the vinyl group or (ii) the aldehyde group was the locus of polymerization.

n H$_2$C=CH–CH=O   ⟶   –(CH$_2$–CH)$_n$– with CHO side group   (i)

                          ⟶   –(O–CH)$_n$– with HC=CH$_2$ side group   (ii)

Depending on the polymerization conditions, different types of polymerization can take place with many other monomers. Furthermore, a name such as poly(vinyl alcohol) refers to a hypothetical source, since this polymer is actually obtained by hydrolysis of poly(vinyl acetate).

Despite these serious deficiencies, source-based nomenclature is still firmly entrenched in the scientific literature. It originated at a time when polymer science was less developed and the structures of most polymers were ill-defined. The significant advances made during the last 50 years in the structure determination of polymers are gradually shifting the emphasis of polymer nomenclature away from starting materials and toward the structure of the synthesized macromolecules.

1.2 The Principles of Structure-Based Nomenclature

Structure-based nomenclature is based on a method of naming the sequence of constitutional or structural units that represent the repeating pattern of the structure of a typical macromolecule in a polymer. The name bears no direct relation to the structure of the (co)monomer(s) used to synthesize the polymer.

2 IUPAC NOMENCLATURE

2.1 Source-Based Nomenclature

2.1.1 Homopolymers

Although source-based names are usually simpler than structure-based names, historically the Commission has not systematically recommended source-based names for homopolymers because they considered that structure-based names were more appropriate for scientific communications [1]. This rationale was based upon two specific points.

(1) A homopolymer that can be represented by a source-based name can usually be given a completely unambiguous structure-based representation, and such a representation should therefore be ideally accompanied by a structure-based name. A source-based name was therefore deemed redundant in these cases.
(2) There are cases in which the simplicity of source-based nomenclature leads to ambiguous names for homopolymers. Two examples illustrate this point.
   (i) A source-based name such as poly(buta-1,3-diene) or polybutadiene does not indicate whether the structure contains 1,2, 1,4-*cis* or 1,4-*trans* units.
   (ii) The polyacrylaldehyde example cited in Section 1.1.

Despite the Commission's long-standing position, the scientific community has continued to use source-based nomenclature for homopolymers such as polystyrene and poly(vinyl acetate) because of their simplicity, convenience and obvious relationship with the monomers from which the homopolymers are prepared. The Commission therefore decided to recommend source-based nomenclature as an alternative official nomenclature for homopolymers in a 2001 publication [2]. Consequently, both source-based and structure-based names are now available for most polymers. The names of monomers in the source-based names may be traditional or semi-systematic, if well established by usage, and not necessarily only those retained in the 1993 *A Guide to IUPAC Nomenclature of Organic Compounds* [3].

The broad principle is to write the source-based name of a homopolymer by combining the prefix 'poly' with the name of the monomer. When the latter consists of more than one word, or any ambiguity is anticipated, the name of the monomer is parenthesized.

The same publication [2] recommended a generic source-based nomenclature, which comprises the optional addition of a polymer class name to the source-based name of the polymer. The addition is recommended when it is necessary to avoid ambiguity or to add clarification.

Example: the source-based name 'poly(vinyloxirane)' is ambiguous; ambiguity about the structure is removed by writing either 'polyalkylene:vinyloxirane' to accompany the source-based representation,

or 'polyether:vinyloxirane' to accompany the source-based representation,

2.1.2 Copolymers
Copolymers are polymers that are derived from more than one species of monomer.

NOMECLATURE

Table 1. IUPAC Nomenclature of Copolymers [4]

Copolymer type	Arrangement of monomeric units	Representation	Connective	Example
unspecified	unknown or unspecified	(A-*co*-B)	-*co*-	poly[styrene-*co*-(methyl methacrylate)]
statistical	obeys known statistical laws	(A-*stat*-B-*stat*-C)	-*stat*-	poly(styrene-*stat*-acrylonitrile-*stat*-butadiene)
random	obeys Bernoullian statistics	(A-*ran*-B)	-*ran*-	poly[ethene-*ran*-(vinyl acetate)]
alternating	alternating	(AB)$_n$	-*alt*-	poly[(ethylene glycol)[a]-*alt*-(terephthalic acid)]
periodic	periodic with respect to at least three monomeric units	(ABC)$_n$ (ABB)$_n$ (AABB)$_n$ (ABAC)$_n$	-*per*-	poly[formaldehyde-*per*-(ethene oxide)-*per*-(ethene oxide)]
block	linear arrangement of blocks	–AAAAA–BBBBB–	-*block*-	polystyrene-*block*-polybutadiene
graft	polymeric side chain different from main chain[b]	–AAAAAAAAAAAA– | B B B B B	-*graft*-	polybutadiene-*graft*-polystyrene

[a] ethane-1,2-diol, [b] main chain (backbone) is specified first in the name.

INTRODUCTION TO POLYMER NOMENCLATURE

Similarly to homopolymers, source-based nomenclature has been applied to copolymers [4]. The principal problem is to define the kind of arrangement in which various types of monomeric units are related to each other. Seven types of separate arrangements have been defined, which are shown in Table 1, where A, B and C represent the names of monomers. The monomer names are linked either through an italicized qualifier or connective (infix), such as '-*co*-', to form the name of the copolymer, as in poly(styrene-*co*-acrylonitrile). The order of citation of the monomers is arbitrary.

The names of copolymers can further be modified to indicate various structural features. For example, the chemical nature of end groups can be specified as α-X-ω-Y-polyA-*block*-polyB, as in α-butyl-ω-carboxy-polystyrene-*block*-polybutadiene.

Subscripts placed immediately after the formula of a monomeric unit or a block designate the degree of polymerization or repetition. On the other hand, mass and mole fractions and relative molecular masses, which in most cases are average quantities, are expressed by placing the corresponding figures after the complete name of the copolymer. The order of citation is as for the monomeric species in the name. Unknown quantities are designated by *a, b*, etc. Table 2 gives some examples.

Table 2. Examples of the IUPAC nomenclature formats for copolymers

Description	Nomenclature Formats
A block copolymer containing 75 mass % of polybutadiene and 25 mass % of polystyrene	polybutadiene-*block*-polystyrene (0.75:0.25 *w*)
A graft copolymer, comprising a polyisoprene backbone grafted with isoprene and acrylonitrile units in an unspecified arrangement that contains 85 mol % of isoprene units and 15 mol % of acrylonitrile units	polyisoprene-*graft*-poly(isoprene-*co*-acrylonitrile) (0.85:0.15 *x*)
A graft copolymer comprising 75 mass % of polybutadiene with a relative molecular mass of 90 000 as the backbone, and 25 mass % of polystyrene in grafted chains with a relative molecular mass of 30 000	polybutadiene-*graft*-polystyrene (75:25 mass %; 90 000:30 000 M_r)
A graft copolymer in which the polybutadiene backbone has a degree of polymerization (DP) of 1 700 and the polystyrene grafts have an unknown DP	polybutadiene-*graft*-polystyrene (1 700:*a* DP)

The complete document [4] should be consulted for names of more complex copolymers, e.g. those having a multiplicity of grafts or having chains radiating from a central atom.

The utility of generic source-based nomenclature [2] can also be illustrated for source-based names of copolymers. As for homopolymers, the addition of a polymer class name is recommended when it is necessary to avoid ambiguity or to add clarification.

NOMENCLATURE

Example: the source-based name poly[(benzene-1,2,4,5-tetracarboxylic 1,2:4,5-dianhydride)-*alt*-(4,4'-oxydianiline)] is ambiguous; ambiguity about the structure is removed by writing poly(amide-acid):[(benzene-1,2,4,5-tetracarboxylic 1,2:4,5-dianhydride)-*alt*-(4,4'-oxydianiline)] to accompany the graphical structure-based representation

and polyimide:[(benzene-1,2,4,5-tetracarboxylic 1,2:4,5-dianhydride)-*alt*-(4,4'-oxydianiline)] to accompany the graphical structure-based representation

In 1997, the principles of source-based nomenclature were extended to non-linear macromolecules and macromolecular assemblies [5], which are classified according to the following terms: cyclic; branched (long- and short-branched); micronetwork; network; blend; semi-interpenetrating polymer network; interpenetrating polymer network; and polymer—polymer complex. The range of italicized qualifiers, listed as connectives in Table 1 for copolymers, was extended by addition of new qualifiers such as '*blend*', '*comb*', '*star*' or '*net*'. Table 3 lists the new qualifiers, together with examples of their use. Quantitative characteristics of a macromolecule or an assembly of macromolecules, such as mass and mole fractions or percentages as well as the degrees of polymerization and relative molecular masses, may be expressed by placing corresponding figures after the complete name. The complete document [5] should be consulted for details and more examples.

2.2 Structure-Based Nomenclature

2.2.1 Regular Single-Strand Organic Polymers

For organic polymers that are regular (i.e. have only one species of constitutional repeating unit (CRU) in a single sequential arrangement) and comprise only single strands, the IUPAC has promulgated a structure-based system of naming polymers [1]. The system consists of naming a polymer as 'poly(constitutional repeating unit)', wherein the repeating unit is named as a divalent organic substituent according to the usual nomenclature rules for organic chemistry. It is important to note that, in structure-based nomenclature, the name of the constitutional repeating unit has no direct relationship to the source from which the unit was prepared. The name is simply that of the identifiable constitutional repeating unit in the polymer, and locants for unsaturation, substituents, etc are dictated by the *structure* of the unit.

The steps involved in naming the CRU are: (1) identification of the unit (taking into account the kinds of atoms in the main chain and the location of substituents); (2) orientation of the unit; and (3) naming of the unit. The CRUs

Table 3. IUPAC Source-Based Nomenclature for Non-Linear Macromolecules and Macromolecular Assemblies [5]

Macromolecule or macromolecular assembly	Qualifier	Example	Comment
cyclic	*cyclo*	*cyclo*-poly(dimethylsiloxane) *cyclo*-poly(styrene-*stat*-α-methylstyrene)	
branched, unspecified	*branch*	*branch*-polystyrene-*v*-divinylbenzene[a]	polystyrene crosslinked with divinylbenzene, insufficient to cause a network to form
short-chain-branched	*sh-branch*	*sh-branch*-polyethene	
long-chain-branched	*l-branch*	*l-branch*-poly(ethyl acrylate)-*v*-(ethane-1,2-diyl dimethacrylate)[a]	
branched with branch points of functionality *f*	*f-branch*[b]		
comb(like)	*comb*	*comb*-poly(styrene-*stat*-acrylonitrile)	both the main chain and side chains are statistical copolymer chains of styrene and acrylonitrile
		polystyrene-*comb*-polyacrylonitrile[c]	main chains of styrene homopolymer; side chains of acrylonitrile homopolymer
		polystyrene-*comb*-[polyacrylonitrile; poly(methyl methacrylate)][c]	main chains of styrene homopolymer; side chains of acrylonitrile homopolymer and methyl methacrylate homopolymer

NOMECLATURE

star	*star*	variegated star copolymer molecule consisting of arm(s) of polyA, arm(s) of polyB and arm(s) of polyC
	star-(polyA; polyB; polyC)	
	star-(polyA-*block*-polyB-*block*-polyC)	star copolymer molecule, each arm of which consists of the same block-copolymer chain
	star-[polystyrene-*block*-poly(methyl methacrylate)]	each arm of the star macromolecule is a block copolymer styrene and methyl methacrylate with a polystyrene block attached to the central unit
star with f arms	f-*star*[b]	
	4-*star*-polystyrene	
	3-*star*-polystyrene-*v*-trichloro(methyl)silane[a]	star macromolecule consisting of arms of polystyrene crosslinked by trichloro(methyl)silane
	star-(polyacrylonitrile; polystyrene) (M_r 100 000:20 000)	star macromolecule consisting of arms of polyacrylonitrile of a total M_r = 100 000 and arms of polystyrene of a total M_r = 20 000
	6-*star*-(polyacrylonitrile (f 3); polystyrene (f 3)) (M_r (arm) 50 000:10 000)	six-armed star macromolecule consisting of three arms of polyacrylonitrile, each of M_r = 50 000, and three arms of polystyrene, each of M_r = 10 000

268

network	*net*	*net*-polybutadiene	network comprising only butadiene constitutional units
		net-poly(phenol-*co*-formaldehyde)	network derived from the random condensation of phenol and formaldehyde molecules
		net-polybutadiene-*v*-sulfur[a]	polybutadiene vulcanized with sulfur
		net-polystyrene-*v*-divinylbenzene	polystyrene crosslinked with divinylbenzene to form a network
		net-poly[(hexane-1,6-diyl diisocyanate)-*alt*-glycerol]	
		net-poly[styrene-*alt*-(maleic anhydride)]-*v*-(ethylene glycol)[a]	alternating copolymer of styrene and maleic anhydride crosslinked with ethane-1,2-diol to form a network
		poly[(ethylene glycol)-*alt*-(maleic anhydride)]-*net*-oligostyrene	unsaturated polyester cured with styrene
micronetwork	μ-*net*	μ-*net*-poly[styrene-*stat*-(vinyl cinnamate)]	crosslinked copolymer micronetwork
polymer blend	*blend*	polystyrene-*blend*-poly(2,6-dimethylphenol)	
		(*net*-polystyrene)-*blend*-(*net*-polybutadiene)	blend of two networks
interpenetrating polymer network (IPN)	*ipn*	(*net*-polybutadiene)-*ipn*-(*net*-polystyrene)	IPN of two networks

NOMECLATURE

		[*net*-poly(styrene-*stat*-butadiene)]-*ipn*-[*net*-poly(ethyl acrylate)]	IPN of two networks
semi-interpenetrating polymer network (SIPN)	*sipn*	(*net*-polystyrene)-*sipn*-poly(vinyl chloride)	SIPN of a polystyrene network and a linear poly(vinyl chloride)
polymer—polymer complex	*compl*	poly(acrylic acid)-*compl*-poly(4-vinylpyridine)	complex of homopolymers
		poly[(methyl methacrylate)-*stat*-(methacrylic acid)]-*compl*-poly(*N*-vinyl-2-pyrrolidone)	complex of a statistical copolymer and a homopolymer

[a] In source-based nomenclature for non-linear macromolecules and macromolecular assemblies, junction units are optionally specified by their source-based names after the name of the macromolecule with the connective (Greek) ν, separated by hyphens [5].
[b] *f* is given a numerical value.
[c] In naming non-linear copolymer molecules having linear chains of two or more types, the italicized connective for the skeletal structure is placed between the source-based names of constituent single strand chains.

INTRODUCTION TO POLYMER NOMENCLATURE

are named systematically according to the IUPAC-recommended nomenclature of organic compounds [3]. Examples of names for some common polymers are given in Table 4. Note that, in this system, parentheses, brackets, braces or a combination of these, are always used to enclose the CRU. For organic names, the IUPAC uses parentheses for the innermost set, then brackets, then braces, thus: $\{[(\ldots)]\}$. If further nesting is required, this cycle is repeated: $(\{[(\ldots)]\})$, $[(\{[(\ldots)]\})]$, $\{[(\{[(\ldots)]\})]\}$, etc.

Table 4. Examples of structure-based and source-based names for some common polymers[a]

Structure	Structure-based name	Source-based name; Traditional name
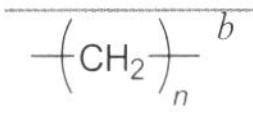 $-(CH_2)_n-$ [b]	poly(methylene)	polyethene; polyethylene[c]
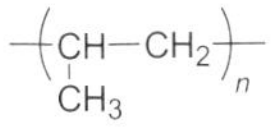 $-(CH(CH_3)-CH_2)_n-$	poly(1-methylethane-1,2-diyl)	polypropene; polypropylene
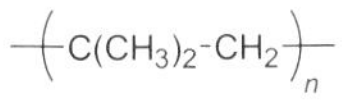 $-(C(CH_3)_2-CH_2)_n-$	poly(1,1-dimethylethane-1,2-diyl)	poly(2-methylpropene); polyisobutylene
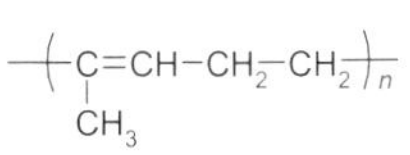 $-(C(CH_3)=CH-CH_2-CH_2)_n-$	poly(1-methylbut-1-ene-1,4-diyl)	polyisoprene
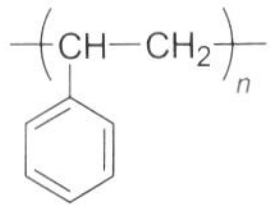 $-(CH(C_6H_5)-CH_2)_n-$	poly(1-phenylethane-1,2-diyl)	polystyrene
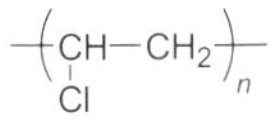 $-(CH(Cl)-CH_2)_n-$	poly(1-chloroethane-1,2-diyl)	poly(vinyl chloride)
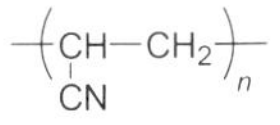 $-(CH(CN)-CH_2)_n-$	poly(1-cyanoethane-1,2-diyl)	polyacrylonitrile
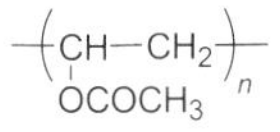 $-(CH(OCOCH_3)-CH_2)_n-$	poly(1-acetoxyethane-1,2-diyl)	poly(vinyl acetate)
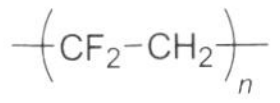 $-(CF_2-CH_2)_n-$	poly(1,1-difluoroethane-1,2-diyl)	poly(vinylidene fluoride)
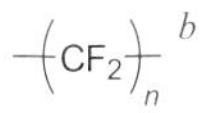 $-(CF_2)_n-$ [b]	poly(difluoromethylene)	poly(tetrafluoroethene); polytetrafluoroethylene
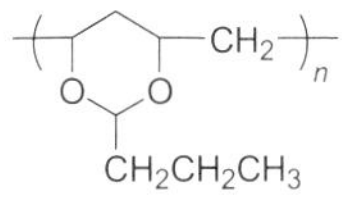 $-(\text{dioxane-CH}_2)_n-$ (2-propyl-1,3-dioxane ring with CH$_2$CH$_2$CH$_3$)	poly[(2-propyl-1,3-dioxane-4,6-diyl)methylene]	—; poly(vinyl butyral)
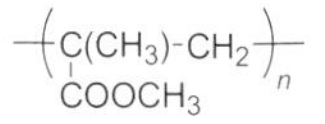 $-(C(CH_3)(COOCH_3)-CH_2)_n-$	poly[1-(methoxycarbonyl)-1-methylethane-1,2-diyl]	poly(methyl methacrylate)
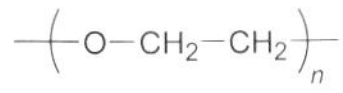 $-(O-CH_2-CH_2)_n-$	poly(oxyethane-1,2-diyl)	poly(ethene oxide); polyoxirane
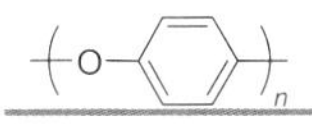 $-(O-C_6H_4)_n-$	poly(oxy-1,4-phenylene)	poly(phenylene oxide)

271

NOMENCLATURE

Structure	Structure-based name	Source-based/other name
–[O–CH₂–CH₂–O–C(=O)–C₆H₄–C(=O)]ₙ–	poly(oxyethane-1,2-diyloxyterephthaloyl)	poly(ethylene terephthalate)
–[NH–C(=O)–(CH₂)₄–C(=O)–NH–(CH₂)₆]ₙ–	poly(iminoadipoyliminohexane-1,6-diyl)	poly[hexamethylenediamine[d]-*alt*-(adipic acid)]; poly(hexamethylene adipamide)
–[CH(–C(=O)–O–C(=O)–)–CH–CH(C₆H₅)–CH₂]ₙ–	poly[(2,5-dioxotetrahydrofuran-3,4-diyl)(1-phenylethane-1,2-diyl)]	poly[(maleic anhydride)-*alt*-styrene]

[a] Ref. 1.
[b] The formulae –(CH₂–CH₂)ₙ– and –(CF₂–CF₂)ₙ– are more often used; they are acceptable owing to past usage and an attempt to retain some similarity to the CRU formulae of homopolymers derived from other vinyl monomers.
[c] The name 'ethylene' should be used only for the divalent group '-CH₂-CH₂-', and not for the monomer 'CH₂=CH₂'; the name for the latter is 'ethene.'
[d] hexane-1,6-diamine.

Structure-based nomenclature can be utilized to name polymers of great complexity, provided only that they be regular and single-strand. Among these are polymers the CRUs of which consist of a series of smaller subunits, polymers with hetero atoms or heterocyclic systems in the main chain, and polymers with substituents on acyclic or cyclic subunits of CRUs. It should be noted that structure-based nomenclature is also applicable to copolymers having a regular structure, regardless of the starting materials used, e.g. poly(oxyethane-1,2-diyloxyterephthaloyl) for poly(ethylene terephthalate) [4].

The basic principles of structure-based nomenclature have been extended beyond regular, single-strand organic polymers to single-strand and quasi-single-strand inorganic and coordination polymers [6], double-strand (ladder and spiro) polymers [7], irregular single-strand organic polymers [8] and other complex systems. Work in this field is continuing.

2.2.2 Inorganic and Coordination Polymers

The nomenclature of regular single-strand inorganic and coordination polymers [6] is governed by the same fundamental principles as that for single-strand organic polymers. The name of such a polymer is that of the CRU prefixed by the terms 'poly', '*catena*' (for linear chains) or other structural descriptor and designations for end-groups. The structural units are named by the nomenclature rules for inorganic and coordination chemistry [9]. Table 5 lists some examples.

2.2.3 Stereochemical Definitions and Notations

The nomenclature of regular polymers can denote stereochemical features if the CRU used is the configurational base unit [10], i.e. a CRU having one or more sites of defined stereoisomerism in the main chain of a polymer molecule [11]. Structure-based names are then derived in the usual fashion. The various stereochemical features that are possible in a polymer must first be defined.

The pioneering work of Natta and co-workers introduced the concept of tacticity, i.e. the orderliness of the succession of configurational repeating units in the main chain of a polymer. For example, in polypropene (polypropylene), possible steric arrangements are (shown in Fischer projections displayed horizontally):

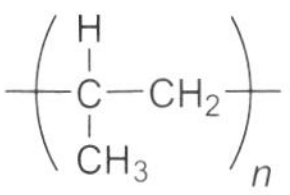

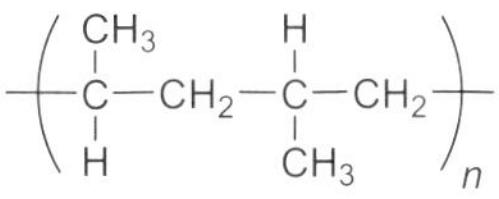

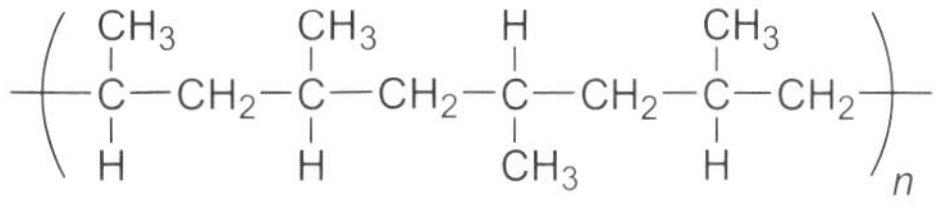

INTRODUCTION TO POLYMER NOMENCLATURE

$$\left(\begin{array}{c} H \\ | \\ C-CH_2 \\ | \\ CH_3 \end{array}\right)_n \quad \left(\begin{array}{cc} CH_3 & H \\ | & | \\ C-CH_2-C-CH_2 \\ | & | \\ H & CH_3 \end{array}\right)_n \quad \left(\begin{array}{cccc} CH_3 & CH_3 & H & CH_3 \\ | & | & | & | \\ C-CH_2-C-CH_2-C-CH_2-C-CH_2 \\ | & | & | & | \\ H & H & CH_3 & H \end{array}\right)_n$$

  isotactic   syndiotactic   atactic

Table 5. Examples of inorganic and coordination polymers[a]

Name	Structure
catena-poly[dimethyltin]	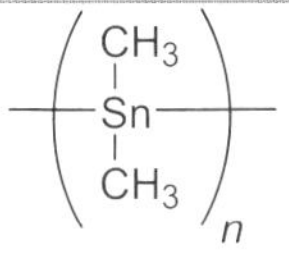 $\left(\begin{array}{c}CH_3\\|\\-Sn-\\|\\CH_3\end{array}\right)_n$
catena-poly[titanium-tri-μ-chlorido]	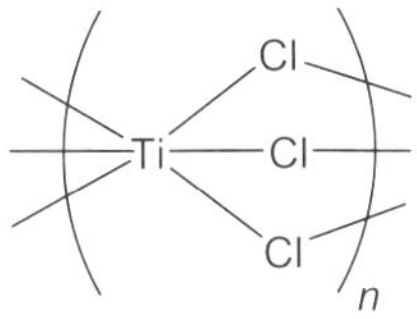 (Ti with three μ-Cl bridges)$_n$
catena-poly[nitrogen-μ-sulfido]	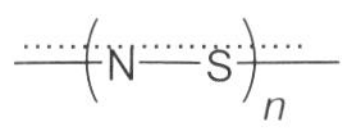 $\left(N\cdots S\right)_n$
catena-poly[silver-μ-(cyanido-*N:C*)]	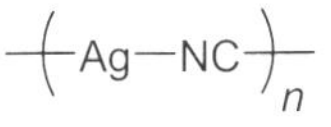 $\left(Ag-NC\right)_n$
catena-poly[(diphenylsilicon)-μ-oxido]	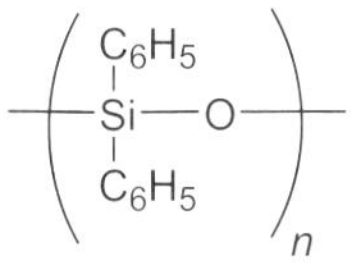 $\left(\begin{array}{c}C_6H_5\\|\\-Si-O-\\|\\C_6H_5\end{array}\right)_n$
α-ammine-ω-(amminedichloridozinc)-*catena*-poly[(amminechloridozinc)-μ-chlorido]	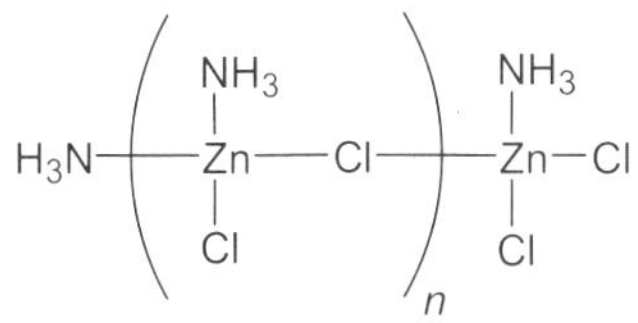 $H_3N-\left(\begin{array}{c}NH_3\\|\\Zn-Cl\\|\\Cl\end{array}\right)_n\begin{array}{c}NH_3\\|\\Zn-Cl\\|\\Cl\end{array}$

[a] If named according to the rules for regular single-strand organic polymers [1], the polymers on lines 1 and 5 would be oriented and named poly(dimethylstannanediyl) and poly[oxy(diphenylsilanediyl)] respectively.

An isotactic polymer has only one species of configurational base unit in a single sequential arrangement and a syndiotactic polymer shows an alternation of configurational base units that are enantiomeric, whereas in an atactic polymer the molecules have equal numbers of the possible configurational base units in a random sequence distribution. This can be generalized as follows in zig-zag and horizontal Fischer projections:

NOMENCLATURE

isotactic:

syndiotactic:

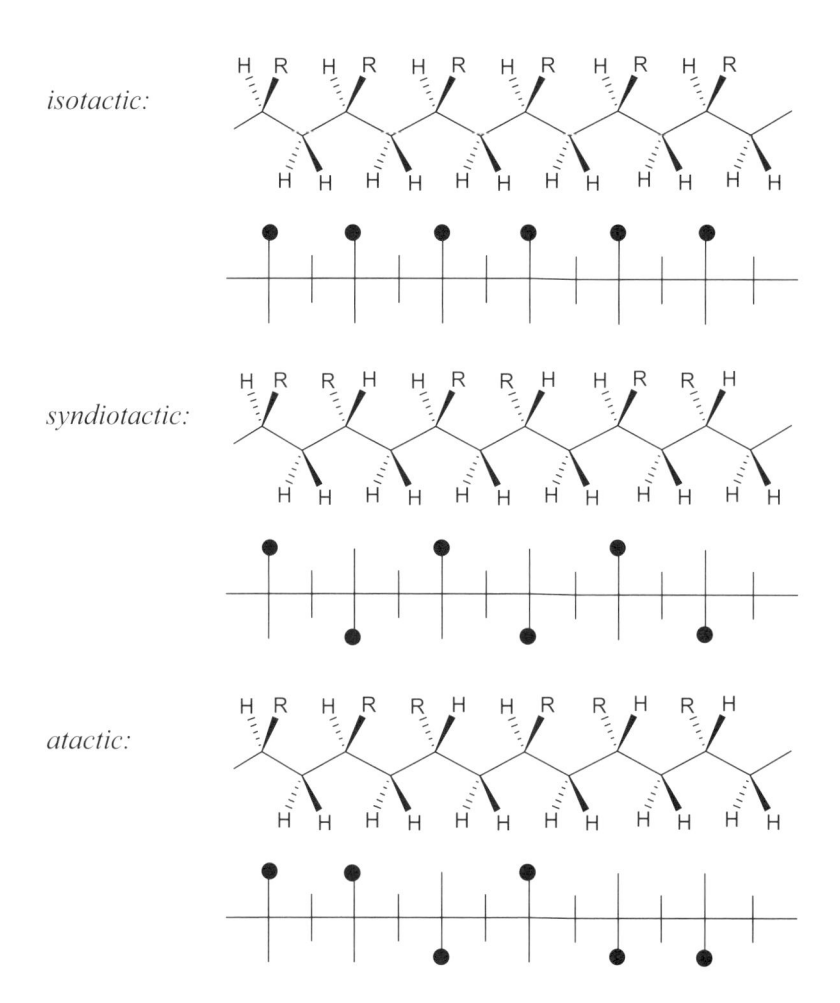

atactic:

Further examples of tactic polymers are

isotactic poly[oxy(1-methylethane-1,2-diyl)]

isotactic poly(methylmethylene)

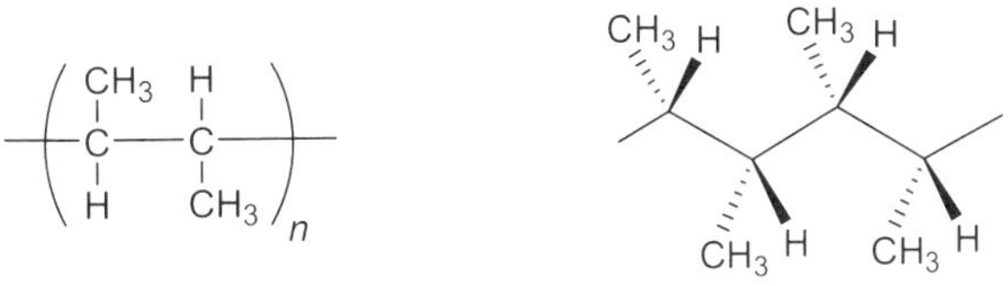

syndiotactic
poly(methylmethylene)

The published IUPAC document [11] should be consulted for more information.

2.2.4 Regular Double-Strand (Ladder and Spiro) Organic Polymers

Structure-based and source-based nomenclature rules have been extended to regular double-strand (ladder and spiro) organic polymers [7]. A double-strand polymer is defined as a polymer the molecules of which are formed by an uninterrupted sequence of rings with adjacent rings having one atom in common (spiro polymer) or two or more atoms in common (ladder polymer).

The structure-based nomenclature rests upon the selection of a preferred CRU [1, 12] of which the polymer is a multiple; the name of the polymer is the name of this repeating unit prefixed by 'poly'. The unit itself is named wherever possible according to the established principles of organic nomenclature [3]. For double-strand polymers, this unit usually is a tetravalent group denoting attachment to four atoms. Since some of these attachments may be double bonds, the unit may be hexavalent or octavalent. Table 6 lists some examples.

Table 6. Examples of structure-based nomenclature for regular double-strand (ladder and spiro) organic polymers

Name	Structure
poly(buta-1,3-diene-1,4:3,2-tetrayl)	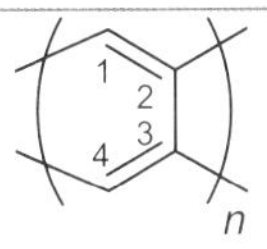
poly(naphthalene-2,3:6,7-tetrayl-6,7-dimethylene)	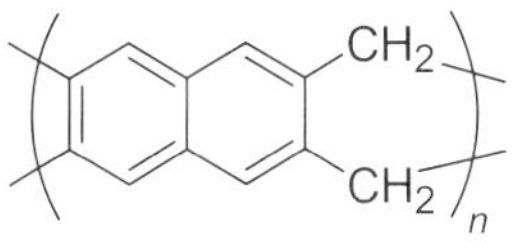
poly[(7,12-dioxo-7,12-dihydrobenzo[*a*]anthracene-3,4:9,10-tetrayl)-9,10-dicarbonyl]	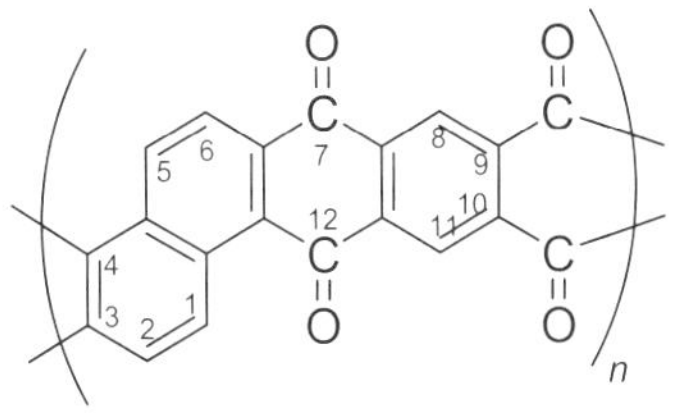
poly(cyclohexane-1,1:4,4-tetrayl-4,4-dimethylene)	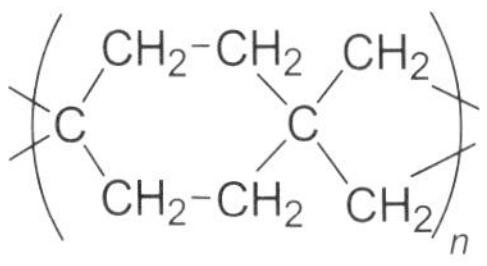

NOMENCLATURE

As with single-strand polymers, substituents that are part of the CRU are denoted by their names prefixed to the name of the subunit to which they are bound (see Table 7).

End-groups are specified by prefixes cited in front of the name of the polymer. The ends designated by α and α' are those attached to the left side of the CRU as drawn; the end groups attached to the right side are designated by ω and ω'. End-groups are named by the nomenclature rules for organic compounds [3]. The designations and names of the end groups proceed in a clockwise direction starting from the lower left: α,α',ω,ω'. Table 7 lists two examples with substituents (lines 1 and 2) and one with both substituents and end-groups (line 3).

Table 7. Examples of nomenclature for regular double-strand (ladder and spiro) organic polymers with substituents and end-groups

Name	Structure
poly[(1,4-dimethyl-7-azabicyclo[2.2.1]heptane-2,3:5,6-tetrayl)-5,6-dicarbonyl]	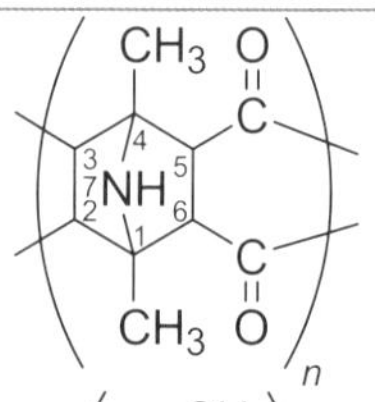
poly(2,4-dimethyl-1,3,5-trioxa-2,4-disilapentane-1,5:4,2-tetrayl)[a]	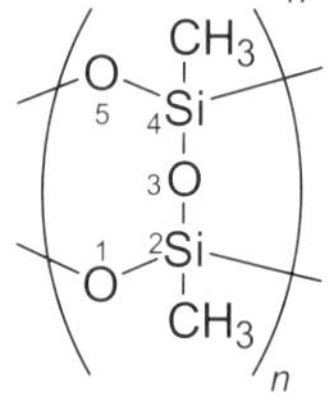
α,α'-dihydro-ω,ω'-dihydroxypoly(2,4-diphenyl-1,3,5-trioxa-2,4-disilapentane-1,5:4,2-tetrayl)[a]	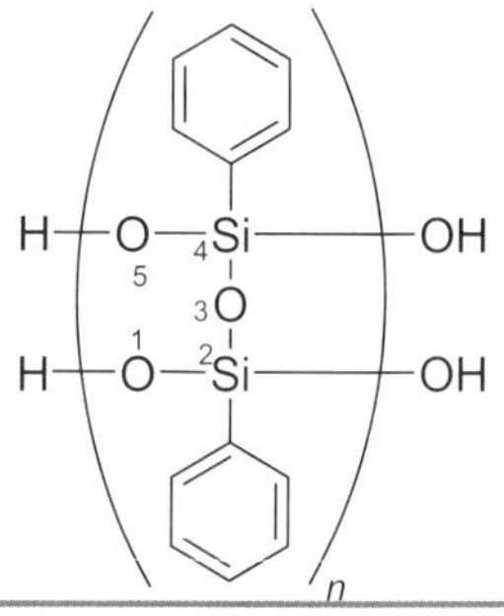

[a] The CRU is named using replacement nomenclature ('a' nomenclature) [3,8]. N.B. In replacement 'a' nomenclature as conventionally applied to acyclic structures with several heteroatoms, terminal heteroatoms are not designated with 'a' prefixes but are named as characteristic groups of the structure, i.e., as hydroxy, amino, carboxylic acid, etc. However, heteroatoms in such positions within the CRUs of ladder or spiro polymer molecules are not terminal units and the structures are not acyclic. Consequently, such atoms are designated with 'a' prefixes, and thereby the simplicity afforded by the application of replacement nomenclature to polymer molecules is enhanced.

Source-based nomenclature identifies the starting monomer(s) from which the double-strand polymer is prepared with addition of an appropriate prefix '*ladder-*' or '*spiro-*'. Examples are:

ladder-poly(diphenyldiacetylene)[*]
ladder-poly(2,5-dichloro-3,6-dihydroxy-1,4-benzoquinone)

---

[*] 1,4-diphenylbuta-1,3-diyne

ladder-poly[(pyromellitic dianhydride)†-alt-(1,2,5,6-tetraaminoanthraquinone)]
spiro-poly[dispiro[3.1.3.1]decane-2,8-bis(carbonyl chloride)]

The rules are complex, and the published IUPAC document [7] should be consulted for more information.

2.2.5 Irregular Single-Strand Organic Polymers

Irregular single-strand organic polymers are single-strand organic polymers that can be described by the repetition of more than one type of constitutional unit or that comprise constitutional units not all connected identically with respect to directional sense. The 1994 system [8] names irregular polymers for which the source-based system for copolymers is inadequate, e.g., polymers that have undergone partial chemical modification, homopolymers having both head-to-tail and head-to-head arrangements of monomeric units, and polymers derived from a single monomer that can provide more than one kind of monomeric unit.

The 1994 system also provides a structure-based alternative to source-based nomenclature for copolymers. Irregular polymers, oligomers or blocks are named by placing the prefix 'poly' or 'oligo', as appropriate, before the structure-based names of the irregularly repeating constitutional units. The names of the irregularly repeating constitutional units are enclosed in parentheses, with the names of the component units separated by oblique strokes.

Block copolymers are named by using dashes (double-length hyphens) for the bonding of blocks with each other and with junction units. With graft and star polymers, the grafts or the arms, respectively, are considered to be substituents to the main chain, and the structure is named in the same way as a regular or irregular polymer. Table 8 lists some examples.

The published IUPAC document [8] should be consulted for more information.

3 ABBREVIATIONS

Because both the source-based and structure-based names of polymers can sometimes be lengthy, abbreviations are frequently used for common industrial polymers. The IUPAC recognizes that there can be advantages in some cases to the use of abbreviations, but urges that each abbreviation be fully defined the first time it appears in a text and that no abbreviations be used in titles of publications. Because of the inherent difficulties in assigning systematic and unique abbreviations to polymeric structures, only a short list has the IUPAC's official sanction [12].

---

† benzene-1,2,4,5-tetracarboxylic 1,2:4,5-dianhydride

NOMECLATURE

Table 8. IUPAC Nomenclature for Irregular Single-Strand Organic Polymers [8]

Name	Structure[a]	Characteristics
poly(1-chloroethane-1,2-diyl/1-phenylethane-1,2-diyl)	$\left(-CH-CH_2-/-CH-CH_2-\right)_n$ with Cl and C_6H_5 substituents	statistical copolymer comprising vinyl chloride and styrene units linked head-to-tail
poly(1-chloroethane-1,2-diyl/2-chloroethane-1,2-diyl)	$\left(-CH-CH_2-/-CH_2-CH-\right)_n$ with Cl substituents	irregular polymer containing vinyl chloride units linked both head-to-tail and head-to-head
poly(chloromethylene/dichloromethylene/methylene)	$\left(-CH-/-C-/-CH_2-\right)$ with Cl substituents	chlorinated polyethylene
poly[1-acetoxyethane-1,2-diyl/1-hydroxyethane-1,2-diyl]	$\left(-CH-CH_2-/-CH-CH_2-\right)_n$ with OCOCH$_3$ and OH	partially hydrolysed head-to-tail poly(vinyl acetate)
poly(oxy-1,4-phenylene)—poly(2-cyanoethane-1,2-diyl)—poly(ethane-1,2-diyloxy)	$\left(-O-\bigcirc-\right)_p\left(CH_2-CH\right)_q\left(CH_2-CH_2-O\right)_r$ with CN	triblock copolymer with blocks linked directly or through unspecified junction units
poly(ethane-1,2-diyloxy)—dimethylsilanediyl—poly(1-chloroethane-1,2-diyl)	$\left(CH_2-CH_2-O\right)_p\text{-Si}(CH_3)_2\text{-}\left(CH-CH_2\right)_q$ with Cl	diblock copolymer in which the blocks are linked by a specific junction unit
poly[methylene/poly(1-phenylethane-1,2-diyl)methylene]	$\left[-CH_2-/-CH-\left(CH_2-CH\right)_p\right]_n$ with C_6H_5	graft copolymer with many graft units of a single type
poly{oxy-1,4-phenylene/oxy-2-[poly(1-chloroethane-1,2-diyl)carbonyloxy]-1,4-phenylene}	$\left[-O-\bigcirc-/-O-\bigcirc-O-C(=O)-\left(CH_2-CH\right)_p-Cl\right]_n$	poly(1,4-phenylene oxide) with a poly(vinyl chloride) block grafted through a carbonyloxy group (a junction unit) in some of the rings
bis[poly(but-2-ene-1,4-diyl)][poly(1-phenylethane-1,2-diyl)][poly(2-phenylethane-1,2-diyl)]silane	$\left(CH_2-CH=CH-CH_2\right)_p\text{Si}\left(CH_2-CH=CH-CH_2\right)_p$ with C_6H_5 and $\left(CH-CH_2\right)_q$, $\left(CH-CH_2\right)_r$	four-armed star polymer with a silicon atom as the central unit and four polymer blocks

[a] The dashes at each end of the formulae of entries 1, 2, 3, 4, 7, and 8 are drawn fully inside the enclosing marks, because the identities of the units at the ends of the chains are unknown.

4 REFERENCES

1. IUPAC. 'Nomenclature of regular single-strand organic polymers (IUPAC Recommendations 2002)', *Pure Appl. Chem.* **74**, 1921-1956 (2002). Reprinted as Chapter 15 this volume.
2. IUPAC. 'Generic source-based nomenclature for polymers (IUPAC Recommendations 2001)', *Pure Appl. Chem.* **73**, 1511-1519 (2001). Errata, *Pure Appl. Chem.* **74**, 2019 (2002). Reprinted as Chapter 21 this volume.
3. IUPAC. *A Guide to IUPAC Nomenclature of Organic Compounds (Recommendations 1993)*, prepared for publication by R. Panico, W. H. Powell and J.-C. Richer (Senior Editor), Blackwell Scientific Publications, Oxford (1993). Errata. *Pure Appl. Chem.* **71**, 1327-1330 (1999).
4. IUPAC. 'Source-based nomenclature for copolymers (Recommendations 1985)', *Pure Appl. Chem.* **57**, 1427-1440 (1985). Reprinted as Chapter 19 this volume.
5. IUPAC. 'Source-based nomenclature for non-linear macromolecules and macromolecular assemblies (IUPAC Recommendations 1997)', *Pure Appl. Chem.* **69**, 2511-2521 (1997). Reprinted as Chapter 20 this volume.
6. IUPAC. 'Nomenclature for regular single-strand and quasi-single-strand inorganic and coordination polymers (Recommendations 1984)', *Pure Appl. Chem.* **57**, 149-168 (1985).
7. IUPAC. 'Nomenclature of regular double-strand (ladder and spiro) organic polymers (IUPAC Recommendations 1993)', *Pure Appl. Chem.* **65**, 1561-1580 (1993). Reprinted as Chapter 16 this volume.
8. IUPAC. 'Structure-based nomenclature for irregular single-strand organic polymers (IUPAC Recommendations 1994)', *Pure Appl. Chem.* **66**, 873-889 (1994). Reprinted as Chapter 17 this volume.
9. IUPAC. *Nomenclature of Inorganic Chemistry. Recommendations 2005* (the 'Red Book'). Prepared for publication by N. G. Connelly and T. Damhus, RSC Publishing, Cambridge, UK (2005).
10. IUPAC. 'Stereochemical definitions and notations relating to polymers (Recommendations 1980)', *Pure Appl. Chem.* **53**, 733-752 (1981). Reprinted as Chapter 2 this volume.
11. IUPAC. 'Glossary of basic terms in polymer science (IUPAC Recommendations 1996)', *Pure Appl. Chem.* **68**, 2287-2311 (1996). Reprinted as Chapter 1 this volume.
12. IUPAC. 'Use of abbreviations for names of polymeric substances (Recommendations 1986)', *Pure Appl. Chem.* **59**, 691-693 (1987). Revised version reprinted as Chapter 22 this volume.

15: Nomenclature of Regular Single-Strand Organic Polymers

CONTENTS

1. Preamble
2. Definitions
3. Fundamental principles
4. Seniority of subunits
    4.1 Heterocyclic rings and ring systems
    4.2 Heteroatom chains
    4.3 Carbocyclic rings and ring systems
    4.4 Acyclic carbon chains
5. Selection of the preferred constitutional repeating unit (CRU)
    5.1 Simple CRUs
    5.2 Complex CRUs
6. Naming the preferred constitutional repeating unit (CRU)
    6.1 Naming subunits
    6.2 Naming the preferred CRU
7. Naming the polymer
8. Polymer chain as a substituent
9. Examples of polymer names
10. References
11. Appendix
    11.1 List of names of common subunits
    11.2 Structure- and source-based names for common polymers

1 PREAMBLE

In 1952, the Subcommission on Nomenclature of the IUPAC Commission on Macromolecules published a report [1] on the nomenclature of polymers that included a method for the systematic naming of linear organic polymers on the basis of structure. A later report [2] dealing with steric regularity utilized that system of nomenclature. When the first report was issued, the skeletal rules were adequate for most needs; indeed, most polymers could at that time be reasonably named on the basis of the real or hypothetical substance used in producing the polymer. In the intervening years, however, the rapid growth of the polymer field had dictated a need for modification and expansion of the earlier rules. The result was a set of rules approved in 1975 [3]. The present report is an updating of the 1975 rules with special attention to developments in the nomenclature of organic chemistry since that time [4, 5].

Originally prepared by a working group consisting of J. Kahovec (Czech Republic), R. B. Fox (USA) and K. Hatada (Japan). Reprinted from *Pure Appl. Chem.* **74**, 1921-1956 (2002).

SINGLE-STRAND ORGANIC POLYMERS

The rules in the present report are designed to name, uniquely and unambiguously, the structures of regular single-strand organic polymers whose repeating structures can be written within the framework of ordinary chemical principles. Although the stereochemistry of polymers is not considered here, examples of names with stereodescriptors are given. A detailed survey of stereochemical notation of polymers is given in a special report [6]. As with organic chemistry nomenclature, this nomenclature describes chemical structures rather than substances. It is realized that a polymeric substance ordinarily may include many structures, and that a complete description of even a single polymer molecule would include an itemization of terminal groups, branching, random impurities, degree of steric regularity, chain imperfections, etc.

Nevertheless, it is useful to think of the macromolecules of a polymer as being represented by a single structure that may itself be hypothetical. To the extent that the structure can be portrayed as a chain of structural repeating units (SRUs) or constitutional repeating units (CRUs) (the terms are synonymous), the structure can be named by the rules in this report; in addition, provision has been made for including end-groups in the name.

The fundamental principles and the basic rules of the structure-based nomenclature are given first, accompanied by detailed extensions and applications. An Appendix contains names of common subunits as well as a list of acceptable source-based names, along with the corresponding structure-based names, of common polymers. There is no strong preference for the use of structure-based names over source-based names for polymers where the latter are clear and unambiguous, but for certain purposes one system of naming may be preferred to the other.

The rules of structure-based nomenclature of regular single-strand organic polymers are of fundamental importance in polymer nomenclature. The names of other kinds of polymers such as double-strand [7] and irregular polymers [8] are based on the principles given in this Report.

After more than two decades of use, many improvements in the 1975 rules [3] have been suggested. As a result, the present new rules are proposed. The new rules do not represent any change in principles. They involve mainly rearrangement of the material, a generalization of basic rules, a clearer presentation, an avoidance of manifold repetition of the same principles at various places and the use of graphical means for the visualization of the principles. Also, some additions are made, such as a rule on naming polymer chain substituents, and several new examples of polymers including those with an 'inorganic' backbone and those where stereodescriptors are essential.

2 DEFINITIONS

2.1 regular polymer
A polymer composed of regular macromolecules, i.e. macromolecules the structure of which essentially comprises the repetition of a single constitutional unit with all units connected identically with respect to directional sense [9].

2.2 single-strand polymer
A polymer the macromolecules of which are single-strand macromolecules, i.e. macromolecules comprising constitutional units connected in such a way that adjacent constitutional units are joined to each other through two atoms, one on each constitutional unit [3, 7, 9].

NOMENCLATURE

2.3   constitutional unit
An atom or group of atoms (with pendant atoms or groups, if any) comprising a part of the essential structure of a macromolecule, an oligomer molecule, a block or a chain [9].

2.4   constitutional repeating unit (CRU)
The smallest constitutional unit, the repetition of which constitutes a regular macromolecule, a regular oligomer molecule, a regular block or a regular chain [9].

2.5   main chain backbone
That linear chain to which all other chains, long or short or both, may be regarded as being pendant [9].

2.6   end-group
A constitutional unit that is an extremity of a macromolecule or oligomer molecule [9].

2.7   subunit
The largest main-chain (backbone) segment of the constitutional repeating unit that can be named as a single unit under organic nomenclature rules [4, 5]. This may be a ring or ring system, a heteroatom or a heteroatom chain or an acyclic carbon chain.

2.8   path length
The path length between two subunits is the number of polymer main chain (backbone) atoms between the two subunits. Where a ring or ring system constitutes all or part of a path between two subunits, the shortest continuous chain of atoms in the ring or ring system is selected.

2.9   seniority
Priority in a set of atoms or groups of atoms according to a prescribed order.

2.10   locant
A numeral or letter that identifies position in a structure.

For the regular single-strand polymer, $E^1{-}(A{-}B{-}C{-}D)_n{-}E^2$ with substituents R^1 on B and R^2 on D,

$-A-B(R^1)-C-D(R^2)-$ is the constitutional repeating unit (CRU)

$-A-$, $-B-$, $-C-$, $-D-$ are subunits

$-B(R^1)-$, $-D(R^2)-$ are substituted subunits

R^1, R^2 are substituents to subunits, and

E^1, E^2 are end-groups

SINGLE-STRAND ORGANIC POLYMERS

3   FUNDAMENTAL PRINCIPLES

This nomenclature method rests upon the selection of a preferred constitutional repeating unit (CRU) of which the polymer molecule is a multiple. Wherever possible, the CRU and subunits are named according to the IUPAC-recommended nomenclature of organic chemistry [4, 5].
In this nomenclature, the steps to be followed when naming a polymer are:

1. Write the structure of the polymer chain. A sufficient portion of the chain should be written to show structure repetition. The portion that repeats is a constitutional repeating unit (CRU).
2. Select the preferred CRU.
3. Name the preferred CRU by citing, from left to right, the names of the subunits, including their substituents if present.
4. Name the polymer.

Structure of the polymer chain

In simple cases, the CRU involves a single subunit. In more complex cases, it is often necessary to draw a large segment of the polymer chain, for example,

$$-\text{OCHCH}_2-\text{OCHCH}_2-\text{OCHCH}_2-\text{OCHCH}_2-\text{OCHCH}_2-$$
$$\hspace{1.3cm}|\hspace{1.5cm}|\hspace{1.5cm}|\hspace{1.5cm}|\hspace{1.5cm}|$$
$$\hspace{1.2cm}\text{Br}\hspace{1.2cm}\text{Br}\hspace{1.2cm}\text{Br}\hspace{1.2cm}\text{Br}\hspace{1.2cm}\text{Br}$$

Selection of the preferred constitutional repeating unit

There are many ways to write the CRU for most chain structures. In simple cases, these units are readily identified. In the polymer chain given above, the possible CRUs are

$$-\text{OCHCH}_2-,\ -\text{CH}_2\text{OCH}-,\ -\text{OCH}_2\text{CH}-,\ -\text{CH}_2\text{CHO}-,\ -\text{CHOCH}_2-,\ -\text{CHCH}_2\text{O}-$$
$$\hspace{0.3cm}|\hspace{1.7cm}|\hspace{1.7cm}|\hspace{1.7cm}|\hspace{1.7cm}|\hspace{1.7cm}|$$
$$\hspace{0.2cm}\text{Br}\hspace{1.6cm}\text{Br}\hspace{1.6cm}\text{Br}\hspace{1.6cm}\text{Br}\hspace{1.6cm}\text{Br}\hspace{1.6cm}\text{Br}$$

To allow construction of a unique name, a single CRU must be selected. The following rules have been designed to specify both seniority among subunits, i.e., the point at which to begin writing the CRU, and the direction along the chain in which to continue to the end of the CRU. The preferred constitutional repeating unit will be one beginning with the subunit of highest seniority (see Section 4). From this subunit, one proceeds toward the subunit next in seniority. In the preceding example, the subunit of highest seniority is an oxygen atom and the subunit next in seniority is a substituted $-\text{CH}_2\text{CH}_2-$ unit. The CRU is written to read from left to right with the chain atoms being numbered consecutively in the same direction. The preferred CRU will therefore be either

$$-\text{OCHCH}_2-\ \ \text{or}\ \ -\text{OCH}_2\text{CH}-$$
$$\hspace{0.5cm}|\hspace{3.2cm}|$$
$$\hspace{0.4cm}\text{Br}\hspace{3.0cm}\text{Br}$$

Naming the preferred constitutional repeating unit

The name of the preferred CRU is formed by citing, in the order in which they appear in the CRU, the names of the subunits within the CRU. In the example, the oxygen atom is called oxy and the $-\text{CH}_2\text{CH}_2-$ (preferred to $-\text{CH}_2-$ because it is larger and can be named as a unit) is called ethane-1,2-diyl; the latter unit substituted with one bromine atom is called 1-bromoethane-1,2-diyl. The preferred CRU is therefore named oxy(1-bromoethane-1,2-diyl).

Further choice in this case is based on the lowest locant* for substitution, so that the preferred CRU is

$$-\underset{\underset{Br}{|}}{O}CHCH_2- \quad \text{rather than} \quad -OCH_2\underset{\underset{Br}{|}}{C}H-$$

1-substitution   2-substitution

Naming the polymer

The name of the polymer is simply the name of the preferred CRU enclosed in curves, square brackets or braces and prefixed by poly. The nesting order of enclosing marks is curves, square brackets, braces, then curves, square brackets, braces, etc., i.e. {[({[( )]})]}. This is well illustrated in Examples 21 and 31 in Section 9.

The $-(\underset{\underset{Br}{|}}{O}CHCH_2)_n-$ polymer is named poly[oxy(1-bromoethane-1,2-diyl)].

4 SENIORITY OF SUBUNITS

Where reference is made in the following rules to the lowest set of locants, this is defined as the set that, when compared term by term with other locant sets, each cited in order of increasing value, has the lowest term at the first point of difference.

Rule 1
The basic order of seniority of subunits is: heterocyclic rings and ring systems > heteroatom chains > carbocyclic rings and ring systems > acyclic carbon chains

The order of seniority of subunits is of primary importance in the generation of polymer names. Further classification according to seniority is based on the nature of the subunits (kind or size or both) and, among identical subunits, (a) on their degree of unsaturation and (b) on their substituents (number, kind and locants). The following criteria are applied consecutively until a decision is reached.

4.1 Heterocyclic rings and ring systems

Rule 2
Among heterocyclic rings and ring systems, the descending order of seniority is:

a. a ring or ring system containing nitrogen;
b. a ring or ring system containing the heteroatom occurring earliest in the order given in Rule 4;
c. a ring or ring system containing the greatest number of rings;

---

* Where reference is made in the following rules to the lowest set of locants, this is defined as the set that, when compared term by term with other possible locant sets, each cited in ascending order, has the lowest term at the first point of difference.

d. a ring or ring system having the largest individual ring at the first point of difference when the rings are represented in decreasing order of size;
e. a ring or ring system having the greatest number of heteroatoms;
f. a ring or ring system containing the greatest variety of heteroatoms;
g. the ring or ring system having the greatest number of heteroatoms earliest in the order given in Rule 4;
h. of the two rings or ring systems of the same size containing the same number and kind of heteroatoms the senior system is that one with the lower locants for the heteroatoms.

Note: This order is a paraphrased extract of that in [10].

Examples of the application of seniority rules among different heterocyclic rings and ring systems are (appropriate rule in parentheses):

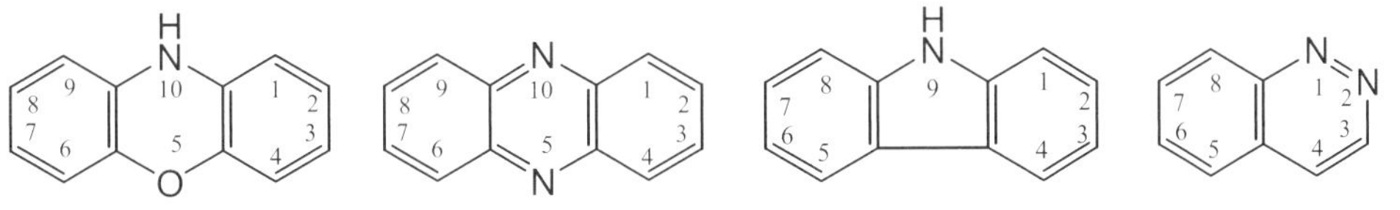

phenoxazine > phenazine > carbazole > cinnoline >
(2b)   (2d)   (2c)   (2h)

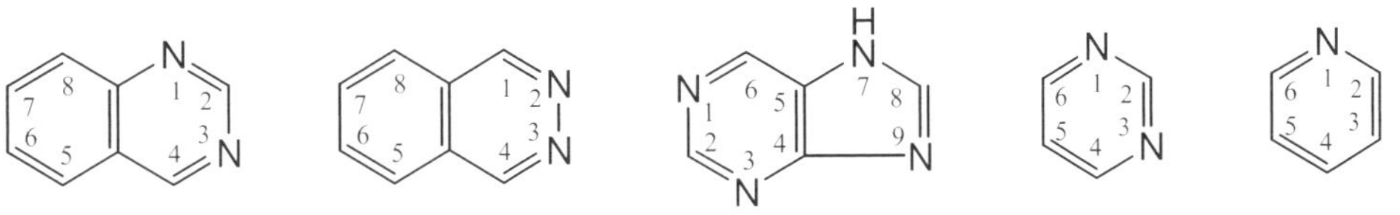

quinazoline > phthalazine > 7H-purine > pyrimidine > pyridine >
(2h)   (2d)   (2c)   (2e)   (2d)

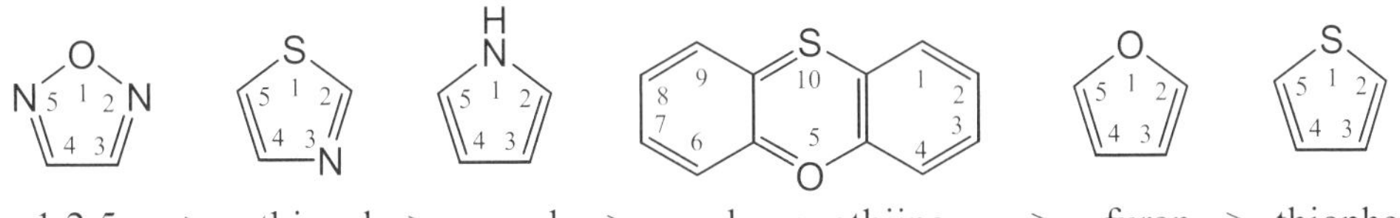

1,2,5- > thiazole > pyrrole > phenoxathiine > furan > thiophene
oxadiazole (2e) (2e)   (2a)   (2c)   (2b)

Rule 3

The order of decreasing seniority within a given heterocyclic ring or ring system is:

a. when rings or ring systems differ only in degree of unsaturation, the senior system is the most unsaturated one;
b. when rings or ring systems of the same degree of unsaturation differ in the positions of double bonds, the senior system is that having the lowest locants for double bonds;
c. in heterocyclic ring assemblies, the assembly of highest seniority is that having lowest locants for the points of attachment between the rings within the assembly, consistent with the fixed numbering of the ring or ring system;
d. a ring or ring system with the lowest locants of free valences;
e. a ring or ring system with the largest number of substituents;

NOMENCLATURE

f. a ring or ring system having substituents with the lowest locants;
g. a ring or ring system in which the substituent first in alphabetical order has the lowest locant.

Examples of the application of seniority rules within a given heterocyclic ring or ring system are (appropriate rule in parentheses):

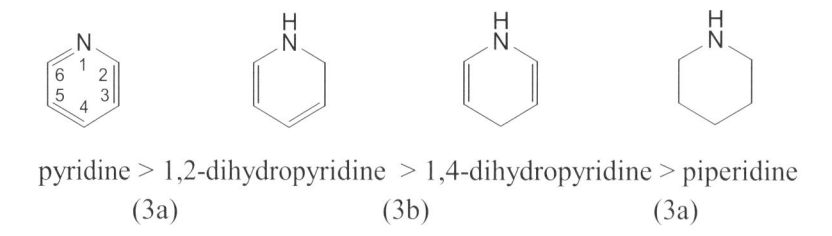

pyridine > 1,2-dihydropyridine > 1,4-dihydropyridine > piperidine
(3a)   (3b)   (3a)

4.2 Heteroatom chains

Rule 4
For the most common heteroatoms, the descending order of seniority is:

$$O > S > Se > Te > N > P > As > Sb > Bi > Si > Ge > Sn > Pb > B > Hg$$

Note: Other heteroatoms may be placed within this order as indicated by their positions in the periodic table [5].

Rule 5
A more substituted single heteroatom is senior to a less substituted single heteroatom of the same kind.

Examples:

Rule 6
Within mono- or disubstituted single heteroatoms, the heteroatom carrying a substituent (substituents) earlier in the alphabet is senior.

Example:

Rule 7
The order of decreasing seniority for chains of heteroatoms of the same kind that have equal length is:

a. when chains differ only in degree of unsaturation, the senior chain is the most unsaturated one;

Note: This rule applies also to single heteroatoms.

b. when chains of the same degree of unsaturation differ in the positions of multiple bonds, the senior chain is that having the lowest locants for double bonds;
c. the chain with the largest number of substituents;
d. the chain having substituents with the lowest locants;
e. the chain in which the substituent first in alphabetical order has the lowest locant.

Examples:

$$-\underset{O}{\overset{O}{\underset{\|}{\overset{\|}{S}}}}-\underset{O}{\overset{O}{\underset{\|}{\overset{\|}{S}}}}-S- \quad > \quad -S-\underset{O}{\overset{O}{\underset{\|}{\overset{\|}{S}}}}-\underset{O}{\overset{O}{\underset{\|}{\overset{\|}{S}}}}- \qquad -N= \;>\; -NH- \qquad -N=N- \;>\; -NHNH-$$

(7d)   (7a)   (7a)

4.3 Carbocyclic rings and ring systems

Rule 8

Among carbocyclic rings and ring systems, the decreasing order of seniority is:

a. a ring system containing the greatest number of rings;
b. the largest ring or a ring system with the largest individual ring at the first point of difference when the rings are represented in decreasing order of size;
c. a ring system having the greatest number of atoms common to the rings.

Note: The criteria for further choice are found in Rule C-14.1 in [4].

Examples of the application of seniority among carbocyclic rings and ring systems are (appropriate rule in parentheses):

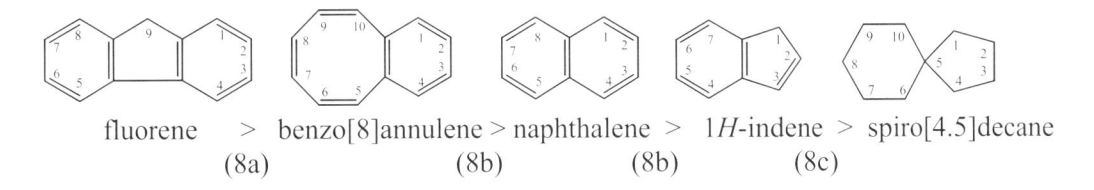

fluorene   >   benzo[8]annulene > naphthalene >   1*H*-indene > spiro[4.5]decane
(8a)   (8b)   (8b)   (8c)

Rule 9

The order of decreasing seniority within a given carbocyclic ring or ring system is:

a. when rings or ring systems differ only in degree of unsaturation, the senior system is the most unsaturated one;
b. when rings or ring systems of the same degree of unsaturation differ in the positions of double bonds, the senior system is that having the lowest locants for double bonds;
c. in carbocyclic ring assemblies, the assembly of highest seniority is that having lowest numbers for the points of attachment between the systems within the assembly, consistent with the fixed numbering of the ring or ring system;
d. a ring or ring system with the lowest locants for free valences;
e. a ring or ring system with the largest number of substituents;
f. a ring or ring system having substituents with the lowest locants;
g. a ring or ring system in which the substituent first in alphabetical order has the lowest locant.

NOMENCLATURE

Examples of the application of seniority rules within a given carbocyclic ring or ring system are (appropriate rule in parentheses):

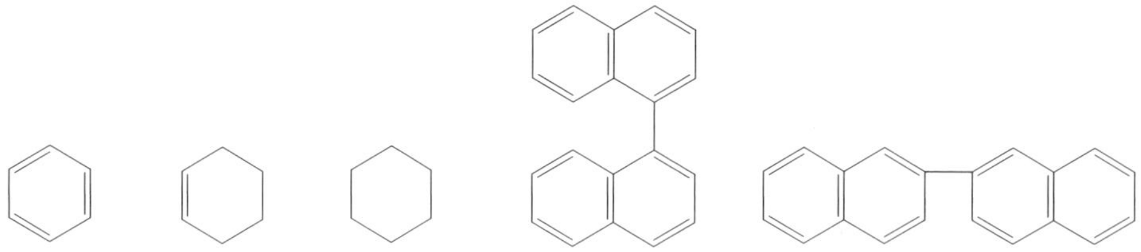

benzene > cyclohexene > cyclohexane   1,1'-binaphthalene > 2,2'-binaphthalene
     (9a)   (9a)   (9c)

naphthalene-1,5-diyl   >   naphthalene-2,6-diyl
                        (9d)

4.4 Acyclic carbon chains

Rule 10

The order of decreasing seniority of acyclic carbon chains of equal length is:

a. when chains differ only in degree of unsaturation, the senior chain is the most unsaturated one;
Note: This rule applies also to single carbon atoms.
b. when chains of the same degree of unsaturation differ in the positions of multiple bonds, the senior chain is that having multiple bonds with the lowest locants;
c. the chain with the largest number of substituents;
d. the chain having substituents with the lowest locants;
e. the chain in which the substituent first in alphabetical order has the lowest locant;
f. the chain with substituents having the same locant is the first in alphabetical order.

Examples:

—CH= > —CH$_2$—   -CH=CHCH$_2$CH$_2$- > -CH$_2$CH=CHCH$_2$- > —(CH$_2$)$_4$—
     (10a)   (10b)   (10a)

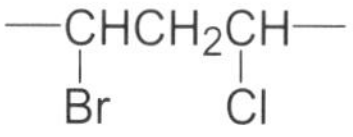

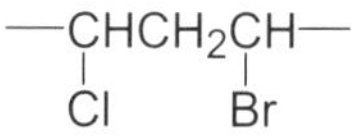

—CHCH$_2$CH—   >   —CHCH$_2$CH—
    | | | |
   Br   Cl   Cl   Br
                   (10e)

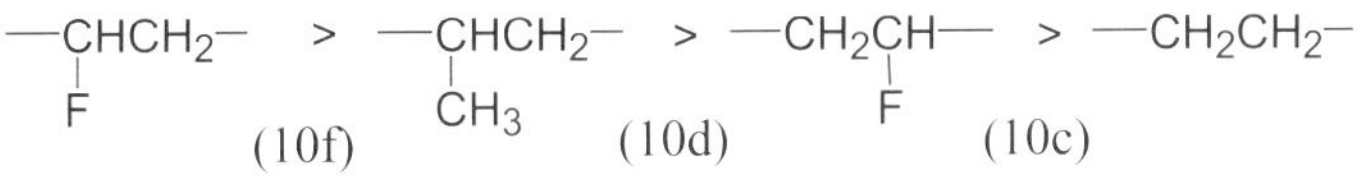

$$\begin{array}{cccc}
\underset{\underset{F}{|}}{-CHCH_2-} > & \underset{\underset{CH_3}{|}}{-CHCH_2-} > & \underset{\underset{F}{|}}{-CH_2CH-} > & -CH_2CH_2- \\
(10f) & (10d) & (10c) &
\end{array}$$

5 SELECTION OF THE PREFERRED CONSTITUTIONAL REPEATING UNIT (CRU)

The CRU in a polymer chain may contain one or more subunits. A simple CRU is one that can be described by single subunit. A complex CRU has at least two subunits.

5.1 Simple CRUs

In selecting the preferred CRU, the steps to be followed in sequence are:

1. Write the structure of a representative portion of the polymer chain;
2. Identify the single subunit and its substituents;
3. Choose the direction so that the locants of free valences of the subunit are as low as possible;
4. Choose the direction so that the locants of the substituents are as low as possible.

Rule 11
For acyclic and monocarbocyclic subunits preference in lowest numbers is given to the carbon atoms through which they are attached to the main chain of the CRU. The point of attachment at the left-hand side of the subunit as written in the CRU must have locant 1.

Examples:

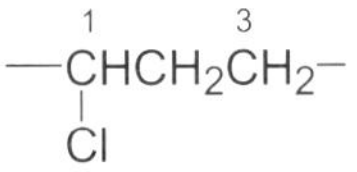

$$\underset{\underset{Cl}{|}}{-\overset{1}{C}HCH_2\overset{3}{C}H_2-}$$

1-chloropropane-1,3-diyl 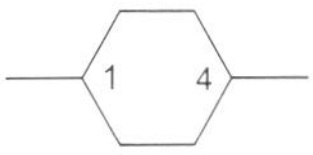   cyclohexane-1,4-diyl

Rule 12
In polycyclic hydrocarbons, bridged hydrocarbons, spiro hydrocarbons, ring assemblies and heterocyclic systems, numbering is fixed for the ring system. The points of attachment of such subunits to the main chain of the CRU should have the lowest permissible locants consistent with the fixed numbering. The same fixed numbering is retained for either direction of progress through the group in generating the subunit name. Where there is a choice, the point of attachment at the left-hand side of the ring as written in the CRU should have the lowest permissible locant.
Examples:

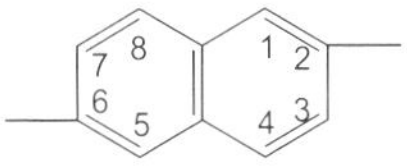

naphthalene-2,6-diyl 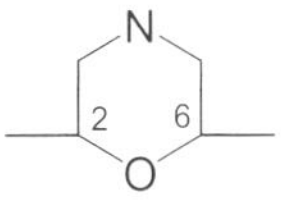   morpholine-2,6-diyl

5.2 Complex CRUs

The factors in the selection of the preferred CRU are in order:

- seniority of subunits
- path length between subunits

In selecting the preferred CRU, the steps to be followed in sequence are:

1. Write the structure of a representative portion of the polymer chain;
2. Identify subunits and substituents;
3. Classify the subunits according to their seniority;
4. Find the shortest path in atoms, irrespective of their nature, from the subunit of the highest seniority to the subunit of the same seniority (Rules 15,16), if present, or of the second highest seniority (Rules 13,14). Where paths of equal length are identified as shortest, the choice depends on the seniority of the remaining subunits and the number and positions of substituents;
5. Orient the structure so that the direction from the most senior subunit to the subunit of the next highest seniority reads left to right;
6. Identify the preferred CRU starting from the highest seniority subunit and moving in the direction determined above.

These steps are further elaborated in Rules 13–16.

Rule 13

The starting point for the preferred CRU is at the subunit of highest seniority (A). Of the two paths leading from the subunit of highest seniority (A) along the main chain (backbone) to both the next subunits of second highest seniority (B), the shorter path is to be followed.

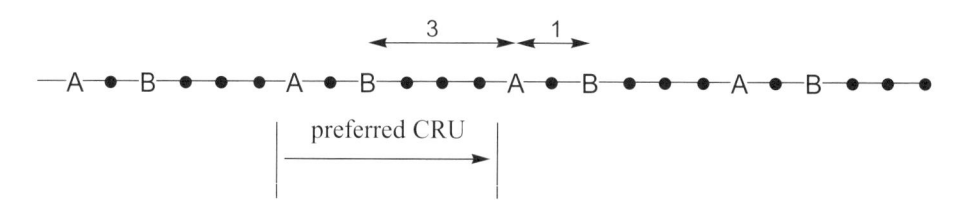

(dots represent subunits of lowest seniority)

(B is closer to A in the preferred CRU, i.e.,

A—•—B—•—•—• > A—•—•—•—B—•—)

Example:

SINGLE-STRAND ORGANIC POLYMERS

(O is senior to a benzene ring; the one-atom path from O to the benzene ring is preferred to the three-atom path.)

Rule 14
When two paths from the starting subunit (A) to both the next subunits of second highest seniority (B) are equally long, the shorter path from the starting subunit A to the subunit of third highest seniority (C) is to be followed.

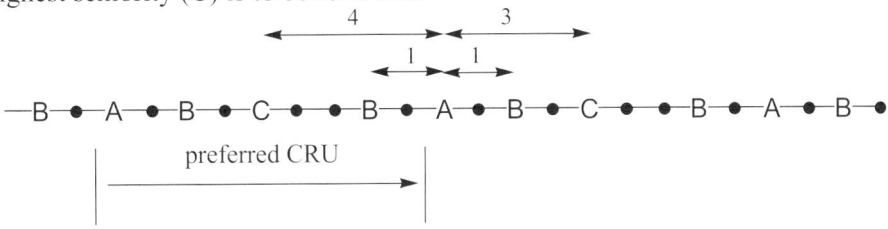

—B•A•B•C••B•A•B•C••B•A•B•

preferred CRU

Example:

·····—CH$_2$SCH$_2$OCH$_2$SCH$_2$NHCH$_2$CH$_2$SCH$_2$OCH$_2$SCH$_2$NHCH$_2$CH$_2$SCH$_2$O—·····
preferred CRU

(O is senior to S which is senior to N; the one-atom path from O to S is to be followed by the six-atom path from S to O in which N is closer to S)

—A–C–B•—A–C–B•—A–C–B•—A–C–B•—A–C–B•—A–C–B•

preferred CRU

Example:

·····—ONHCH$_2$SCH$_2$CH$_2$ONHCH$_2$SCH$_2$CH$_2$ONHCH$_2$SCH$_2$CH$_2$ONHCH$_2$SCH$_2$CH$_2$O—·····
preferred CRU

(O is senior to S which is senior to N; of the two equal paths from O to S, the one that traverses N is preferred to that which traverses CH$_2$)

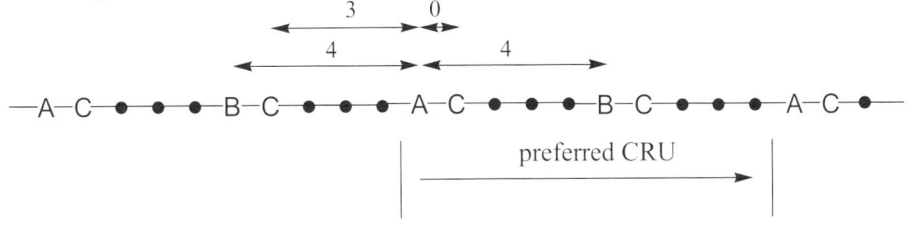

—A–C••—B–C•••—A–C•••—B–C•••—A–C•—

preferred CRU

(C is closer to A in the preferred CRU; both B's are equally distant from A)

NOMENCLATURE

Example:

—O—⟨C₆H₄⟩—CH₂CH₂CH₂NH—⟨C₆H₄⟩—CH₂CH₂CH₂O—⟨C₆H₄⟩—CH₂CH₂CH₂NH—

preferred CRU

(O is senior to N which is senior to a benzene ring; of the two equal paths from O to N, the one that traverses the benzene ring earlier is preferred.)

Rule 15

When two identical subunits of the highest seniority (A) are present in a CRU in the main chain (backbone), the shorter path between the identical subunits is to be followed. The starting point is chosen in such a way that the shorter path to the subunit of second highest seniority (B) is followed. If the paths are equal, the paths to subunits of third highest seniority (C) are considered in the sense of Rule 14.

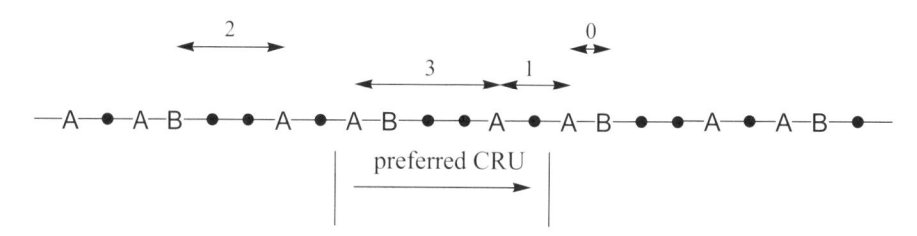

—A—•—A–B—•—•—A—•—A–B—•—•—A—•—A–B—•—•—A—•—A–B—•—

preferred CRU

(B is closer to both A subunits in the preferred CRU)

Example:

----—O—CH₂—O—SO—CH₂CH₂—O—CH₂—O—SO—CH₂CH₂—O—CH₂—O—SO—CH₂CH₂—----

preferred CRU

(O is senior to S; the one-atom path from O to O is preferred to the three-atom path from O to O; S is closer to O in the preferred CRU)

—•—A—•—A–C–B—•—A—•—A–C–B—•—A—•—A–C–B—•—A—•—A–C—

preferred CRU

(of the two equal paths from —A—•—A— to B, the one that traverses C is preferred to that which traverses —•—.)

SINGLE-STRAND ORGANIC POLYMERS

Example:

```
----—OCH₂O–NHCH₂SCH₂CH₂OCH₂O-NHCH₂SCH₂CH₂OCH₂O-NHCH₂SCH₂CH₂O—----
              |      preferred CRU      |
```

(O is senior to S, which is senior to N; the one-atom path from O to O is preferred to the five-atom path; N is closer to O in the preferred CRU)

Rule 16
When three or more identical subunits of the highest seniority (A) are present in a CRU in the main chain (backbone), the starting point and direction are chosen in such a way that the shortest path through all the subunits A results. If there is a choice, the CRU with the shorter path to the subunits of second or third highest seniority (B or C) is selected.

Examples:

```
----—OCH₂CH₂OCH₂CH₂OCH₂CH₂CH₂CH₂OCH₂CH₂OCH₂CH₂OCH₂CH₂CH₂CH₂—----
     |        preferred CRU           |
```

(shortest path through all oxygens, either direction)

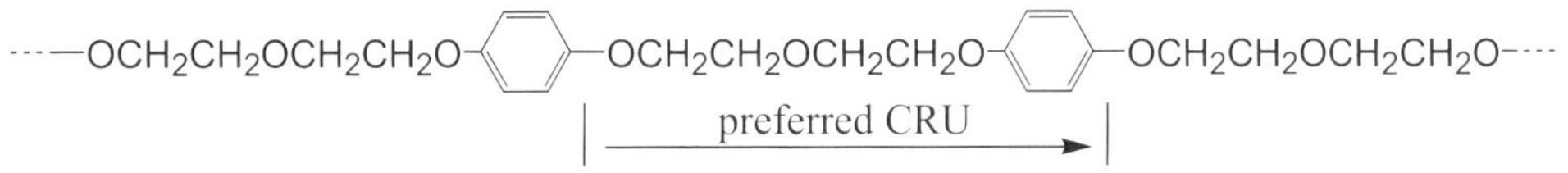

```
----—OCH₂CH₂OCH₂CH₂O-⟨phenyl⟩-OCH₂CH₂OCH₂CH₂O-⟨phenyl⟩-OCH₂CH₂OCH₂CH₂O----
                    |       preferred CRU       |
```
(shortest path through all oxygens)

```
------ONHCH₂OCH₂CH₂OCH₂CH₂ONHCH₂OCH₂CH₂OCH₂CH₂ONHCH₂OCH₂CH₂O----
                         |    preferred CRU    |
```

(starting O and direction determined by N)

Rule 17
If a choice is possible between a divalent and a higher-valent CRU, the number of free valences is minimized only after all other orders of seniority have been observed.

Example:

–CH=CH– (not =CH–CH=)

6 NAMING THE PREFERRED CONSTITUTIONAL REPEATING UNIT (CRU)

6.1 Naming subunits

Rule 18
The subunits and substituted subunits, which are parenthesized or bracketed, are named by organic nomenclature rules [4,5] with the following exceptions:

NOMENCLATURE

1. Alkyl-substituted acyclic carbon chain subunits are named as such, not as single-chain units to differentiate between the length of the acyclic carbon chain which is part of the main chain (backbone) and an acyclic carbon chain substituent on that backbone. This is an exception from the rules in [5].
2. Similarly, heterocyclic or carbocyclic ring assemblies, if not in the main chain (backbone), are not named as such, but as substituted rings or ring systems.

Examples:

$$-CH-CH_2-$$
$$\quad\;\;|$$
$$(CH_2)_9CH_3$$

1-decylethane-1,2-diyl
(not dodecane-2,1-diyl according to ref. 5)

$$-CHCH_2CH_2-$$
$$\;\;|$$
$$CH_3$$

1-methylpropane-1,3-diyl
(not butane-3,1-diyl according to ref. 5)

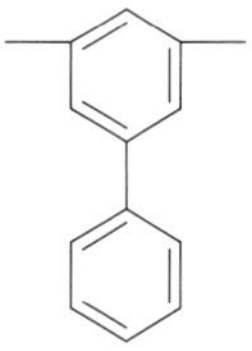

5-phenyl-1,3-phenylene
(not biphenyl-3,5-diyl)

1. A list of names of common subunits is in Appendix 11.1.

6.2 Naming the preferred CRU

Rule 19
The name of the CRU is formed from the names of its subunits, including substituents (substituted subunits), and cited in order from left to right as they appear in the CRU.

Examples:

$-OCH_2CH_2-$ oxyethane-1,2-diyl

$-OCH-$
$\;\;\;|$
$\;\;C_6H_5$ oxy(phenylmethylene)

$-OCHCH_2-$
$\;\;\;|$
$\;\;COOCH_3$ oxy[1-(methoxycarbonyl)ethane-1,2-diyl]

$\quad CH_3$
$\quad\;|$
$-N^+-CH_2CH_2-$
$\quad\;|$
$\quad CH_3\;\;Br^-$ (dimethyliminio)ethane-1,2-diyl bromide

7 NAMING THE POLYMER

Rule 20
Polymers (or oligomers) are named with the prefix poly (or oligo) followed in parentheses or brackets by the name of the CRU. If the name of the repeating unit is 'ABC', the corresponding polymer (or oligomer) name is

SINGLE-STRAND ORGANIC POLYMERS

$\mathrm{+(ABC)}_{\overline{n}}$ poly(ABC) or oligo(ABC)

Note: Where it is desired to specify the chain length in an oligomer, the appropriate Greek prefix (deca, docosa, etc.) may be used.

Example:

$\mathrm{+(OCH_2CH_2)}_{10}$ deca(oxyethane-1,2-diyl)

Rule 21

End-groups may be specified by prefixes placed ahead of the name of the polymer. The end-group designated by α is that attached to the left-hand side of the CRU written as described in the preceding rules and the other end-group is designated by ω; the end-groups are cited in that order. If there is a choice, the end-group with the name starting earlier in the alphabet should be cited first.

Examples:

Cl$_3$C—(C$_6$H$_4$—CH$_2$)$_n$—Cl   α-(trichloromethyl)-ω-chloropoly(1,4-phenylenemethylene)

H—(OCH$_2$CH$_2$)$_n$—OCH$_3$ α-hydro-ω-methoxypoly(oxyethane-1,2-diyl)

(not α-methyl-ω-hydroxypoly(oxyethane-1,2-diyl); alphabetical order of end-groups decides)

CH$_3$O(CH$_2$CH$_2$O)$_n$CO(CH$_2$)$_4$CO(OCH$_2$CH$_2$)$_n$OCH$_3$

α,α'-adipoylbis[ω-methoxypoly(oxyethane-1,2-diyl)]

α,α',α''-benzene-1,3,5-triyltris[poly(1-phenylethane-1,2-diyl)]

(a three-star polymer consisting of a central branch point and three single-strand chains, wherein the benzene ring is the end-group linking the three chains)

8 POLYMER CHAIN AS A SUBSTITUENT

Rule 22

If a regular single-strand chain is linked to a constitutional unit of the main chain (backbone) of a polymer molecule or to a low-molecular-weight structure, either directly or through an intervening unit, it is considered a substituent of the constitutional unit or structure. In naming the polymeric substituent, the actual bonding relations are reflected in the name of the CRU.

NOMENCLATURE

Examples:

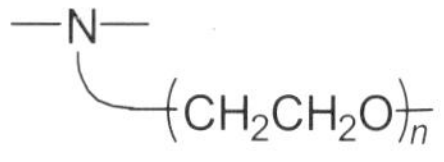

[poly(oxyethane-1,2-diyl)]imino (a polymer-substituted imino subunit)

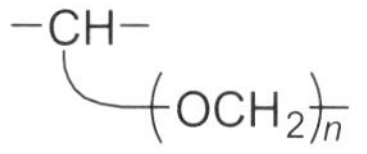

[poly(methyleneoxy)]methylene (a polymer-substituted methylene subunit)

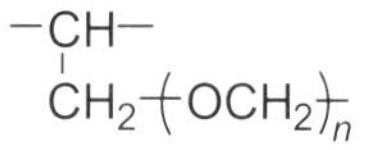

{[poly(methyleneoxy)]methyl}methylene (a polymer-substituted substituted methylene subunit)

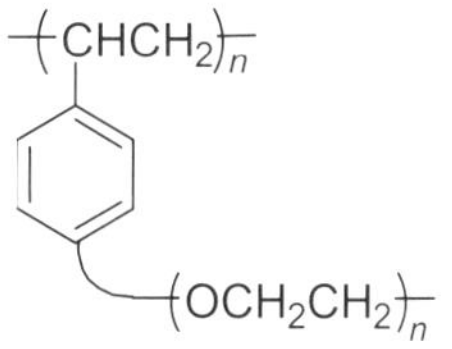

poly(1-{4-[poly(oxyethane-1,2-diyl)]phenyl}ethylene) (a polymer-substituted CRU)

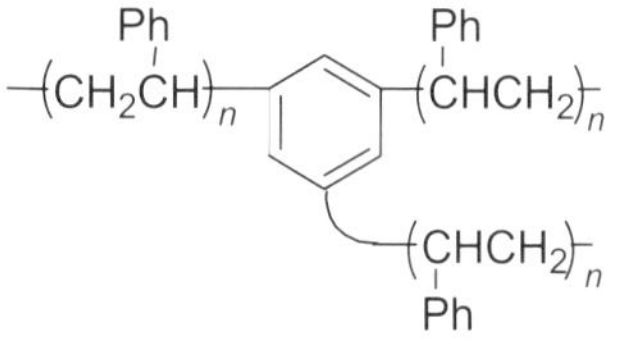

1,3,5-tris[poly(2-phenylethane-1,2-diyl)]benzene (a polymer-substituted low-molecular-weight compound)

9   EXAMPLES OF POLYMER NAMES

To illustrate the present rules for naming various kinds of polymers, examples of polymers are given in this section. The key steps in the naming and the corresponding rule numbers are also given.

Example 1:

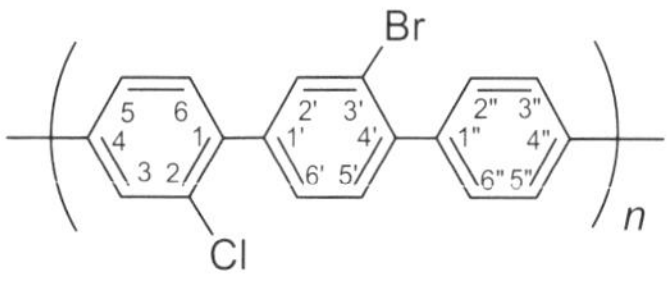

poly(3'-bromo-2-chloro[1,1':4',1''-terphenyl]-4,4''-diyl)
(Rules 9c, 9f, 12)

Example 2:

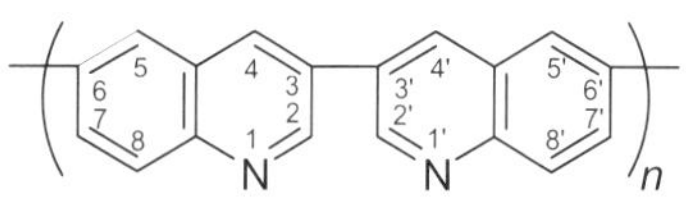

poly([3,3'-biquinoline]-6,6'-diyl)
(Rules 3c, 12)

SINGLE-STRAND ORGANIC POLYMERS

Example 3:

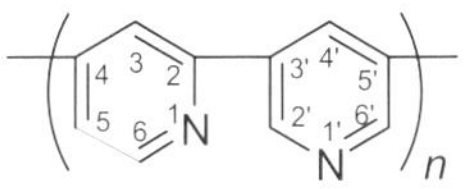

poly([2,3′-bipyridine]-4,5′-diyl)
(Rules 3c,12)

Example 4:

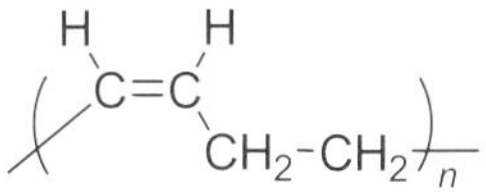

poly[(Z)-but-1-ene-1,4-diyl]
(double bond takes lowest locant; Rule 10b)

Example 5:

−(CH=CH)$_n$−

poly(ethene-1,2-diyl)
(divalent CRU is preferred to =CH−CH=$_n$, poly(ethanediylidene); Rule 17)

Example 6:

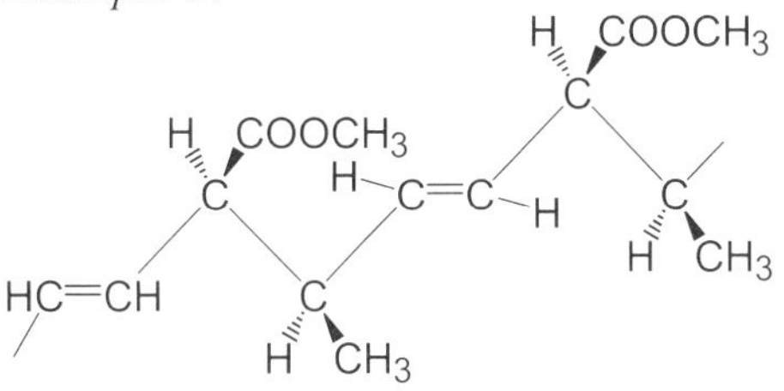

and/or

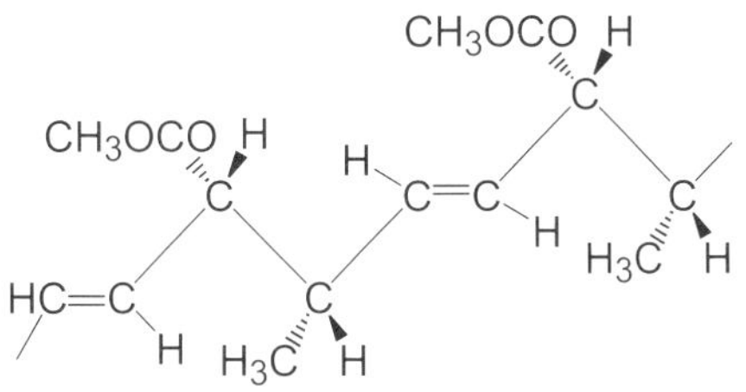

diisotactic poly[*threo*-(E)-3-(methoxycarbonyl)-4-methylbut-1-ene-1,4-diyl]
(double bond takes lowest locant; Rule 10b)

297

NOMENCLATURE

Example 7:

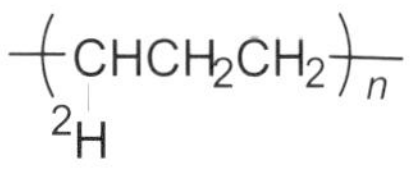

$\text{+CHCH}_2\text{CH}_2\text{+}_n$ with ^2H substituent   poly[(1-^{2}H$_1$)propane-1,3-diyl]
(lowest locant for ^{2}H; Rule 10d)

Example 8:

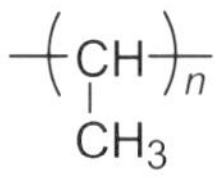

+CH+_n with CH$_3$ substituent   poly(methylmethylene)
(Rule 18)

Example 9:

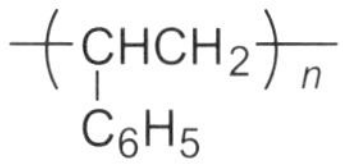

$\text{+CHCH}_2\text{+}_n$ with C$_6$H$_5$ substituent   poly(1-phenylethane-1,2-diyl)
(Rule 10d)

Example 10:

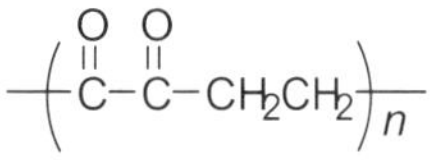

$\text{+C(=O)-C(=O)-CH}_2\text{CH}_2\text{+}_n$ poly(1,2-dioxobutane-1,4-diyl)
(Rule 10d)

Example 11:

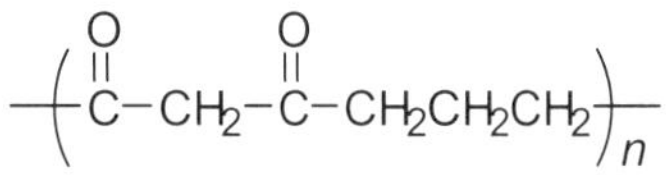

$\text{+C(=O)-CH}_2\text{-C(=O)-CH}_2\text{CH}_2\text{CH}_2\text{+}_n$ poly(1,3-dioxohexane-1,6-diyl)
(Rule 10d)

Example 12:

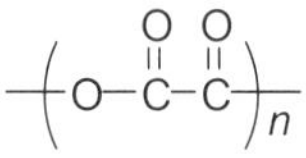

+O-C(=O)-C(=O)+_n poly(oxyoxalyl)
(O is senior; Rule 1)

Example 13:

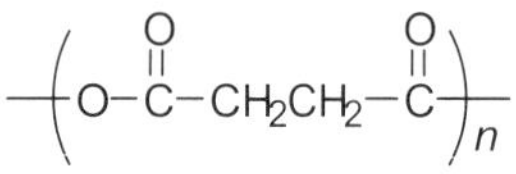

$\text{+O-C(=O)-CH}_2\text{CH}_2\text{-C(=O)+}_n$ poly(oxysuccinyl)
(O is senior; Rule 1)

Example 14:

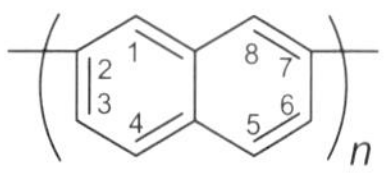

(naphthalene-2,7-diyl repeat unit, numbered 1-8)

poly(naphthalene-2,7-diyl)
(lower free-valence locant on the left; Rule 12)

Example 15:

poly(2*H*-furo[3,2-*b*]pyran-2,6-diyl)
(lower free-valence locant on the left; Rule 12)

Example 16:

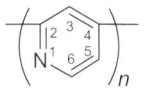

poly(pyridine-2,4-diyl)
(lower locant on the left; Rule 12)

Example 17:

poly(sodium 1-carboxylatoethane-1,2-diyl)
(lower locant for substituent; Rule 10d)

Example 18:

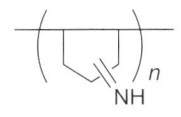

poly(*x*-iminocyclopentane-1,2-diyl) [lower free-valence locant on the left; Rule 12; the *x* is required to differentiate the structure from the following:

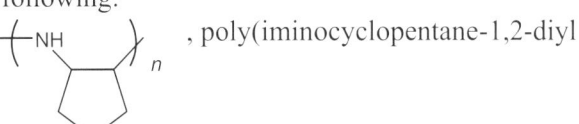

, poly(iminocyclopentane-1,2-diyl

Example 19:

poly(pyridine-3,5-diylpiperidine-2,4-diyl)
(pyridine is senior to piperidine; Rule 3a)

NOMENCLATURE

Example 20:

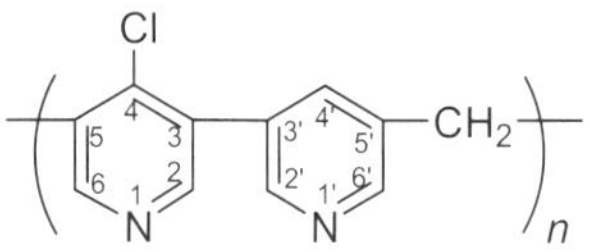

poly[(4-chloro-[3,3'-bipyridine]-5,5'-diyl)methylene]
(ring assembly is senior to acyclic carbon chain; Rules 1,3c,3e,12)

Example 21:

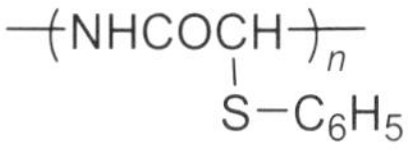

poly{imino[1-oxo-2-(phenylsulfanyl)ethane-1,2-diyl]}
(heteroatom is senior to acyclic carbon chain; in the chain the substituent earlier in the alphabetical order has a lower locant; Rules 1,10e)

Example 22:

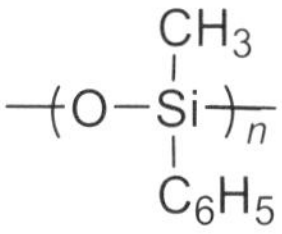

poly[oxy(methylphenylsilanediyl)]
(O is senior to Si; Rule 4) or poly(methylphenylsiloxane)

Example 23:

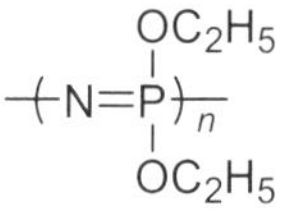

poly[nitrilo(diethoxy-λ^5-phosphanylylidene)]
(N is senior to P; Rule 4) or poly(diethoxyphosphazene)

Example 24:

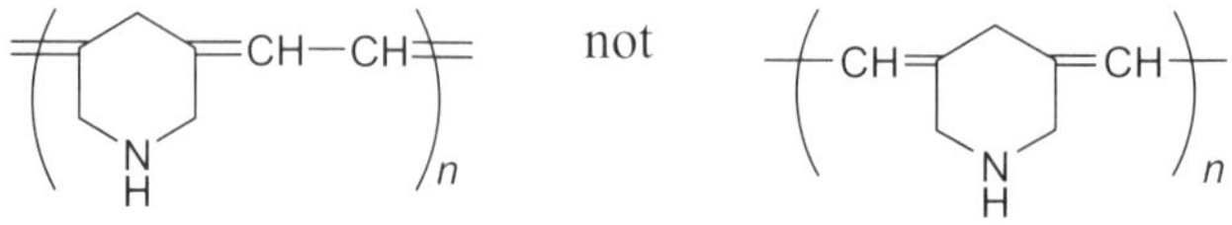

poly(piperidine-3,5-diylideneethanediylidene)
(piperidine ring is senior to acyclic carbon chain; Rule 1)

Example 25:

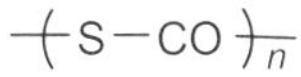

$+S-CO+_n$ poly(sulfanediylcarbonyl)
(S is senior; Rule 1)

Example 26:

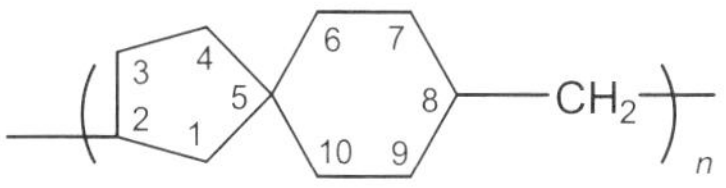

poly(spiro[4.5]decane-2,8-diylmethylene)
(ring system is senior to acyclic carbon chain; lower free-valence locant on the left;
 Rules 1,12)

Example 27:

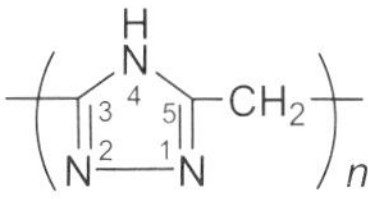

poly(4*H*-1,2,4-triazole-3,5-diylmethylene)
(heterocycle is senior to acyclic carbon chain; lower free-valence locant on the left;
 Rules 1,12)

Example 28:

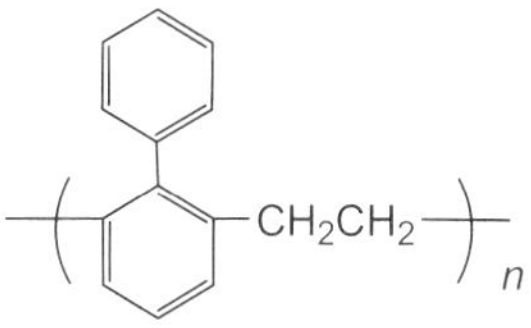

poly[(2-phenyl-1,3-phenylene)ethane-1,2-diyl]
(ring system is senior to acyclic carbon chain; lower free-valence locant on the left;
 Rules 1,12,18)

Example 29:

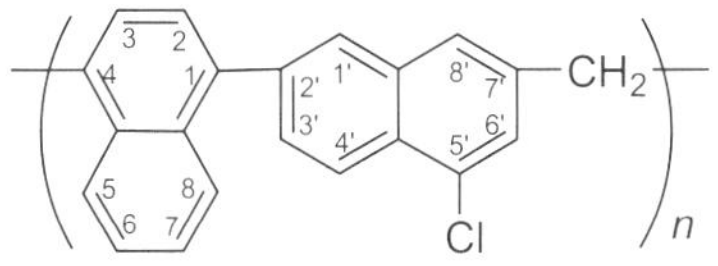

poly[(5'-chloro[1,2'-binaphthalene]-4,7'-diyl)methylene]
(ring system is senior to acyclic carbon chain; Rules 9c,12)

NOMENCLATURE

Example 30:

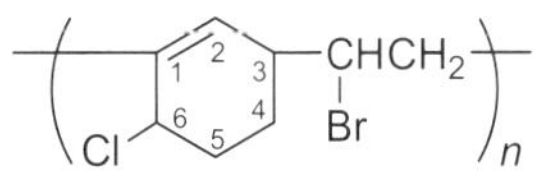

poly[(6-chlorocyclohex-1-ene-1,3-diyl)(1-bromoethane-1,2-diyl)]
(ring is senior to acyclic carbon chain; lower free valence locant on the left; lower locant for Br in ethane-1,2-diyl; Rules 1,9b,10d,11)

Example 31:

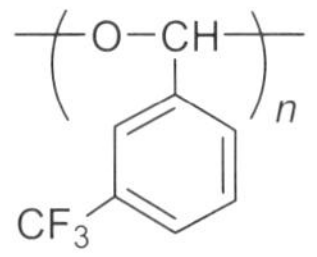

poly(oxy{[3-(trifluoromethyl)phenyl]methylene})
(O is senior to methylene; Rule 1)

Example 32:

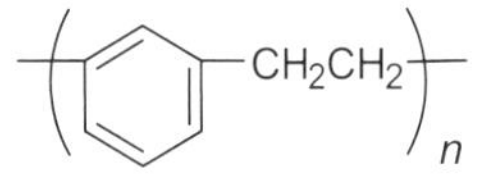

poly(1,3-phenyleneethane-1,2-diyl)
(ring is senior to acyclic carbon chain; Rules 1,11)

Example 33:

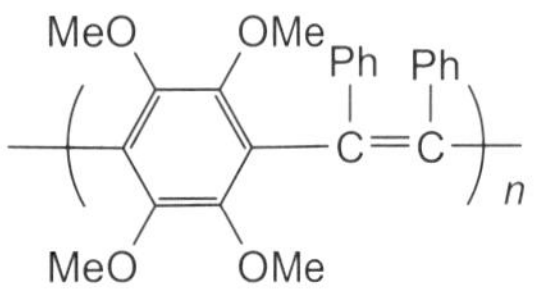

poly[(tetramethoxy-1,4-phenylene)(1,2-diphenylethene-1,2-diyl)]
(ring is senior to acyclic carbon chain; Rule 1)

Example 34:

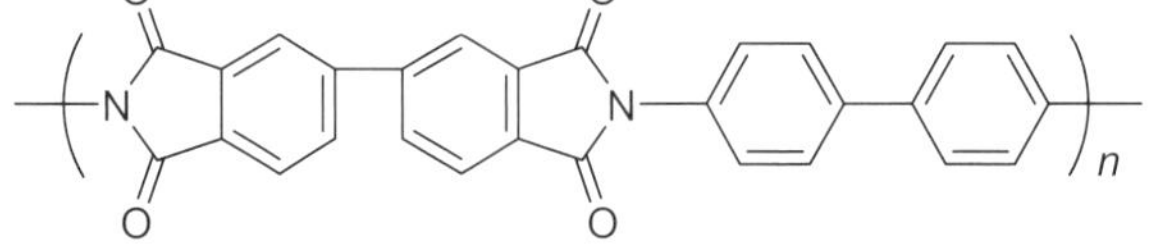

poly{(1,1',3,3'-tetraoxo[5,5'-biisoindoline]-2,2'-diyl)biphenyl-4,4'-diyl}
(heterocyclic system is senior to carbocyclic system; Rules 1,3c,12)

SINGLE-STRAND ORGANIC POLYMERS

Example 35:

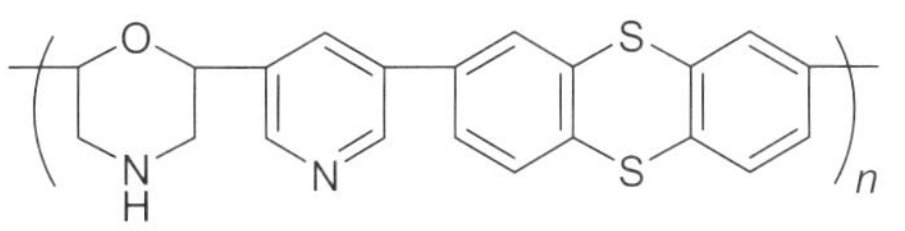

poly(morpholine-2,6-diylpyridine-3,5-diylthianthrene-2,8-diyl)
(nitrogen heterocycles are senior to non-nitrogen heterocycle; ring with larger number of heteroatoms is senior; Rules 2a,2e,12)

Example 36:

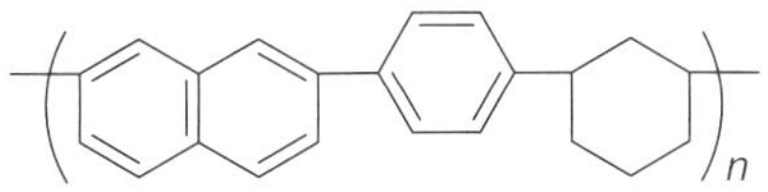

poly(naphthalene-2,7-diyl-1,4-phenylenecyclohexane-1,3-diyl)
(two-ring system is senior to single-ring systems; benzene as least hydrogenated is senior to cyclohexane; Rules 8a,9a,11,12)

Example 37:

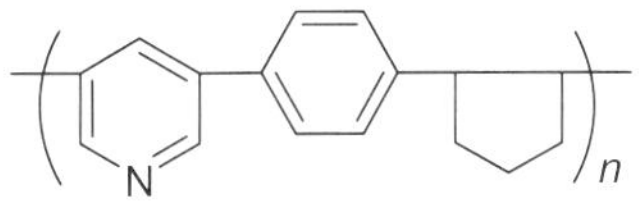

poly(pyridine-3,5-diyl-1,4-phenylenecyclopentane-1,2-diyl)
(heterocycle is senior to carbocycles; larger carbocycle is senior to smaller; Rules 1,8b,11,12)

Example 38:

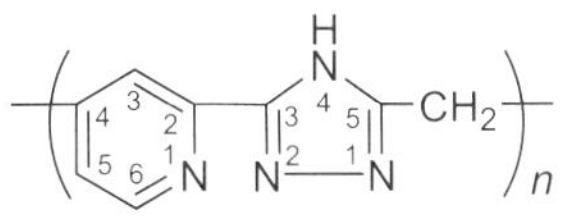

poly(pyridine-4,2-diyl-4*H*-1,2,4-triazole-3,5-diylmethylene)
(heterocycles are senior to acyclic carbon chain; larger nitrogen ring; Rules 1,2d,12)

Example 39:

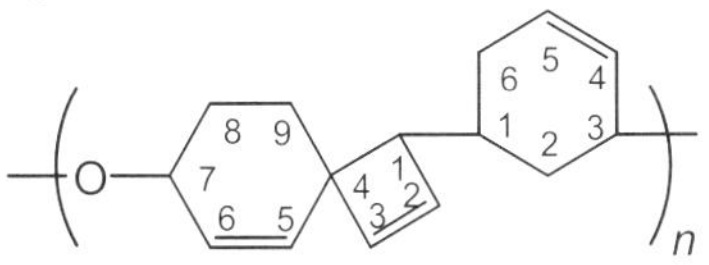

poly(oxyspiro[3.5]nona-2,5-diene-7,1-diylcyclohex-4-ene-1,3-diyl)
(O is senior; shorter path to the senior ring system; Rules 1,8a,11,12)

NOMENCLATURE

Example 40:

poly(piperidine-4,2-diyloxymethylene)
(heterocycle is senior to heteroatom; shorter path between them; Rules 1,12)

Example 41:

poly(piperidine-2,4-diyloxymethylene)
(heterocycle is senior to heteroatom; shorter path between them; Rules 1,12)

Example 42:

poly(pyridine-3,5-diylmethyleneoxy-1,4-phenylene)
(pyridine is senior to O; shorter path between them; Rules 1,11,12)

Example 43:

poly[imino(1-chloro-2-oxoethane-1,2-diyl)(4-nitro-1,3-phenylene)(3-bromopropane-1,3-diyl)]
(heteroatom is senior to carbocycle; shorter path between them; Rules 1,11,12)

Example 44:

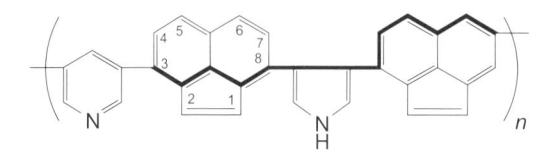

poly(pyridine-3,5-diylacenaphthylene-3,8-diylpyrrole-3,4-diylacenaphthylene-3,7-diyl)
(pyridine is senior to pyrrole; shorter path between them; heavy line denotes path followed: Rules 2d,12)

Example 45:

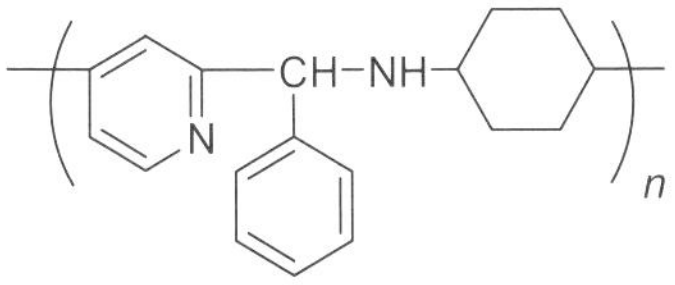

poly[pyridine-4,2-diyl(phenylmethylene)iminocyclohexane-1,4-diyl]
(heterocycle is senior to heteroatom; shorter path between them; Rules 1,11,12)

Example 46:

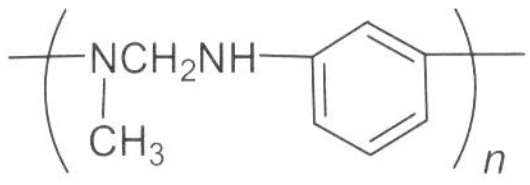

poly[(methylimino)methyleneimino-1,3-phenylene]
(heteroatoms are senior to carbocycle; shorter path through both N; substituted heteroatom is senior to the same unsubstituted heteroatom; Rules 1,5,11,15)

Example 47:

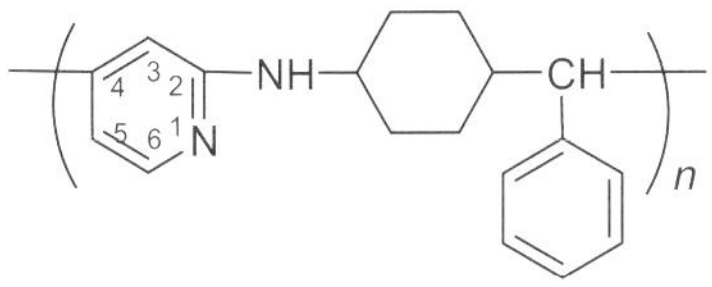

poly[pyridine-4,2-diyliminocyclohexane-1,4-diyl(phenylmethylene)]
(heterocycle is senior to heteroatom; shorter path between them; Rules 1,11,12)

Example 48:

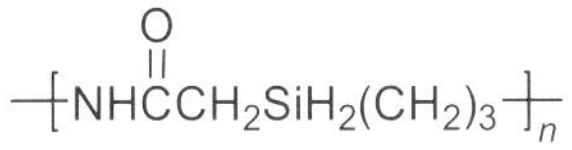

poly[imino(1-oxoethane-1,2-diyl)silanediylpropane-1,3-diyl]
(N is senior to Si; shorter path; Rule 4)

Example 49:

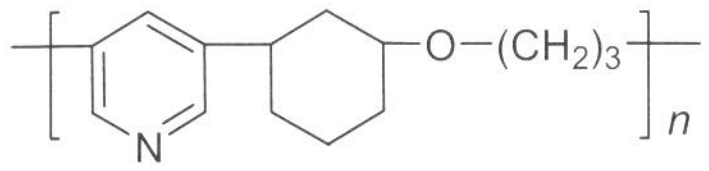

poly(pyridine-3,5-diylcyclohexane-1,3-diyloxypropane-1,3-diyl)
(of two equal paths of three atoms between heterocycle and heteroatom, the path through the carbocycle is preferred to the path through the aliphatic chain; Rules 11,12,14)

NOMENCLATURE

Example 50:

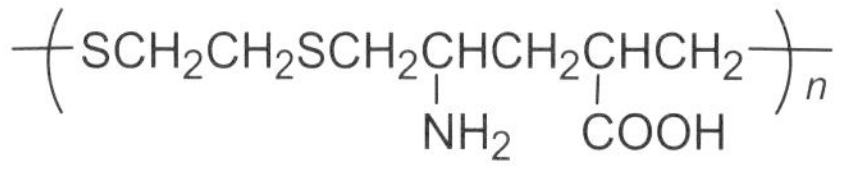

$$-\!\!\left(\text{SCH}_2\text{CH}_2\text{SCH}_2\underset{\underset{\text{NH}_2}{|}}{\text{CH}}\text{CH}_2\underset{\underset{\text{COOH}}{|}}{\text{CH}}\text{CH}_2\right)_{\!\!n}\!\!-$$

poly[sulfanediylethane-1,2-diylsulfanediyl(2-amino-4-carboxypentane-1,5-diyl)]
(shorter path from S to S; direction determined by lower locant for the substituent earlier in the alphabetical order; Rules 10e,15)

Example 51:

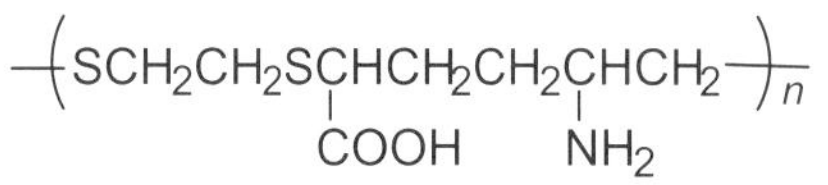

$$-\!\!\left(\text{SCH}_2\text{CH}_2\underset{\underset{\text{COOH}}{|}}{\text{SCH}}\text{CH}_2\text{CH}_2\underset{\underset{\text{NH}_2}{|}}{\text{CH}}\text{CH}_2\right)_{\!\!n}\!\!-$$

poly[sulfanediylethane-1,2-diylsulfanediyl(4-amino-1-carboxypentane-1,5-diyl)]
(shorter path through both S; lowest locants for substituents; Rules 10d,15)

Example 52:

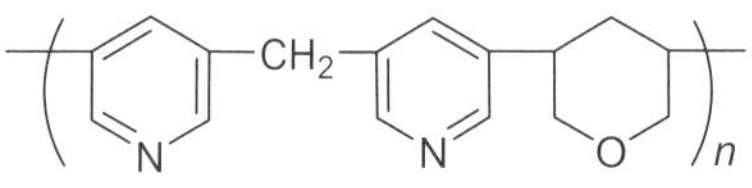

poly(pyridine-3,5-diylmethylenepyridine-3,5-diyltetrahydropyran-3,5-diyl)
(shorter path between pyridine subunits; Rules 12,15)

Example 53:

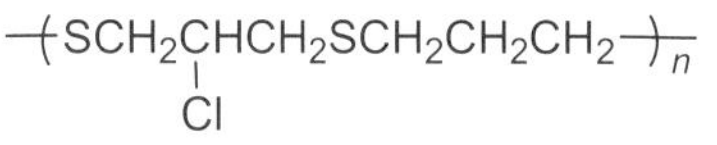

$$-\!\!\left(\text{SCH}_2\underset{\underset{\text{Cl}}{|}}{\text{CH}}\text{CH}_2\text{SCH}_2\text{CH}_2\text{CH}_2\right)_{\!\!n}\!\!-$$

poly[sulfanediyl(2-chloropropane-1,3-diyl)sulfanediylpropane-1,3-diyl]
(substituted acyclic carbon chain is senior to unsubstituted; Rules 10c,15)

Example 54:

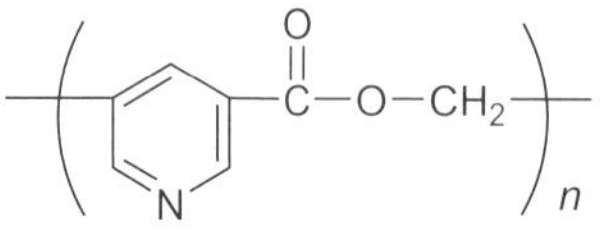

poly(pyridine-3,5-diylcarbonyloxymethylene)
(heterocycle is senior to heteroatom; substituted acyclic carbon chain is senior to unsubstituted one; Rules 1,10c)

Example 55:

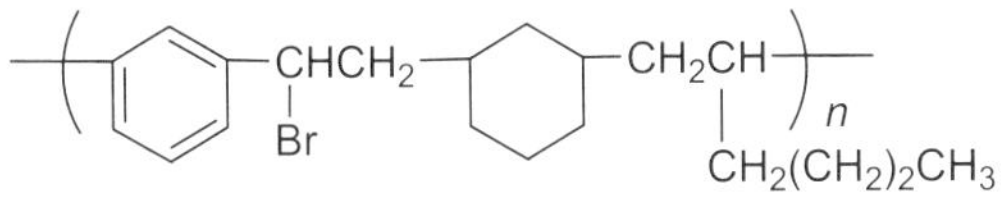

<chemical structure>

poly[1,3-phenylene(1-bromoethane-1,2-diyl)cyclohexane-1,3-diyl(2-butylethane-1,2-diyl)]
(least hydrogenated ring is senior; the direction is determined by the alphabetical order of the substituents in the carbon chain; Rules 9a,10e,11)

Example 56:

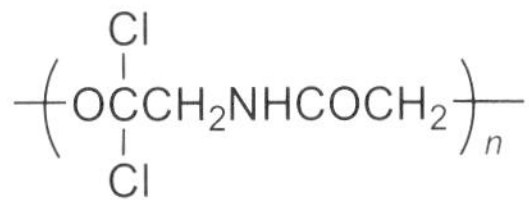

<chemical structure>

poly[oxy(1,1-dichloroethane-1,2-diyl)imino(1-oxoethane-1,2-diyl)]
(O is senior to N; of the two two-carbon chains the one with the larger number of substituents is senior; Rules 4,10c)

Example 57:

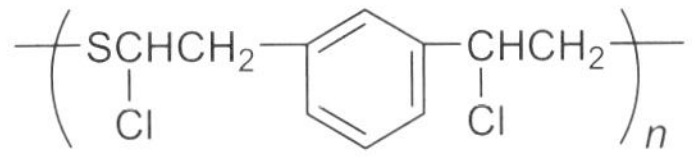

<chemical structure>

poly[sulfanediyl(1-chloroethane-1,2-diyl)(1,3-phenylene)(1-chloroethane-1,2-diyl)]
(heteroatom is senior to carbocycle; of equal chains with the same substituents that with lower locant is senior; Rules 1,10d)

Example 58:

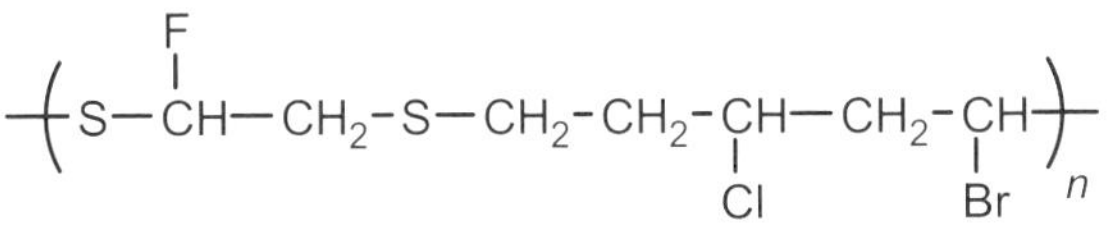

<chemical structure>

poly[sulfanediyl(1-fluoroethane-1,2-diyl)sulfanediyl(5-bromo-3-chloropentane-1,5-diyl)]
(shorter path through both S; direction determined by the lower locant of the fluoro substituent; Rules 10d,15)

Example 59:

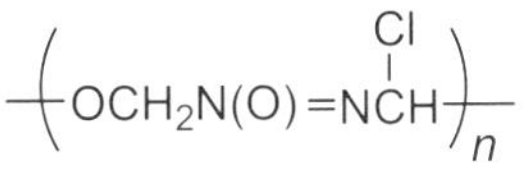

<chemical structure>

poly[oxymethylene-*ONN*-azoxy(chloromethylene)]
(O is senior to N; direction in the group –N(O)=N– is indicated by the prefix *ONN*; Rules 4,7d)

NOMENCLATURE

Example 60:

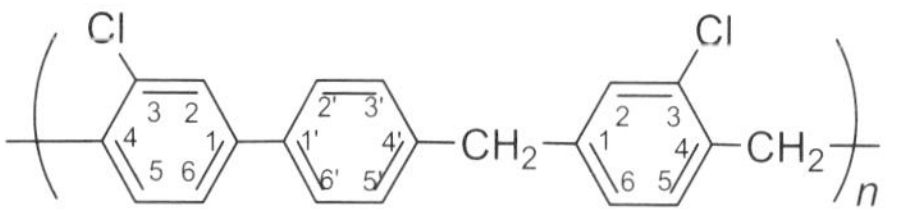

poly[(3-chlorobiphenyl-4,4'-diyl)methylene(3-chloro-1,4-phenylene)methylene]
(direction determined by the lower locant for the chloro substituent in biphenyl; Rules 9c, 9f, 16)

Example 61:

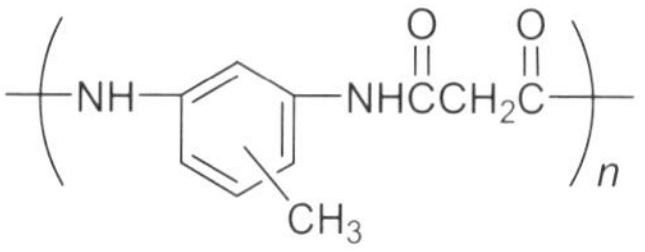

poly[imino(x-methyl-1,3-phenylene)iminomalonyl]
benzene is senior to acyclic carbon chain; the path from N to N through the ring can go in either direction because of the absence of a specific locant for the methyl group; Rules 1, 11, 15)

Example 62:

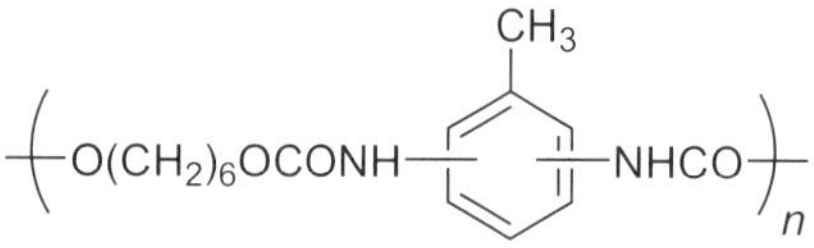

poly[oxyhexane-1,6-diyloxycarbonylimino(methylphenylene)iminocarbonyl]
(path between the two oxygens through the acyclic carbon chain is shorter, since for the benzene ring with unknown positions of its attachment to other atoms the longest possible path of four carbon atoms is being considered; Rule 15)

Example 63:

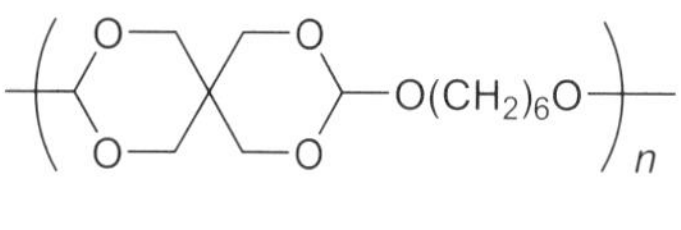

poly(2,4,8,10-tetraoxaspiro[5.5]undecane-3,9-diyloxyhexane-1,6-diyloxy)
(heterocyclic ring system is senior to heteroatoms; Rules 1, 12)

SINGLE-STRAND ORGANIC POLYMERS

Example 64:

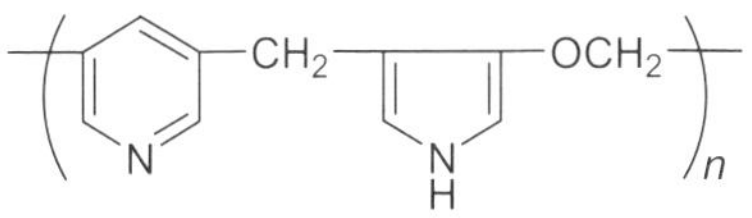

$+\!\!\left(\!\!\begin{array}{c}\text{—CH}_2\!-\!\end{array}\!\!\right)_n$

poly(pyridine-3,5-diylmethylenepyrrole-3,4-diyloxymethylene)
(pyridine is senior to pyrrole; shorter paths; Rules 2d,12)

Example 65:

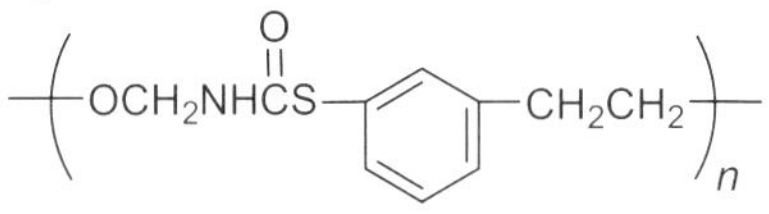

poly(oxymethyleneiminocarbonylsulfanediyl-1,3-phenyleneethane-1,2-diyl)
(O is senior to S; shorter path; Rules 4,11)

Example 66:

$+(\text{ONHCH}_2\text{NHNHCH}_2)_n$

poly(oxyiminomethylenehydrazine-1,2-diylmethylene)
(O is senior to N; shortest path between O and N; Rule 4)

Example 67:

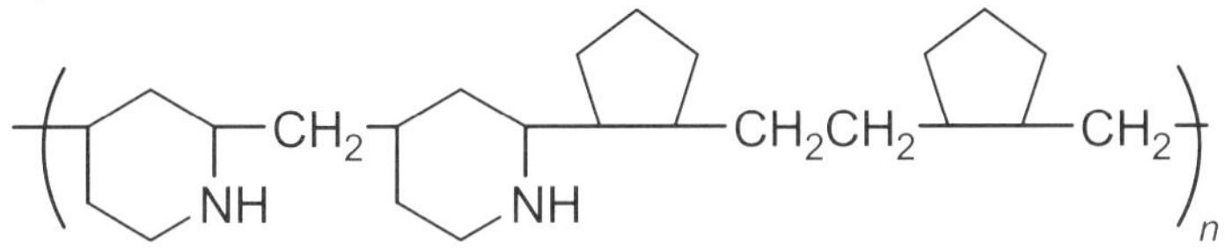

poly(piperidine-4,2-diylmethylenepiperidine-4,2-diylcyclopentane-1,2-diylethane-1,2-diylcyclopentane-1,2-diylmethylene)
(shorter path between two identical senior heterocycles; shorter path from heterocycle to carbocycle; Rules 1,11,12,15)

Example 68:

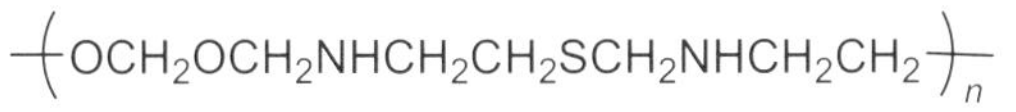

$+(\text{OCH}_2\text{OCH}_2\text{NHCH}_2\text{CH}_2\text{SCH}_2\text{NHCH}_2\text{CH}_2)_n$

poly(oxymethyleneoxymethyleneiminoethane-1,2-diylsulfanediylmethylene-iminoethane-1,2-diyl) or poly(1,3-dioxa-8-thia-5,10-diazadodecane-1,12-diyl)[*]
(O is senior to S; shorter path from O to O to N; Rules 4,15)

---

[*] In Examples 68, 69 and 75, alternative names that use replacement ('a') nomenclature are presented [5]. As conventionally applied to acyclic structures with multiple heteroatoms, terminal heteroatoms are not designated with 'a' prefixes but are named as characteristic groups of the structure, i.e., as hydroxy, amino, carboxylic acid, etc. However, heteroatoms in such positions within the constitutional units of polymer molecules are not terminal units. Consequently, such atoms are designated with 'a' prefixes, and thereby the simplicity afforded by the application of replacement nomenclature to polymers is enhanced.

NOMENCLATURE

Example 69:

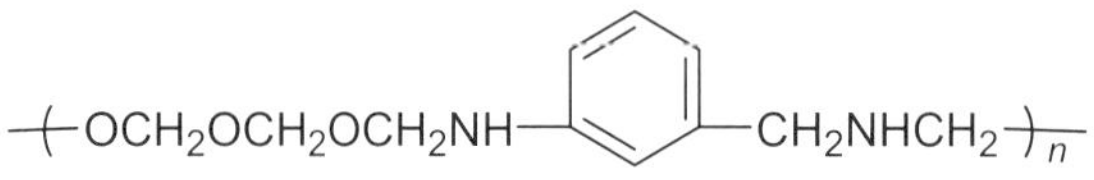

poly(oxymethyleneoxymethyleneoxymethyleneimino-1,3-phenylene-methyleneiminomethylene) or
poly(1,3,5-trioxa-7-azaheptane-1,7-diyl-1,3-phenylene-2-azapropane-1,3-diyl)*
(O is senior; the shorter path through all oxygens to the ring has been taken; Rules 4,11,16)

Example 70:

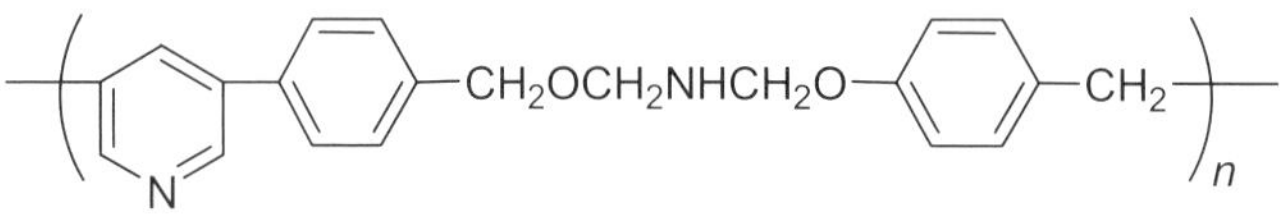

poly(pyridine-3,5-diyl-1,4-phenylenemethyleneoxymethyleneiminomethyleneoxy-1,4-phenylenemethylene)
(heterocycle is senior; shorter path from pyridine to benzene; Rules 1,11,12)

Example 71:

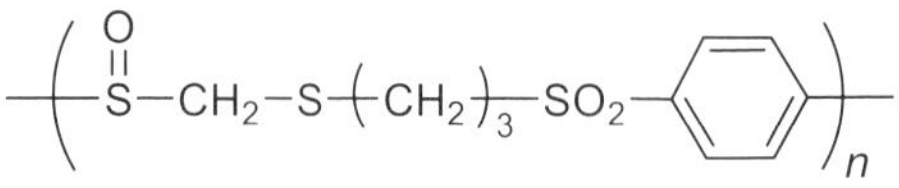

poly(sulfinylmethylenesulfanediylpropane-1,3-diylsulfonyl-1,4-phenylene)
(shortest path between the heteroatoms; Rule 16)

Example 72:

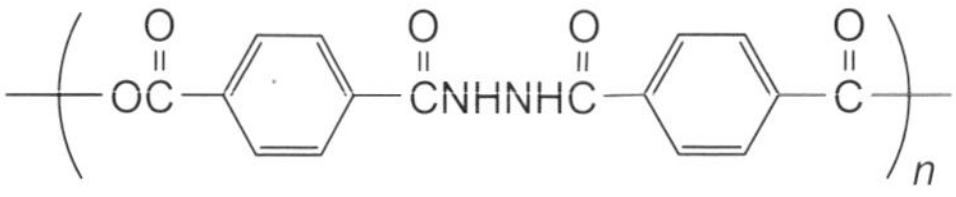

poly(oxyterephthaloylhydrazine-1,2-diylterephthaloyl)
 (O is senior to N; Rule 4)

Example 73:

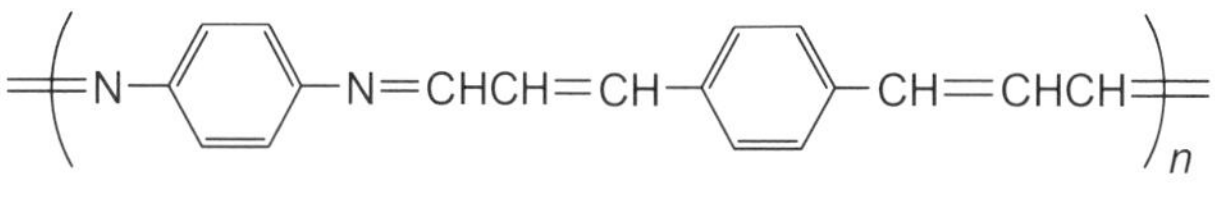

poly(nitrilo-1,4-phenylenenitriloprop-2-en-3-yl-1-ylidene-1,4-phenyleneprop-1-en-1-yl-3-ylidene)
(N is senior; shorter path between N; Rules 1,15)

Example 74:

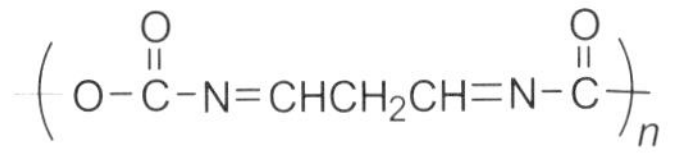

$$\left(O-\overset{O}{\underset{\|}{C}}-N=CHCH_2CH=N-\overset{O}{\underset{\|}{C}} \right)_n$$

poly(oxycarbonylnitrilopropane-1,3-diylidenenitrilocarbonyl)
(O is senior to N; the direction of bonding in unsymmetrical nitrilo subunits, =N– or –N=, is indicated by the endings of the names of the adjacent subunits in the CRU; Rule 3)

Example 75:

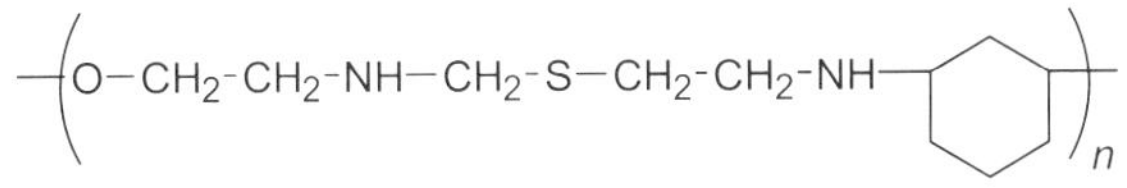

$$\left(O-CH_2-CH_2-NH-CH_2-S-CH_2-CH_2-NH-\text{[cyclohexane]} \right)_n$$

poly(oxyethane-1,2-diyliminomethylenesulfanediylethane-1,2-diyliminocyclohexane-1,3-diyl)
(O is senior to S; shorter path; Rules 4,11)
poly(1-oxa-6-thia-4,9-diazanonane-1,9-diylcyclohexane-1,3-diyl)*

Example 76:

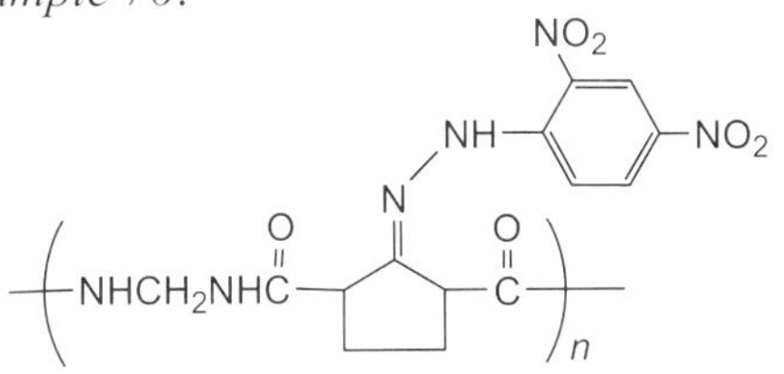

$$\left(NHCH_2NH\overset{O}{\underset{\|}{C}}-\underset{\substack{\|\\N\\\|\\NH-\text{[2,4-dinitrophenyl]}}}{\text{[cyclopentane]}}-\overset{O}{\underset{\|}{C}} \right)_n$$

poly(iminomethyleneiminocarbonyl{2-[(2,4-dinitrophenyl)hydrazono]cyclopentane-1,3-diyl}carbonyl)
(either path through both N to ring; Rule 15)

Example 77:

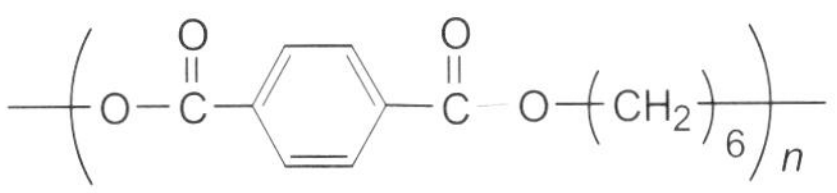

$$\left(O-\overset{O}{\underset{\|}{C}}-\text{[benzene]}-\overset{O}{\underset{\|}{C}}-O-(CH_2)_6 \right)_n$$

poly(oxyterephthaloyloxyhexane-1,6-diyl)
(O is senior to benzene; shorter path between them; Rules 1,15)

---

* see footnote on page 308.

Example 78:

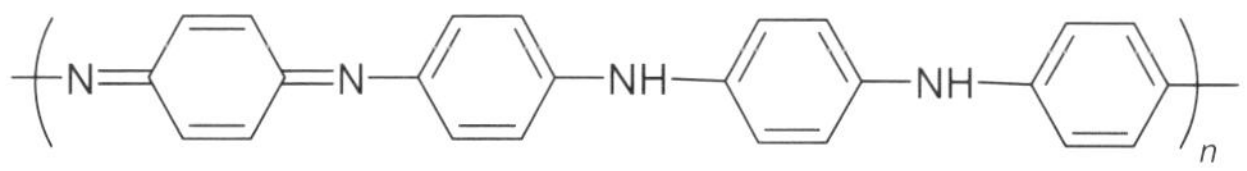

poly(nitrilocyclohexa-2,5-diene-1,4-diylidenenitrilo-1,4-phenyleneimino-1,4-phenyleneimino-1,4-phenylene)
(–N= is most senior; Rule 7a)

Example 79:

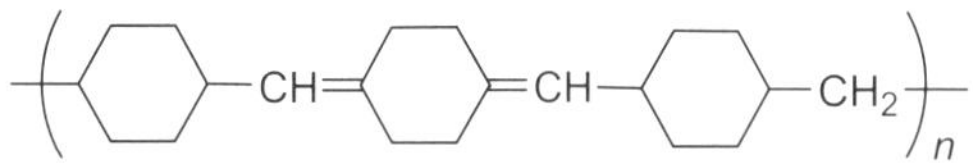

poly(cyclohexane-1,4-diylmethanylylidenecyclohexane-1,4-diylidenemethanylylidenecyclohexane-1,4-diylmethylene)
(cyclohexane is most senior; the path to the next cyclohexane goes through a senior acyclic carbon atom –CH=; the number of free valences is minimized; Rules 10a,17)

10 REFERENCES

1. IUPAC. 'Report on nomenclature in the field of macromolecules', *J. Polym. Sci.* **8**, 257-277 (1952).
2. IUPAC. 'Report on nomenclature dealing with steric regularity in high polymers', *Pure Appl. Chem.* **12**, 643-656 (1966).
3. IUPAC. 'Nomenclature of regular single-strand organic polymers (Rules approved 1975)', *Pure Appl. Chem.* **48**, 373-385 (1976).
4. IUPAC. *Nomenclature of Organic Chemistry,* Sections A, B, C, D, E, F and H, Pergamon Press, Oxford (1979).
5. IUPAC. *A Guide to IUPAC Nomenclature of Organic Compounds,* Blackwell Scientific Publications, Oxford (1993).
6. IUPAC. 'Stereochemical definitions and notations relating to polymers (Recommendations 1980)', *Pure Appl. Chem.* **53**, 733-752 (1981). Reprinted as Chapter 2 this volume.
7. IUPAC. 'Nomenclature of regular double-strand (ladder and spiro) organic polymers (IUPAC Recommendations 1993)', *Pure Appl. Chem.* **65**, 1561-1580 (1993). Reprinted as Chapter 16 this volume.
8. IUPAC. 'Structure-based nomenclature for irregular single-strand organic polymers (IUPAC Recommendations 1993)', *Pure Appl. Chem.* **66**, 873-889 (1994). Reprinted as Chapter 17 this volume.
9. IUPAC. 'Glossary of basic terms in polymer science (IUPAC Recommendations 1996), *Pure Appl. Chem.* **68**, 2287-2311 (1996). Reprinted as Chapter 1 this volume.
10. IUPAC. 'Nomenclature of fused and bridged fused ring systems (IUPAC Recommendations 1998)', *Pure Appl. Chem.* **70**, 143-216 (1998).

SINGLE-STRAND ORGANIC POLYMERS

11 APPENDIX

11.1 List of names of common subunits

(The use of subunit names denoted with an asterisk (*) are not acceptable [5]. However, it should be noted that those entries also marked † are retained IUPAC names for groups other than the ones being indicated here.)

adipoyl*	see hexanedioyl
azo*	see diazenediyl
azoimino*	see triazene-1,3-diyl
azoxy	–N(O)=N– or –N=N(O)–
benzoylimino	$C_6H_5CON<$
benzylidene*†	see phenylmethylene
biphenyl-3,5-diyl	see 5-phenyl-1,3-phenylene
biphenyl-4,4'-diyl	
butanedioyl	–COCH$_2$CH$_2$CO–
butane-1,1-diyl	see propylmethylene
butane-1,4-diyl	–(CH$_2$)$_4$–
butylidene*†	see propylmethylene
but-1-ene-1,4-diyl	–CH=CHCH$_2$CH$_2$–
carbonimidoyl	–C(=NH)–
carbonothioyl	–CS–
carbonyl	–CO–
cyclohexane-1,1-diyl	
cyclohexane-1,4-diyl	
cyclohexylidene*†	see cyclohexane-1,1-diyl
decanedioyl	–CO(CH$_2$)$_8$CO–
diazenediyl	–N=N–
diazoamino*	see triaz-1-ene-1,3-diyl
dimethylmethylene	(CH$_3$)$_2$C<
dioxy*	see peroxy
diphenylmethylene	(C$_6$H$_5$)$_2$C<
disulfanediyl	–SS–
dithio*	see disulfanediyl
ethanedioyl	see oxalyl
ethane-1,1-diyl*	see methylmethylene
ethane-1,2-diyl	–CH$_2$CH$_2$–
ethanediylidene	=CHCH=
ethene-1,2-diyl	–CH=CH–
ethylene*	see ethane-1,2-diyl
ethylidene*†	see methylmethylene
glutaryl*	see pentanedioyl
hexamethylene*	see hexane-1,6-diyl

NOMENCLATURE

hexanedioyl	$-CO(CH_2)_4CO-$
hexane-1,6-diyl	$-(CH_2)_6-$
hydrazine-1,2-diyl	$-NHNH-$
hydrazo*	see hydrazine-1,2-diyl
hydroxyimino	$HO-N<$
imino	$-NH-$
iminio	$-NH^+-$
isophthaloyl	—CO—⌬—CO— (1,3-benzene)
isopropylidene*†	see dimethylmethylene
malonyl	$-COCH_2CO-$
methanylylidene	$-CH=$
methylene	$-CH_2-$
1-methylethane-1,1-diyl*	see dimethylmethylene
1-methylethane-1,2-diyl	$-CH(CH_3)CH_2-$
methylidyne ($-CH=$)*†	see methanylylidene
methylmethylene	$CH_3CH<$
methylylidene*	see methanylylidene
naphthalene-1,8-diyl	(naphthalene with bonds at 1,8)
nitrilo	$-N=$
oxalyl	$-COCO-$
oxy	$-O-$
pentamethylene*	see pentane-1,5-diyl
pentanedioyl	$-COCH_2CH_2CH_2CO-$
pentane-1,5-diyl	$-(CH_2)_5-$
peroxy	$-OO-$
1,4-phenylene	(1,4-benzene)
phenylmethylene	$C_6H_5CH<$
5-phenyl-1,3-phenylene	(1,3-phenylene with phenyl at 5)
phthaloyl	(1,2-benzene with two CO)
piperidine-1,4-diyl	(piperidine with bonds at N and 4)
propanedioyl	see malonyl

SINGLE-STRAND ORGANIC POLYMERS

propane-1,3-diyl	–(CH$_2$)$_3$–
propane-2,2-diyl	see dimethylmethylene
propylene*	see 1-methylethane-1,2-diyl
propylmethylene	CH$_3$CH$_2$CH$_2$CH<
silanediyl	–SiH$_2$–
silylene	see silanediyl
succinyl	see butanedioyl
sulfanediyl	–S–
sulfinyl	–SO–
sulfonyl	–SO$_2$–
thio*†	see sulfanediyl
terephthaloyl	—CO—C$_6$H$_4$—CO—
tetramethylene*	see butane-1,4-diyl
thiocarbonyl*	see carbonothioyl
triazene-1,3-diyl	–N=N–NH–
trimethylene*	see propane-1,3-diyl
vinylene*	see ethene-1,2-diyl
vinylidene	see methylidenemethylene

11.2 Structure- and source-based names for common polymers

The Commission recognized that a number of common polymers have semisystematic or trivial source-based names that are well established by usage; it is not intended that they be immediately supplanted by the structure-based names. Nevertheless, it is hoped that for scientific communication the use of semisystematic or trivial source-based names for polymers will be kept to a minimum.

For the following idealized structural representations, the semisystematic or trivial source-based names given are approved for use in scientific work; the corresponding structure-based names are given as alternative names. Equivalent names for close analogues of these polymers [e.g. other alkyl ester analogues of poly(methyl acrylate)] are also acceptable.

Structure	Source-based name (preferred given first)	Structure-based name
—(CH$_2$)$_n$— [a]	polyethene polyethylene[b]	poly(methylene)

—(CHCH$_2$)$_n$—
      |
     CH$_3$
	polypropene
	polypropylene[b]
		poly(1-methylethane-1,2-diyl)

CH$_3$ | —(CCH$_2$)$_n$— | CH$_3$	poly(2-methylpropene) polyisobutylene[b]	poly(1,1-dimethylethane-1,2-diyl)
—(CH=CH—CH$_2$—CH$_2$)$_n$—	poly(buta-1,3-diene)	poly(but-1-ene-1,4-diyl)

NOMENCLATURE

Structure	Common name	Systematic name
$-(CH_2-C(CH_3)=CH-CH_2)_n-$	polybutadiene polyisoprene	poly(2-methylbut-2-ene-1,4-diyl)
$-(CHCH_2)_n-$ with C_6H_5	poly(ethenylbenzene) polystyrene[b]	poly(1-phenylethane-1,2-diyl)
$-(CHCH_2)_n-$ with CN	Polyacrylonitrile[b]	poly(1-cyanoethane-1,2-diyl)
$-(CHCH_2)_n-$ with OH	poly(vinyl alcohol)	poly(1-hydroxyethane-1,2-diyl)
$-(CHCH_2)_n-$ with OCOCH$_3$	poly(vinyl acetate)	poly(1-acetoxyethane-1,2-diyl)
$-(CHCH_2)_n-$ with Cl	poly(vinyl chloride)	poly(1-chloroethane-1,2-diyl)
$-(CF_2CH_2)_n-$	poly(1,1-difluoroethene) poly(vinylidene difluoride)[b]	poly(1,1-difluoroethane-1,2-diyl)
$-(CF_2)_n-$ [a]	poly(tetrafluoroethene) poly(tetrafluoroethylene)[b]	poly(difluoromethylene)
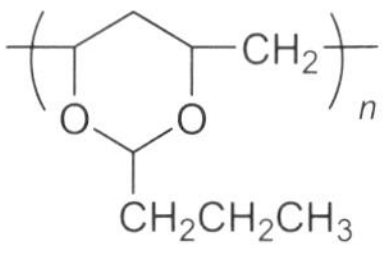 (dioxane ring with CH$_2$CH$_2$CH$_3$)$_n$	poly(vinyl butyral)[b]	poly[(2-propyl-1,3-dioxane-4,6-diyl)methylene]
$-(CHCH_2)_n-$ with COOCH$_3$	poly(methyl acrylate)	poly[1-(methoxycarbonyl)ethane-1,2-diyl]
$-(C(CH_3)CH_2)_n-$ with COOCH$_3$	poly(methyl methacrylate)	poly[1-(methoxycarbonyl)-1-methylethane-1,2-diyl]
$-(OCH_2)_n-$	polyformaldehyde	poly(oxymethylene)
$-(OCH_2CH_2)_n-$	poly(ethylene oxide)	poly(oxyethane-1,2-diyl)
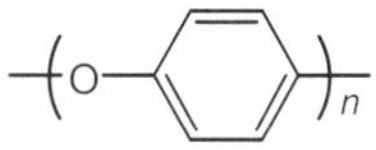 $-(O-C_6H_4)_n-$	poly(phenylene oxide)	poly(oxy-1,4-phenylene)
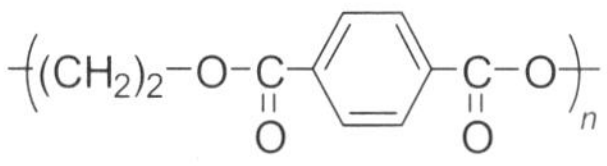 $-((CH_2)_2-O-CO-C_6H_4-CO-O)_n-$	poly(ethane-1,2-diyl terephthalate)	poly(oxyethane-1,2-diyloxy-terephthaloyl)

${-}$(NHCO(CH$_2$)$_5$)$_n{-}$	poly(hexano-6-lactam) poly(ε-caprolactam)	poly[imino(1-oxohexane-1,6-diyl)]
${-}$(NHCH$_2$CH$_2$)$_n{-}$	polyaziridine poly(ethylenimine)	poly(iminoethane-1,2-diyl)

[a] The formulae $-(CH_2CH_2)_n-$ and $-(CF_2CF_2)_n-$ are more often used; they are acceptable due to past usage and an attempt to retain some similarity to the CRU formulae of homopolymers derived from other ethene derivatives.
[b] Traditional names.

16: Nomenclature of Regular Double-Strand (Ladder and Spiro) Organic Polymers

CONTENTS

1. Preamble
2. Definitions
3. Structure-based nomenclature
   3.1 Fundamental principles
   3.2 Identification of the preferred constitutional repeating unit
   3.3 Orientation of the constitutional repeating unit
   3.4 Naming of the preferred constitutional repeating unit
   3.5 Naming of the polymer
   3.6 Polymers constituted of repeatedly fused or spiro carbocycles
   3.7 Polymers constituted of repeatedly fused or spiro carbocyclic systems
   3.8 Polymers constituted of repeatedly fused or spiro heterocycles
   3.9 Polymers constituted of repeatedly fused or spiro heterocyclic systems
   3.10 Substituents
   3.11 End-groups
4. Source-based nomenclature
5. Acknowledgment
6. References

1 PREAMBLE

In 1976, the IUPAC Commission on Macromolecular Nomenclature published rules on the nomenclature of regular single-strand organic polymers. These were later revised [1]. A regular single-strand polymer is one which can be described by constitutional repeating units with only two terminals composed of one atom each.

In 1985, the Commission jointly with the IUPAC Commission on Nomenclature of Inorganic Chemistry published rules on the nomenclature for regular single-strand and quasi-single-strand inorganic and coordination polymers [2]. A quasi-single-strand polymer is one in which the molecules can be described by constitutional repeating units connected to adjacent constitutional repeating unit (or to an end-group) through a single atom at one terminal and through two atoms at the other terminal.

In 1985, the Commission also published rules on source-based nomenclature for copolymers [3] which by definition are polymers derived from more than one species of monomer [4].

Originally prepared by a Working Group consisting of W. V. Metanomski (USA), R. E. Bareiss (FRG), J. Kahovec (Czech Republic), K. L. Loening (USA), L. Shi (PRC) and V. P. Shibaev (Russia). Reprinted from *Pure Appl. Chem.* **65**, 1561-1580 (1993).

DOUBLE-STRAND ORGANIC POLYMERS

In this report, both the structure-based and source-based nomenclature rules are extended to regular double-strand (ladder and spiro) organic polymers. Biopolymers, however, such as DNA are not considered here. Rules for quasi-single-strand coordination polymers, resembling spiro polymers yet not covalently bonded, are *not* included.

2 DEFINITIONS

2.1 regular polymer
Polymer, the molecules of which essentially comprise the repetition of a single constitutional unit with all units connected identically with respect to directional sense.

2.2 constitutional unit
Atom or group of atoms (with pendant atoms or groups, if any) comprising a part of the essential structure of a macromolecule, an oligomer molecule, a block, or a chain [4].

2.3 constitutional repeating unit
Smallest constitutional unit, the repetition of which constitutes a regular macromolecule, a regular oligomer molecule, a regular block, or a regular chain [4].

2.4 ring
Arrangement of atoms bound to each other in a cyclic manner (a closed chain).

2.5 ring system
Single ring or a combination of rings linked one to another by atoms common to both.

2.6 double-strand polymer
Polymer the molecules of which consist of an uninterrupted sequence of rings with adjacent rings having one atom in common (spiro polymer) or two or more atoms in common (ladder polymer).

Examples:

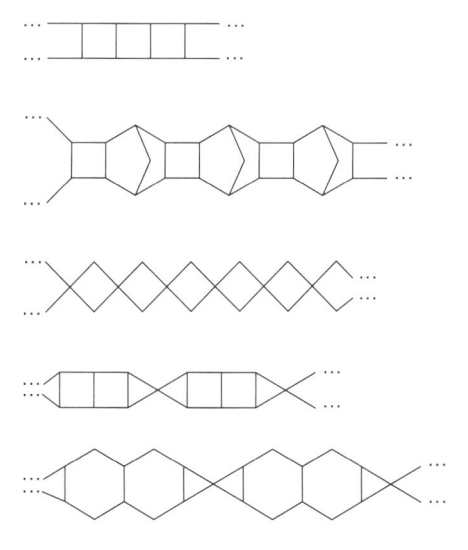

NOMENCLATURE

Note: Partial ladder (imperfect ladder, block ladder) polymers [5], in which the sequence of rings is interrupted and a divalent constitutional repeating unit can be identified, are *not* double-strand polymers. They are named as single-strand polymers.

Example:

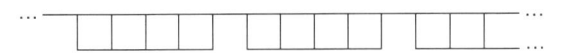

On the other hand, occasional irregular deviations from ideality can be ignored in constructing graphical representations for double-strand polymers and in assigning corresponding names.
Specific examples of double-strand polymers:

  a. constituted of repeatedly fused carbocycles

  (i)

  b. constituted of repeatedly fused or spiro carbocyclic systems

  (ii)

  (iii)

  c. constituted of repeatedly fused heterocycles

  (iv)

  d. constituted of repeatedly fused or spiro heterocyclic systems

  (v)

  (vi)

DOUBLE-STRAND ORGANIC POLYMERS

(vii) 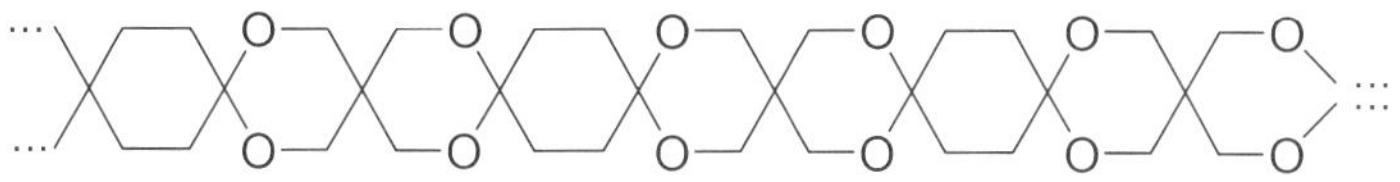 [structure]

(viii) [structure]

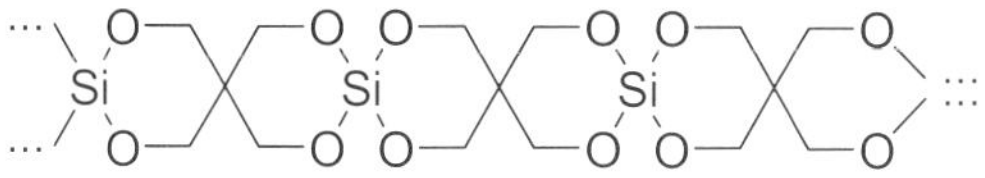

(ix) 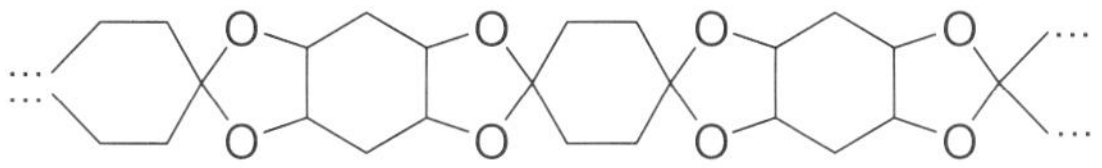 [structure]

Reviews on double-strand, ladder and spiro polymers published in the literature [5, 6] have not addressed their nomenclature in any detailed fashion.

3 STRUCTURE-BASED NOMENCLATURE

3.1 Fundamental principles

This nomenclature rests upon the selection of a preferred *constitutional repeating unit* (CRU) [1, 4] of which the polymer is a multiple; the name of the polymer is the name of this repeating unit prefixed by 'poly'. The unit itself is named wherever possible according to the established principles of organic nomenclature [7]. For double-strand polymers, this unit usually is a tetravalent group denoting attachment to four atoms. Since some of these attachments may be double bonds, the unit may be hexavalent or octavalent.

In using this nomenclature, the steps to be followed in sequence are the same as those for regular single-strand polymers:

(1) *identify* the preferred constitutional repeating unit,

(2) *orient* the constitutional repeating unit,

(3) *name* the constitutional repeating unit, and

(4) *name* the polymer.

Identification and orientation must precede naming of the polymer.

3.2 Identification of the preferred constitutional repeating unit

Examples given in Section 2 above have illustrated the repeating nature of the constitutional units within each polymer structure. Some of the ways to break the same structures into constitutional repeating units are illustrated below.

(i) 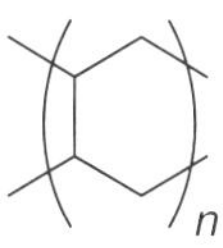 [structure] or [structure]

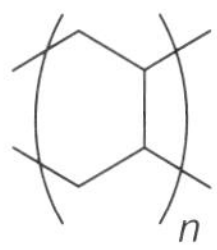

NOMENCLATURE

(ii) [structures]

(iii) [structures]

(iv) [structures]

(v) [structures]

or [structure]

(vi) [structures]

(vii) [structures]

(viii) [structures]

(ix) [structures]

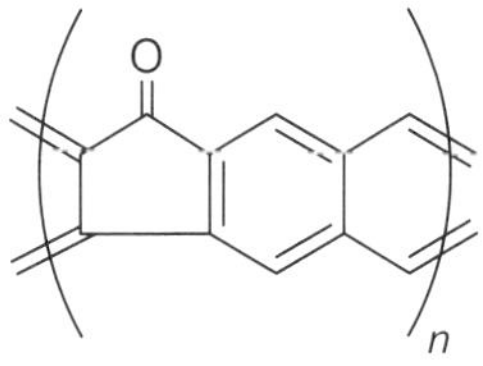

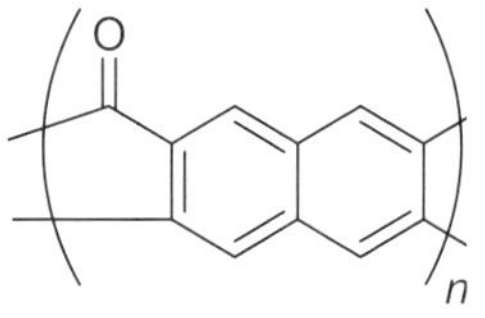

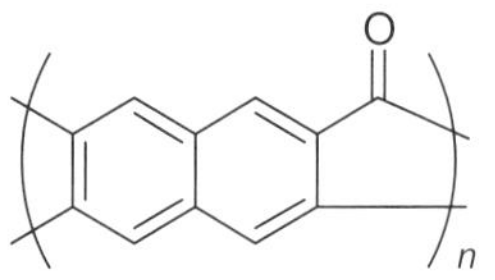

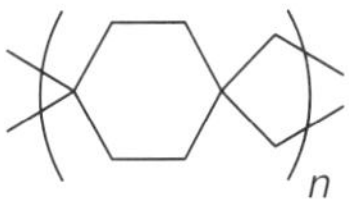

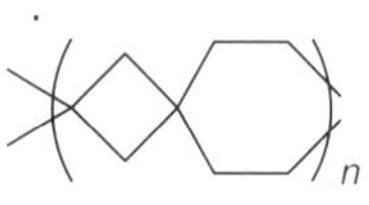

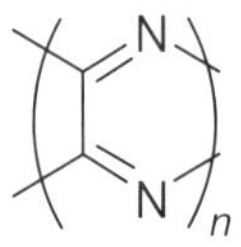

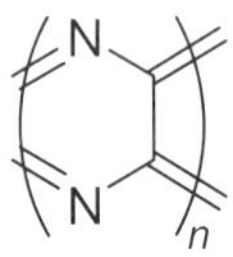

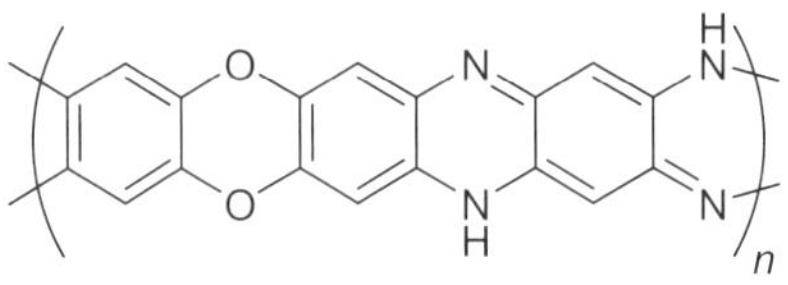

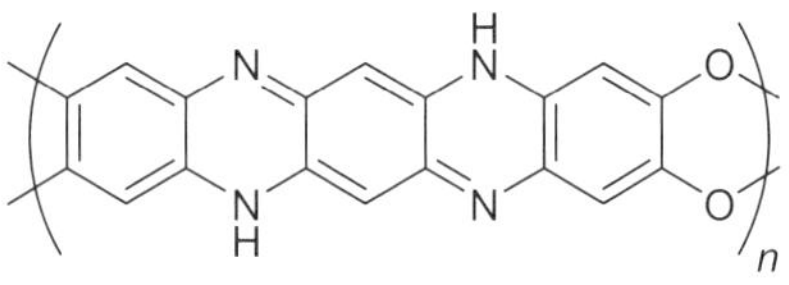

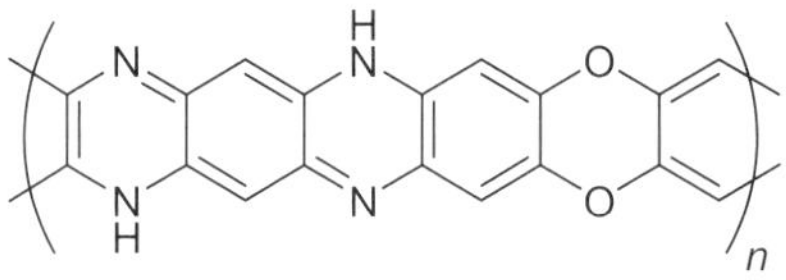

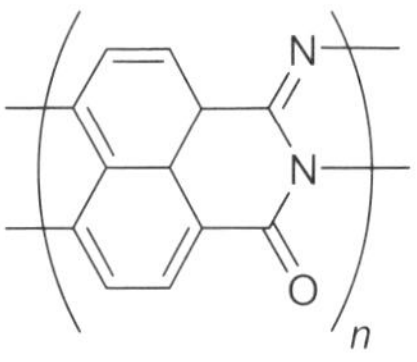

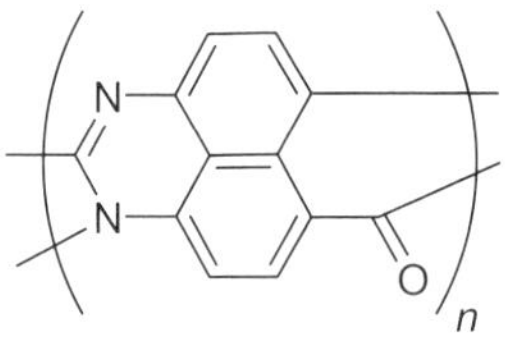

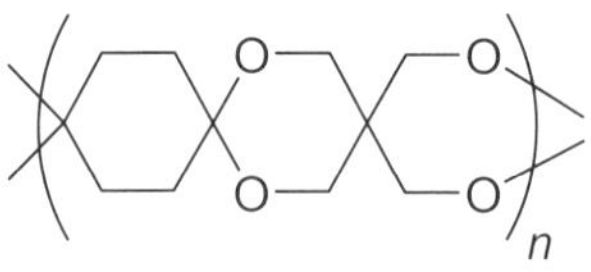

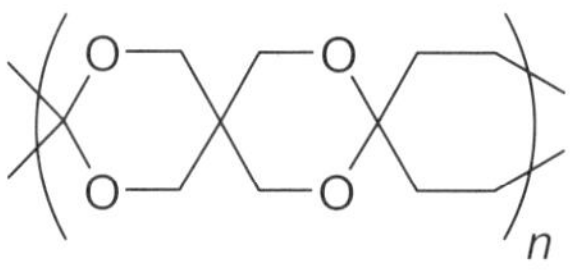

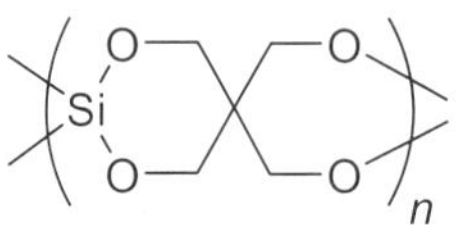

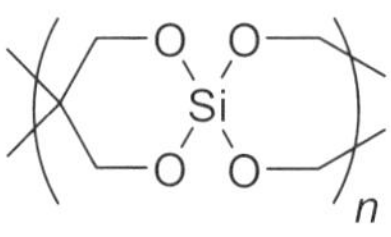

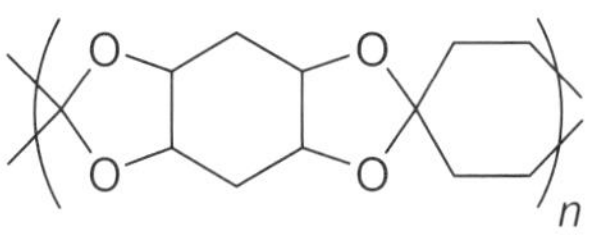

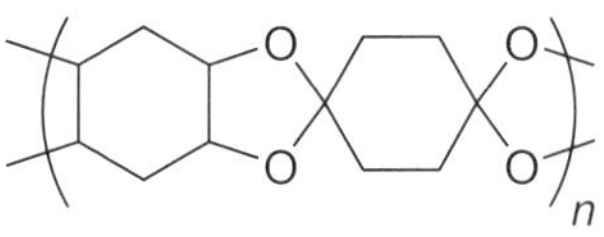

DOUBLE-STRAND ORGANIC POLYMERS

To allow construction of a unique name, a single *preferred* constitutional repeating unit must be selected.

Rule 1
The polymer chain is interrupted by breaking a non-aromatic ring (aromatic rings, i.e. those having a chemistry typified by benzene [8] are broken only when no other rings are present) by observing the following criteria in decreasing order of priority:

(1) *minimize* the number of free valences of the constitutional repeating unit,

(2) *maximize* the number of most preferred hetero atoms in the ring system,

(3) *retain* the most preferred ring system,

(4) *choose* the longest chain for acyclic constitutional repeating unit.

Seniority of ring systems [1] is decided by applying the following criteria, successively, until a decision is reached:

(1) A heterocycle is senior to any carbocycle, regardless of the number of rings

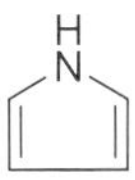

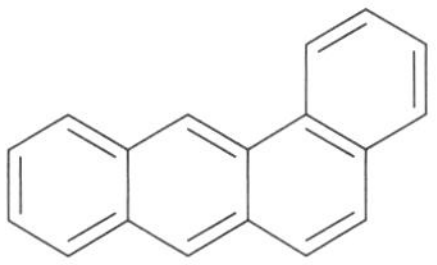

(2) For heterocycles, the criteria are based on the nature and position of the hetero atom [1]

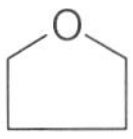

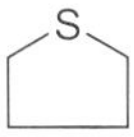

(3) Larger number of rings

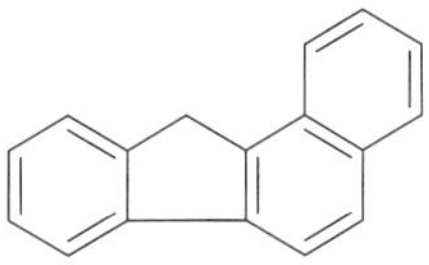

(four rings) senior to (three rings)

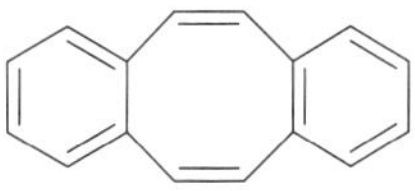

(4) Larger individual ring at first point of difference

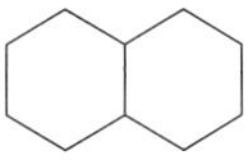

(6,6) senior to (6,5)

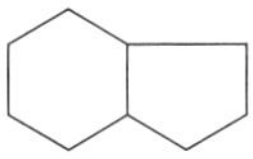

323

NOMENCLATURE

senior to

(8,5)   (7,7)

(5) Larger number of atoms in common among rings

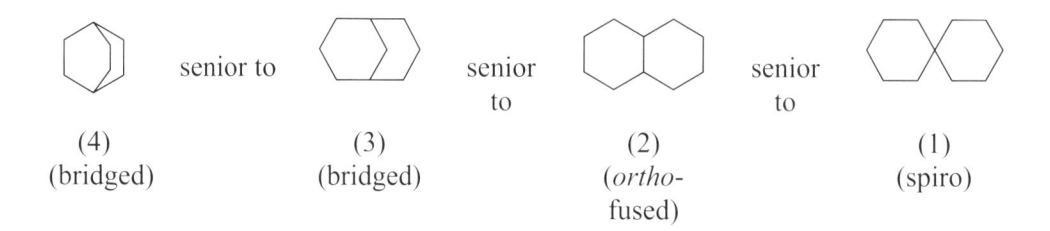

(4) senior to (3) senior to (2) senior to (1)
(bridged)   (bridged)   (*ortho*-fused)   (spiro)

For further rules and additional details, [1] needs to be consulted.

3.3 Orientation of the constitutional repeating unit

Rule 2
The acyclic constitutional repeating unit that results from breaking the ring is oriented in such a way that the lowest free valence locant is at the lower left.

Example:

is preferred to

1,4:3,2-   2,3:4,1-

If there is a further choice, the acyclic unit is oriented to give the lower locant number on the occasion of the first difference.

Example:

is preferred to

1,2:3,1-   1,3:3,2-

Rule 3
If the preferred constitutional repeating unit contains a ring system with fixed numbering [7], it is oriented (e.g. by rotation) in such a way that the lowest free valence locant is at the lower left and next locants in the ascending order proceed in a clockwise direction.
Example:

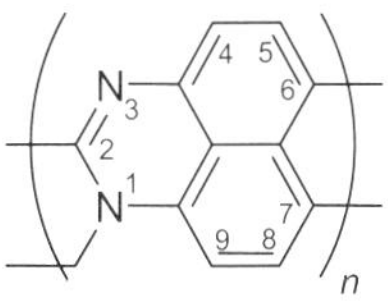

DOUBLE-STRAND ORGANIC POLYMERS

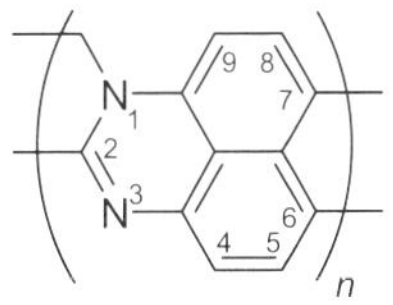

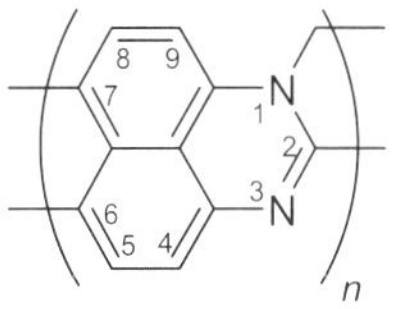

Whenever orienting the ring with the lowest free valence locant at the lower left results in an anticlockwise order of ascending locants, the orientation is changed to the next lowest free valence locant at the lower left to preserve the ascending order of locants in the clockwise direction.

Example:

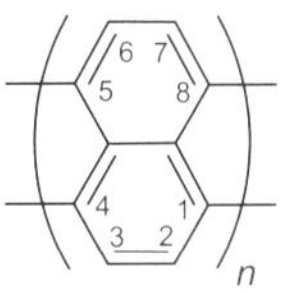

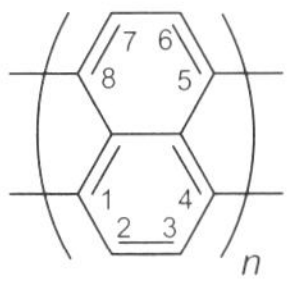

Rule 4
If the preferred constitutional repeating unit consists of a ring system and acyclic subunits, it is oriented in such a way that the ring system is on the left side and the acyclic subunits are on the right of the ring system. The ring system is oriented according to Rule 3.

Example:

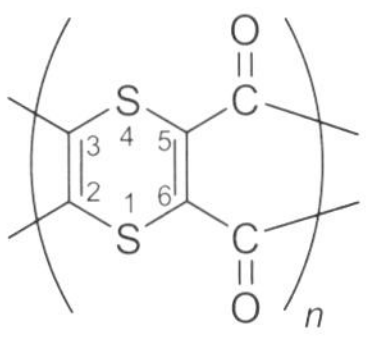

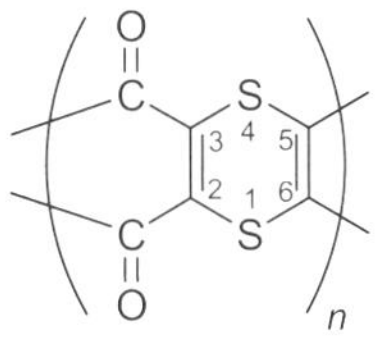

3.4 Naming of the preferred constitutional repeating unit

Rule 5
The preferred constitutional repeating unit is named as drawn, i.e. from left to right. The name of an acyclic constitutional repeating unit is based on the longest chain with the lowest free valence locant at the lower left with respect to the left parenthesis. The name of a ring system is formed according to the principles of naming carbocyclic and heterocyclic ring systems. The names are followed by an appropriate suffix, e.g. '2,3:5,6-tetrayl', '2,3:9-triyl-10-ylidene', '2,3-diyl:6,7-diylidene'. The free valence locants are always placed just in front of the corresponding suffix and are cited in the order:

lower left, upper left: upper right, lower right

(i.e., in a clockwise direction), the left locants being separated from the right locants by a colon. Then the names of acyclic subunits (broken ring fragments) are cited in the order:

325

NOMENCLATURE

upper right, lower right, preceded by the corresponding locants (the locants refer to the ring system positions where the acyclic subunits are attached).

Note: Locants are placed immediately before that part of the name to which they relate, e.g. ethane-1,2:2,1-tetrayl (not 1,2:2,1-ethanetetrayl) or buta-1,3-diene-1,4:3,2-tetrayl (not 1,3-butadiene-1,4:3,2-tetrayl) [7].

Examples:

1.

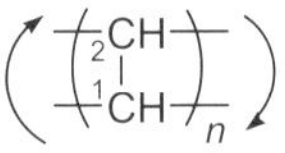

poly(ethane-1,2:2,1-tetrayl)
(not ethane-1,2:1,2-diylidene)

2.

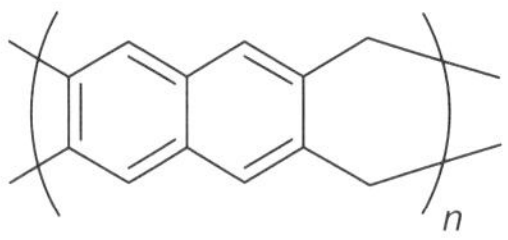

poly(naphthalene-2,3:6,7-tetrayl-6,7-dimethylene)

3.5 Naming of the polymer

Rule 6

The polymer is named with the prefix 'poly' followed by the name of the preferred constitutional repeating unit in parentheses or brackets.

3.6 Polymers constituted of repeatedly fused or spiro carbocycles

The constitutional repeating unit resulting from a ladder or spiro polymer consisting of a repeating *carbocycle* is a tetravalent acyclic group.

Examples:

1.

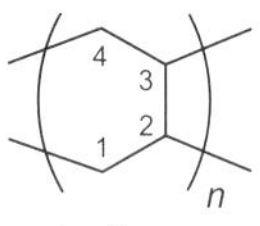

poly(butane-1,4:3,2-tetrayl)

2.

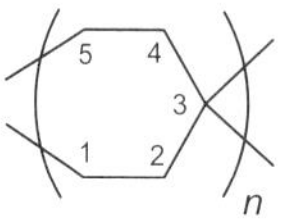

not

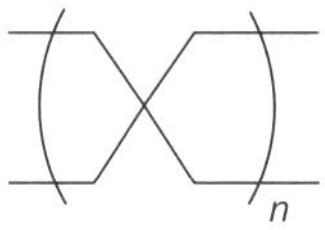

poly(pentane-1,5:3,3-tetrayl)
(the longest straight chain is preferred)

DOUBLE-STRAND ORGANIC POLYMERS

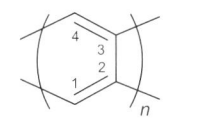

3.

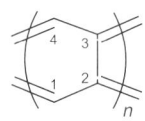

poly(buta-1,3-diene-1,4:3,2-tetrayl)
(when an aromatic ring is broken, double bonds are fixed
and free valences in the total constitutional repeating unit
are minimized)

3.7 Polymers constituted of repeatedly fused or spiro carbocyclic systems

The constitutional repeating unit resulting from a ladder or spiro polymer consisting of a repeating *polycarbocyclic* (multi-ring) system is a tetravalent group or a combination of a tetravalent (multivalent) cyclic group with one or two divalent (multivalent) acyclic groups.

Examples:

1.

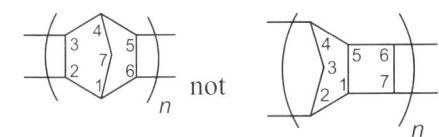

poly(bicycle[2.2.1]heptane-2,3:5,6-tetrayl)
(the preferred two-ring system is 5,5 not 4,5)

2.

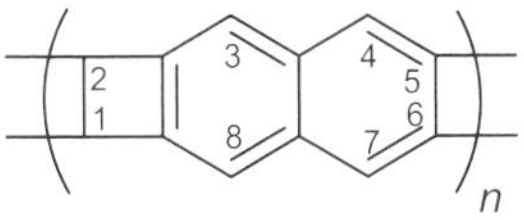

poly(1,2-dihydrocyclobuta[*b*]naphthalene-1,2:5,6-tetrayl)

3.

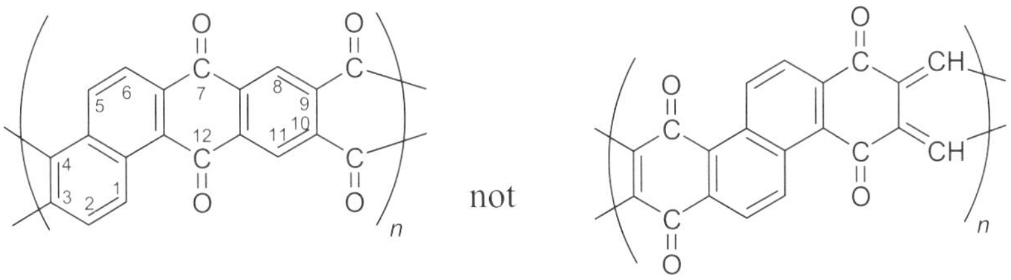

poly[(7,12-dioxo-7,12-dihydrobenzo[*a*]anthracene-3,4:9,10-tetrayl)-9,10-dicarbonyl]

(unless there is no choice, the aromatic ring is not broken)

4.

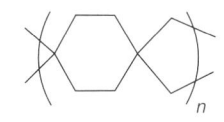

poly[(cyclohexane-1,1:4,4-tetrayl)-4,4-dimethylene]

NOMENCLATURE

5.

poly[(1,3-dioxotrispiro[3.1.1.3.1.1]tridecane-2,2:10,10-tetrayl)-10,10-dimethylene]

3.8 Polymers constituted of repeatedly fused or spiro heterocycles

The constitutional repeating unit resulting from a ladder or spiro polymer consisting of a repeating *heterocycle* is viewed as a tetravalent (multivalent) acyclic carbon group in which some carbon atoms are replaced by hetero atoms. The constitutional repeating unit is oriented in such a way that hetero atoms have lowest locants and that the hetero atom of highest seniority [1] has lower locant than that of next seniority. The constitutional repeating unit is then named by replacement nomenclature [7].[*]

Examples:

1.

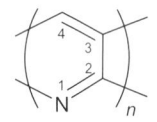

poly(1-azabuta-1,3-diene-1,4:3,2-tetrayl)

2.

poly(1-oxa-4-azabut-2-ene-1,4:3,2-tetrayl)

3.

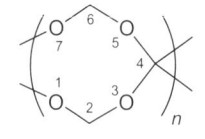

poly(1,3,5,7-tetraoxaheptane-1,7:4,4-tetrayl)

---

[*] In replacement 'a' nomenclature as conventionally applied to acyclic structures with multiple heteroatoms, terminal heteroatoms are not designated with 'a' prefixes but are named as characteristic groups of the structure, i.e., as hydroxy, amino, carboxylic acid, etc. However, heteroatoms in such positions within the constitutional units of ladder or spiro polymer molecules are not terminal units and the structures are not necessarily acyclic. Consequently, such atoms are designated with 'a' prefixes, and thereby the simplicity afforded by the application of replacement nomenclature to polymer molecules is enhanced.

DOUBLE-STRAND ORGANIC POLYMERS

4.

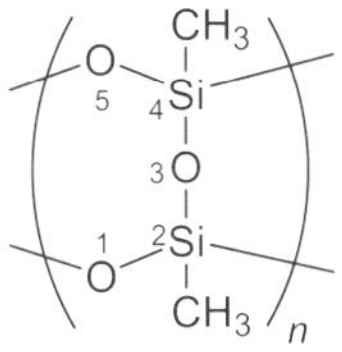

poly(2,4-dimethyl-1,3,5-trioxa-2,4-disilapentane-1,5:4,2-tetrayl)

5.

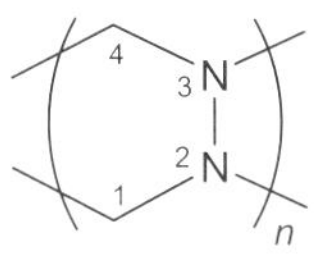

poly(2,3-diazabutane-1,4:3,2-tetrayl)

3.9 Polymers constituted of repeatedly fused or spiro heterocyclic systems

The constitutional repeating unit resulting from a ladder or spiro polymer consisting of a repeating *heterocyclic* ring system is a combination of a tetravalent (multivalent) ring system group with one or two bivalent (multivalent) acyclic groups. If there is a choice, the constitutional repeating unit is oriented in such a way that the acyclic subunit with the terminal atom of higher seniority is at the upper right.

Examples:

1.

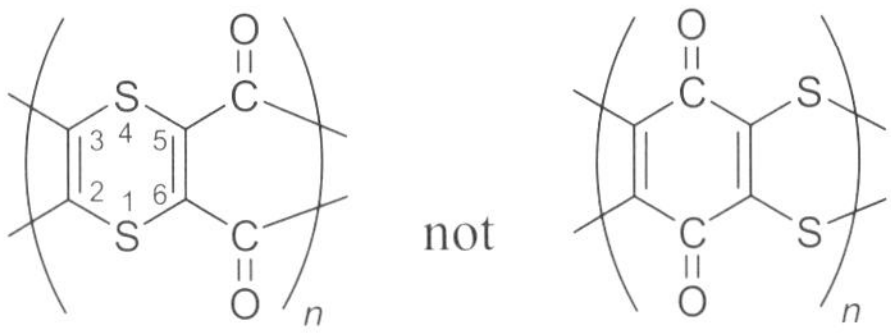

poly[(1,4-dithiine-2,3:5,6-tetrayl)-5,6-dicarbonyl]
(the carbocyclic ring is broken in preference to the heterocyclic ring to keep the hetero atoms in the ring system)

2.

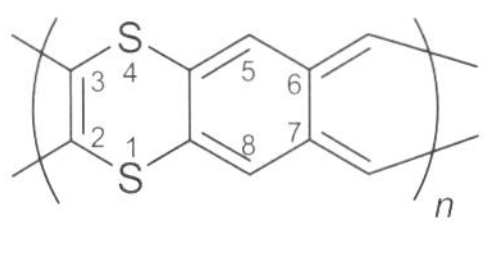

poly[(1,4-benzodithiine-2,3-diyl:6,7-diylidene)-6,7-dimethanylylidene]

3.

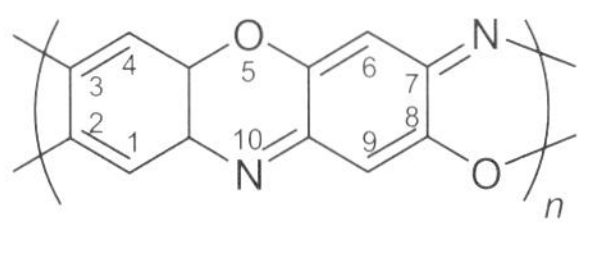

poly(7H-phenoxazine-2,3-diyl:7-ylidene-8-yl-7-nitrilo-8-oxy)
(the heterocyclic ring is broken in preference to the carbocyclic aromatic ring to minimize the number of free valences; the citation of bivalent groups preserves the order 'upper right, lower right' in preference to the seniority of hetero atoms)

NOMENCLATURE

4.

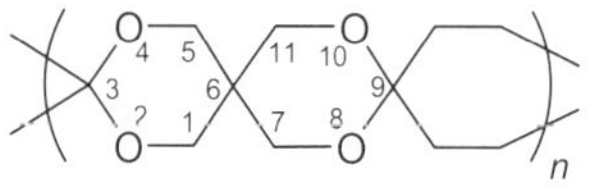

poly(2,4,8,10-tetraoxaspiro[5.5]undecane-3,3:9,9-tetrayl-9,9-diethane-1,2-diyl)
(all hetero atoms are in the ring system)

5.

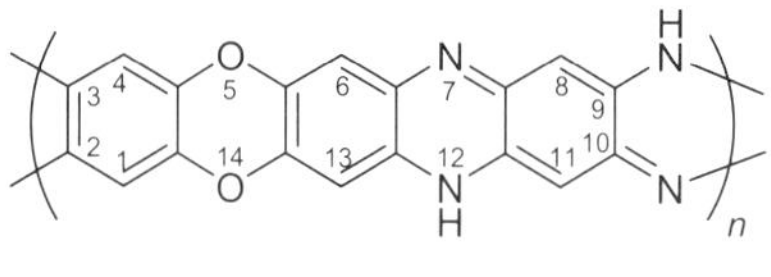

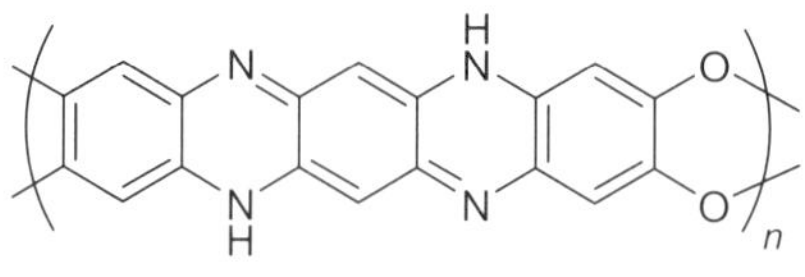

poly([1,4]benzodioxino[2,3-*b*]phenazine-2,3:9-triyl-10(12*H*)-ylidene-9-imino-10-nitrilo)
(the number of preferred hetero atoms is maximized in the ring system; the locants '2,3:9-triyl-10(12*H*)-ylidene' are preferred to '2,3:10-triyl-9(7*H*)-ylidene' (lower locants for free valences)

6.

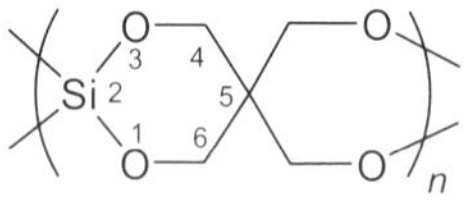

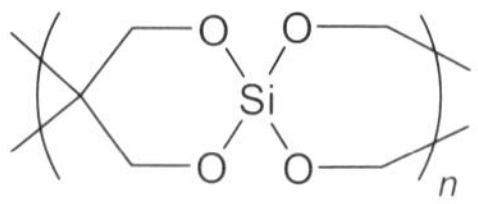

poly[1,3-dioxa-2-silacyclohexane-2,2:5,5-tetrayl-5,5-bis(methyleneoxy)]
(fixed ring numbering decides the orientation)

7.

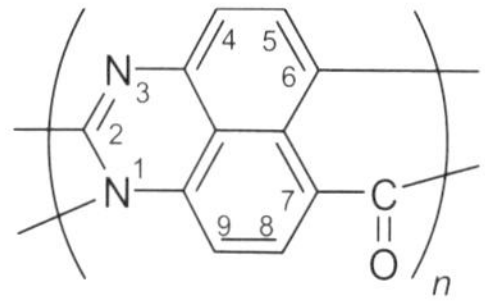

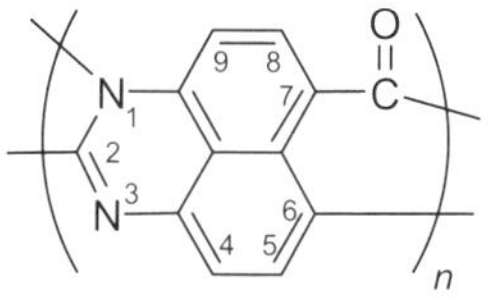

poly(1*H*-perimidine-1,2:6,7-tetrayl-7-carbonyl)
(lower free valence locant is at the lower left)

8.

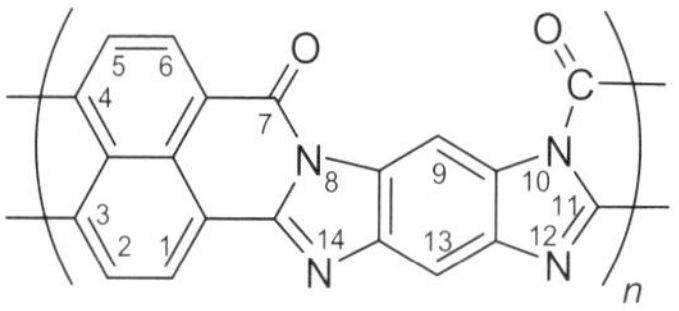

poly[(7-oxo-7*H*,10*H*-benzo[*de*]imidazo[4′,5′:5,6]benzimidazo[2,1-*a*]isoquinoline-3,4:10,11-tetrayl)-10-carbonyl]

DOUBLE-STRAND ORGANIC POLYMERS

9.

poly[(1,7,11,16-tetramethyl-9,18-dioxo-9,9a,10,12,13,14,15,17,17a,18-decahydro-10,17:12,15-diepoxyphenanthro[1,2-*a*]pentacene-3,4:13,14-tetrayl)-13,14-dicarbonyl]

10.

poly[(hexahydrobenzo[1,2-*d*:4,5-*d'*]bis[1,3]dioxole-2,2:6,6-tetrayl)-6,6-diethane-1,2-diyl]
(fused ring system is preferred to spiro ring system, because of larger number of atoms in common among rings)

11.

poly[(5,8-dioxo-2,3,5,8,10,11-hexahydrobenzo[*lmn*]diimidazo[2,1-*b*:1',2'-*j*][3,8]phenanthroline-2,2:11,11-tetrayl)-11,11-diethane-1,2-diyl]

(fused ring system is retained intact and spiro ring broken)

3.10 Substituents
Substituents that are part of the constitutional repeating unit are denoted by means of prefixes affixed to the name of the subunit to which they are bound.

Examples:

1.

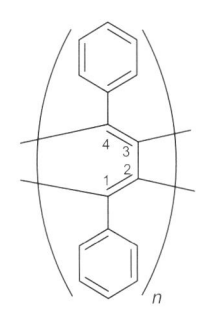

poly(1,4-diphenylbuta-1,3-diene-1,4:3,2-tetrayl)

NOMENCLATURE

2.

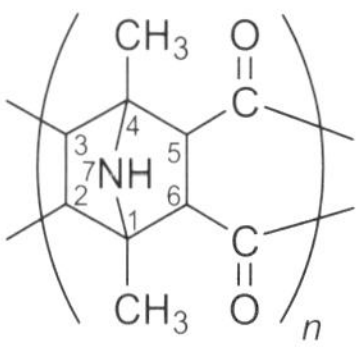

poly[(1,4-dimethyl-7-azabicyclo[2.2.1]heptane-2,3:5,6-tetrayl)-5,6-dicarbonyl]

3.

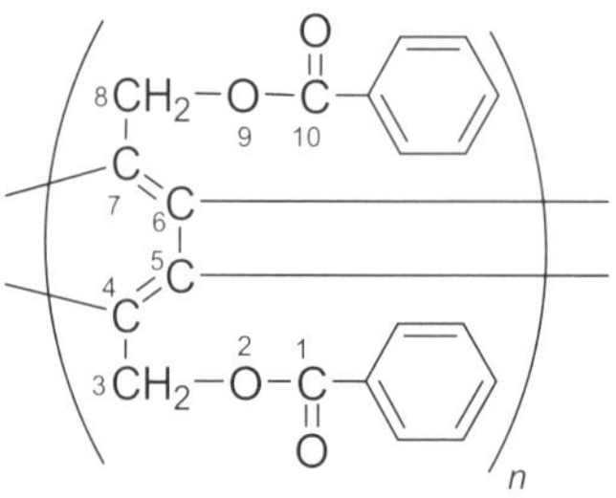

poly(1,10-dioxo-1,10-diphenyl-2,9-dioxadeca-4,6-diene-4,7:6,5-tetrayl)

4.

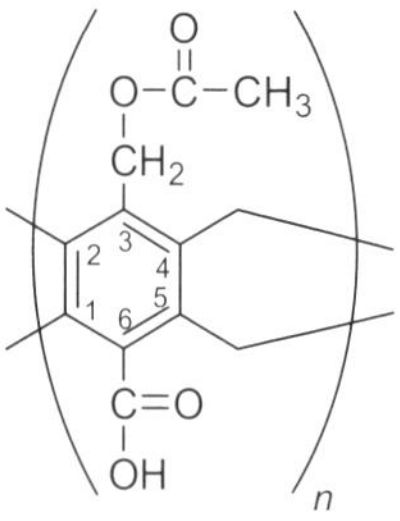

poly{[3-(acetoxymethyl)-6-carboxybenzene-1,2:4,5-tetrayl]-4,5-dimethylene}
(for substituted structures, free valences have preference over substituents when competing for low locants)

3.11 End groups

Rule 7

End groups are specified by prefixes cited in front of the name of the polymer. The ends designated by α and α' are those attached to the left side of the constitutional repeating unit as drawn and named by the preceding rules; the end groups attached to the right side are designated by ω and ω'. The designations and names of the end groups proceed in a clockwise direction starting from the lower left: α,α',ω,ω'.

DOUBLE-STRAND ORGANIC POLYMERS

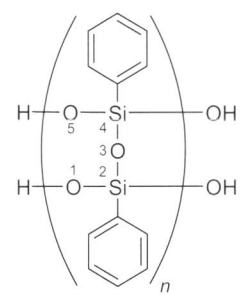

Examples:

1.

α,α′-dihydro-ω,ω′-dihydroxypoly(2,4-diphenyl-1,3,5-trioxa-2,4-disilapentane-1,5:4,2-tetrayl)

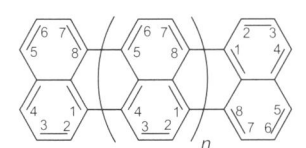

2.

α,α′:ω,ω′-bis(naphthalene-1,8-diyl)poly(naphthalene-4,5:8,1-tetrayl)
(in order to preserve the ascending order of locants in the clockwise direction, the lowest possible free valence locant of the constitutional repeating unit placed at the lower left is 4 rather than 1)

4   SOURCE-BASED NOMENCLATURE

For many polymers, the structural identity of all the constitutional repeating units and of their sequential arrangement is not known, yet some indication is available as to their general characterization such as ladder or spiro polymers.

At times, it is also useful to identify the starting monomers especially for such complex polymers as double-strand polymers, the synthesis of which is often a multi-step reaction involving condensation, cyclization and crosslinking.

It may also be useful at times to provide both source- and structure-based names for the same polymer. For a couple of examples below (6 and 11), references within this document are given to the corresponding constitutional repeating units and their names.

The source-based nomenclature identifies the starting material(s) from which the ladder or spiro polymer is prepared. It is derived from the nomenclature system for copolymers [3]. The system is based on the following principles:

(1) A homopolymer is described by the prefix 'poly' followed by citation of the name of the monomer used. For monomer names consisting of two or more words, and for names containing substituents, parentheses or brackets are used.

(2) A copolymer, including a polymer obtained from two complementary monomers, is described by the prefix 'poly' followed by citation in parentheses or brackets of the names of the monomers used. For monomer names consisting of two or more words, and for names containing substituents, additional parentheses or brackets are

NOMENCLATURE

used. The monomer names are linked through a connective (infix) '-*alt*-', '-*per*-' or '-*co*-'.

Note: '-*co*-' is used for cases where it is not possible to establish alternating or periodic structure.

(3) The specification of the type of structure in a polymer is shown by an italicized prefix '*ladder*-' or '*spiro*-' preceding 'poly'.

(4) No seniority rule is provided for the order of citation of the monomer names.

(5) Semisystematic or trivial names for monomers, which are well established by usage, are allowed.

Examples:

1. *ladder*-poly(methyl vinyl ketone)

2. *ladder*-poly(1,4-diphenyldiacetylene)

3. *ladder*-poly(2,5-dichloro-3,6-dihydroxy-1,4-benzoquinone)

4. *spiro*-poly[dispiro[3.1.3.1]decane-2,8-bis(carbonyl chloride)]

5. *ladder*-poly[(pyromellitic dianhydride[*])-*alt*-(1,2,5,6-tetraaminoanthraquinone)]

6. *ladder*-poly[(naphthalene-1,4,5,8-tetracarboxylic acid)-*alt*-(1,2,4,5-tetraaminobenzene)]

Note: For the corresponding structure-based representation, see example 8 in Section 3.9.

7. *ladder*-poly{pyromellitonitrile[†]-*alt*-[9,9-bis(4-aminophenyl)fluorene]}

8. *ladder*-poly[(1,2,4,5-tetramethylidenecyclohexane)-*alt*-1,4-benzoquinone]

9. *ladder*-poly[(2,5-diaminoterephthalic acid)-*alt*-(2,5-dicyanoterephthaloyl chloride)]

10. *spiro*-poly(cyclohexane-1,4-dione-*alt*-pentaerythritol[‡])

Note: For the corresponding structure-based representation, see example 4 in Section 3.9.

---

[*] benzene-1,2,4,5-tetracarboxylic 1,2:4,5-dianhydride
[†] benzene-1,2,4,5-tetracarbonitrile
[‡] 2,2-bis(hydroxymethyl)propane-1,3-diol

5 ACKNOWLEDGMENT

The Commission acknowledges the first structure-based nomenclature rules for organic ladder and spiro polymers proposed by the Nomenclature Committee of the Division of Polymer Chemistry of the American Chemical Society [9].

6 REFERENCES

1. IUPAC. 'Nomenclature of regular single-strand organic polymers (IUPAC Recommendations 2002)', *Pure Appl. Chem.* **74**, 1921-1956 (2002). Reprinted as Chapter 15 this volume.
2. IUPAC. 'Nomenclature for regular single-strand and quasi-single-strand inorganic and coordination polymers (Recommendations 1984)', *Pure Appl. Chem.* **57**, 149-168 (1985). Reprinted as Chapter 16 this volume.
3. IUPAC. 'Source-based nomenclature for copolymers (Recommendations 1985)', *Pure Appl. Chem.* **57**, 1427-1440 (1985). Reprinted as Chapter 20 this volume.
4. IUPAC. 'Glossary of basic terms in polymer science (IUPAC Recommendations 1996)', *Pure Appl. Chem.* **68**, 2287-2311 (1996). Reprinted as Chapter 1 this volume.
5. W. J. Bailey. 'Ladder and Spiro Polymers' in *Encyclopedia of Polymer Science and Technology,* H. F. Mark, N. G. Gaylord, N. M. Bikales (Eds.), Wiley-Interscience, New York (1968), **8**, 97-120. 'Ladder and Spiro Polymers' in *Encyclopedia of Polymer Science and Engineering*, H. F. Mark, N. M. Bikales, C. G. Overberger, G. Menges, J. I. Kroschwitz (Eds.), 2nd ed., Wiley-Interscience, New York (1990), Index Vol., 158-245.
6. C. G. Overberger and J. A. Moore. 'Ladder Polymers', *Adv. Polym. Sci.,* **7**, 113-150 (1970).
7. *A Guide to IUPAC Nomenclature of Organic Compounds, Recommendations 1993*, R. Panico, W. H. Powell, J.-C. Richer (Eds.), Blackwell Scientific Publications (1993). Corrigenda, *Pure Appl. Chem.* **71**, 1327-1330 (1999).
8. IUPAC. 'Glossary of terms used in physical organic chemistry (Recommendations 1982)', *Pure Appl. Chem.* **55**, 1281-1371 (1983).
9. American Chemical Society. 'A structure-based nomenclature for linear polymers', *Macromolecules* **1**, 193-198 (1968).

17: Structure-Based Nomenclature for Irregular Single-Strand Organic Polymers

CONTENTS

Preamble
Basic principles
Rules for naming irregular polymers
Additional examples
References

PREAMBLE

Previous reports from this Commission have presented structure-based systems for naming regular single-strand organic polymers [1, 2]. These are polymers whose molecules can be described by only a single type of constitutional repeating unit in a single sequential arrangement [3]. A source-based system for naming copolymers, which are mostly irregular, has also been published [4].

This document describes a structure-based nomenclature system for irregular single-strand organic polymers, i.e., single-strand organic polymers that can be described by the repetition of more than one type of constitutional unit or that comprise constitutional units not all connected identically with respect to directional sense. The new system names irregular polymers for which the source-based system is inadequate (e.g., polymers that have undergone partial chemical modification, homopolymers having both head-to-tail and head-to-head arrangements of monomeric units, and polymers derived from a single monomer that can provide more than one kind of monomeric unit) and in addition it offers a structure-based alternative to source-based nomenclature for copolymers [4].

BASIC PRINCIPLES

Irregular polymers or blocks are named by placing the prefix 'poly' before the structure-based names of the constitutional units, collectively enclosed in parentheses or brackets, with the individual constitutional units separated by oblique strokes. A typical name derived from this new system is

poly(A/B),

which represents an irregular polymer consisting of the constitutional units A and B.

Originally prepared by a Working Group consisting of R. B. Fox (USA), N. M. Bikales (USA), K. Hatada (Japan) and J. Kahovec (Czech Republic). Reprinted from *Pure Appl. Chem.* **66**, 873-889 (1994).

IRREGULAR SINGLE-STRAND ORGANIC POLYMERS

Block copolymers in which the sequential arrangement of regular blocks is known are named by using dashes for the bonding of blocks with each other and with junction units:

poly(A)—X_A—poly(B) —X_B—poly(C) —X_C...

where A, B, C, ... are the names of constitutional units and X_A, X_B, X_C ... are the names of junction units.

The procedure for structure-based naming of irregular single-strand polymers is as follows:

1. Write the structure of the polymer chain based on available information.
2. Select and orient the minimum number of constitutional units necessary to represent the polymer structure [1]; where there is a choice, selection is based on the seniority rules of [1]*. Structures are written to be read from left to right.
3. To be certain that the selected constitutional units are the correct ones, combine their individual structures in all possible ways to form a polymer chain. Incorrect constitutional units will give chain segments that do not correspond to the polymer structure or description as written.
4. Name the constitutional units according to the rules of organic [5, 6] and structure-based polymer [1] nomenclatures†.
5. Write the name of the polymer as described in the following rules.

Note: The fact that specific structural information, such as the location of substituents, is missing can be conveyed in the polymer name through the symbols ?-, *x*- or locant numerals.

RULES FOR NAMING IRREGULAR POLYMERS

Rule 1
Polymers with an irregular arrangement of constitutional units are named

poly(A/B/C/...)

---

* The seniority of constitutional units is heterocyclic rings > hetero atoms in a chain > carbocyclic rings > acyclic carbon chains. Unsaturation is senior to saturation. Within these groups, seniority runs in the sequence: rings with nitrogen > rings with other hetero atoms > ring systems with the largest number of rings > the largest ring > a ring system having the greatest number of hetero atoms > a ring system containing the greatest variety of hetero atoms > a ring system having the greatest number of hetero atoms highest in the order given in the following list. Among hetero atoms, the order of seniority is O > S > Se > Te > N > P > As > Sb > Bi > Si > Ge > Sn > Pb > B > Hg with other atoms placed in the order according to their positions in the periodic table. The seniority in carbocyclic ring systems follows the order: the largest number of rings > the largest ring > the lowest locant numbers for points of attachment. Other things being equal, seniority is based on substituents in the following order: the acyclic chain with the largest number of substituents > the chain having substituents with lowest locants > the alphabetical order of substituents. Further details and examples are given in [1].

† Rules of organic nomenclature are under continuous development. In this document, many of the changes recommended in the latest [6] organic rules have been incorporated. Nevertheless, [6] does provide for flexibility, and it allows the use of many triaditional names; where the older names are retained, it is because of common usage.

NOMENCLATURE

where A, B, C,... are the structure-based names of the appropriate constitutional units, separated by oblique strokes. The preferred order for the citation of constitutional units is alphabetical.

Note 1: The oblique strokes specify that the sequential arrangement of the constitutional units is unknown.

Note 2: The constitutional unit can also be a regular or irregular block.

Example 1.1

A statistical copolymer consisting of units derived from styrene (phenylethene) and vinyl chloride (chloroethene) joined head-to-tail:

$$-CH(C_6H_5)-CH_2-CH(Cl)-CH_2-CH(Cl)-CH_2-CH(C_6H_5)-CH_2-CH(C_6H_5)-CH_2-CH(Cl)-CH_2-$$

constitutional units: $-CH(Cl)-CH_2-$, $-CH(C_6H_5)-CH_2-$

name: poly(l-chloroethane-1,2-diyl/l-phenylethane-1,2-diyl)*

Example 1.2

An irregular polymer derived from vinyl chloride (chloroethene), the constitutional units of which are linked both head-to-tail and head-to-head:

$$-CH(Cl)-CH_2-CH(Cl)-CH_2-CH_2-CH(Cl)-CH(Cl)-CH_2-CH_2-CH(Cl)-CH_2-CH(Cl)-$$

constitutional units: $-CH(Cl)-CH_2-$, $-CH_2-CH(Cl)-$

name: poly(l-chloroethane-1,2-diyl/2-chloroethane-1,2-diyl)

Example 1.3

A polyamide derived from adipic acid† and a mixture of hexane-l,6-diamine and butane-l,4-diamine:

$$-NH-(CH_2)_4-NHCO-(CH_2)_4-CONH-(CH_2)_6-NHCO-(CH_2)_4-CONH-(CH_2)_6-CO-(CH_2)_4-CONH-(CH_2)_4-NHCO-(CH_2)_4-CONH-$$

constitutional units: -NH-(CH$_2$)$_4$-NHCO-(CH$_2$)$_4$-CO-, -NH-(CH$_2$)$_6$-NHCO-(CH$_2$)$_4$-CO-

name: poly(iminobutane-1,4-diyliminohexanedioyl/iminohexane-1,6-diyliminohexanedioyl)

---

* Footnote to the Second Edition: To accord with this name and with the rules of Chapters 18 and 19, the graphical representation of a statistical copolymer comprising constitutional units derived from styrene and vinyl chloride joined head-to-tail should be written as follows:

$$\left(-CH(Cl)-CH_2- \;/\; -CH(C_6H_5)-CH_2- \right)_n$$

The same rationale may be applied to a number of the structural representations in this chapter.

† hexanedioic acid

IRREGULAR SINGLE-STRAND ORGANIC POLYMERS

Note: The constitutional units are selected on the basis of seniority of their constituent parts, and they are cited alphabetically.

Example 1.4
A polymer consisting of constitutional units derived from the 1,4 and 1,2 polymerization of buta-1,3-diene:

$$-CH=CH-CH_2-CH_2-CH-CH_2-CH=CH-CH_2-CH_2-CH=CH-CH_2-CH_2-CH-CH_2-$$
$$\qquad\qquad\qquad\qquad\quad |\qquad\qquad\qquad\qquad\qquad\qquad\qquad\qquad\qquad\qquad\qquad\quad |$$
$$\qquad\qquad\qquad\qquad\quad CH=CH_2\qquad\qquad\qquad\qquad\qquad\qquad\qquad\qquad\qquad\quad CH=CH_2$$

constitutional units: $-CH=CH-CH_2-CH_2-$, $-CH-CH_2-$
$$\qquad\qquad\qquad\qquad\qquad\qquad\qquad\qquad\qquad\qquad\qquad\qquad\qquad\qquad\quad |$$
$$\qquad\qquad\qquad\qquad\qquad\qquad\qquad\qquad\qquad\qquad\qquad\qquad\qquad\quad CH=CH_2$$

name: poly(but-1-ene-1,4-diyl/1-vinylethane-1,2-diyl)

Note: The choice of the constitutional units is dictated by the seniority rules of [1], i.e., but-1-ene-1,4-diyl is senior to but-2-ene-1,4-diyl.

Example 1.5
A polymer consisting of constitutional units derived from 1,4 polymerization and both head-to-head and head-to-tail 1,2 polymerization of buta-1,3-diene:

$$\qquad\qquad\qquad\qquad\qquad\qquad CH=CH_2$$
$$\qquad\qquad\qquad\qquad\qquad\qquad |$$
$$-CH_2-CH=CH-CH_2-CH_2-CH-CH-CH_2-CH_2-CH=CH-CH_2-$$
$$\qquad\qquad\qquad\qquad\qquad\qquad\qquad\quad |$$
$$\qquad\qquad\qquad\qquad\qquad\qquad\qquad\quad CH=CH_2$$

constitutional units: $-CH_2-CH=CH-CH_2-$, $-CH-CH_2-$, $-CH_2-CH-$
$$\qquad\qquad\qquad\qquad\qquad\qquad\qquad\qquad\qquad\qquad\qquad\quad |\qquad\qquad\qquad\qquad |$$
$$\qquad\qquad\qquad\qquad\qquad\qquad\qquad\qquad\qquad\qquad\quad CH=CH_2\qquad\qquad CH=CH_2$$

name: poly(but-2-ene-1,4-diyl/1-vinylethane-1,2-diyl/2-vinylethane-1,2-diyl)

Note: But-1-ene-1,4-diyl would not be selected as a constitutional unit to represent the structure given above because its use would also require the use of the constitutional units of but-2-ene-1,4-diyl, vinylmethylene and methylene, thus violating the principle of minimizing the number of constitutional units. In addition, the combination of the selected constitutional units, like methylene, generates chain segments that do not correspond to the polymer structure as written, such as propane-1,3-diyl.

Example 1.6
A chlorinated polyethylene:

$$\qquad\qquad\qquad\qquad\quad Cl$$
$$\qquad\qquad\qquad\qquad\quad |$$
$$-CH_2-CH-CH_2-C-CH-CH_2-CH_2-CH-CH-CH_2-$$
$$\qquad\qquad\quad |\qquad\qquad |\quad |\qquad\qquad\qquad\quad |\quad |$$
$$\qquad\qquad\quad Cl\qquad\quad Cl\ Cl\qquad\qquad\qquad Cl\ Cl$$

$$\qquad\qquad\qquad\qquad\qquad\qquad\qquad\qquad\quad Cl$$
$$\qquad\qquad\qquad\qquad\qquad\qquad H\qquad\qquad\quad |$$
constitutional units: $-C-$, $-C-$, $-CH_2-$
$$\qquad\qquad\qquad\qquad\qquad\qquad |\qquad\qquad\quad |$$
$$\qquad\qquad\qquad\qquad\qquad\qquad Cl\qquad\qquad\quad Cl$$

name: poly(chloromethylene/dichloromethylene/methylene)

NOMENCLATURE

Example 1.7
A chlorinated poly(vinyl chloride); any methylene units are necessarily separated by an odd number of carbon atoms:

$$-\underset{\underset{Cl}{|}}{\overset{\overset{Cl}{|}}{C}}-CH_2-\underset{\underset{Cl}{|}}{CH}-CH_2-\underset{\underset{Cl}{|}}{\overset{\overset{Cl}{|}}{C}}-\underset{\underset{Cl}{|}}{\overset{\overset{Cl}{|}}{C}}-\underset{\underset{Cl}{|}}{CH}-\underset{\underset{Cl}{|}}{CH}-\underset{\underset{Cl}{|}}{CH}-CH_2-\underset{\underset{Cl}{|}}{\overset{\overset{Cl}{|}}{C}}-\underset{\underset{Cl}{|}}{CH}-\underset{\underset{Cl}{|}}{CH}-CH_2-\underset{\underset{Cl}{|}}{CH}-\underset{\underset{Cl}{|}}{\overset{\overset{Cl}{|}}{C}}-$$

constitutional units:

$$-\underset{\underset{Cl}{|}}{CH}-CH_2-, \quad -\underset{\underset{Cl}{|}}{\overset{\overset{Cl}{|}}{C}}-CH_2-, \quad -\underset{\underset{Cl}{|}}{CH}-\underset{\underset{Cl}{|}}{CH}-, \quad -\underset{\underset{Cl}{|}}{\overset{\overset{Cl}{|}}{C}}-\underset{\underset{Cl}{|}}{\overset{\overset{Cl}{|}}{C}}-,$$

$$-\underset{\underset{Cl}{|}}{\overset{\overset{Cl}{|}}{C}}-\underset{\underset{}{|}}{CH}-, \quad -\underset{\underset{Cl}{|}}{CH}-\underset{\underset{Cl}{|}}{\overset{\overset{Cl}{|}}{C}}-$$

name: poly(1-chloroethane-1,2-diyl/1,1-dichloroethane-1,2-diyl/1,2-dichloroethane-1,2-diyl/tetrachloroethane-1,2-diyl/1,1,2-trichloroethane-1,2-diyl/1,2,2-trichloroethane-1,2-diyl)

Example 1.8
A partially hydrolyzed head-to-tail poly(vinyl acetate):

$$-\underset{\underset{OCOCH_3}{|}}{CH}-CH_2-\underset{\underset{OH}{|}}{CH}-CH_2-\underset{\underset{OCOCH_3}{|}}{CH}-CH_2-\underset{\underset{OH}{|}}{CH}-CH_2-\underset{\underset{OH}{|}}{CH}-CH_2-\underset{\underset{OCOCH_3}{|}}{CH}-CH_2-$$

constitutional units: $\quad -\underset{\underset{OCOCH_3}{|}}{CH}-CH_2-, \quad -\underset{\underset{OH}{|}}{CH}-CH_2-$

name: poly(1-acetoxyethane-1,2-diyl/1-hydroxyethane-1,2-diyl)

Example 1.9
A polymer that consists of an irregular sequence of blocks of polyacrylonitrile, polystyrene and the irregular block corresponding to Example 1.4:

constitutional units:

$$\left(-\underset{\underset{CN}{|}}{CH}-CH_2-\right)_p, \quad \left(-\underset{\underset{C_6H_5}{|}}{CH}-CH_2-\right)_q,$$

$$\left(-CH_2-CH=CH-CH_2- \;/\; -\underset{\underset{CH=CH_2}{|}}{CH}-CH_2-\right)_r$$

name: poly[poly(but-1-ene-1,4-diyl/1-vinylethane-1,2-diyl)/poly(1-cyanoethane-1,2-diyl)/poly(1-phenylethane-1,2-diyl)]

Rule 2
Irregular polymers with regular or irregular blocks and junction units in a specific sequential arrangement are named by linking the names of the blocks and junction units with dashes:

IRREGULAR SINGLE-STRAND ORGANIC POLYMERS

poly(A)—X_A—poly(B)—X_B—poly(C)—X_C—...

where A, B, C, ... are the structure-based names of the appropriate constitutional units and X_A, X_B, X_C... are the structure-based names of the junction units. The order of citation of the block names corresponds to the order of the blocks in the chain as written from left to right [4]. If there are no junction units, or they are not known, the polymer is named

poly(A)—poly(B)—poly(C)—...

A polymer consisting of repeated specific sequences is named

poly[poly(A)—X_A—poly(B)—X_B—poly(C)—X_C—...].

Note: Since the Rules for seniority distinguishing between blocks have not been defined, use of seniority rules of [1] for blocks is regarded as one possibility. An alternative is shown in the note to Ex. 2.1.

Example 2.1
A triblock copolymer consisting of a sequence of three blocks joined directly or through unspecified junction units:

name: poly(oxy-1,4-phenylene)-poly(2-cyanoethane-1,2-diyl)-poly(ethane-1,2-diyloxy)

Note: If the graphic representation of the structure is reversed, another possible name is

name: poly(oxyethane-1,2-diyl)-poly(1-cyanoethane-1,2-diyl)-poly(1,4-phenyleneoxy)

Example 2.2
A diblock copolymer in which the blocks are joined by a specific junction unit, as in the polymer:

name: poly(ethane-1,2-diyloxy)-dimethylsilanediyl-poly(1-chloroethane-1,2-diyl)

NOMENCLATURE

Example 2.3
A segmented polyurethane consisting of poly(oxybutane-1,4-diyl) blocks joined through junction units derived from 1,3- and 1,4-phenylene diisocyanate and arranged alternately:

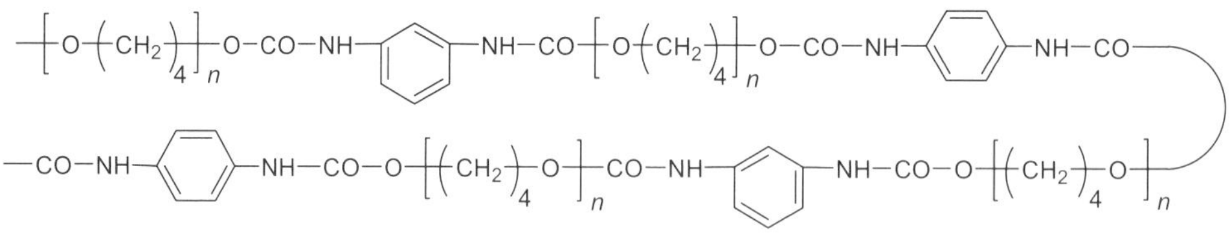

[structure diagram]

constitutional units: [structure diagrams]

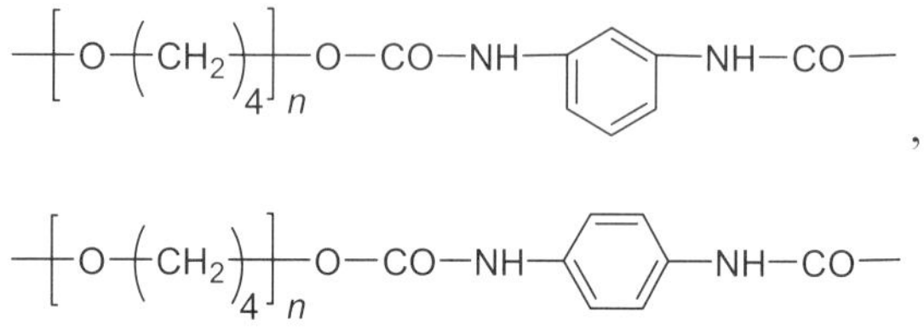

name: poly[poly(oxybutane-1,4-diyl)-oxycarbonylimino-1,3-phenyleneiminocarbonyl-
poly(oxybutane-1,4-diyl)-oxycarbonylimino-l,4-phenyleneiminocarbonyl]

Rule 3
Irregular polymers in which polymer or oligomer chains are attached to the main chain (as in graft copolymers) are named as follows. The attached polymer or oligomer chains are considered to be substituents to the main chain and named in the same way as regular polymers [1] or irregular polymers (cf. Rules 1 and 2) but without the suffix -yl. The atom in the attached chains nearest to the point of attachment to the main chain is given the locant 1.

Note 1: A graft copolymer with poly(A) blocks grafted to the main chain of a regular polymer consisting of constitutional units Z is named

   poly[Z/poly(A)Z]

   where poly(A) is the substituent on constitutional unit Z.

Note 2: A graft copolymer having several kinds of grafts attached to the main chain of a regular polymer consisting of constitutional units Z is named

   poly[Z/poly(A)Z/poly(B)Z/poly(C)Z/...]

   where poly(A), poly(B), poly(C), ... are substituents on the constitutional unit Z.

Example 3.1
A graft copolymer with many of one type of graft unit:

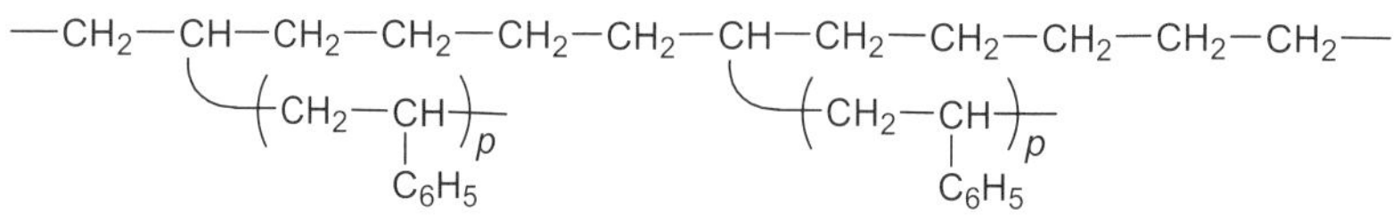

[structure diagram]

IRREGULAR SINGLE-STRAND ORGANIC POLYMERS

constitutional units: —CH$_2$— , —CH—$\underset{\substack{|\\C_6H_5}}{(\text{CH}_2\text{—CH})_p}$

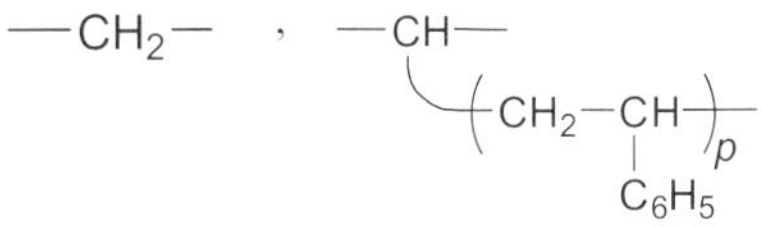

name: poly[methylene/poly(2-phenylethane-1,2-diyl)methylene]

Example 3.2
A graft copolymer with many of one type of graft unit, itself a copolymer consisting of two kinds of constitutional units, 1-chloroethane-1,2-diyl and 1-phenylethane-1,2-diyl:

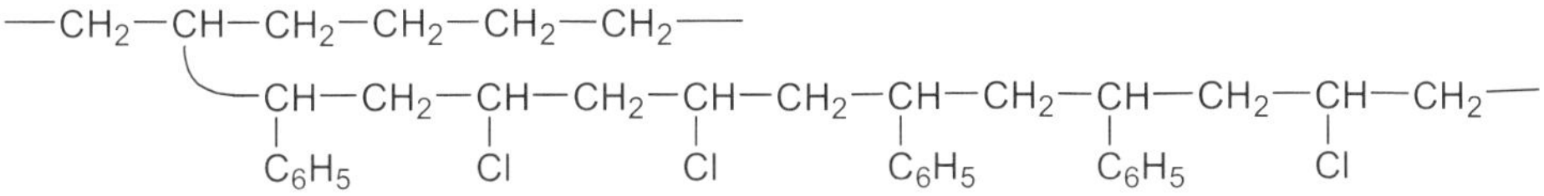

—CH$_2$—CH—CH$_2$—CH$_2$—CH$_2$—CH$_2$—
  └CH—CH$_2$—CH—CH$_2$—CH—CH$_2$—CH—CH$_2$—CH—CH$_2$—CH—CH$_2$—
     | | | | | |
    C$_6$H$_5$ Cl   Cl   C$_6$H$_5$ C$_6$H$_5$ Cl

name: poly[methylene/poly(1-chloroethane-1,2-diyl/1-phenylethane-1,2-diyl)methylene]

Note: The graft is the copolymer specified in Ex. 1.1.

Example 3.3
A graft copolymer with two types of graft units:

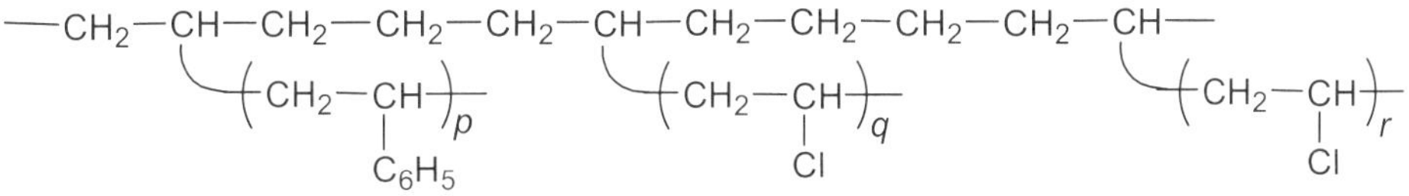

—CH$_2$—CH—CH$_2$—CH$_2$—CH$_2$—CH—CH$_2$—CH$_2$—CH$_2$—CH$_2$—CH—
       | | |
  (CH$_2$—CH)$_p$ (CH$_2$—CH)$_q$ (CH$_2$—CH)$_r$
       | | |
      C$_6$H$_5$ Cl   Cl

constitutional units: —CH$_2$— , —CH—$\underset{\substack{|\\C_6H_5}}{(\text{CH}_2\text{—CH})_p}$, —CH—$\underset{\substack{|\\Cl}}{(\text{CH}_2\text{—CH})_q}$

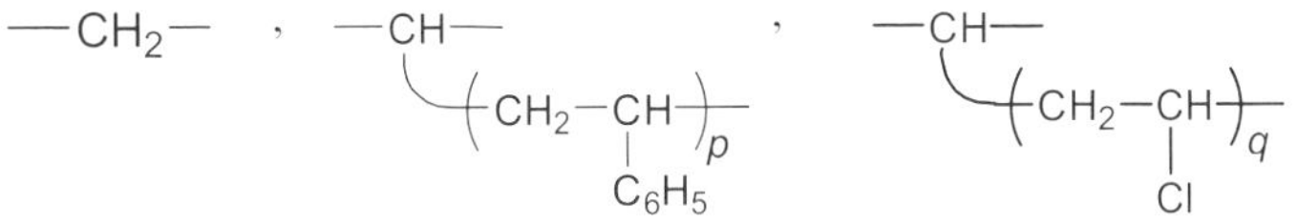

name: poly[methylene/poly(2-chloroethane-1,2-diyl)methylene/poly(2-phenylethane-1,2-diyl)methylene]

Rule 4
Irregular polymers in which polymer or oligomer chains are attached to the main chain through a constitutional unit (linking unit) different from the constitutional unit of the side chain are named as follows. The polymer or oligomer side chains are considered to be substituents to the linking unit and named as in Rule 3. The side chain together with the linking unit, taken as a whole, is considered to be a substituent to the main chain. In forming the name, the atom in the linking unit nearest to the point of attachment to the main chain is given the locant 1.

Note 1: A graft copolymer having the structure

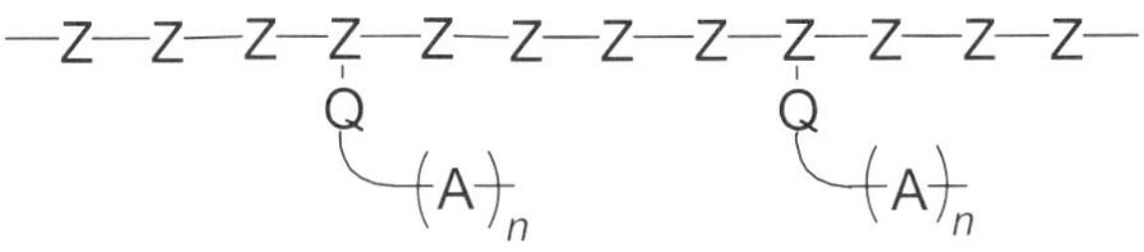

—Z—Z—Z—Z—Z—Z—Z—Z—Z—Z—Z—Z—
         | |
         Q   Q
          (A)$_n$ (A)$_n$

NOMENCLATURE

i.e., with poly(A) blocks attached through linking unit Q to a regular polymer consisting of constitutional units –Z-, is named

poly{Z/[poly(A)Q]Z}.

Note 2: A graft copolymer having the structure

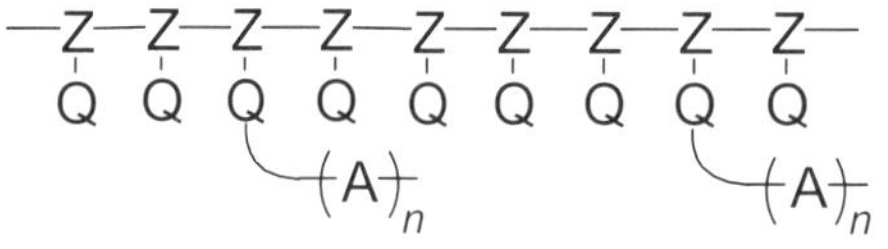

$$\begin{array}{c} -Z-Z-Z-Z-Z-Z-Z-Z-Z- \\ QQQQQQQQQ \\ (A)_n (A)_n \end{array}$$

i.e., with poly(A) blocks attached to substituent Q in a regular polymer consisting of constitutional units –(Q)Z-, is named

poly{(Q)Z/[poly(A)Q]Z}.

Example 4.1
A polystyrene with polymethacrylate grafts on some of the phenyl groups:

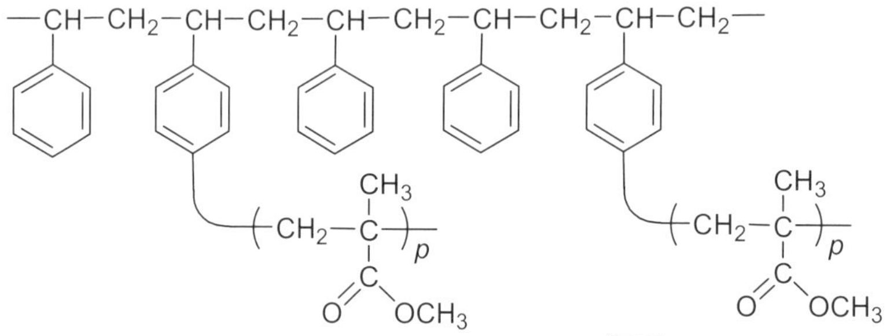

constitutional unit: —CH—CH$_2$— with C$_6$H$_5$ link unit: —C$_6$H$_4$—

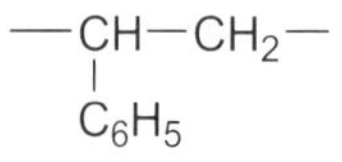

graft: —(CH$_2$—C(CH$_3$)(CO-OCH$_3$))$_p$—

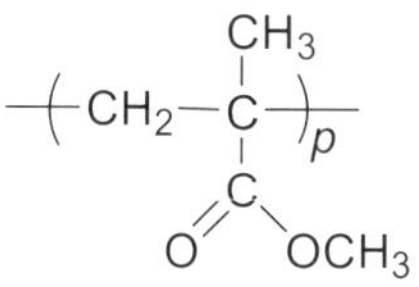

name: poly(1-phenylethane-1,2-diyl/1-(4-{poly[2-(methoxycarbonyl)-2-methylethane-1,2-diyl]}phenyl)ethane-1,2-diyl

Example 4.2
A poly(phenylene oxide) with a poly(vinyl chloride) block grafted through a carbonyloxy group to the 2-position in some of the rings:

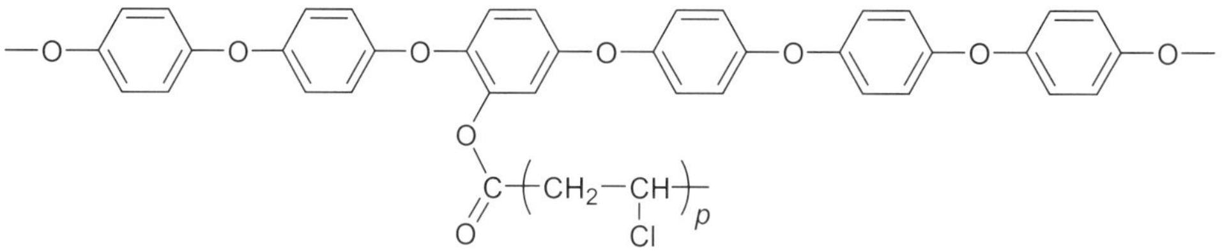

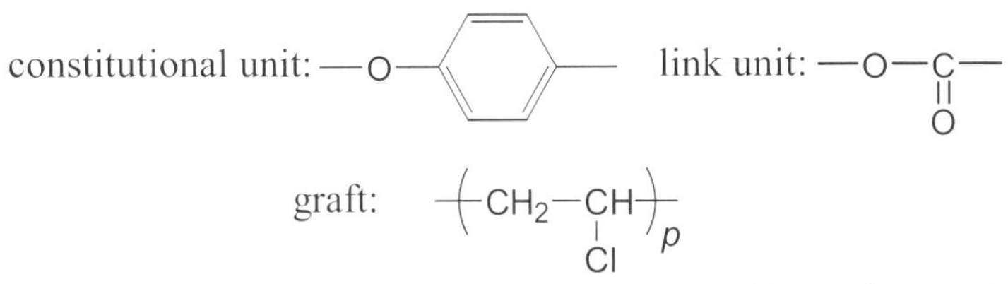

IRREGULAR SINGLE-STRAND ORGANIC POLYMERS

constitutional unit: —O—⟨C₆H₄⟩—   link unit: —O—C(=O)—

graft: —(CH₂—CHCl)ₚ—

name: poly[oxy-l,4-phenylene/oxy({[poly(2-chloroethane-1,2-diyl)carbonyl]oxy}-l,4-phenylene)]

Rule 5
In irregular polymers in which a central unit is substituted with three or more blocks, i.e., star polymers, the blocks are treated as substituents to a single central unit. A star polymer having identical blocks as its arms is named

   m[poly(A)]X

where m represents a Greek multiplicative prefix (tris, tetrakis...) denoting the number of poly(A) substituents on the central unit X. If the number of arms is undefined, the star polymer is named

   n-kis[poly(A)]X.

A star polymer having different blocks as its arms is named

   x-[poly(A)]-y-[poly(B)]-z-[poly(C)]...X

where x, y, z... are locants for positions on the central unit X. The atom in the blocks nearest to the central unit is given the locant 1.

Example 5.1
A three-armed star polymer in which polystyrene blocks are linked to a benzene ring at specified positions:

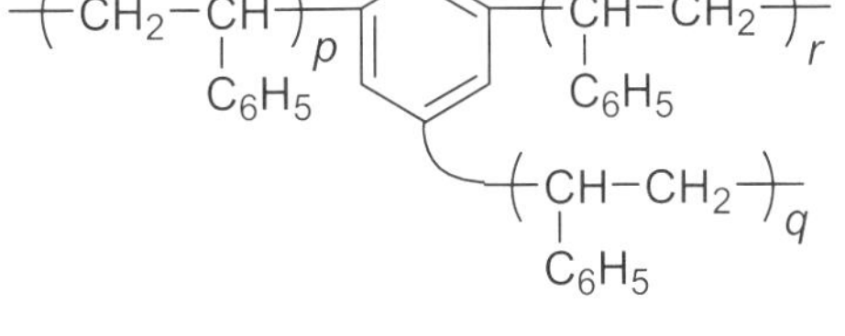

—(CH₂—CH(C₆H₅))ₚ—⟨benzene⟩—(CH—CH₂(C₆H₅))ᵣ—
                              |
                   —(CH—CH₂(C₆H₅))_q—

name: 1,3,5-tris[poly(1-phenylethane-1,2-diyl)]benzene

NOMENCLATURE

Example 5.2
A six-armed star polymer consisting of ethane substituted with polystyrene [poly(1-phenyl-ethane-1,2-diyl)] blocks:

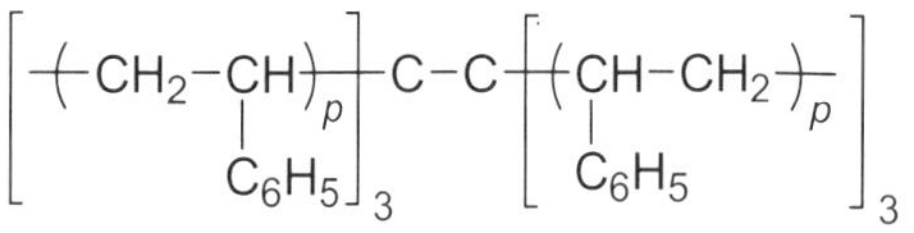

$$\left[+\!\!\left(\!\!\begin{array}{c}CH_2-CH\\|\\C_6H_5\end{array}\!\!\right)_{\!\!p}\!\!\right]_{\!3}\!\!-C-C-\!\left[\!\!\left(\!\!\begin{array}{c}CH-CH_2\\|\\C_6H_5\end{array}\!\!\right)_{\!\!p}\!\!\right]_{\!3}$$

name: hexakis[poly(1-phenylethane-1,2-diyl)]ethane

Example 5.3
A four-armed star polymer consisting of silane substituted with different block polymer chains:

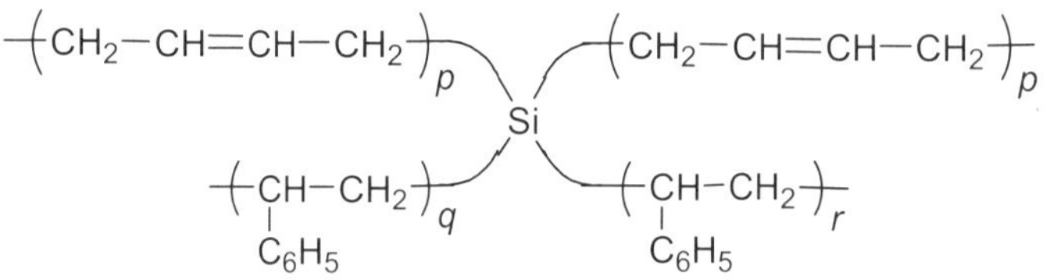

$$+(CH_2-CH=CH-CH_2)_p \quad +(CH_2-CH=CH-CH_2)_p$$
$$\diagdown Si \diagup$$
$$+(CH-CH_2)_q \quad +(CH-CH_2)_r$$
$$\;\;|\qquad\qquad\qquad\;\;|$$
$$C_6H_5 \qquad\qquad\quad C_6H_5$$

name: bis[poly(but-2-ene-1,4-diyl)][poly(1-phenylethane-1,2-diyl)][poly(2-phenylethane-1,2-diyl)] silane

Example 5.4
A three-armed star polymer consisting of silane substituted with poly(vinyl chloride), poly(ethylene oxide) and polystyrene chains, the last of which is linked to the central unit through an oxygen atom.

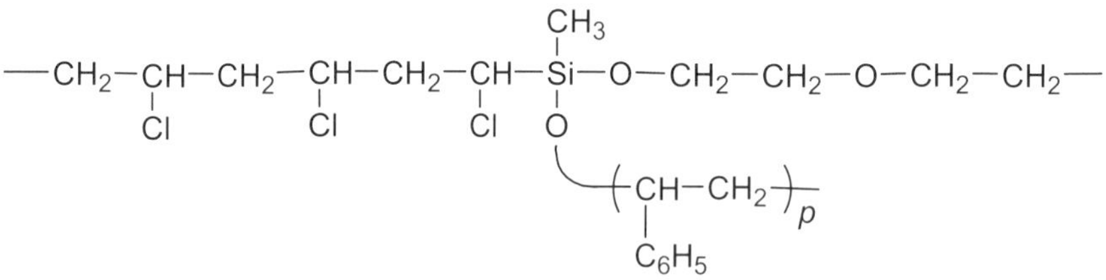

$$-CH_2-\underset{Cl}{CH}-CH_2-\underset{Cl}{CH}-CH_2-\underset{\underset{O}{|}}{CH}-\underset{\underset{|}{CH_3}}{Si}-O-CH_2-CH_2-O-CH_2-CH_2-$$
$$\qquad\qquad\qquad\qquad\qquad\qquad\quad\;\;+(CH-CH_2)_p$$
$$\qquad\qquad\qquad\qquad\qquad\qquad\qquad\;\;|$$
$$\qquad\qquad\qquad\qquad\qquad\qquad\qquad C_6H_5$$

name: methyl[poly(1-chloroethane-1,2-diyl)][poly(oxyethane-1,2-diyl)]{[poly(1-phenyl-ethane-1,2-diyl)]oxy}silane

Rule 6
Names of end groups are prefixed to the name of the polymer and denoted by the Greek letters α and ω. When bonding of the end groups to specific constitutional units is unknown, the irregular polymer is named

α-R-ω-R'-poly(A/B/...)

where R and R' are the names of the end groups. To specify bonding between end groups and constitutional units, the end group is combined with the attached constitutional unit prior to naming.

IRREGULAR SINGLE-STRAND ORGANIC POLYMERS

Example 6.1
The copolymer described in Ex. 1.1 with specified end groups:
name: α-chloro-ω-(trichloromethyl)-poly(1-chloroethane-1,2-diyl/1-phenylethane-1,2-diyl).

Example 6.2
The copolymer described in Ex. 6.1 in which the trichloromethyl group is connected to the 2-position of a 1-phenylethane-1,2-diyl unit and the chlorine to the 1-position of a 1-chloroethane-1,2-diyl unit:

name: α-(2,2-dichloroethyl)-ω-(3,3,3-trichloro-1-phenylpropyl)-poly(1-chloroethane-1,2-diyl/1-phenylethane-1,2-diyl)

Example 6.3
The graft copolymer in which the grafts have specified end groups:

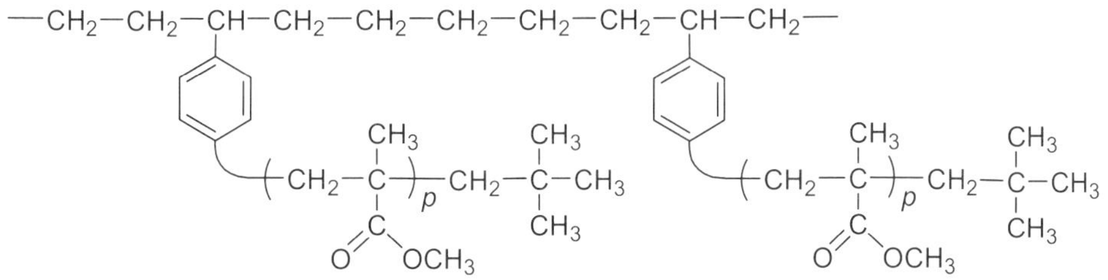

name: poly[ethane-1,2-diyl/l-(4-{ω-2,2'-dimethylpropyl-poly[2-(methoxycarbonyl)-2-methylethane-1,2-diyl]}phenyl)ethane-1,2-diyl]

Note: The free ends of the polymeric substituents, the right hand ends as represented here, are considered to be the ω-positions in accordance with the convention described in Section 2.2.4 of Chapter 1, the other ends, the α ends, being linked to the parent structure.

Rule 7
Specification with regard to mass fractions, mole fractions and molar masses is handled as in source-based copolymer nomenclature [4].

Example 7.1
Chlorinated poly(vinyl chloride) with a mass fraction of chlorine of 0.65:

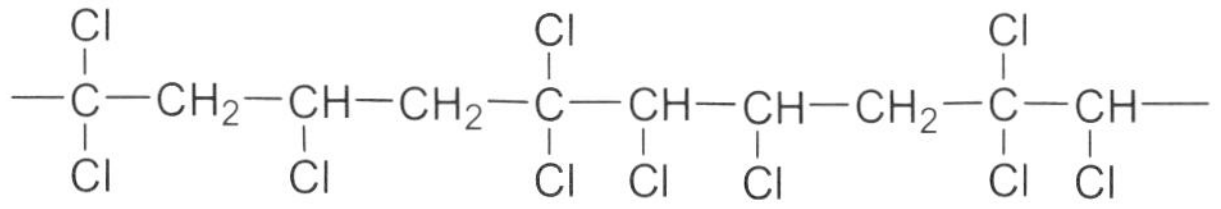

constitutional units: —CH— , —C— , —CH$_2$—
                          | |
                          Cl   Cl
                                    |
                                    Cl

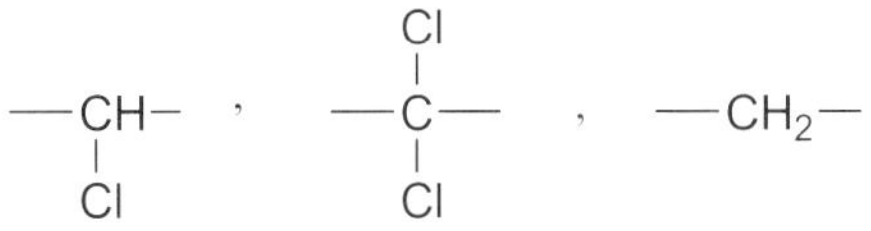

name: poly(chloromethylene/dichloromethylene/methylene) (65 mass % Cl)

347

NOMENCLATURE

ADDITIONAL EXAMPLES

Example 8.1
A diblock copolymer consisting of a block derived from the 1,4 and 1,2 polymerization of buta-1,3-diene and a block of polystyrene:

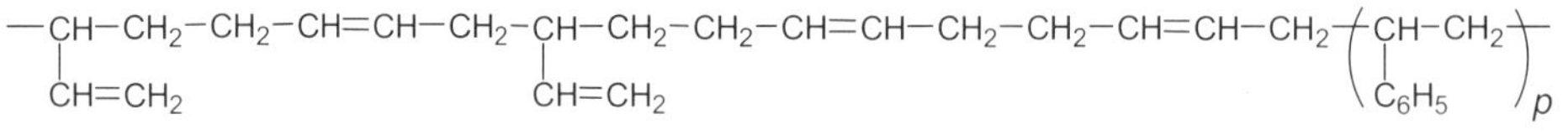

$$-\underset{\underset{CH=CH_2}{|}}{CH}-CH_2-CH_2-CH=CH-CH_2-\underset{\underset{CH=CH_2}{|}}{CH}-CH_2-CH_2-CH=CH-CH_2-CH_2-CH=CH-CH_2-\left(\underset{\underset{C_6H_5}{|}}{CH}-CH_2\right)_p$$

constitutional units: $-CH_2-CH=CH-CH_2-$, $-\underset{\underset{CH=CH_2}{|}}{CH}-CH_2-$, $-\underset{\underset{C_6H_5}{|}}{CH}-CH_2-$

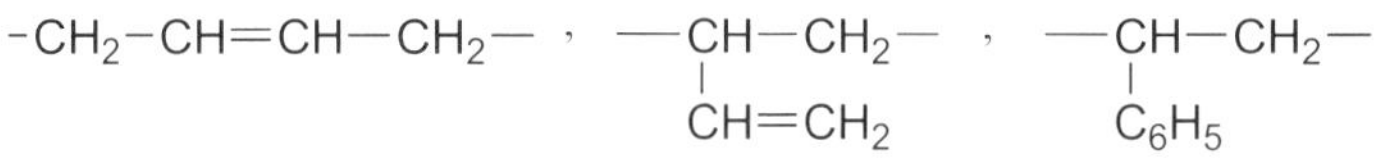

name: poly(but-2-ene-1,4-diyl/1-vinylethane-1,2-diyl)-poly(1-phenylethane-1,2-diyl).

Example 8.2
Poly(ethane-1,2-diyl/1-hydroxyethane-1,2-diyl) grafted with polystyrene [poly(1-phenylethane-1,2-diyl)] at hydroxymethylene units:

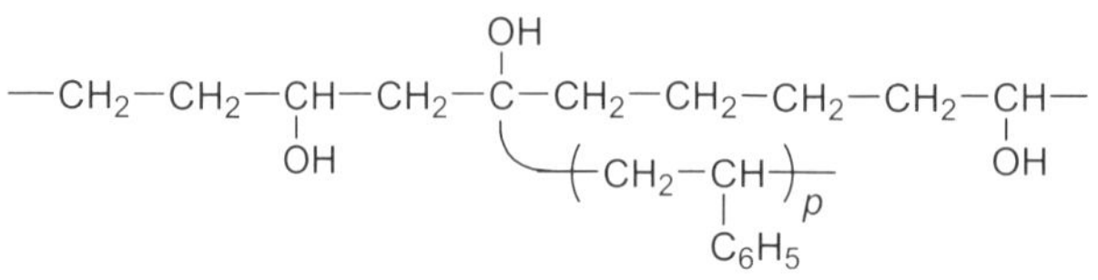

$$-CH_2-CH_2-\underset{\underset{OH}{|}}{CH}-CH_2-\underset{\underset{\left(CH_2-\underset{\underset{C_6H_5}{|}}{CH}\right)_p}{|}}{\overset{\overset{OH}{|}}{C}}-CH_2-CH_2-CH_2-CH_2-\underset{\underset{OH}{|}}{CH}-$$

constitutional units: $-CH_2-CH_2-$, $-\underset{\underset{OH}{|}}{CH}-CH_2-$, $-\underset{\underset{\left(CH_2-\underset{\underset{C_6H_5}{|}}{CH}\right)_p}{|}}{\overset{\overset{OH}{|}}{C}}-CH_2-$

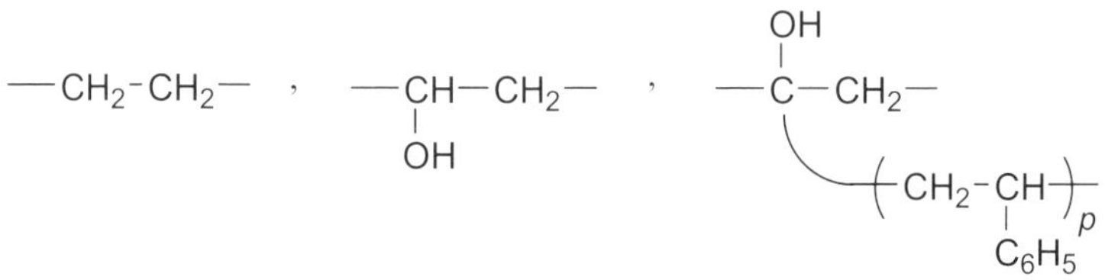

name: poly{ethane-1,2-diyl/1-hydroxyethane-1,2-diyl/1-hydroxy-1-[poly(2-phenylethane-1,2-diyl)]ethane-1,2-diyl}

Note: If the grafting point is not known, the graft copolymer is named

poly{ethane-1,2-diyl/1-hydroxyethane-1,2-diyl/*x*-[poly(2-phenylethane-1,2-diyl)]ethane-1,2-diyl/*x*-[poly(2-phenylethane-1,2-diyl)]-1-hydroxyethane-1,2-diyl}.

Example 8.3
A diblock copolymer consisting of a block of polyethylene grafted with poly(vinyl chloride) and a block of polystyrene, the structure of which is:

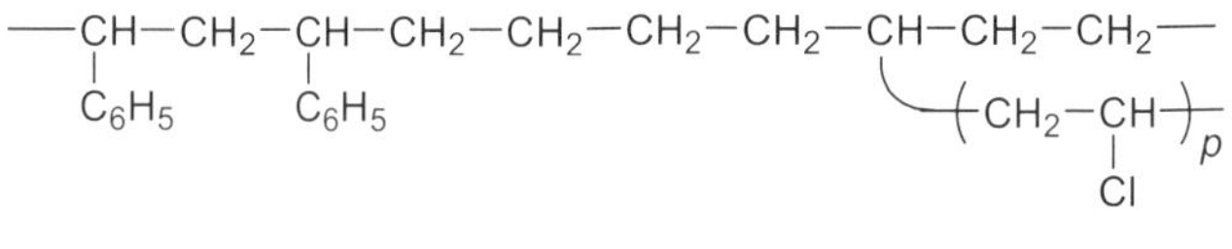

$$-\underset{\underset{C_6H_5}{|}}{CH}-CH_2-\underset{\underset{C_6H_5}{|}}{CH}-CH_2-CH_2-CH_2-CH_2-\underset{\underset{\left(CH_2-\underset{\underset{Cl}{|}}{CH}\right)_p}{|}}{CH}-CH_2-CH_2-$$

constitutional units: $-\underset{\underset{C_6H_5}{|}}{CH}-CH_2-$, $-\underset{\underset{\left(CH_2-\underset{\underset{Cl}{|}}{CH}\right)_p}{|}}{CH}-CH_2-$, $-CH_2-CH_2-$

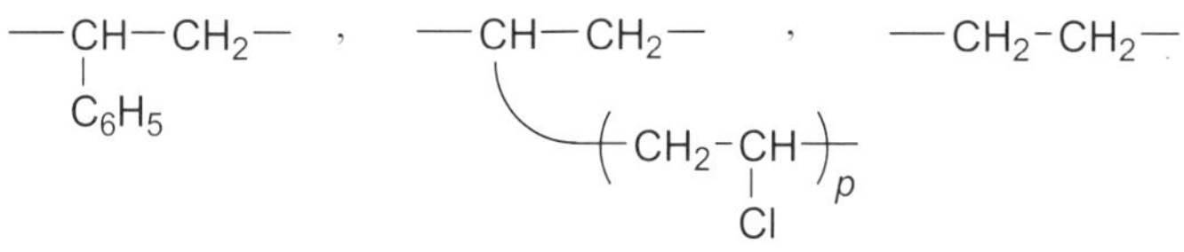

IRREGULAR SINGLE-STRAND ORGANIC POLYMERS

name: poly(1-phenylethane-1,2-diyl)-poly[ethane-1,2-diyl/1-poly(1-chloroethane-1,2-diyl)ethene/1-poly(2-chloroethane-1,2-diyl)ethene/2-poly(1-chloroethane-1,2-diyl)ethene/2-poly(2-chloroethane-1,2-diyl)ethene]

REFERENCES

1. IUPAC. 'Nomenclature of regular single-strand organic polymers (IUPAC Recommendations 2002)', *Pure Appl. Chem.* **74**, 1921-1956 (2002). Reprinted as Chapter 15 this volume.
2. IUPAC. 'Nomenclature for regular single-strand and quasi-single-strand inorganic and coordination polymers (Rules Approved 1984)', *Pure Appl. Chem.* **57**, 149-168 (1985).
3. IUPAC. 'Glossary of basic terms in polymer science (IUPAC Recommendations 1996)', *Pure Appl. Chem.* **68**, 2287-2311 (1996). Reprinted as Chapter 1 this volume.
4. IUPAC. 'Source-based nomenclature for copolymers (Rules approved 1985)', *Pure Appl. Chem.* **57**, 1427-1440 (1985). Reprinted as Chapter 19 this volume.
5. IUPAC. 'Nomenclature of Organic Chemistry (1979 Edition)', Pergamon Press, Oxford.
6. IUPAC. 'A Guide to IUPAC Nomenclature of Organic Compounds (Recommendations 1993)', R. Panico, W. H. Powell, J.-C. Richer, (Eds.), Blackwell Scientific Publications, Oxford, 1993.

18: Graphic Representations (Chemical Formulae) of Macromolecules

CONTENTS

Preamble
1. General rules
2. Regular polymers
3. Irregular polymers
4. Copolymers
References

PREAMBLE

Graphic representations (chemical formulae) of macromolecules are used extensively in the scientific literature on polymers including IUPAC documents on macromolecular nomenclature. This document establishes rules for the unambiguous representation of macromolecules by chemical formulae. The rules apply principally to synthetic macromolecules. Insofar as is possible, these rules are consistent with the formulae given in IUPAC documents [2-4] and they also cover the presentation of formulae for irregular macromolecules [5], copolymer molecules [1, 6] and star macromolecules.

In comparison with chemical formulae of low-molecular-weight compounds, the graphical representations of which have been addressed in a recent IUPAC document [7], chemical formulae of polymers must additionally reflect the multiplicity of constitutional units in a macromolecule and the various possibilities for connecting the constitutional units in a macromolecule.

Throughout the text the term *constitutional unit* [1] is taken to include both *constitutional repeating unit* [1] and *monomeric unit* [1]; one of these types of unit should be used wherever possible and appropriate.

As a general rule, chemical formulae for macromolecules should be written only in those cases where the structures of the constitutional units are known. A given structure may, however, be written in various ways to emphasize specific structural features; such alternative structures need not necessarily reflect the order of citation dictated by structure-based nomenclature [2].

1   GENERAL RULES

Rule 1.1: The formula representation of constitutional units shall be in accordance with usage in organic [8] and inorganic [9] chemistry, and with IUPAC rules for the nomenclature of polymers [2-6].

Prepared by a Working Group consisting of R. E. Bareiss (FRG), J. Kahovec (Czech Republic) and P. Kratochvíl (Czech Republic). Reprinted from *Pure Appl. Chem.* **66**, 2469-2482 (1994).

Rule 1.2: Consistent with the structure of the macromolecule, the order of citation of constitutional units within the formulae is arbitrary and, hence, need not comply with that given in [2].

Rule 1.3: To make the formulae more concise, dashes representing chemical bonds may be omitted. At the ends of constitutional repeating units and monomeric units, dashes must be attached.
Note: The absence of one or more of the dashes from a chiral or prochiral atom, or of dashes from atoms linked by a double bond, signifies lack of knowledge about the configuration of the corresponding site of stereoisomerism or lack of intention to specify it [1, 3].

Rule 1.4: Side-groups or substituents written on the same line as the backbone of the macromolecule and consisting of more than one atom symbol are set between enclosing marks, usually parentheses.

Rule 1.5: Enclosing marks together with subscript letters denote multiplicity of the enclosed constitutional units. The enclosing marks are parentheses (round brackets) or (square) brackets and can be used at random, except for inorganic polymers, for which exclusive use of parentheses is recommended for this purpose, in order to avoid confusion with (square) brackets, which denote coordination structures.

Rule 1.6: The subscript letters n, p, q, r, etc. denote multiplicities of polymeric sequences, whereas the subscript letters a, b, c, etc. denote multiplicities of oligomeric sequences. The subscripts should be printed in italic type.

Rule 1.7: The formulae of end groups, if known, may be attached to the bonds at the ends of the constitutional units, but placed outside the enclosing marks.

Rule 1.8: Specifications about mass fractions (w), mole fractions (x), molar masses (M), relative molecular masses (M_r), degrees of polymerization (DP) or the average values of the latter three quantities, may be expressed by placing the corresponding values in parentheses after the formula of the macromolecule in a manner analogous to that recommended for the naming of copolymers [6].

Applications of the general rules are illustrated in the following sections.

2   REGULAR POLYMERS

Rule 2.1: The formula of a **regular polymer** ([1], Definition 2.15) with the **constitutional repeating unit** ([1], Definition 1.15) —R— is given as:

$-(\text{R})_n-$ or $-[\text{R}]_n-$

and in cases where the end groups E′ and E″ are known:

$\text{E}'-(\text{R})_n-\text{E}''$ or $\text{E}'-[\text{R}]_n-\text{E}''$

Note: The chemical bonds connecting the constitutional repeating units are represented by dashes drawn across the enclosing marks.

NOMENCLATURE

Examples[*][†]

2-E1: poly(1-phenylethane-1,2-diyl)
polystyrene

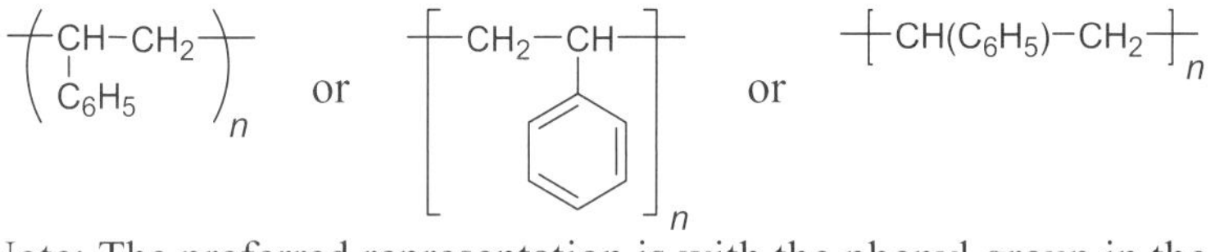

$$\left(\begin{matrix} CH-CH_2 \\ | \\ C_6H_5 \end{matrix}\right)_n \quad \text{or} \quad \left[\begin{matrix} CH_2-CH \\ | \\ C_6H_5 \end{matrix}\right]_n \quad \text{or} \quad \left[CH(C_6H_5)-CH_2\right]_n$$

Note: The preferred representation is with the phenyl group in the 1- position, i.e. that which corresponds to the name poly(1-phenylethane-1,2-diyl) [2].

2-E2: syndiotactic poly(1-phenylethane-1,2-diyl)
syndiotactic polystyrene

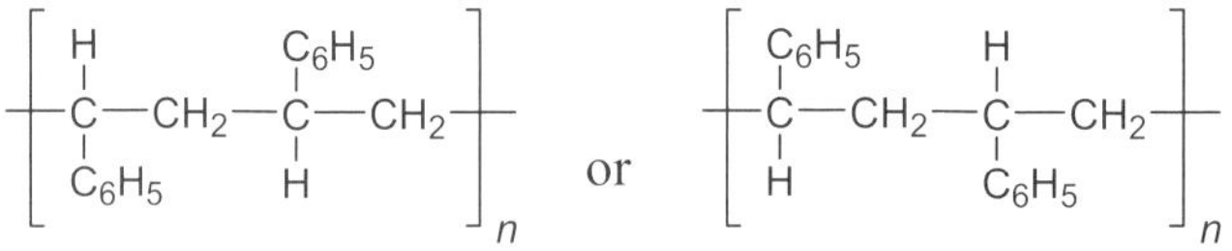

$$\left[\begin{matrix} H & & C_6H_5 \\ | & & | \\ C-CH_2-C-CH_2 \\ | & & | \\ C_6H_5 & & H \end{matrix}\right]_n \quad \text{or} \quad \left[\begin{matrix} C_6H_5 & & H \\ | & & | \\ C-CH_2-C-CH_2 \\ | & & | \\ H & & C_6H_5 \end{matrix}\right]_n$$

Note: Analogous formulae can be drawn for other tactic macromolecules [3].

2-E3: poly(but-1-ene-1,4-diyl)
1,4-polybutadiene

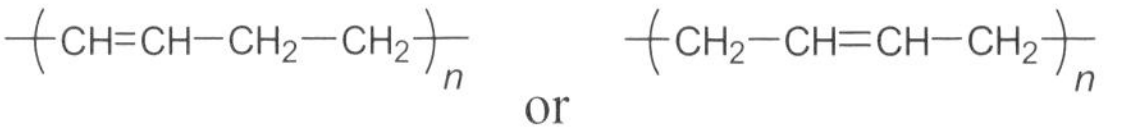

$$[CH=CH-CH_2-CH_2]_n \quad \text{or} \quad [CH_2-CH=CH-CH_2]_n$$

Note: The preferred representation is with the double bond in the 1-position, i.e., that which corresponds to the name poly(but-1-ene-1,4-diyl) [2].

2-E4: poly[(E)-1-methylbut-1-ene-1,4-diyl]
trans-1,4-polyisoprene

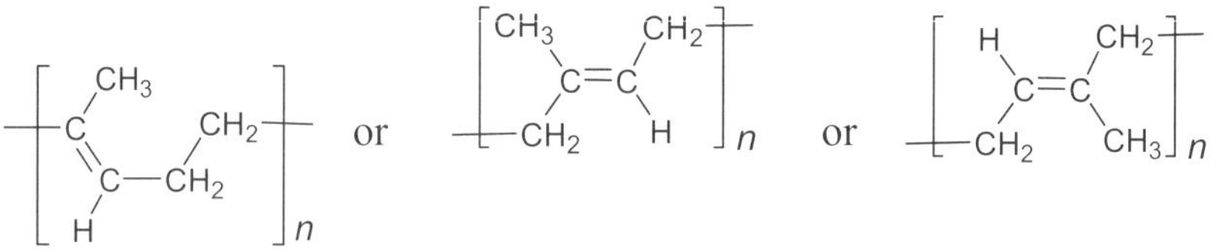

$$\left[\begin{matrix} CH_3 \\ C \\ \| \\ C-CH_2 \\ | \\ H \end{matrix} CH_2\right]_n \quad \text{or} \quad \left[\begin{matrix} CH_3 & CH_2 \\ C=C \\ CH_2 & H \end{matrix}\right]_n \quad \text{or} \quad \left[\begin{matrix} H & CH_2 \\ C=C \\ CH_2 & CH_3 \end{matrix}\right]_n$$

Note: The preferred representation is with the double bond in the 1-position, i.e. that which corresponds to the name poly[(E)-1-methylbut-1-ene-1,4-diyl] [2].

---

[*] Structure-based names [2-4, 5, 9] are given first, followed by source-based names or traditional names [2-4, 9], if these exist.
[†] The formulae of the constitutional repeating units depicted are understood to be non-exclusive.

GRAPHIC REPRESENTATIONS

2-E5: poly[1-(methoxycarbonyl)-1-methylethane-1,2-diyl]
poly(methyl methacrylate)

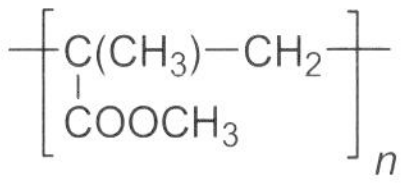

$$-[C(CH_3)(COOCH_3)-CH_2]_n-$$

2-E6: poly(oxyethane-1,2-diyloxyterephthaloyl)
poly(ethylene terephthalate)

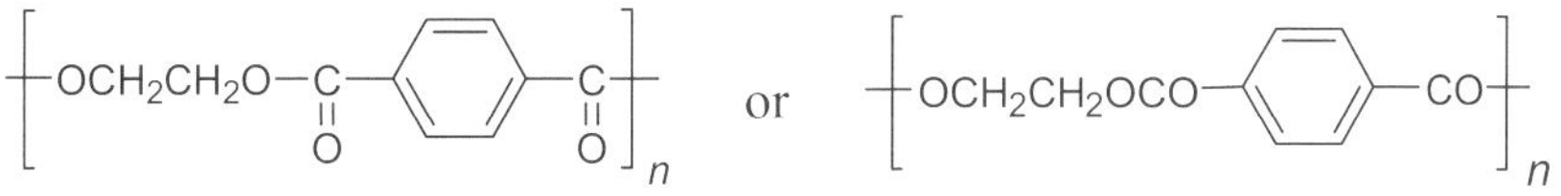

$$-[OCH_2CH_2O-CO-C_6H_4-CO]_n- \text{ or } -[OCH_2CH_2OCO-C_6H_4-CO]_n-$$

2-E7: poly[imino(1-oxohexane-1,6-diyl)]
poly(ε-caprolactam), poly(hexano-6-lactam)

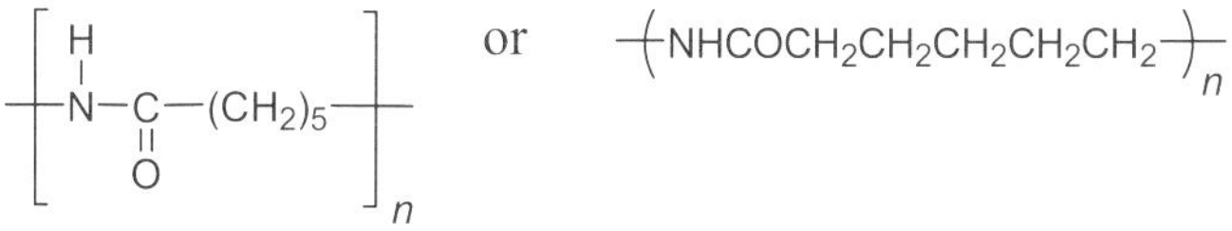

$$-[N(H)-C(=O)-(CH_2)_5]_n- \text{ or } -(NHCOCH_2CH_2CH_2CH_2CH_2)_n-$$

2-E8: poly[sulfanediyl-(R,R)-1,2-dimethylethane-1,2-diyl] or poly[sulfanediyl-(S,S)-1,2-dimethylethane-1,2-diyl]
poly[(R,S)-2,3-dimethylthiirane], poly(cis-2,3-dimethylthiirane)

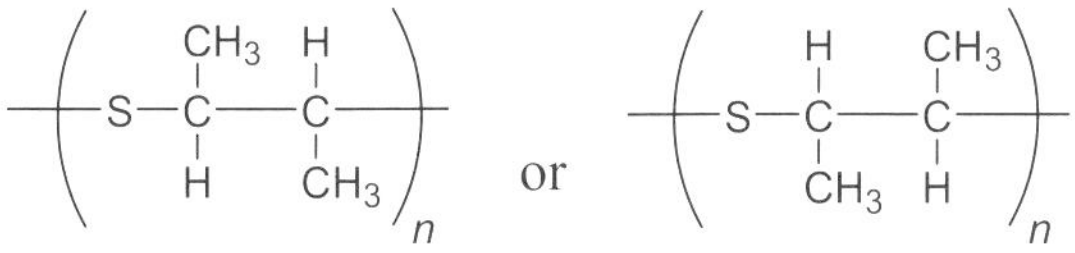

(structures of two stereoisomeric repeat units shown)

Note: Enantioselective polymerization of cis-(R,S)-2,3-dimethylthiirane can produce poly[sulfanediyl-(R,R)-1,2-dimethylethane-1,2-diyl] or its enantiomer.

2-E9: poly{1-[({5-[(4′-cyanobiphenyl-4-yl)oxy]pentyl}oxy)carbonyl]ethane-1,2-diyl}

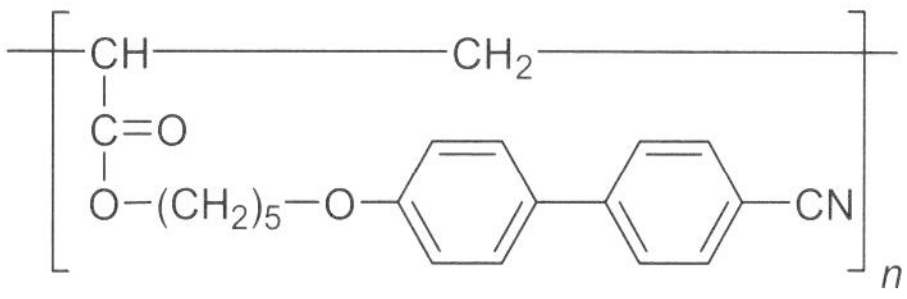

$$-[CH(C(=O)O-(CH_2)_5-O-C_6H_4-C_6H_4-CN)-CH_2]_n-$$

2-E10: α-hydro-ω-hydroxypoly(oxyethane-1,2-diyl)
poly(ethylene glycol)

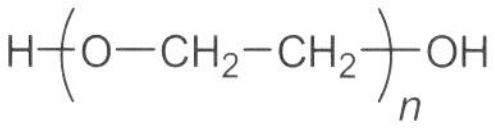

$$H-(O-CH_2-CH_2)_n-OH$$

353

NOMENCLATURE

2-E11: poly(5-oxaspiro[3.5]nonane-2,7-diyl)

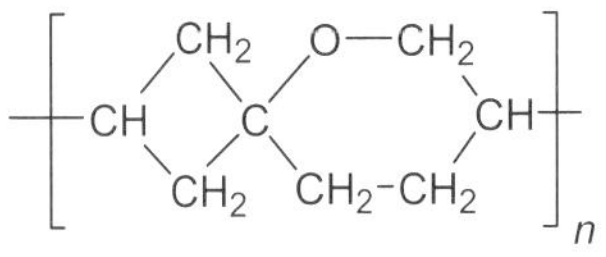

2-E12: poly[imino(2-isobutyl-1-oxoethane-1,2-diyl)imino(1-oxoethane-1,2-diyl)]
poly(glycylleucine)

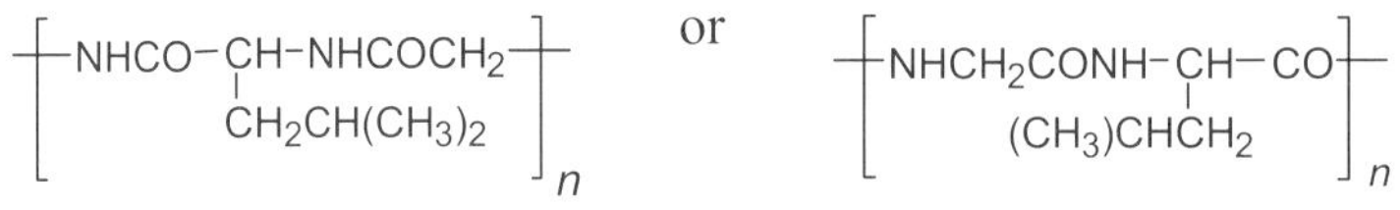

2-E13: poly(cyclopentane-1,3-diylethene-1,2-diyl)
polynorbornene or poly(8,9,10-trinorborn-2-ene)

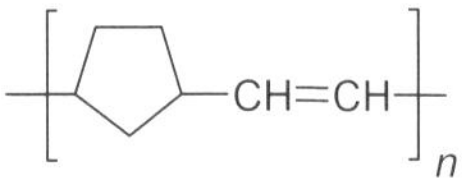

2-E14: poly(but-1-ene-1,4:3,2-tetrayl) [10]
ladder-poly(methyl vinyl ketone) [10]

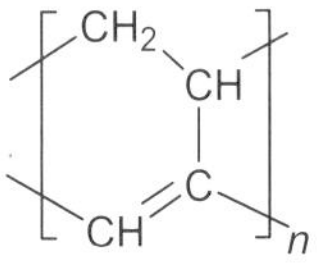

Note: The source-based name identifies the starting monomer of this ladder polymer, the synthesis of which comprises a multistep reaction involving condensation and cyclization.

2-E15: *catena*-poly[(diphenylsilicon)-μ-oxo] [4] or
poly[oxy(diphenylsilanediyl)] [2, 7]
poly(diphenylsiloxane)

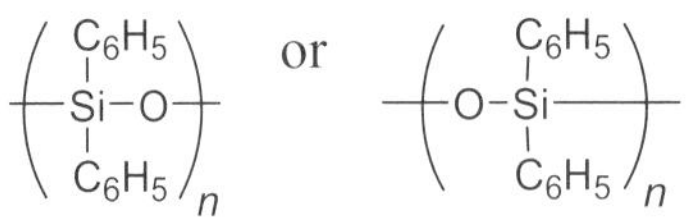

2-E16: *catena*-poly[(diethoxophosphorus)-μ-nitrido] [4]
poly[nitrilo(diethoxy-λ^5-phosphanylylidene)] [2]
poly(diethoxyphosphazene)

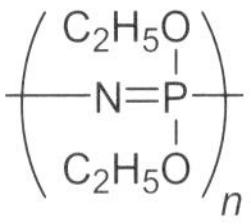

GRAPHIC REPRESENTATIONS

2-E17: *catena*-poly{caesium [cuprate-tri-μ-chloro](1-) or
catena-poly{caesium [cuprate(II)-tri-μ-chloro])

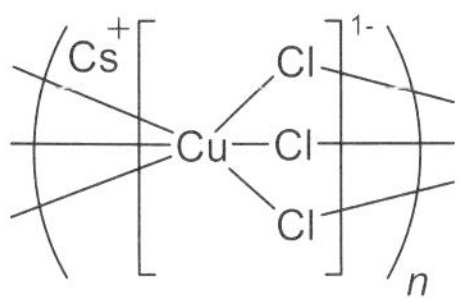

$$\left(Cs^+ \left[\begin{array}{c} Cl \\ | \\ Cu-Cl \\ | \\ Cl \end{array} \right]^{1-} \right)_n$$

2-E18: α-ammine-ω-(amminedichlorozinc)-*catena*-poly[(amminechlorozinc)-μ-chloro]

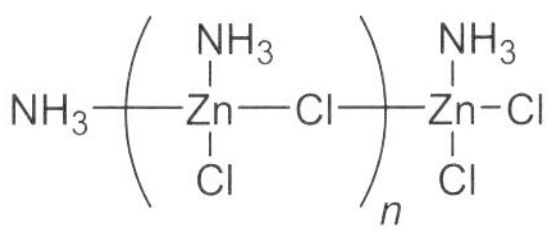

$$NH_3 \left(\begin{array}{c} NH_3 \\ | \\ Zn-Cl \\ | \\ Cl \end{array} \right)_n \begin{array}{c} NH_3 \\ | \\ Zn-Cl \\ | \\ Cl \end{array}$$

3   IRREGULAR POLYMERS [1, 5]

Rule 3.1: The formula of an **irregular polymer** ([1], Definition 3.2) or **irregular block** ([1], Definition 3.16) comprised of the constitutional units —U—, —V—, —W—, etc., is given as:

$$(—U—/—V—/—W—/...)_n \quad \text{or} \quad [—U—/—V—/—W—/...]_n$$

Note 1: The sequence of dots denotes the presence of further constitutional units.
Note 2: The order of the constitutional units in the formula is arbitrary.
Note 3: The oblique stroke drawn between the constitutional units means that the sequential arrangement of these units is irregular or unknown.
Note 4: The dashes at each end of the formula are drawn fully inside the enclosing marks, because they do not necessarily denote terminal chemical bonds of the macromolecules.
Note 5: The validity of the constitutional units selected should always be checked by arranging their formulae with repetition, thus forming formulae of longer sequences. In this way it is possible to exclude combinations of constitutional units that do not occur in the macromolecules.

*Examples**

3-E1: poly(1-chloroethane-1,2-diyl/2-chloroethane-1,2-diyl)

(an irregular polymer derived from vinyl chloride, the units of which are joined both head-to-tail and head-to-head):

---

* Structure-based names are given first [5] followed, if necessary, by an explanation (in parentheses) and a formula segment of the macromolecule. Then the monomeric or constitutional units necessary to describe the complete structure are given followed by the proposed formula.

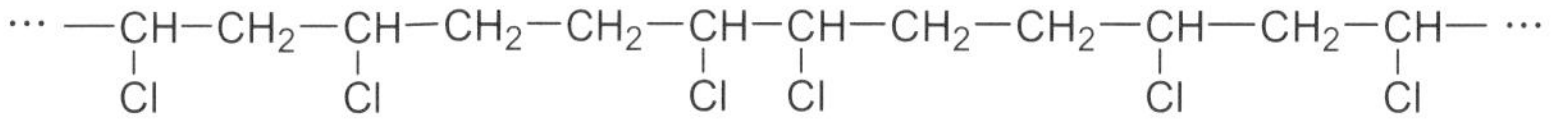

NOMENCLATURE

$$\cdots -\underset{\underset{Cl}{|}}{CH}-CH_2-\underset{\underset{Cl}{|}}{CH}-CH_2-CH_2-\underset{\underset{Cl}{|}}{CH}-\underset{\underset{Cl}{|}}{CH}-CH_2-CH_2-\underset{\underset{Cl}{|}}{CH}-CH_2-\underset{\underset{Cl}{|}}{CH}-\cdots$$

with monomeric units: $-\underset{\underset{Cl}{|}}{CH}-CH_2-$, $-CH_2-\underset{\underset{Cl}{|}}{CH}-$

$$\left(-\underset{\underset{Cl}{|}}{CH}-CH_2-\ /\ -CH_2-\underset{\underset{Cl}{|}}{CH}-\right)_n$$

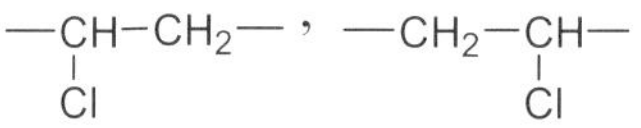

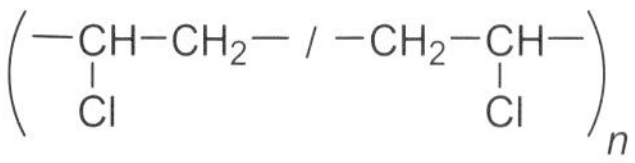

Note: The sequence of dots denotes the continuation of the macromolecular chain.

3-E2: poly(chloromethylene/methylene)
(chlorinated polyethene that does not contain dichloromethylene units):

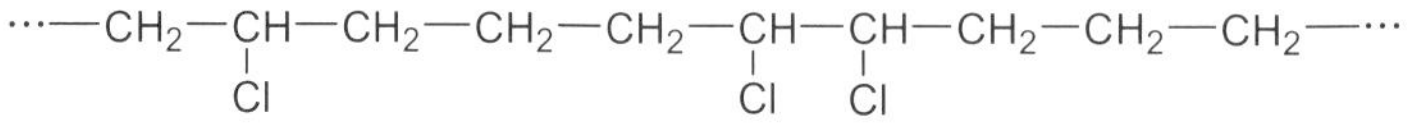

$$\cdots -CH_2-\underset{\underset{Cl}{|}}{CH}-CH_2-CH_2-CH_2-\underset{\underset{Cl}{|}}{CH}-\underset{\underset{Cl}{|}}{CH}-CH_2-CH_2-CH_2-\cdots$$

with constitutional units:

$-\underset{\underset{Cl}{|}}{CH}-$, $-CH_2-$

$$\left(-\underset{\underset{Cl}{|}}{CH}-\ /\ -CH_2-\right)_n$$

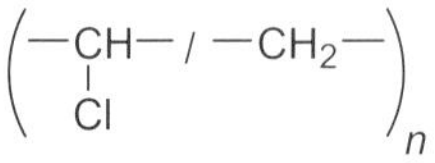

See note to 3-El.

3-E3: poly(chloromethylene/dichloromethylene/methylene)
(chlorinated poly(vinyl chloride) with a mass fraction of chlorine of 0.65):

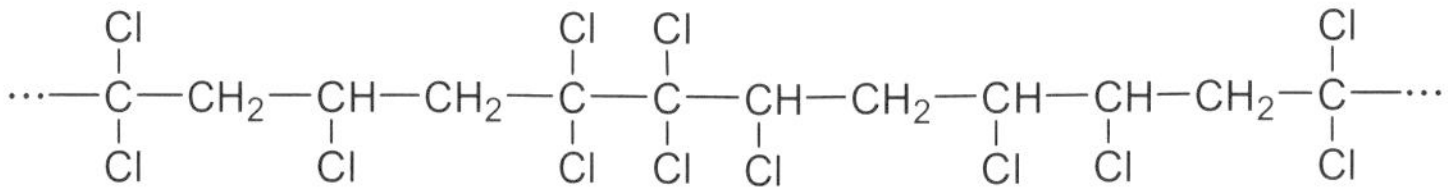

$$\cdots -\underset{\underset{Cl}{|}}{\overset{\overset{Cl}{|}}{C}}-CH_2-\underset{\underset{Cl}{|}}{CH}-CH_2-\underset{\underset{Cl}{|}}{\overset{\overset{Cl}{|}}{C}}-\underset{\underset{Cl}{|}}{\overset{\overset{Cl}{|}}{C}}-\underset{\underset{Cl}{|}}{CH}-CH_2-\underset{\underset{Cl}{|}}{CH}-\underset{\underset{Cl}{|}}{CH}-CH_2-\underset{\underset{Cl}{|}}{\overset{\overset{Cl}{|}}{C}}-\cdots$$

with constitutional units: $-\underset{\underset{Cl}{|}}{CH}-$, $-\underset{\underset{Cl}{|}}{\overset{\overset{Cl}{|}}{CH}}-$, $-CH_2-$

$$\left(-\underset{\underset{Cl}{|}}{CH}-\ /\ -\underset{\underset{Cl}{|}}{\overset{\overset{Cl}{|}}{C}}-\ /\ -CH_2-\right)_n \quad (w(Cl) = 0.65)$$

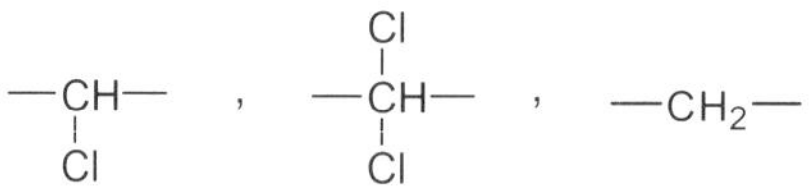

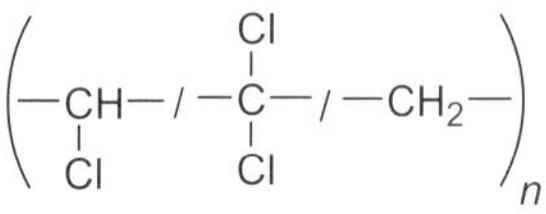

See note to 3-El.

3-E4: poly(but-2-ene-1,4-diyl/1-vinylethane-1,2-diyl/2-vinylethane-1,2-diyl)
(irregular polymer comprising units derived from 1,4- and 1,2-additions in the polymerization of buta-1,3-diene):

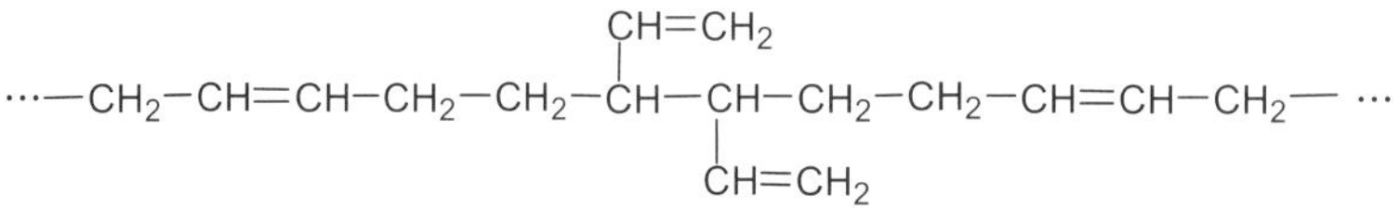

$$\cdots -CH_2-CH=CH-CH_2-CH_2-\underset{\underset{CH=CH_2}{|}}{\overset{\overset{CH=CH_2}{|}}{CH}}-CH-CH_2-CH_2-CH=CH-CH_2-\cdots$$

with monomeric units:

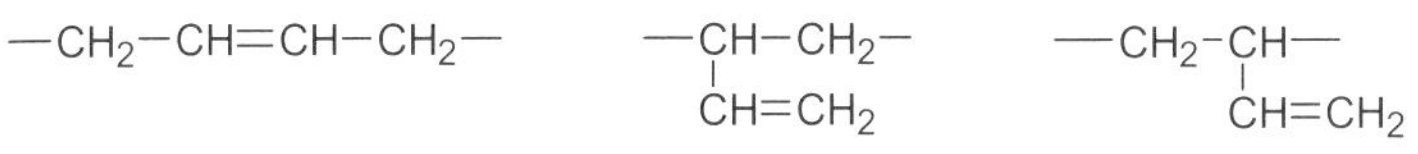

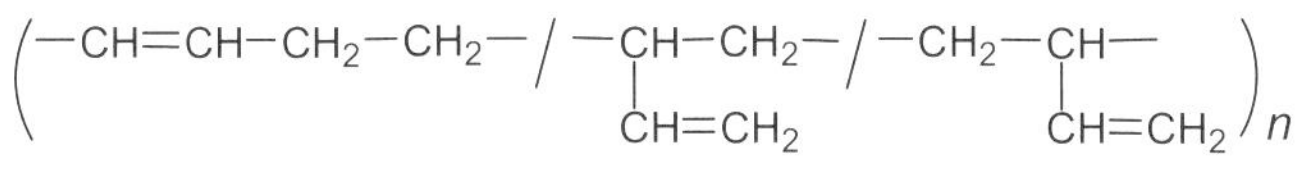

—CH$_2$—CH=CH—CH$_2$—   —CH—CH$_2$—   —CH$_2$—CH—
                          | |
                          CH=CH$_2$ CH=CH$_2$

$\left(\text{—CH=CH—CH}_2\text{—CH}_2\text{—} / \text{—CH—CH}_2\text{— } / \text{—CH}_2\text{—CH—} \atop \qquad\qquad\qquad\qquad\qquad \text{CH=CH}_2 \qquad\qquad \text{CH=CH}_2 \right)_n$

See note to 3-E1.

3-E5: poly(carbonyl/hydroperoxymethylene/methylene/vinylene) (oxidized polyethene):

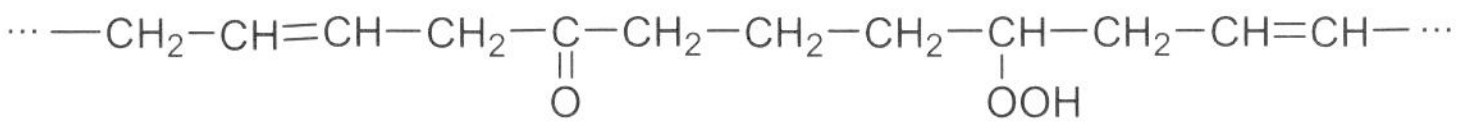

···—CH$_2$—CH=CH—CH$_2$—C—CH$_2$—CH$_2$—CH$_2$—CH—CH$_2$—CH=CH—···
                        || |
                        O   OOH

with constitutional units:

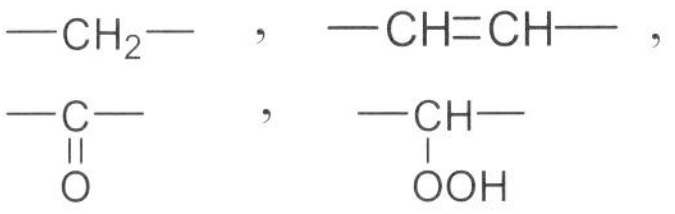

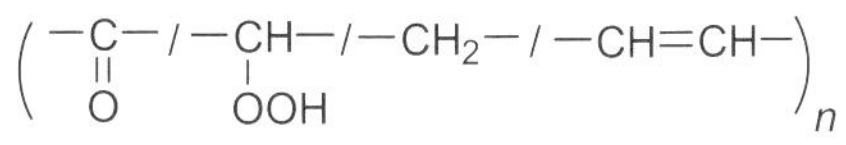

—CH$_2$— , —CH=CH— ,

—C— , —CH—
|| |
O   OOH

$\left(\text{—C—} / \text{—CH—} / \text{—CH}_2\text{—} / \text{—CH=CH—} \atop \;\;\text{||} \qquad \;\;\text{|} \right)_n$
$\qquad\;\; \text{O} \qquad \text{OOH}$

See note to 3-E1.

3-E6: poly(iminohexane-1,6-diyliminohexanedioyl/iminobutane-1,4-diyliminohexanedioyl)
(polyamide derived from adipoyl chloride and a mixture of hexane-1,6-diamine and butane-1,4-diamine):

- - —NHCO—(CH$_2$)$_4$-CONH—(CH$_2$)$_4$-NHCO—(CH$_2$)$_4$-CONH—(CH$_2$)$_6$-NHCO—(CH$_2$)$_4$-CONH—(CH$_2$)$_6$—

with constitutional units:

—NH—(CH$_2$)$_4$—NHCO—(CH$_2$)$_4$—CO— , —NH—(CH$_2$)$_6$—NHCO—(CH$_2$)$_4$—CO—

$\left[\text{—NH—(CH}_2\text{)}_4\text{—NH—C—(CH}_2\text{)}_4\text{—C—} / \text{—NH—(CH}_2\text{)}_6\text{—NH—C—(CH}_2\text{)}_4\text{—C—} \atop \qquad\qquad\qquad\qquad\; \text{||} \qquad\qquad\;\; \text{||} \qquad\qquad\qquad\qquad\qquad\; \text{||} \qquad\qquad\;\; \text{||} \right]_n$
$\qquad\qquad\qquad\qquad\qquad\quad \text{O} \qquad\qquad\;\; \text{O} \qquad\qquad\qquad\qquad\qquad\qquad \text{O} \qquad\qquad\;\; \text{O}$

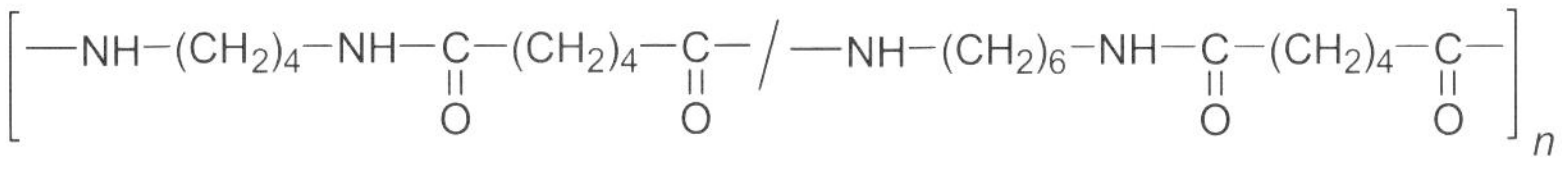

See note to 3-E1.

3-E7: poly[poly(1-cyanoethane-1,2-diyl)/poly(1-phenylethane-1,2-diyl)/poly(1-vinylethane-1,2-diyl)] (irregular polymer which consists of regular blocks of polyacrylonitrile, polystyrene and 1,2-polybutadiene with an unspecified sequential arrangement of the blocks $+\text{CH(CN)—CH}_2+_p$, $+\text{CH(C}_6\text{H}_5\text{)—CH}_2+_q$ and $+\text{CH(CH=CH}_2\text{)—CH}_2+_r$):

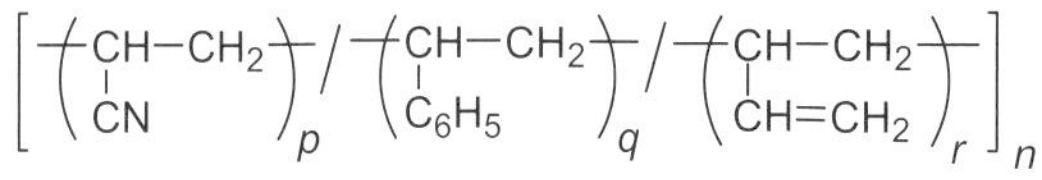

NOMENCLATURE

$$\left[\left(\begin{array}{c}\text{CH}-\text{CH}_2\\|\\\text{CN}\end{array}\right)_p \Big/ \left(\begin{array}{c}\text{CH}-\text{CH}_2\\|\\\text{C}_6\text{H}_5\end{array}\right)_q \Big/ \left(\begin{array}{c}\text{CH}-\text{CH}_2\\|\\\text{CH}=\text{CH}_2\end{array}\right)_r\right]_n$$

Note: The choice of the constitutional units is dictated by Rule 1.1 and the rules of [2], e.g. 1-cyanoethane-1,2-diyl is preferred to 2-cyanoethane-1,2-diyl.

4 COPOLYMERS

4.1 Alternating and Periodic Copolymers

Rule 4.1.1: Alternating and periodic copolymers, as far as possible, are treated as regular polymers.

Rule 4.1.2: Pseudoperiodic copolymers, e.g. those in which only some of the constitutional units occur regularly, are treated as irregular polymers (see Ex. 4.1-E1) or as unspecified copolymers (see example 4.1-E3).

Examples[*]

4.1-E1: poly[styrene-*alt*-(maleic anhydride)]
poly[(2,5-dioxotetrahydrofuran-3,4-diyl)(1-phenylethane-1,2-diyl)]

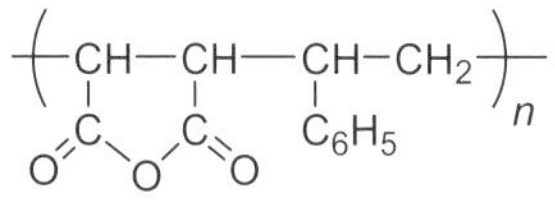

Note: In cases in which both 1-phenylethane-1,2-diyl and 2-phenylethane-1,2-diyl units are present in measurable amounts, the formula for an irregular polymer is used:

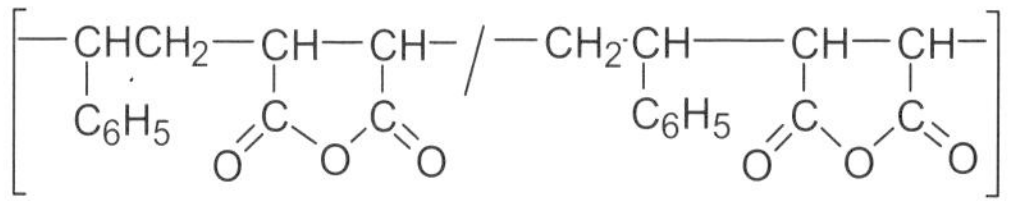

Note: The preferred representations for these constitutional units are with the heterocyclic rings leftmost [2].

---

[*] Source-based (co)polymer names [2, 6] are given first, followed by structure-based names [2], if these exist, before the proposed formula.

GRAPHIC REPRESENTATIONS

4.1-E2: poly(1,3,6-trioxaoctane)*
       poly(oxymethyleneoxyethane-1,2-diyloxyethane-1,2-diyl)

$$+(OCH_2OCH_2CH_2OCH_2CH_2)_n$$

4.1-E3: poly[(ethane-1,2-diol)-*alt*-(terephthalic acid; isophthalic acid)]
       poly[(ethylene terephthalate)-*co*-(ethylene isophthalate)]

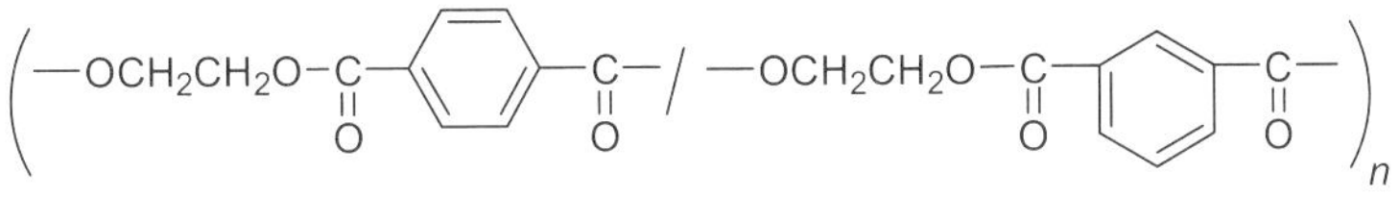

$$\left(-OCH_2CH_2O-\underset{O}{\overset{\|}{C}}-\bigcirc-\underset{O}{\overset{\|}{C}}-\Big/-OCH_2CH_2O-\underset{O}{\overset{\|}{C}}-\bigcirc-\underset{O}{\overset{\|}{C}}-\right)_n$$

4.2 Statistical, Random and Unspecified Copolymers

Rule 4.2.1: Statistical, random and unspecified copolymers are treated as irregular polymers.

Examples[†]

4.2-E1: poly(styrene-*stat*-buta-1,3-diene)

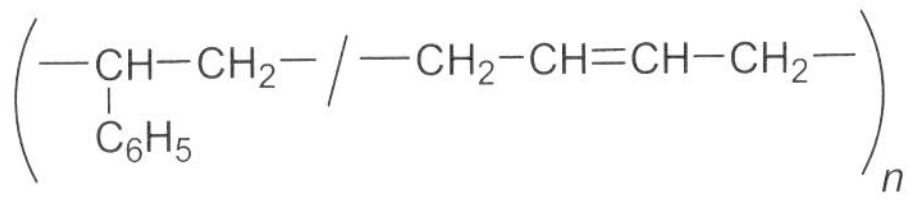

$$\left(\begin{array}{c}-CH-CH_2-\\ |\\ C_6H_5\end{array}\Big/-CH_2-CH=CH-CH_2-\right)_n$$

Note: The buta-1,3-diene is exclusively incorporated by 1,4-addition, which cannot be reflected in the source-based name.

4.2-E2: poly[(6-aminohexanoic acid)-*stat*-(7-aminoheptanoic acid)]

$$\left[-NH(CH_2)_5CO-\Big/-NH(CH_2)_6CO-\right]_n$$

Note: The preferred representations for the constitutional units from 6-aminohexanoic acid and 7-aminoheptanoic acid are -NH-CO-(CH$_2$)$_5$- and -NH-CO-(CH$_2$)$_6$-, respectively [2].

4.2-E3: poly[ethene-*ran*-(vinyl acetate)]

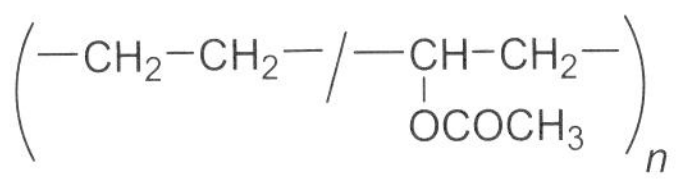

$$\left(-CH_2-CH_2-\Big/\begin{array}{c}-CH-CH_2-\\ |\\ OCOCH_3\end{array}\right)_n$$

---

* In replacement nomenclature as conventionally applied to acyclic structures with multiple heteroatoms, terminal heteroatoms are not designated with 'a' prefixes but are named as characteristic groups of the structure, i.e., as hydroxy, amino, carboxylic acid, etc. However, heteroatoms in such positions within the constitutional units of polymer molecules are not terminal units. Consequently, such atoms are designated with 'a' prefixes, and thereby the simplicity afforded by the application of replacement nomenclature to polymer molecules is enhanced.
† Source-based copolymer names [6] are given, followed by the proposed formula.

NOMENCLATURE

4.2-E4: poly[(4-hydroxybenzoic acid)-*co*-hydroquinone-*co*-(terephthalic acid)]

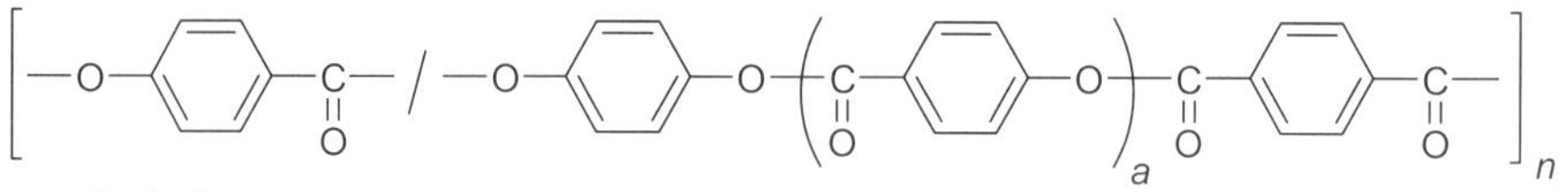

[structure]

$a = 0, 1, 2, ...$

4.2-E5: poly[styrene-*co*-(methyl methacrylate)] (75 : 25 mass %; 10^5 $\overline{M}_{r,n}$)

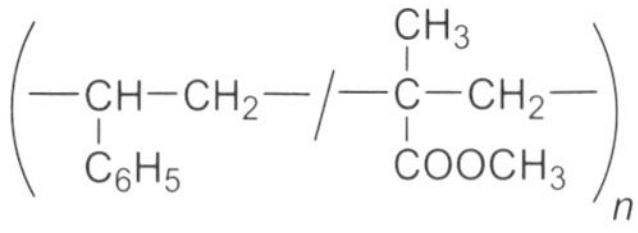

[structure]

4.2-E6: α-methyl-ω-hydroxy-poly[(ethylene oxide)-*co*-(propylene oxide)*]

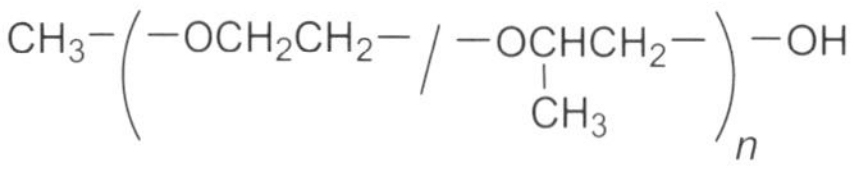

[structure]

Note: The dashes for the bonds to the end groups are not drawn through the parentheses of the polymer formula, because this formula does not specify which end group is attached to which monomeric unit.

4.2-E7: α-butyl-ω-carboxy-poly[styrene-*co*-(4-chlorostyrene)]
α-(1-phenylhexyl)-ω-[2-carboxy-2-(4-chlorophenyl)ethyl]poly[styrene-*co*-(4-chlorostyrene)] [6]

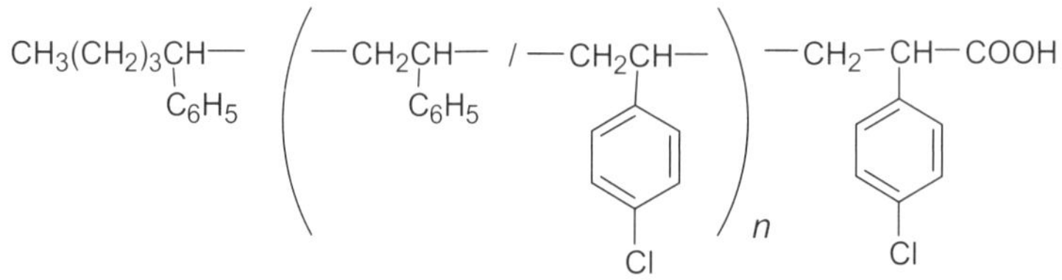

[structure]

Note 1: The preferred representations for the constitutional units are -CH(C_6H_5)-CH_2- and -CH(4Cl-C_6H_4)-CH_2-[2]; use of the preferred representations would result in α-pentyl and ω-[1-carboxy-1-(4-chlorophenyl) methyl] end-groups.
Note 2: See note to 4.2-E6.
Note 3: The formula is more specific than the first of the two given names in that it specifies that the butyl end group is linked to the 2-position of a 1-phenylethyl group and the carboxy group is attached to the 2-position of a 2-(4-chlorophenyl)ethyl group.

---

* According to 'A Guide to IUPAC Nomenclature of Organic Compounds (Recommendations 1993)' the traditional name 'propylene oxide' should be replaced by 'methyloxirane'.

GRAPHIC REPRESENTATIONS

4.3 Block Copolymers

Rule 4.3.1: The formulae of **block copolymers** ([1], Definition 3.35) consisting of a sequence of **regular blocks** ([1], Definition 3.15) and, if known, **junction units** ([6], Definition Rule 5.5) in *known* sequential arrangement are written as, e.g.,

$$-(\!(A)\!)_p-(\!(B)\!)_q-(\!(C)\!)_r-\cdots$$

$$-(\!(A)\!)_p-X-(\!(B)\!)_q-(\!(C)\!)_r-\cdots$$

$$-(\!(A)\!)_p-X-(\!(B)\!)_q-Y-(\!(C)\!)_r-\cdots$$

where A, B, C, etc., are the constitutional repeating units of the regular blocks $-(\!(A)\!)_p$, $-(\!(B)\!)_q$, $-(\!(C)\!)_r$, etc., and X, Y, etc. are junction units which are not considered to be parts of the blocks.

Note: The sequence of dots denotes the presence of further constitutional units or blocks or both.

Rule 4.3.2: The formulae of block copolymers consisting of sequences of regular blocks and, if known, junction units in an *unknown* sequential arrangement are written with the formulae of blocks and junction units separated by oblique strokes. Thus,

$$[-(\!(A)\!)_p / -(\!(B)\!)_q / -(\!(C)\!)_r /\ldots]_n$$

represents a block copolymer consisting of an unknown sequence of regular blocks $-(\!(A)\!)_p$, $-(\!(B)\!)_q$, $-(\!(C)\!)_r$, etc., and

$$[-(\!(A)\!)_p-X- / -(\!(B)\!)_q-Y- / -(\!(C)\!)_r-]_n$$

represents a block copolymer consisting of an unknown sequence of regular blocks $-(\!(A)\!)_p$, $-(\!(B)\!)_q$, $-(\!(C)\!)_r$ with junction units X and Y, attached to $-(\!(A)\!)_p$ and $-(\!(B)\!)_q$, respectively.

The sequence of dots denotes the presence of further constitutional units or blocks or both.

Rule 4.3.3: The formulae of block copolymers consisting of a sequence of irregular blocks are written as:

$$(-A- / -B- /\ldots)_p (-U- / -V- /\ldots)_q \ldots$$

NOMENCLATURE

where A, B, etc. and U, V, etc. are the constitutional repeating units of the irregular blocks

$(-A-/-B-/...)_p, (-U-/-V-/...)_q$, etc.

Note 1: The sequence of dots denotes the presence of further constitutional units or blocks or both.

Note 2: The bonds emanating from the first and last constitutional or junction unit of an irregular block are written within the enclosing marks when it is not known to which of the units of the irregular block the other blocks or the end groups of the polymer are attached. Thus,

$+(A)_p(-B-/-C-)_q+(D)_r$

represents a block copolymer consisting of an irregular block $(-B-/-C-)_q$ (see Rules 3.1 and 1.6) between and connected to the regular blocks $+(A)_p$, $+(D)_r$ and

$E'-(-A-/-B-)_p-X+(C)_q-E''$

represents a block copolymer consisting of an irregular block $(-A-/-B-)_p$ linked on one end via a junction unit —X— to the regular block $+(C)_q$ with the end group E″ and on the other end to the end group E′.

Examples[*]

4.3-E1: oligostyrene-*block*-octakis(methyl acrylate)

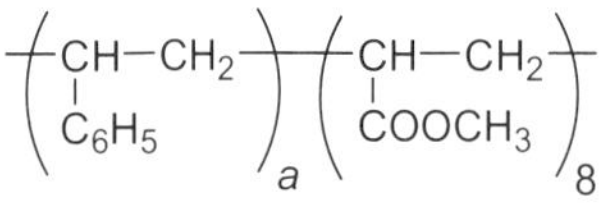

$$\left(\begin{array}{c}\text{CH}-\text{CH}_2\\|\\\text{C}_6\text{H}_5\end{array}\right)_a\left(\begin{array}{c}\text{CH}-\text{CH}_2\\|\\\text{COOCH}_3\end{array}\right)_8$$

4.3-E2: polystyrene-*block*-1,4-polybutadiene-*block*-polystyrene

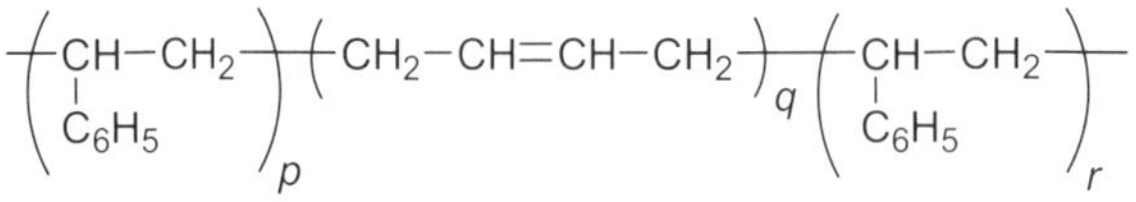

$$\left(\begin{array}{c}\text{CH}-\text{CH}_2\\|\\\text{C}_6\text{H}_5\end{array}\right)_p\left(\text{CH}_2-\text{CH}=\text{CH}-\text{CH}_2\right)_q\left(\begin{array}{c}\text{CH}-\text{CH}_2\\|\\\text{C}_6\text{H}_5\end{array}\right)_r$$

---

[*] Source-based copolymer names [6] are given, followed by proposed formula.

4.3-E3: tris[polystyrene-*block*-1,4-oligobutadiene-*block*-poly(methyl methacrylate)]

$$\left[\left(\begin{array}{c}-\text{CH}-\text{CH}_2-\\|\\\text{C}_6\text{H}_5\end{array}\right)_p\left(-\text{CH}_2-\text{CH}=\text{CH}-\text{CH}_2-\right)_a\left(\begin{array}{c}\text{CH}_3\\|\\-\text{C}-\text{CH}_2-\\|\\\text{COOCH}_3\end{array}\right)_q\right]_3$$

4.3-E4: poly[poly(methyl methacrylate)-*block*-polystyrene-*block*-poly(methyl acrylate)]

$$\left[\left(\begin{array}{c}\text{CH}_3\\|\\-\text{C}-\text{CH}_2-\\|\\\text{COOCH}_3\end{array}\right)_p/\left(\begin{array}{c}-\text{CH}-\text{CH}_2-\\|\\\text{C}_6\text{H}_5\end{array}\right)_q/\left(\begin{array}{c}-\text{CH}-\text{CH}_2-\\|\\\text{COOCH}_3\end{array}\right)_r\right]_n$$

4.3-E5: poly(styrene-*stat*-buta-1,3-diene)-*block*-polystyrene-*block*-1,2-polybutadiene

$$\left(\begin{array}{c}-\text{CH}-\text{CH}_2-\\|\\\text{C}_6\text{H}_5\end{array}/-\text{CH}_2-\text{CH}=\text{CH}-\text{CH}_2-\right)_p\left(\begin{array}{c}-\text{CH}-\text{CH}_2-\\|\\\text{C}_6\text{H}_5\end{array}\right)_q\left(\begin{array}{c}-\text{CH}-\text{CH}_2-\\|\\\text{CH}=\text{CH}_2\end{array}\right)_r$$

See note to 4.2-E1.

4.3-E6: polystyrene-*block*-dimethylsilanediyl-*block*-1,4-polybutadiene

$$\left(\begin{array}{c}-\text{CH}-\text{CH}_2-\\|\\\text{C}_6\text{H}_5\end{array}\right)_p\begin{array}{c}\text{CH}_3\\|\\-\text{Si}-\\|\\\text{CH}_3\end{array}\left(-\text{CH}_2-\text{CH}=\text{CH}-\text{CH}_2-\right)_q$$

4.4 Graft Copolymers

Rule 4.4.1: The formula of a graft copolymer consisting of a polymeric backbone of monomeric units A to which an unknown number of blocks of monomeric units B (grafts) are linked at **known** sites to some of the monomeric units A, is written as:

$$\left[-\text{A}-/-\text{A}'-\underset{(\text{B})_p}{|}\right]_n \quad \text{or} \quad \left[-\text{A}'-/-\text{A}-\underset{(\text{B})_p}{|}\right]_n$$

where A′ denotes the monomeric unit A **modified** by the substitution with the graft.

Rule 4.4.2: The formula of a graft copolymer consisting of a polymeric backbone of monomeric units A to which an unknown number of blocks of monomeric units B (grafts) are linked at **unknown** sites to some of the monomeric units A, is written as:

cont.

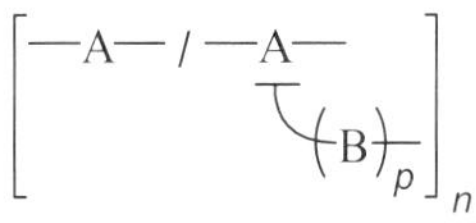

NOMENCLATURE

$$\left[\begin{array}{c} -A-/-A- \\ \diagdown(B)_p \end{array} \right]_n \quad \text{or} \quad \left[\begin{array}{c} -A-/-A- \\ \diagdown(B)_p \end{array} \right]_n$$

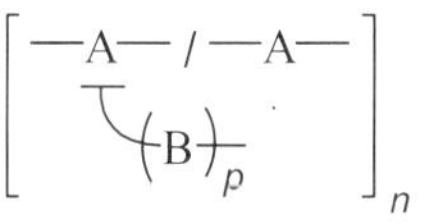

where the horizontal line under the **unmodified** monomeric unit A denotes that the point of attachment of the graft is not known.

Note: If the backbone of a graft copolymer is itself a copolymer consisting of monomeric units A, B, and it is not known to which of the monomeric units A or B, nor at which site of the individual units A or B, the grafts $-(C)_p-$ are attached, the formula is written as:

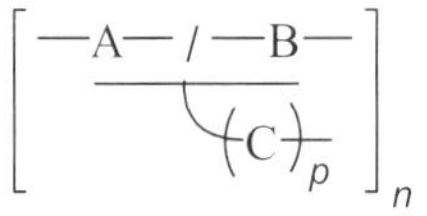

$$\left[\begin{array}{c} -A-/-B- \\ \diagdown(C)_p \end{array} \right]_n$$

Rule 4.4.3: When known, the average number of grafts per copolymer molecule *(i)* is given in parentheses following the formula

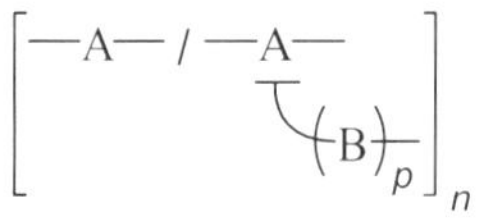

$$\left[\begin{array}{c} -A-/-A- \\ \diagdown(B)_p \end{array} \right]_n \qquad (i \text{ grafts per molecule})$$

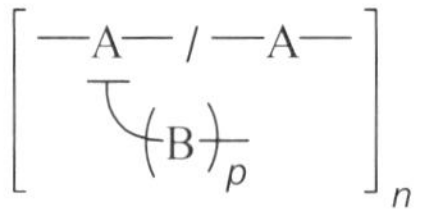

$$\left[\begin{array}{c} -A-/-A- \\ \diagdown(B)_p \end{array} \right]_n \qquad (i \text{ grafts per molecule})$$

See Rules 4.4.1 (for A′) and 4.4.2.

*Examples**

4.4-E1: 1,4-polybutadiene-*graft*-polystyrene
(a) polystyrene blocks grafted at unknown sites to but-2-ene-1,4-diyl units)

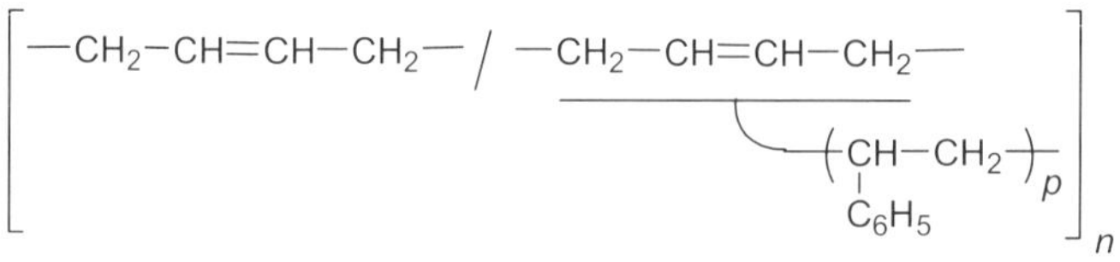

$$\left[\begin{array}{c} -CH_2-CH=CH-CH_2-/-CH_2-CH=CH-CH_2- \\ \diagdown(CH-CH_2)_p \\ | \\ C_6H_5 \end{array} \right]_n$$

(b) polystyrene blocks grafted to known sites of the but-2-ene-1,4-diyl units with the two ends of the backbone chain bearing one chloro and one trichloromethyl end group and the free ends of the polystyrene grafts bonded to hydrogen end groups)

---

* Source-based copolymer names [6] are given first, followed by an explanation in parentheses before the proposed formula. Structure-based names [5] are given for star copolymers in the notes to examples 4.4-E6 and 4.4-E7

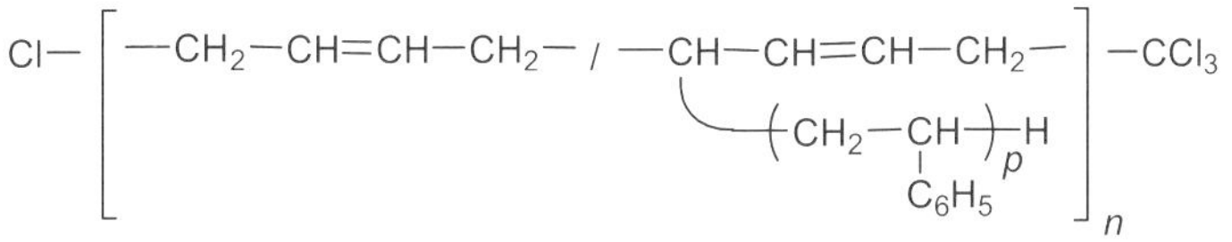

GRAPHIC REPRESENTATIONS

$$\text{Cl} - \left[\begin{array}{c} -\text{CH}_2-\text{CH}=\text{CH}-\text{CH}_2-\,/\,-\text{CH}-\text{CH}=\text{CH}-\text{CH}_2- \\ \underset{}{\smile}(\text{CH}_2-\underset{|}{\text{CH}})_p\text{H} \\ \text{C}_6\text{H}_5 \end{array} \right]_n -\text{CCl}_3$$

Note: The preferred representation for the buta-1,3-diene constitutional unit is with the double bond in the 1- position, i.e. that which corresponds to the name but-1-ene-1,4-diyl [2].

4.4-E2: 1,4-polybutadiene-*block*-(polystyrene-*graft*-oligoacrylonitrile)
(oligoacrylonitrile grafts linked to a 1,4-polybutadiene-polystyrene two-block copolymer at known sites of some of the monomeric units from styrene)

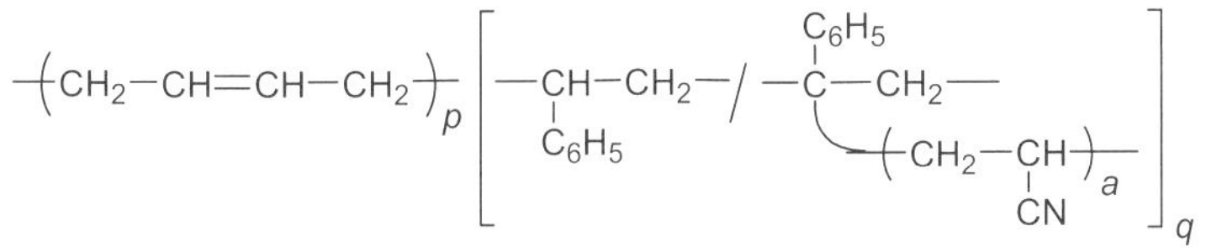

$$-(\text{CH}_2-\text{CH}=\text{CH}-\text{CH}_2)_p\left[\begin{array}{c} \text{C}_6\text{H}_5 \\ -\text{CH}-\text{CH}_2-\,/\,-\overset{|}{\text{C}}-\text{CH}_2- \\ | \\ \text{C}_6\text{H}_5\smile(\text{CH}_2-\underset{|}{\text{CH}})_a \\ \text{CN} \end{array} \right]_q$$

4.4-E3 : poly(1,3-butadiene-*stat*-styrene)-*graft*-polyacrylonitrile
(polyacrylonitrile grafted to a statistical buta-1,3-diene-styrene copolymer at unspecified sites)

$$\left[\begin{array}{c} \text{C}_6\text{H}_5 \\ -\text{CH}_2-\text{CH}=\text{CH}-\text{CH}_2-\,/\,-\overset{|}{\text{CH}}-\text{CH}_2- \\ \overline{} \\ \smile(\text{CH}_2-\underset{|}{\text{CH}})_p \\ \text{CN} \end{array} \right]_n$$

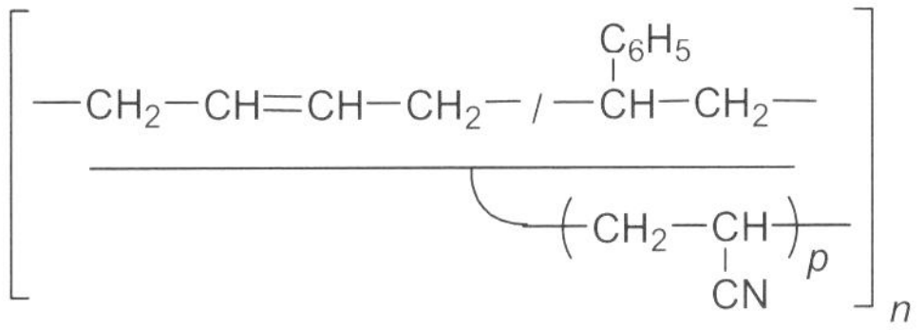

See note to 4.2-E1.

4.4-E4: polystyrene-*block*-[1,4-polybutadiene-*graft*-poly(styrene-*co*-acrylonitrile)]
(copolymer from styrene and acrylonitrile grafted to a 1,4-polybutadiene-polystyrene two-block copolymer at unspecified sites of some of the but-2-ene-1,4-diyl units)

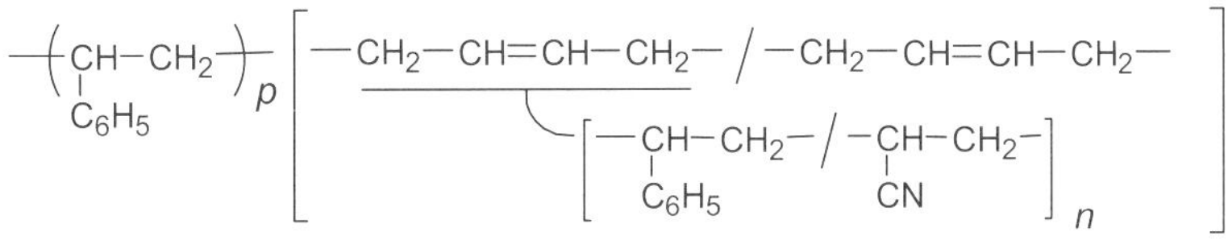

$$-(\underset{\underset{\text{C}_6\text{H}_5}{|}}{\text{CH}}-\text{CH}_2)_p\left[\begin{array}{c} -\text{CH}_2-\text{CH}=\text{CH}-\text{CH}_2-\,/\,-\text{CH}_2-\text{CH}=\text{CH}-\text{CH}_2- \\ \overline{} \\ \left[-\underset{\underset{\text{C}_6\text{H}_5}{|}}{\text{CH}}-\text{CH}_2-\,/\,-\underset{\underset{\text{CN}}{|}}{\text{CH}}-\text{CH}_2-\right] \end{array} \right]_n$$

4.4-E5: polyacrylonitrile-tris(-*graft*-polystyrene)
(three polystyrene grafts per molecule linked to a polyacrylonitrile backbone at unspecified sites of monomeric units from acrylonitrile)

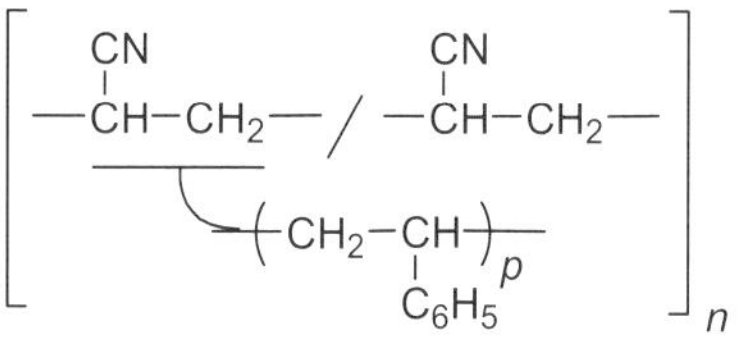

NOMENCLATURE

$$\left[\begin{array}{c} \underset{|}{\text{CN}} \\ -\text{CH}-\text{CH}_2- \\ \end{array} \middle/ \begin{array}{c} \underset{|}{\text{CN}} \\ -\text{CH}-\text{CH}_2- \\ \underset{|}{\big(}\text{CH}_2-\underset{|}{\text{CH}}\underset{C_6H_5}{\big)_p} \end{array} \right]_n$$ (3 grafts per macromolecule)

4.4-E6: deca(buta-1,3-diene)-*block*-(methylsilanetriyl-*graft*-polystyrene)-*block*-pentadeca(buta-1,3-diene)
(star copolymer, consisting of one polystyrene and two oligo(buta-1,3-diene) chains attached to a central methylsilane unit)

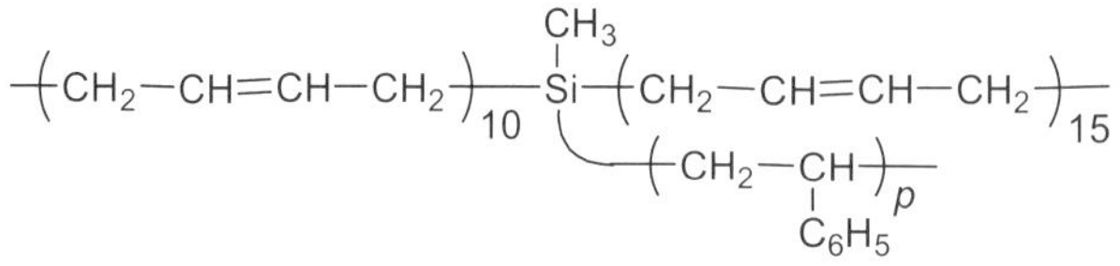

Note: The buta-1,3-diene is exclusively incorporated by 1,4-addition which cannot be reflected in the source-based name. The structure-based name (see [5]) is: [deca(but-2-ene-1,4-diyl)]methyl[pentadeca(but-2-ene-1,4-diyl)][poly(2-phenylethylene)]silane

4.4-E7: polystyrene-*block*-{silanetetraylbis[-*graft*-poly(buta-1,3-diene)]}-*block*-polystyrene, or poly(buta-1,3-diene)-*block*-[silanetetraylbis(-*graft*-polystyrene)]-*block*-poly(buta-1,3-diene)
(star copolymer, consisting of two polystyrene and two poly(buta-1,3-diene) chains attached to a central Si atom)

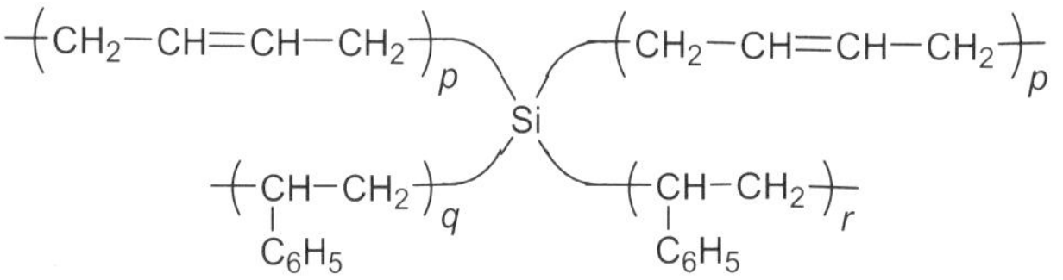

Note: The buta-1,3-diene is exclusively incorporated by 1,4-oligomerization which cannot be reflected in the source-based name. The structure-based name is: bis[poly(but-2-ene-1,4-diyl)][poly(1-phenylethane-1,2-diyl)][poly(2-phenylethane-1,2-diyl)]silane.

REFERENCES

1. IUPAC. 'Glossary of basic terms in polymer science (IUPAC Recommendations 1996)', *Pure Appl. Chem.* **68**, 2287-2311 (1996). Reprinted as Chapter 1 this volume.
2. IUPAC. 'Nomenclature of regular single-strand organic polymers (IUPAC Recommendations 2002)', *Pure Appl. Chem.* **74**, 1921-1956 (2002). Reprinted as Chapter 15 this volume.
3. IUPAC. 'Stereochemical definitions and notations relating to polymers (Recommendations 1980)', *Pure Appl. Chem.* **53**, 733-752 (1981). Reprinted as Chapter 2 this volume.

4. IUPAC. 'Nomenclature for regular single-strand and quasi-single-strand inorganic and coordination polymers (Recommendations 1984)', *Pure Appl. Chem.* **57**, 149-168 (1985).
5. IUPAC. 'Structure-based nomenclature for irregular single-strand organic polymers (IUPAC Recommendations 1994)', *Pure Appl. Chem.* **66**, 873-889 (1994). Reprinted as Chapter 17 this volume.
6. IUPAC. 'Source-based nomenclature for copolymers (Recommendations 1985)', *Pure Appl. Chem.* **57**, 1427-1440 (1985). Reprinted as Chapter 19 this volume.
7. IUPAC. 'Graphical Representation Standards for Chemical Structure Diagrams (Recommendations 2008),' *Pure Appl. Chem.* **80**, 277-410 (2008).
8. IUPAC. *A Guide to IUPAC Nomenclature of Organic Compounds (Recommendations 1993),* R. Panico, W. H. Powell and J.-C. Richer, (Eds.), Blackwell Scientific Publications, Oxford (1993).
9. IUPAC. *Nomenclature of Inorganic Chemistry, IUPAC Recommendations 2005* (the 'Red Book'), N. G. Connelly, T, Damhus, R. M. Hartshorn and A. T. Hutton, (Eds.), RSC Publishing, Cambridge (2005).
10. IUPAC. 'Nomenclature of regular double-strand (ladder and spiro) organic polymers (IUPAC Recommendations 1993)', *Pure Appl. Chem.* **65**, 1561-1580 (1993). Reprinted as Chapter 16 this volume.

19: Source-Based Nomenclature for Copolymers

CONTENTS

Preamble
Basic concept
Classification and definition of copolymers
   1. Copolymers with an unspecified arrangement of monomeric units
   2. Statistical copolymers
   3. Alternating copolymers
   4. Other types of periodic copolymers
   5. Block copolymers
   6. Graft copolymers
   7. Polymers made by polycondensation or related polymerization
   8. Specification with regard to mass fractions, mole fractions, molar masses and degrees of polymerization
Appendix: alternative nomenclature for copolymers
References

PREAMBLE

Copolymers have gained considerable importance both in scientific research and in industrial applications. A consistent and clearly defined system for naming these polymers would, therefore, be of great utility. The nomenclature proposals presented here are intended to serve this purpose by setting forth a system for designating the types of monomeric-unit sequence arrangements in copolymer molecules.

In principle, a comprehensive structure-based system of naming copolymers would be desirable. However, such a system presupposes a knowledge of the structural identity of all the constitutional units as well as their sequential arrangements within the polymer molecules; this information is rarely available for the synthetic polymers encountered in practice. For this reason, the proposals presented in this Report embody an essentially *source-based nomenclature system*.

Application of this system should not discourage the use of structure-based nomenclature whenever the copolymer structure is fully known and is amenable to treatment by the rules for single-strand polymers [1, 2]. It is intended that the present nomenclature system supersede the previous recommendations published in 1952 [3].

Originally prepared by a working group consisting of W. Ring (FRG), I. Mita (Japan), A. D. Jenkins (UK) and N. M. Bikales (USA). Reprinted from *Pure Appl. Chem.* **57**, 1427-1440 (1985).

COPOLYMERS

BASIC CONCEPT

The nomenclature system presented here is designed for copolymers. By definition, copolymers are polymers that are derived from more than one species of monomer [4]. Various classes of copolymers are discussed, which are based on the characteristic sequence arrangements of the monomeric units within the copolymer molecules. Generally, the names of monomers are used to specify monomeric units; the latter can be named using the trivial, semi-systematic or systematic form. The classes of copolymers are as follows:

unspecified	Rule 1.1
statistical	Rule 2.1
random	Rule 2.2
alternating	Rule 3.1
periodic	Rule 4.1
block	Rule 5.1
graft	Rule 6.1

In those cases where copolymer molecules can be described by only one species of constitutional unit in a single sequential arrangement, copolymers are regular polymers [4] and can, therefore, be named on a structure basis [1, 2]. Examples will be quoted later in the text.

Polymers having monomeric units differing in constitutional or configurational features, but derived from a single monomer, are not regarded as copolymers, in accordance with the basic definitions [4]. Examples of such polymers, which are not copolymers, are:

(1) polybutadiene with mixed sequences of 1,2- and 1,4-units; (2) poly(methyloxirane), also known as poly(propylene oxide), obtained through polymerization of a mixture of the two enantiomers, *R* and *S*, and containing both *R* and *S* monomeric units.

The nomenclature system presented here can, however, also be applied to such pseudo-copolymers. Polymers having monomeric units differing in constitutional features, but derived from a homopolymer by chemical modification, can be named in the same way, e.g.,

(3) partially hydrolysed poly(vinyl acetate) containing both ester and alcohol units.

CLASSIFICATION AND DEFINITION OF COPOLYMERS

A systematic source-based nomenclature for copolymers must identify the constituent monomers and provide a description of the sequence arrangement of the different types of monomeric units present. According to the present proposals, these objectives are achieved by citing the names of the constituent monomers after the prefix 'poly', and by placing between the names of each pair of monomers an italicized connective to denote the kind of arrangement by which those two types of monomeric units are related in the structure. Seven types of sequence arrangement are listed below, together with the corresponding connectives and examples, in which A, B and C represent the names of monomers.

NOMENCLATURE

Type	Connective	Example
unspecified	-*co*-	poly(A-*co*-B)
statistical	-*stat*-	poly(A-*stat*-B)
random	-*ran*-	poly(A-*ran*-B)
alternating	-*alt*-	poly(A-*alt*-B)
periodic	-*per*-	poly(A-*per*-B-*per*-C)
block	-*block*-	polyA-*block*-polyB
graft	-*graft*-	polyA-*graft*-polyB

Each of these types of copolymer is considered in more detail below. When the chemical nature of the end groups is to be specified, the name of the copolymer (as described above) is preceded by the systematic names of the terminal units. The prefix α or ω refers to the terminal unit attached to the left or right, respectively, of the structure, as written.

Example:

α-X-ω-Y-poly(A-*co*-B).

The citation of A, B and C is not intended to reflect an order of seniority, unless such seniority is specified in the rules. As a result, more than one name is often possible.

1   Copolymers with an unspecified arrangement of monomeric units

Rule 1.1

An unspecified sequence arrangement of monomeric units is represented by

(A-*co*-B)

and the corresponding copolymer has the name

poly(A-*co*-B)

Example: An unspecified copolymer of styrene and methyl methacrylate is named

poly[styrene-*co*-(methyl methacrylate)].

2   Statistical copolymers

Statistical copolymers are copolymers in which the sequential distribution of the monomeric units obeys known statistical laws; e.g. the monomeric-unit sequence distribution may follow Markovian statistics of zeroth (Bernoullian), first, second or a higher order. Kinetically, the elementary processes leading to the formation of a statistical sequence of monomeric units do not necessarily proceed with equal *a priori* probability. These processes can lead to various types of sequence distribution comprising those in which the arrangement of monomeric units tends towards alternation, tends towards clustering of like units, or exhibits no ordering tendency at all [5]. In simple binary copolymerization, the nature of this sequence distribution can be indicated by the numerical value of a function either of the reactivity ratios or of the related run number [5, 6].

The term statistical copolymer is proposed here to embrace a large proportion of those copolymers that are prepared by simultaneous polymerization of two or more monomers in admixture. Such copolymers are often described in the literature as 'random copolymers', but this is almost always an improper use of the term random and such practice should be abandoned.

Rule 2.1
A statistical sequence arrangement of monomeric units is represented by

(A-*stat*-B), (A-*stat*-B-*stat*-C), etc.

where -*stat*- indicates that the statistical sequence distribution with regard to A, B, C, etc, units is considered to be known. Statistical copolymers are named

poly(A-*stat*-B), poly(A-*stat*-B-*stat*-C), etc.

Examples:

poly(styrene-*stat*-butadiene)

poly(styrene-*stat*-acrylonitrile-*stat*-butadiene)

Random copolymers. A random copolymer is a special case of a statistical copolymer. It is a statistical copolymer in which the probability of finding a given monomeric unit at any given site in the chain is independent of the nature of the neighbouring units at that position (Bernoullian distribution). In other words, for such a copolymer, the probability of finding a sequence ...ABC... of monomeric units A, B, C..., i.e. $P(...ABC...]$, is given by

$$P[...ABC...] = P[A] \cdot P[B] \cdot P[C] = \prod_{i=A,B,C} P[i]$$

where $P[A]$, $P[B]$, $P[C]$, etc. are the unconditional probabilities of the occurrence of various monomeric units. As already noted above, the term 'random' should not be used for statistical copolymers except in this narrow sense.

Some authors use the term 'random' to denote the Bernoullian case further restricted by the condition that the monomeric units be present in exactly equal amounts [7].

Rule 2.2
A random sequence arrangement of monomeric units is represented by

(A-*ran*-B), (A-*ran*-B-*ran*-C), etc.

where -*ran*- indicates a random sequence distribution with regard to A, B, C, etc, units. Random copolymers are named

poly(A-*ran*-B), poly(A-*ran*-B-*ran*-C), etc.

NOMENCLATURE

Example:

poly[ethene-*ran*-(vinyl acetate)]

3 Alternating copolymers

An alternating copolymer is a copolymer comprising two species of monomeric units distributed in alternating sequence. The arrangement

-ABABABAB-   or   (AB)$_n$

thus represents an alternating copolymer.

Rule 3.1

An alternating sequence arrangement of monomeric units is represented by

(A-*alt*-B)

and the corresponding alternating copolymer is named

poly(A-alt-B).

Example:

poly(styrene-*alt*-(maleic anhydride)]

Alternating sequence arrangements can form constitutionally regular structures and may, in those cases, also be named utilizing the structure-based nomenclature for regular single-strand organic polymers. The example above would be then named

poly[(2,5-dioxotetrahydrofuran-3,4-diyl)(1-phenylethane-1,2-diyl)]

4 Other types of periodic copolymers

In addition to alternating polymers, other structures are known in which the monomeric units appear in an ordered sequence. Examples are:

-ABCABCABC-	or	(ABC)$_n$
-ABBABBABB-	or	(ABB)$_n$
-AABBAABBAABB-	or	(AABB)$_n$
-ABACABACABAC-	or	(ABAC)$_n$

Rule 4.1

A periodic sequence arrangement of monomeric units is represented by

(A-*per*-B-*per*-C)
(A-*per*-B-*per*-B)
(A-*per*-A-*per*-B-*per*-B)
(A-*per*-B-*per*-A-*per*-C), etc.

and the corresponding periodic copolymers are named

poly(A-*per*-B-*per*-C)
poly(A-*per*-B-*per*-B)
poly(A-*per*-A-*per*-B-*per*-B)
poly(A-*per*-B-*per*-A-*per*-C), etc, respectively.

If these polymers are regular. they can also be named according to the structure-based nomenclature for regular single-strand organic polymers [1].

Example: The binary monomer mixture consisting of formaldehyde and ethylene oxide might yield the periodically sequenced copolymer

$-(-CH_2-O-CH_2-CH_2-O-CH_2-CH_2-O-)_n$

which is named

poly[formaldehyde-*per*-(ethylene oxide)-*per*-(ethylene oxide)]

or
poly[formaldehyde-*alt*-bis(ethylene oxide)]

or, alternatively,
poly(oxymethyleneoxyethyleneoxyethylene).

Rule 4.2

If copolymer structures comprise several types of periodic sites, only some of which are always occupied by particular species of monomeric units (A, B...), and sites of the other types are occupied by two or more types of monomeric unit (U, V...) in irregular arrangement, the names of the monomers in the latter sites are embraced by parentheses and are separated by semicolon(s).

Examples:

1. The copolymer with the sequence arrangement

    -AUAVAVAUAVAUAU- is named poly[A-*alt*-(U;V)].

2. The copolymer with the sequence arrangement

    -AUBUAVBUAVBVAUBVAUBU- is named poly[A-*per*-(U;V)-*per*-B-*per*-(U;V)].

5   Block copolymers

A block polymer is a polymer comprising molecules in which there is a linear arrangement of blocks, a block being defined as a portion of a polymer molecule in which the monomeric units have at least one constitutional or configurational feature absent from the adjacent portions [4]. In a block copolymer, the distinguishing feature is constitutional, i.e. each of the blocks comprises units derived from a characteristic species of monomer.

NOMENCLATURE

In the sequence arrangements

-AAAAAAAA-BBBBBBBBBB-
-AAAAAAAA-BBBBBBBBBB-AAAAAAAA-
-AABABAAABB-AAAAAAAA-BBBBBBBBBB-

the sequences -AAAAAAAA-, -BBBBBBBBBB- and -AABABAAABB- are blocks.

Rule 5.1
A block sequence arrangement is represented by A_k-*block*-B_m, A_k-*block*-(A-*stat*-B), etc. and the corresponding polymers are named

polyA-*block*-polyB, polyA-*block*-poly(A-*stat*-B), etc., respectively.

If no ambiguity arises, a long dash may be used to designate block connections, as follows:

polyA—polyB.

For complex cases, use of -*block*- rather than the long dash is always encouraged. The order of citation of the block names corresponds to the order of succession of the blocks in the chain as written from left to right.

Examples: In the following examples, the subscripts k, m, ...represent different multiplicity of the monomeric units for different blocks. They may be indeterminate or specific (see *Rule* 5.3). In each case, the first line gives a representation of the block sequence arrangement, the second the corresponding name and the third an illustration of a specific case.

A_k-*block*-B_m
polyA-*block*-polyB
polystyrene-*block*-polybutadiene

A_k-*block*-B_m-*block*-A_k
polyA-*block*-polyB-*block*-poly A
polystyrene-*block*-polybutadiene-*block*-polystyrene

(A-*stat*-B)-*block*-A_k-*block*-B_m
poly(A-*stat*-B)-*block*-polyA-*block*-polyB
poly(styrene-*stat*-butadiene)-*block*-polystyrene-*block*-polybutadiene
A_k-*block*-B_n-*block*-C_m
polyA-*block*-polyB-*block*-polyC
polystyrene-*block*-polybutadiene-*block*-poly(methyl methacrylate)

Rule 5.2
Where a succession of blocks, such as -A_k-B_n-C_m- is repeated, the appropriate multiplying prefix is used.

Examples:

(A_k-*block*-B_n-*block*-C_m)$_3$
tris(polyA-*block*-polyB-*block*-polyC)

$(A_k\text{-}block\text{-}B_n\text{-}block\text{-}C_m)_p$
poly(polyA-*block*-polyB-*block*-polyC).

Rule 5.3
When it is possible to specify the chain length of a block, the appropriate Greek prefix (e.g. hecta for 100) may be used rather than poly. Although short sequence lengths are not strictly embraced within the definition of 'block', the same device may usefully be employed by using the general prefix 'oligo' or the appropriate specific prefix (e.g. tri).

Examples:

$A_c\text{-}block\text{-}B_8$
oligoA-*block*-octaB
$(A_c\text{-}block\text{-}B_k\text{-}block\text{-}C_3)_n$
poly(oligoA-*block*-polyB-*block*-triC)
$(A_c\text{-}block\text{-}B_k)_4$
tetrakis(oligoA-*block*-poly B)

where c is a small integer corresponding to the degree of polymerization of the oligomeric sequence.

Rule 5.4
Those block copolymers, derived from more than two monomers, that also exhibit statistical block sequence arrangements are named according to the principles of *Rule* 2.1.

Example: The statistical sequence arrangement

$-A_k\text{-}block\text{-}B_m\text{-}block\text{-}C_n\text{-}block\text{-}B_m\text{-}block\text{-}A_k\text{-}block\text{-}C_n\text{-}$

is named poly(polyA-*stat*-polyB-*stat*-polyC).

Rule 5.5
In the name of block copolymers with blocks connected by way of junction units, X, that are not part of the blocks, the name of the junction unit is inserted in the appropriate place. The connective, -*block*-, may be omitted. Thus,

$A_k\text{-}block\text{-}X\text{-}block\text{-}C_m$ or $A_k\text{-}X\text{-}C_m$

is named polyA-*block*-X-*block*-polyC or polyA—X—polyC.

The same designations can be applied to block polymers.

Examples:

polystyrene-*block*-dimethylsilanediyl-*block*-polybutadiene
or
polystyrene—dimethylsilanediyl—polybutadiene

and

NOMENCLATURE

polystyrene-*block*-dimethylsilanediyl-*block*-polystyrene
or
polystyrene—dimethylsilanediyl polystyrene.

Rule 5.6
A block copolymer wherein A_k and B_m blocks, connected through junctions X, are distributed in statistical manner in the polymer molecules, as in

-A_k-*block*-X-*block*-B_m-*block*-X-*block*-B_m-*block*-X-*block*-A_k-

is named poly[(polyA-*block*-X)-*stat*-(polyB-*block*-X)].

A block copolymer wherein A_k and B_m blocks and junction units X are all distributed in statistical manner, as in

-A_k-*block*-X-*block*-B_m-*block*-A_k-*block*-B_m-*block*-X-*block*-B_m-*block*-X-*block*-A_k-

is named poly(polyA-*stat*-X-*stat*-polyB).

6 Graft copolymers

A graft polymer is a polymer comprising molecules with one or more species of block connected to the main chain as side chains, these side chains having constitutional or configurational features that differ from those in the main chain [4]. In a graft copolymer the distinguishing feature of the side chains is constitutional, i.e. the side chains comprise units derived from at least one species of monomer different from those which supply the units of the main chain.

Rule 6.1
The simplest case of a graft copolymer can be represented by A_k-*graft*-B_m or the arrangement

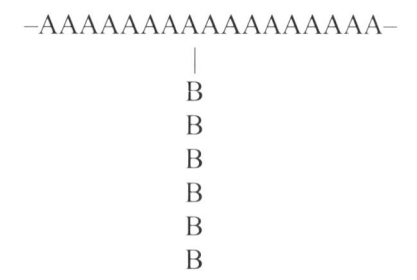

```
    –AAAAAAAAAAAAAAAAA–
            |
            B
            B
            B
            B
            B
            B
```

and the corresponding name is polyA-*graft*-polyB where the monomer named first (A in this case) is that which supplied the backbone (main chain) units, while that named second (B) is in the side chain(s).

Examples:
Each of the following examples presents in order, a representation of the graft sequence arrangement, the corresponding name and an illustration of a specific case.

COPOLYMERS

1. A_k-*graft*-B_m

   polyA-*graft*-polyB

   polybutadiene-*graft*-polystyrene
   (polystyrene grafted to polybutadiene)

2. (A_k-*block*-B_m)-*graft*-C_n

   (polyA-*block*-polyB)-*graft*-polyC

   (polybutadiene-*block*-polystyrene)-*graft*-polyacrylonitrile
   (polyacrylonitrile grafted to a polybutadiene-polystyrene block copolymer at unspecified sites)

3. (A-*stat*-B)-*graft*-C_n

   poly(A-*stat*-B)-*graft*-polyC

   poly(butadiene-*stat*-styrene)-*graft*-polyacrylonitrile
   (polyacrylonitrile grafted to a statistical butadiene-styrene copolymer at unspecified sites)

4. A_k-*block*-(B_m-*graft*-C_n)

   polyA-*block*-(polyB-*graft*-polyC) or (polyB-*graft*-polyC)-*block*-polyA

   polybutadiene-*block*-(polystyrene-*graft*-polyacrylonitrile) or
   (polystyrene-*graft*-polyacrylonitrile)-*block*-polybutadiene
   (polystyrene-polybutadiene block copolymer with polyacrylonitrile grafted to the polystyrene block)

5. [A_k-*graft*-(B-*co*-C)]-*block*-B_m

   [polyA-*graft*-poly(B-*co*-C)]-*block*-polyB or polyB-*block*-[polyA-*graft*-poly(B-*co*-C)]

   [polybutadiene-*graft*-poly(styrene-*co*-acrylonitrile)]-*block*-polystyrene or
   polystyrene-*block*-[polybutadiene-*graft*-poly(styrene-*co*-acrylonitrile)]
   (polybutadiene-polystyrene block copolymer with a styrene-acrylonitrile copolymer with an unspecified sequence arrangement of monomeric units grafted to the polybutadiene block).

Rule 6.2
If more than one type of graft chain is attached to the backbone, semicolons are used to separate the names of the grafts or their symbolic representations.

Example:

A_k-*graft*-(B_m; C_n)
polyA-*graft*-(polyB; polyC)

377

NOMENCLATURE

polybutadiene-*graft*-[polystyrene; poly(methyl methacrylate)]
(polystyrene and poly(methyl methacrylate) chains grafted to a polybutadiene backbone).

Rule 6.3
Graft copolymers with known numbers of graft chains are named using numeric prefixes (mono, bis, tris, etc).

Example:

A_k(-*graft*-B_m)$_3$

polyA-tris(-*graft*-polyB)

polybutadiene-tris(-*graft*-polystyrene)
(three polystyrene grafts per polybutadiene backbone).

If the precise site of grafting is known, it can be specified.

Example:

A_{10}-*block*-(X-*graft*-B_m)-*block*-A_{15}

decaA-*block*-(X-*graft*-poly B)-*block*-pentadecaA

decabutadiene-*block*-(methylsilanetriyl-*graft*-polystyrene)-*block*-pentadecabutadiene.

The system of naming graft copolymers is also applicable, in principle, to 'star copolymers' [*], where chains having different constitutional or configurational features are linked through a central moiety.

Examples:

1. A_k-*block*-[X-(*graft*-B_m)$_2$]-*block*-A_k or B_m-*block*-[X-(*graft*-A_k)$_2$]-*block*-B_m

    polyA-*block*-[X-bis(-*graft*-polyB)]-*block*-polyA or
    polyB-*block*-[X-bis(-*graft*-polyA)]-*block*-polyB

    polystyrene-*block*-[silanetetraylbis(-*graft*-polybutadiene)]-*block*-polystyrene or
    polybutadiene-*block*-[silanetetraylbis(-*graft*-polystyrene)]-*block*-polybutadiene
    (two polystyrene and two polybutadiene chains attached to a central Si atom)

2. A_k-*block*-[X-*graft*-(B_m; C_n)]-*block*-D_p

    polyA-*block*-[X-*graft*-poly B; polyC)]-*block*-polyD
    polystyrene-*block*-[silanetetrayl-*graft*-(polybutadiene; polyisoprene)]-*block*-poly(methyl methacrylate)

---

[*] Footnote for the Second Edition: Star copolymers can be named by using the prefix '*star*-' as in *star*-(polyA; ;poly: B; polyC) (see Chapter 20).

COPOLYMERS

(a polybutadiene chain, a polyisoprene chain, a polystyrene chain and a poly(methyl methacrylate) chain attached to the same central Si atom).

In the absence of a seniority rule, several other names are possible.

7 Polymers made by polycondensation or related polymerization

The nomenclature system for copolymers is also applicable to polymers made by polycondensation of more than one monomeric species, or, more generally, by polymerization of more than one monomeric species where molecules of all sizes (i.e. monomers, oligomers, polymers) can react with each other. One can distinguish the case of polymers made by polycondensation of homopolymerizable monomers from that of polymers made by polycondensation of complementary ingredients that do not usually separately homopolymerize. Rigorous application of the source-based definition of a copolymer [4] embraces polymers such as poly(ethylene terephthalate) or poly(hexane-1,6-diyladipamide)[*] (which are commonly regarded as homopolymers) because two ingredients are, in each case, the usual starting materials of polymerization. If polymers of this type have constitutionally regular structures and are regular polymers, the nomenclature for regular single-strand organic polymers can also be used [1].

This applies, for example, to the polymer derived from terephthalic acid and ethylene glycol, which by source-based copolymer nomenclature would be named as poly[(ethylene glycol)-*alt*-(terephthalic acid)], if in fact the polymer has been prepared by a polycondensation starting with terephthalic acid and ethylene glycol. However, if the starting material is the partial ester, $HOCH_2CH_2OCOC_6H_4COOH$, the appropriate source-based name is that of a homopolymer, whereas use of the starting material bis(2-hydroxyethyl) terephthalate, $HOCH_2CH_2OCOC_6H_4COOCH_2CH_2OH$, extensively used industrially, would suggest the name poly[bis(2-hydroxyethyl) terephthalate]. Regardless of the starting materials used, the structure-based name is poly(oxyethyleneoxyterephthaloyl). The traditional name poly(ethylene terephthalate) is also permitted, because it is so well established in the literature.

For all such polymers made by polycondensation of two complementary difunctional ingredients (or 'monomers'), which can readily be visualized as reacting on a 1:1 basis to give an 'implicit monomer', the homopolymerization of which would give the actual product, the structure-based nomenclature may be suitable insofar as such a polymer is regular and can be represented as possessing a single constitutional repeating unit. It is to be noted that this is applicable only to cases where the mole ratio of the ingredients is 1:1 and the ingredients are exclusively difunctional.

The introduction of a third component into the reaction system necessitates the use of copolymer nomenclature which can logically be developed from the foregoing rules, as the examples below illustrate. The copolymer derived from reaction of ethylene glycol with a mixture of terephthalic and isophthalic acids would be named:

poly{[(ethylene glycol)-*alt*-(terephthalic acid)]-*co*-[(ethylene glycol)-*alt*-(isophthalic acid)]}, poly[(ethylene terephthalate)-*co*-(ethylene isophthalate)] or
poly[(ethylene glycol)-*alt*-(terephthalic acid; isophthalic acid)]

---

[*] poly(*N*,*N*'-hexane-1,6-diylhexanediamide)

NOMENCLATURE

A polymer derived from the polycondensation of a single actual monomer, the molecules of which terminate in two different complementary functional groups (e.g. 6-aminohexanoic acid) is, by definition, a (regular) homopolymer. When two different monomers of this type react together, the product is a copolymer that can be named in appropriate fashion. For example, if 6-aminohexanoic acid is copolycondensed with 7-aminoheptanoic acid, leading to a statistical distribution of monomeric units, the product is named poly[(6-aminohexanoic acid)-*stat*-(7-aminoheptanoic acid)].

8   Specification with regard to mass fractions, mole fractions, molar masses and degrees of polymerization

Whereas subscripts placed immediately after the formula of the monomeric unit or the block designate the degree of polymerization or repetition, mass and mole fractions and molar masses – which in most cases are average quantities – may be expressed by placing corresponding figures after the complete name or symbol of the copolymer. The order of citation is the same as for the monomer species in the name or symbol of the copolymer. Unknown quantities can be designated by a, b, etc.

Although this scheme can be extended to complicated cases, it is recommended that its application be restricted to simple cases; any treatment of more complicated systems should be explained in the text.

Rule 8.1
Mass fractions, or mass percentages, for the monomeric units are placed in parentheses after the copolymer name, followed by the symbol 'w', or the phrase 'mass %', respectively. The order of citation in the parentheses is the same as in the name.

Examples:

1. polybutadiene-*graft*-polystyrene (0.75:0.25 w) or polybutadiene-*graft*-polystyrene (75:25 mass %) (a graft copolymer containing 75 mass % of polybutadiene and 25 mass % of grafted polystyrene)

2. polybutadiene-*graft*-poly(styrene-*stat*-acrylonitrile) (0.75:a:b w) or
   polybutadiene-*graft*-poly(styrene-*stat*-acrylonitrile) (75:a:b mass %)
   (a graft copolymer containing 75 mass % of butadiene units as backbone and unknown quantities in statistical arrangement of styrene and acrylonitrile units in the grafted chains).

Rule 8.2
Mole fractions, or mole percentages, for the monomeric units are placed in parentheses after the copolymer name, followed by the symbol 'x', or the phrase 'mol %', respectively. The order of citation in the parentheses is the same as in the name.

Example:

polybutadiene-*graft*-polystyrene (0.85:0.15 x)   or
polybutadiene-*graft*-polystyrene (85:15 mol %)
(a graft copolymer containing 85 mol % of butadiene units and 15 mol % of styrene units).

COPOLYMERS

Rule 8.3
The molar mass, relative molecular mass or degree of polymerization may be included in the scheme of *Rules* 8.1 and 8.2 by adding the corresponding figures, followed by the symbol M, M_r or DP, respectively.

Examples:

1. polybutadiene-*graft*-polystyrene (75:25 mass %; 90 000:30 000 M_r) (a graft copolymer consisting of 75 mass % of butadiene units with a relative molecular mass of 90 000 as the backbone, and 25 mass % of styrene units in grafted chains with a relative molecular mass of 30 000)

2. polybutadiene-*graft*-polystyrene (1700:290 DP) (a graft copolymer consisting of a polybutadiene backbone with a degree of polymerization of 1700 to which polystyrene with a degree of polymerization of 290 is grafted)

APPENDIX: ALTERNATIVE NOMENCLATURE FOR COPOLYMERS

In the first edition of this book, an alternative nomenclature system was described, which was based on the principle that a copolymer can be described by the prefix 'copoly' followed by citation of the names of the monomers as copoly(A/B), copoly(A/B/C), etc., where A, B and C represent the names of the monomers employed. However, after two decades this system has been scarcely adopted for copolymer naming in the literature, the appendix is therefore omitted from this book.

REFERENCES

1. IUPAC. 'Nomenclature of regular single-strand organic polymers (IUPAC Recommendations 2002)', *Pure Appl. Chem.* **74**, 1921-1956 (2002). Reprinted as Chapter 15, this volume.
2. IUPAC. 'Nomenclature for regular single-strand and quasi-single-strand inorganic and coordination polymers (Recommendations 1984)', *Pure Appl. Chem.* **57**, 149-168 (1985).
3. IUPAC. 'Report on nomenclature in the field of macromolecules', *J. Polym. Sci.* **8**, 257-277 (1952).
4. IUPAC. 'Glossary of basic terms in polymer science (IUPAC Recommendations 1996)', *Pure Appl. Chem.* **68**, 2287-2311 (1996). Reprinted as Chapter 1, this volume.
5. G. E. Ham. In *Encyclopedia of Polymer Science and Technology*, H. F. Mark, N. G. Gaylord, N. M. Bikales (Eds.), Wiley-Interscience, New York (1966), **4**, 165.
6. H. J. Harwood and W. M. Ritchey. 'The characterization of sequence distribution in copolymers', *J. Polym. Sci., Polym. Lett. Ed.* **2**, 601 (1964).
7. H. J. Harwood. 'Characterization of the structures of diene polymers by NMR', *Rubber Chem. Technol.* **55**, 769 (1982).

20: Source-Based Nomenclature for Non-Linear Macromolecules and Macromolecular Assemblies

CONTENTS

Preamble
1. Definitions
2. General principles
3. Non-linear homopolymer molecules
4. Non-linear copolymer molecules
5. Macromolecular assemblies
6. Quantitative specifications
7. References

PREAMBLE

The recommendations for macromolecular nomenclature published so far [1] have dealt with regular single-strand organic macromolecules [2, 3], regular single-strand and quasi-single-strand inorganic and coordination macromolecules [4], copolymer macromolecules [5], regular double-strand organic macromolecules [6] and irregular single-strand organic macromolecules [7]. As far as possible, the established principles of organic and inorganic nomenclature have been followed [8, 9]. The present document is intended to extend the system to non-linear macromolecules such as branched and cyclic macromolecules, micronetworks, polymer networks, and to macromolecular assemblies held together by non-covalent bonds or forces, such as polymer blends, interpenetrating polymer networks and polymer complexes.

Conventionally, the word 'polymer' as a noun is ambiguous in meaning. It is commonly used to refer to both polymer substances and polymer molecules. Henceforth, 'macromolecule' is used for individual molecules and 'polymer' is used to denote a substance composed of macromolecules [10]. 'Polymer' may also be employed unambiguously as an adjective according to accepted usage, e.g. 'polymer blend', 'polymer molecule'.

For non-linear macromolecules, the skeletal structure should be reflected in the name. Non-linear macromolecular structures and molecular assemblies are classified using the following terms:

Originally prepared by a Working Group consisting of J. Kahovec (Czech Republic), P. Kratochvíl (Czech Republic), A. D. Jenkins (UK), I. Mita (Japan), I. M. Papisov (Russia), L. H. Sperling (USA) and R. F. T. Stepto (UK). Reprinted from *Pure & Appl. Chem.,* **69**, 2511-2521 (1997).

NON-LINEAR MACROMOLECULES

Cyclic
Branched
Short-branched
Long-branched
Micronetwork
Network
Blend
Semi-interpenetrating polymer network
Interpenetrating polymer network
Polymer-polymer complex

Definitions of some of the terms used in this document can be found in previous nomenclature reports [10,11].

1 DEFINITIONS

1.1 chain
Whole or part of a macromolecule, an oligomer molecule or a block, comprising a linear or branched sequence of constitutional units between two boundary constitutional units, each of which may be either an end-group, a *branch point* or an otherwise-designated characteristic feature of the macromolecule.
Note 1: Except in linear single-strand macromolecules, the definition of the chain may be somewhat arbitrary.
Note 2: A cyclic macromolecule has no end-groups but may nevertheless be regarded as a chain.
Note 3: Any number of *branch points* may be present between the boundary units.
Note 4: Where appropriate, definitions relating to 'macromolecule' may also be applied to 'chain'.

1.2 linear chain
Chain with no *branch points* intermediate between the boundary units.

1.3 branched chain
Chain with at least one *branch point* intermediate between the boundary units.

1.4 main chain
backbone
That *linear chain* to which all other chains, long or short or both, may be regarded as being pendant.
Note: Where two or more chains could equally be considered to be the main chain, that one is selected which leads to the simplest representation of the molecule.

1.5 subchain
Arbitrarily chosen contiguous sequence of constitutional units in a *chain*.
Note: The term 'subchain' may be used to define designated subsets of the constitutional units in a *chain*.

NOMENCLATURE

**1.6 side-chain
branch
pendant chain**
Oligomeric or polymeric offshoot from a macromolecular *chain*.
Note 1: An oligomeric side-chain may be termed a short-chain branch.
Note 2: A polymeric side-chain may be termed a long-chain branch.

1.7 branch unit
Constitutional unit containing a *branch point*.
Note: A branch unit from which f *linear chains* emanate may be termed an f-functional branch unit, e.g. five-functional branch unit. Alternatively, the terms trifunctional, tetrafunctional, pentafunctional, etc., may be used, e.g. pentafunctional branch unit.

1.8 branch point
Point on a *chain* at which a *branch* is attached.
Note 1: A branch point from which f *linear chains* emanate may be termed an f-functional branch point, e.g. five-functional branch point. Alternatively, the terms trifunctional, tetrafunctional, pentafunctional, etc. may be used, e.g. pentafunctional branch point.
Note 2: A branch point in a *network* may be termed a junction point.

1.9 crosslink
Small region in a macromolecule from which at least four chains emanate, and which is formed by reactions involving sites or groups on *existing* macromolecules or by interactions between *existing* macromolecules.
Note 1: The small region can be an atom, a group of atoms or a number of *branch points* connected by bonds, groups of atoms or oligomeric chains.
Note 2: In the majority of cases, a crosslink is a covalent structure but the term is also used to describe sites of weaker chemical interactions, portions of crystallites and even physical interactions and entanglements.

1.10 network
Highly ramified structure in which essentially each constitutional unit is connected to each other constitutional unit and to the macroscopic phase boundary by many paths through the structure, the number of such paths increasing with the average number of intervening constitutional units; the paths must on average be co-extensive with the structure.
Note 1: Usually, and in all systems that exhibit rubber elasticity, the number of distinct paths is very high, but, in most cases, some constitutional units exist that are connected by a single path only.
Note 2: Modified from [12]. The definition proposed here is a generalization to cover both polymeric networks and networks comprised of particles.
Note 3: If the permanent paths through the structure of a network are all formed by covalent bonds, the term covalent network may be used.
Note 4: The term physical network may be used if the permanent paths through the structure of a network are not all formed by covalent bonds, but, at least in part, by physical interactions, such that removal of the interactions leaves individual molecules or macromolecules or a non-network macromolecule.

1.11 micronetwork
Polymer network that has dimensions of the order of 1 nm to 1 µm.

NON-LINEAR MACROMOLECULES

1.12 macrocycle
Cyclic macromolecule or a macromolecular cyclic portion of a macromolecule.
Note 1: See Note 2 to Definition 1.1.
Note 2: In the literature, the term 'macrocycle' is sometimes used for molecules of low relative molecular mass that would not be considered 'macromolecules'.

1.13 comb macromolecule
Macromolecule comprising a main *chain* with multiple trifunctional *branch points* from each of which a linear side-chain emanates.
Note 1: If the *subchains* between the *branch points* of the main *chain* and the terminal *subchains* of the main *chain* are identical with respect to constitution and degree of polymerization, and the side-chains are identical with respect to constitution and degree of polymerization, the macromolecule is termed a regular comb macromolecule.
Note 2: If at least some of the *branch points* are of functionalities greater than three, the macromolecule may be termed a brush macromolecule.

1.14 star macromolecule
Macromolecule containing a single *branch point* from which *linear chains* (arms) emanate.
Note 1: A star macromolecule with n *linear chains* (arms) attached to the *branch point* is termed an n-star macromolecule, e.g. five-star macromolecule.
Note 2: If the arms of a star macromolecule are identical with respect to constitution and degree of polymerization, the macromolecule is termed a regular star macromolecule.
Note 3: If different arms of a star macromolecule are composed of different monomeric units, the macromolecule is termed a variegated star macromolecule.

1.15 polymer blend
Macroscopically homogeneous mixture of two or more different species of polymer.
Note 1: See Fig. 1a.
Note 2: In most cases, blends are homogeneous on scales larger than several times visual, optical wavelengths.
Note 3: For polymer blends, no account is taken of the miscibility or immiscibility of the constituent polymers, i.e. no assumption is made regarding the number of phases present.
Note 4: The general use of the term 'polymer alloy' as a synonym for a polymer blend is discouraged.

1.16 semi-interpenetrating polymer network (SIPN)
Assembly of macromolecules comprising one or more polymer *networks* and one or more linear or branched polymers characterized by the penetration on a molecular scale of at least one of the *networks* by at least some of the linear or branched macromolecules.
Note 1: See Fig. 1b.
Note 2: A SIPN is distinguished from an *IPN* because the constituent linear or branched macromolecules can, in principle, be separated from the constituent *polymer network(s)* without breaking chemical bonds; it is a *polymer blend*.

1.17 network polymer
Polymer composed of one or more *networks*.

NOMENCLATURE

1.18 interpenetrating polymer network (IPN)
Assembly of macromolecules comprising two or more *networks* that are at least partially interlaced on a molecular scale but not covalently bonded to each other and cannot be separated unless chemical bonds are broken.
Note 1: See Fig. 1c.
Note 2: A mixture of two or more preformed polymer *networks* is not an IPN.

1.19 polymer-polymer complex
Complex, at least two components of which are different polymers.

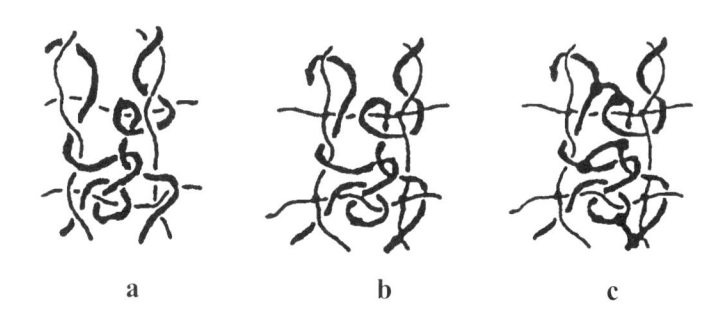

Fig. 1. Schematic representation of macromolecular assemblies: ——— polymer chain 1; ━━━ polymer chain 2; • *branch point*. (a) polymer blend, (b) semi-interpenetrating polymer network, (c) interpenetrating polymer network

2 GENERAL PRINCIPLES

The following procedure is employed to name both non-linear macromolecules and macromolecular assemblies:

1. The topology of the structure of the non-linear macromolecule or modes of combination of the constituent species of macromolecules are ascertained.
2. For each constituent linear *subchain* in a non-linear macromolecule or for each species of macromolecule, a source-based name is assigned according to the rules for naming homopolymer or copolymer molecules [2, 5].
3. The names of the constituent *subchains* in a non-linear macromolecule or those of the macromolecules are then combined together with the appropriate prefix or connective(s), or both.

The following italicized qualifiers can be used as both prefixes (e.g. *blend-, net-*) and connectives (e.g. *-blend-, -net-*), separated by (a) hyphen(s) from the constituent name(s), to designate the skeletal structure of non-linear macromolecules or macromolecular assemblies:

cyclic	*cyclo*
branched, unspecified	*branch*
short-chain-branched	*sh-branch*
long-chain-branched	*l-branch*
branched with *branch points* of functionality f	*f-branch* (f is given a numerical value)
comb(like)	*comb*

NON-LINEAR MACROMOLECULES

star	*star*
star with *f* arms	*f-star* (*f* is given a numerical value)
network	*net*
micronetwork	*μ-net*
polymer blend	*blend*
interpenetrating polymer network	*ipn*
semi-interpenetrating polymer network	*sipn*
polymer-polymer complex	*compl*

3 NON-LINEAR HOMOPOLYMER MOLECULES

Rule 1
In naming non-linear homopolymer molecules, the italicized prefix for the skeletal structure of the macromolecule is placed before the source-based name of the constituent *linear chain*.

Examples:

1.1 *sh-branch*-polyethylene
1.2 4-*star*-polystyrene
1.3 *cyclo*-poly(dimethylsiloxane)
1.4 *net*-polybutadiene

Rule 2
Crosslink, the *branch units* of *star macromolecules* and other junction units [5] are optionally specified by their source-based names after the name of the macromolecule with the connective (Greek) *v*, separated by hyphens.
Note: When the content of the crosslinking monomer is high, the macromolecule is treated as a copolymer molecule.

Examples:

2.1 *net*-polybutadiene-*v*-sulfur
    (polybutadiene vulcanized with sulfur)

2.2 *net*-polystyrene-*v*-divinylbenzene
    (polystyrene crosslinked with divinylbenzene to form a *network*)

2.3 *branch*-polystyrene-*v*-divinylbenzene
    (polystyrene crosslinked with divinylbenzene, insufficient to cause a *network* to form)

2.4 *l-branch*-poly(ethyl acrylate)-*v*-(ethane-1,2-diyl dimethacrylate)

2.5 3-*star*-polystyrene-*v*-trichloro(methyl)silane

NOMENCLATURE

4 NON-LINEAR COPOLYMER MOLECULES

4.1 Copolymer molecules comprising a single species of linear chain

Rule 3

In naming non-linear copolymer molecules comprising linear *subchains* of the same monomeric units in a single type of skeletal structure, the italicized prefix for the skeletal structure is placed before the source-based name of the constituent linear *subchains*. In the case of *star macromolecules* with block copolymer arms, the block named first after the prefix emanates from the *branch point*.

Examples:

3.1 *cyclo*-poly(styrene-*stat*-α-methylstyrene)

3.2 *star*-[polystyrene-*block*-poly(methyl methacrylate)]
 (each arm of the *star macromolecule* is a block copolymer chain with a polystyrene block attached to the central unit)

3.3 *μ-net*-poly[styrene-*stat*-(vinyl cinnamate)]
 (crosslinked copolymeric *micronetwork*)

3.4 *net*-poly(phenol-*co*-formaldehyde)

3.5 *comb*-poly(styrene-*stat*-acrylonitrile)
 (both the *main chain* and side-chains are statistical copolymer chains of styrene and acrylonitrile)

3.6 *net*-poly[(hexane-1,6-diyl diisocyanate)-*alt*-glycerol]

3.7 *net*-poly[styrene-*alt*-(maleic anhydride)]-*v*-(ethylene glycol)
 (alternating copolymer of styrene and maleic anhydride crosslinked with ethylene glycol to form a *network*)

3.8 *star*-(polyA-*block*-polyB-*block*-polyC)
 (star copolymer molecule, each arm of which consists of the same block-copolymer chain, see Fig. 2a).

Rule 4

In naming non-linear copolymer molecules having linear *subchains* of two or more types, the italicized connective for the skeletal structure is placed between the source-based names of the types of constituent linear *subchains*. In the case of branched and comb-like macromolecules, the *linear chain* named before the connective is that which forms the *main chain*, whereas that (those) named after the connective forms (form) the side-chain(s). The names of different species of side-chain are separated by semicolons. In the case of variegated *star macromolecules* the prefix is placed before the name of the macromolecule with the different species of arms separated by semicolons.

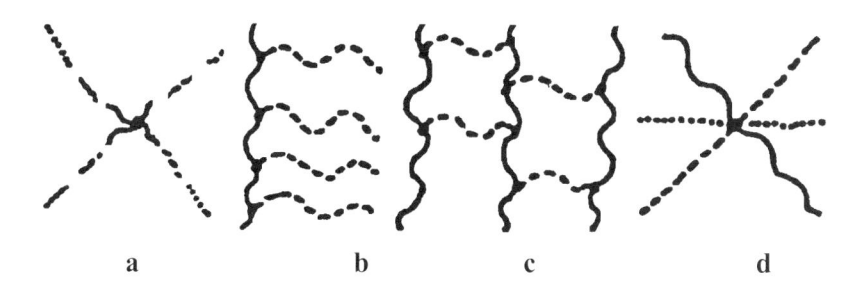

NON-LINEAR MACROMOLECULES

Fig. 2. Schematic representation of non-linear copolymer molecules ——— polymer chain 1, - - - - - - polymer chain 2, · · · · · · polymer chain 3, • *branch point* (a), (d) star macromolecules, (b) comb macromolecule, (c) network macromolecule

4.2 Copolymer molecules comprising a variety of species of chains

Examples:

1.1 polystyrene-*comb*-polyacrylonitrile
(equivalent to polystyrene-*graft*-polyacrylonitrile, see Fig. 2b; -*graft*- is recommended for these cases. However, *comb* cannot be replaced by *graft* if the former is a prefix, cf. Example 3.5)

1.2 polystyrene-*comb*-[polyacrylonitrile; poly(methyl methacrylate)]
(a *comb macromolecule* with polyacrylonitrile and poly(methyl methacrylate) side-chains)

1.3 poly[(ethylene glycol)-*alt*-(maleic anhydride)]-*net*-oligostyrene
(unsaturated polyester cured with styrene, see Fig. 2c)

1.4 poly(methyl methacrylate)-*net*-poly(ethylene oxide)

1.5 *star-(*polyA; polyB; polyC)
(a variegated star copolymer molecule consisting of arm(s) of polyA, arm(s) of polyB and arm(s) of polyC, see Fig. 2d)

Rule 5

The name of a crosslinking monomer molecule having two or more different types of polymerizable groups, each serving as a monomeric unit for one of two or more different *linear chains*, is cited with the name of each of the chains with the symbol *v*.

Note: When the proportion of the crosslinking monomer having two or more different polymerizable groups is a significant fraction of the total, the constituent chains are treated as copolymer chains.

NOMENCLATURE

Examples:

5.1 {poly(butyl vinyl ether)-*v*-[2-(vinyloxy)ethyl methacrylate]}-*net*-{poly(methyl methacrylate)-*v*-[2-(vinyloxy)ethyl methacrylate]}
(poly(butyl vinyl ether) and poly(methyl methacrylate) chains crosslinked mutually with 2-(vinyloxy)ethyl methacrylate to form a *network*)

5.2 (poly{(butyl vinyl ether)-*co*-[2-(vinyloxy)ethyl methacrylate]})-*net*-(poly{(methyl methacrylate)-*co*-[2-(vinyloxy)ethyl methacrylate]})
(a copolymer of butyl vinyl ether and 2-vinyloxyethyl methacrylate formed into a *network* with a copolymer of methyl methacrylate and 2-vinyloxyethyl methacrylate)

5.3 (*branch*-{poly(butyl vinyl ether)-*v*-[2-(vinyloxy)ethyl methacrylate]})-*v*-(*branch*-{poly(methyl methacrylate)-*v*-[2-(vinyloxy)ethyl methacrylate]})
(poly(butyl vinyl ether) and poly(methyl methacrylate) chains crosslinked mutually with 2-(vinyloxy)ethyl methacrylate), insufficient to cause a *network* to form)

5 MACROMOLECULAR ASSEMBLIES

Assemblies of macromolecules include polymer blends, *semi-interpenetrating polymer networks*, network polymers, *interpenetrating polymer networks* and polymer—polymer complexes.

Rule 6
Assemblies of macromolecules held together by non-covalent bonds are named by combination of the names of the constituent macromolecules together with an italicized connective between them.

Examples:

6.1 polystyrene-*blend*-poly(2,6-dimethylphenol)

6.2 (*net*-polystyrene)-*blend*-(*net*-polybutadiene)
(blend of two *networks*)

6.3 (*net*-polystyrene)-*sipn*-poly(vinyl chloride)
(SIPN of a polystyrene *network* and a linear poly(vinyl chloride), see Fig. 1b)

6.4 (*net*-polybutadiene)-*ipn*-(*net*-polystyrene)
(IPN of two *networks*, see Fig. 1c)

6.5 [*net*-poly(styrene-*stat*-butadiene)]-*ipn*-[*net*-poly(ethyl acrylate)]
(IPN of two *networks*)

6.6 poly(acrylic acid)-*compl*-poly(4-vinylpyridine)
(complex of two species of linear macromolecules)

6.7 poly[(methyl methacrylate)-*stat*-(methacrylic acid)]-*compl*-poly(*N*-vinyl-2-pyrrolidone)
(complex of statistical copolymer and homopolymer molecules)

6 QUANTITATIVE SPECIFICATIONS

Quantitative characteristics of a macromolecule or an assembly of macromolecules, such as mass and mole fractions or percentages as well as the degrees of polymerization and molar masses, may be expressed by placing corresponding figures after the complete name. The order of citation is the same as for the monomer species in the name. Some characteristics cannot be defined for all types of macromolecules and assemblies dealt with in this document, e.g. molar mass of a *network*.

Rule 7
Mass or mole fractions or percentages of the monomeric units are placed in parentheses after the name of the macromolecule or assembly of macromolecules in parentheses after the name of the macromolecule or assembly of macromolecules with the symbol 'w', or 'x', respectively, followed by the mass or mole fractions or percentages. The order of citation in the parentheses is the same as that of the monomers in the name; the individual values are separated by colons. Unknown quantities can be designated by a, b, etc., or hyphens.
Note: For examples, see [5], Rule 8.1.

Rule 8
For simple systems, the molar mass, relative molecular mass or degree of polymerization may be included in the scheme of Rule 7 by the symbol M, M or DP, respectively, followed by the corresponding numerical values, separated by colons. Symbols are qualified if quantities refer to parts of macromolecules or assemblies, e.g., M (block), M_r (arm), M (network chain), etc.

Note 1: Average quantities are denoted by the symbol $\overline{M}$, $\overline{M_r}$ or $\overline{DP}$
Note 2: M (network chain) is often denoted M_c. Similarly, M_r (network chain) might be denoted $M_{r,c}$.

Examples:

8.1 *net*-polystyrene-*v*-divinylbenzene (w 98:2%)
    (polystyrene crosslinked with 2% divinylbenzene)

8.2 *star*-(polyacrylonitrile; polystyrene) (M_r 100 000:20 000)
    (*star macromolecule* consisting of arms of polyacrylonitrile of a total $M_r = 100\,000$ and arms of polystyrene of a total $M_r = 20\,000$)

8.3 *star*-(polyacrylonitrile; polystyrene) (M_r (arm) 50 000:10 000)
    (*star macromolecule* consisting of arms of polyacrylonitrile each of $M_r = 50\,000$ and arms of polystyrene each of $M_r = 10\,000$; the macromolecule could be identical to that of example 8.2 with two arms of polyacrylonitrile and two arms of polystyrene)

Rule 9
The functionality of a monomer, oligomer or polymer molecule (or their mixtures) or of a *branch point* may be included in the scheme of Rule 7 by adding the symbol 'f' and the figure corresponding to the functionality in parentheses after the name of the monomer, oligomer or polymer molecule or *branch point*.

NOMENCLATURE

Note: If a mixture of monomer, oligomer or polymer molecules is present with components of the mixture bearing the same type of chemical functionality, an average value of f for the mixture may be quoted in parentheses following the list of components in the mixture.

Examples:

9.1 *net*-poly(glycerol; pentaerythritol* (*f* 3.5)-*co*-(dimethylphenylene diisocyanate)]
(*network* formed by the reaction of the glycerol/pentaerythritol mixture of average functionality 3.5 with dimethylphenylene diisocyanate)

9.2 *star*-polystyrene-*v*-trichloro(methyl)silane (*f* 2.8)

9.3 6-*star*-(polyacrylonitrile (*f* 3); polystyrene (*f* 3)) (*M*$_r$ (arm) 50 000:10 000)
(a six-armed *star macromolecule* consisting of three arms of polyacrylonitrile, each of M_r = 50 000, and three arms of polystyrene, each of M_r = 10 000)

9.4 4-*star*-(polyacrylonitrile-*block*-polystyrene) (*M*$_r$ (arm) 50 000)
(a four-armed *star macromolecule*, each arm of which consists of the same copolymeric block of M_r = 50 000)

9.5 *net*-polystyrene ($\overline{M}_{r,c}$ 10000)
(network chains of average M_r = 10 000)

7   REFERENCES

1. IUPAC. *Compendium of Macromolecular Nomenclature*, W. V. Metanomski (Ed.), Blackwell Scientific Publications, Oxford (1991).
2. IUPAC. 'Nomenclature of regular single-strand organic polymers (IUPAC Recommendations 2002)', *Pure Appl. Chem.* **74**, 1921-1956 (2002). Reprinted as Chapter 15 this volume.
3. IUPAC. 'A classification of linear single-strand polymers', *Pure Appl. Chem.* **61**, 243-254 (1989).
4. IUPAC. 'Nomenclature for regular single-strand and quasi-single-strand inorganic and coordination polymers (Recommendations 1984)', *Pure Appl. Chem.* **57**, 149-168 (1985).
5. IUPAC. 'Source-based nomenclature for copolymers (Recommendations 1985)', *Pure Appl. Chem.* **57**, 1427-1440 (1985). Reprinted as Chapter 19 this volume.
6. IUPAC. 'Nomenclature of regular double-strand (ladder and spiro) organic polymers (IUPAC Recommendations 1993)', *Pure Appl. Chem.* **65**, 1561-1580 (1993). Reprinted as Chapter 16 this volume.
7. IUPAC. 'Structure-based nomenclature for irregular single-strand organic polymers (IUPAC Recommendations 1994)', *Pure Appl. Chem.* **66**, 873-889 (1994). Reprinted as Chapter 17 this volume.

---

* 2,2-bis(hydroxymethyl)propane-1,3-diol

8. *A Guide to IUPAC Nomenclature of Organic Compounds, Recommendations 1993* (the 'Blue Book'), R. Panico, W. H. Powell and J.-C. Richer (Eds.), Blackwell Scientific Publications (1993). Corrigenda, *Pure Appl. Chem.*, **71**, 1327-1330 (1999).
9. IUPAC. *Nomenclature of Inorganic Chemistry,* RSC Publishing, Cambridge (2005).
10. IUPAC. 'Glossary of basic terms in polymer science (IUPAC Recommendations 1996)', *Pure Appl. Chem.* **68**, 2287-2311 (1996). Reprinted as Chapter 1 this volume.
11. IUPAC. 'Definitions of terms relating to individual macromolecules, their assemblies and dilute polymer solutions (Recommendations 1988)', *Pure Appl. Chem.* **61**, 211-241 (1989). Reprinted as Chapter 3 this volume.
12. IUPAC. Compendium of Chemical Terminology, 2nd ed. (the 'Gold Book'). Compiled by A. D. McNaught and A. Wilkinson. Blackwell Scientific Publications, Oxford (1997). XML on-line corrected version: http://goldbook.iupac.org (2006-) created by M. Nic, J. Jirat, B. Kosata; updates compiled by A. Jenkins.

8 ALPHABETICAL INDEX TO DEFINITIONS

backbone	1.4	micronetwork	1.11
branch	1.6	network	1.10
branch point	1.8	network polymer	1.17
branch unit	1.7	pendant chain	1.6
branched chain	1.3	polymer blend	1.15
chain	1.1	polymer-polymer complex	1.19
comb macromolecule	1.13	semi-interpenetrating polymer	
crosslink	1.9		
interpenetrating polymer network (IPN)	1.18	network (SIPN)	1.16
		side-chain	1.6
linear chain	1.2	star macromolecule	1.14
macrocycle	1.12	subchain	1.5
main chain	1.4		

21: Generic Source-Based Nomenclature for Polymers

CONTENTS

1. Preamble
2. Source-based nomenclature for homopolymers
3. Generic source-based nomenclature of polymers
4. Further applications of generic source-based names of polymers
5. References

1 PREAMBLE

The IUPAC Commission on Macromolecular Nomenclature has published three documents [1, 2, 3] on the structure-based nomenclature for organic polymers which enable most polymers, except networks, to be named. The Commission has also produced two documents [4, 5] on the source-based nomenclature of linear copolymers and non-linear polymers. In general, source-based names are simpler and less rigorous than structure-based names. However, there are cases in which the simplicity of the source-based nomenclature leads to ambiguous names for polymers. For example, the condensation of a dianhydride (A) with a diamine (B) gives first a polyamide-acid, which can be cyclized to a polyimide; however, both products have the same name poly(A-*alt*-B) according to current source-based nomenclature. If the class name of the polymer 'amide-acid' or 'imide' is incorporated in the name, differentiation is easily accomplished. Even in cases where only a single product is formed, use of the class name (generic name) may help to clarify the structure of the polymer, especially if it is very complex.

Examples of ambiguous names exist also for homopolymers. The source-based name 'polybutadiene' does not indicate whether the structure is 1,2-, 1,4-*cis*- or 1,4-*trans*-; supplementary information is needed to distinguish between the possibilities.

It is the objective of the present document to introduce a generic nomenclature system to solve these problems, and to yield better source-based names.

Most trivial names, such as polystyrene, are source-based names. Hitherto, the Commission has not systematically recommended source-based names for homopolymers because it considered that the more rigorous structure-based names were more appropriate for scientific communications. However, since the publication of 'Nomenclature of Regular Single-Strand Organic Polymers' in 1976, scientists, in both industry and academia, have continued to use trivial names. Even the Commission itself adopted (1985) a source-based nomenclature for copolymers because of its simplicity and practicality. Based on these facts, the Commission has now decided to recommend source-based

---

Originally prepared by a Working Group consisting of R. E. Bareiss (Germany), R. B. Fox (USA), K. Hatada (Japan), K. Horie (Japan), A. D. Jenkins (UK), J. Kahovec (Czech Republic), P. Kubisa (Poland), E. Maréchal (France), I. Meisel (Germany), W. V. Metanomski (USA), I. Mita (Japan), R. F. T. Stepto (UK), and E. S. Wilks (USA). Reprinted from *Pure Appl. Chem.* **73**, 1511-1519 (2001).

GENERIC SOURCE-BASED NOMENCLATURE

nomenclature as an alternative official nomenclature for homopolymers. In this document, the rules for generating source-based names for homopolymers are described. Consequently, source-based and structure-based names are available for most polymers.

Names of the monomers in the source-based names of polymers should preferably be systematic but they may be trivial if well established by usage. Names of the organic groups, as parts of constitutional repeating units (CRU) in structure-based names, are those based on the principles of organic nomenclature and recommended by the 1993 'A Guide to IUPAC Nomenclature of Organic Compounds' [6].

2 SOURCE-BASED NOMENCLATURE FOR HOMOPOLYMERS

Rule 1
The source-based name of a homopolymer is made by combining the prefix 'poly' with the name of the monomer. When the latter consists of more than one word, or any ambiguity is anticipated, the name of the monomer is parenthesized.

Example 1.1

Source-based name: polystyrene
Structure-based name: poly(1-phenylethane-1,2-diyl)

Example 1.2

Source-based name: poly(vinyl chloride)
Structure-based name: poly(1-chloroethane-1,2-diyl)

3 GENERIC SOURCE-BASED NOMENCLATURE OF POLYMERS

3.1 Fundamental Principles

The basic concept for generic source-based nomenclature of polymers is very simple; just add the polymer class name to the source-based name of the polymer. Addition of the polymer class name is frequently optional; in some cases, the addition is necessary to avoid ambiguity or to clarify. However, the addition is undesirable if it fails to add clarification.

The system presented here can be applied to almost all homopolymers, copolymers and others, such as networks. However, generic source-based nomenclature should not be considered as a third nomenclature system to be added to the other two systems of nomenclature; it must be considered as an auxiliary system and a simple extension of current source-based nomenclature. When the generic part of the name is eliminated from the name of a polymer, the well-established source-based name remains.

NOMENCLATURE

3.2 General Rules

Rule 2
A generic source-based name of a polymer has two components in the following sequence; (1) a polymer class (generic) name (polyG) followed by a colon and (2) the actual or hypothetical monomer name(s) (A, B, etc...), always parenthesized in the case of a copolymer. In the case of a homopolymer parentheses are introduced when it is necessary to improve clarity.

polyG:A   polyG:(B)   polyG:(A-*co*-B)   polyG:(A-*alt*-B)

Note 1: The polymer class name (generic name) describes the most appropriate type of functional group or heterocyclic ring system.
Note 2: All the rules given in the two prior documents on source-based nomenclature [4,5] can be applied to the present nomenclature system, with the addition of the generic part of the name.
Note 3: A polymer may have more than one name; this usually occurs when it can be prepared in more than one way.
Note 4: If a monomer or a pair of complementary monomers can give rise to more than one polymer, or if the polymer is obtained through a series of intermediate structures, the use of generic nomenclature is essential (see examples 2.1, 2.3, and 2.4).
Note 5: In a structure-based name, a dash denotes a junction between blocks (see example 2.1 in [3]).
Note 6: It is assumed that this reaction is limited to only one graft for each CRU.

Example 2.1

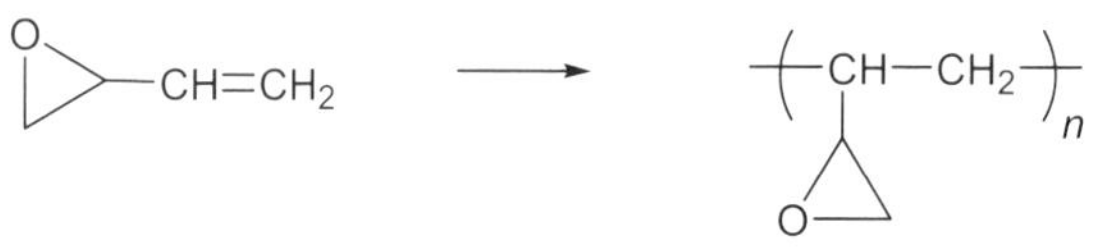

Generic source-based names:
  I. polyalkylene:vinyloxirane
  II. polyether:vinyloxirane

Source-based name:
   poly(vinyloxirane)
   (I and II have the same source-based name.)

Structure-based names:
  I. poly(1-oxiranylethane-1,2-diyl)
  II. poly[oxy(1-vinylethane-1,2-diyl)]

GENERIC SOURCE-BASED NOMENCLATURE

Example 2.2

Generic Source-based name: polyoxadiazole:(4-cyanobenzonitrile *N*-oxide)
Structure-based name: poly(1,2,4-oxadiazole-3,5-diyl-1,4-phenylene)

Example 2.3

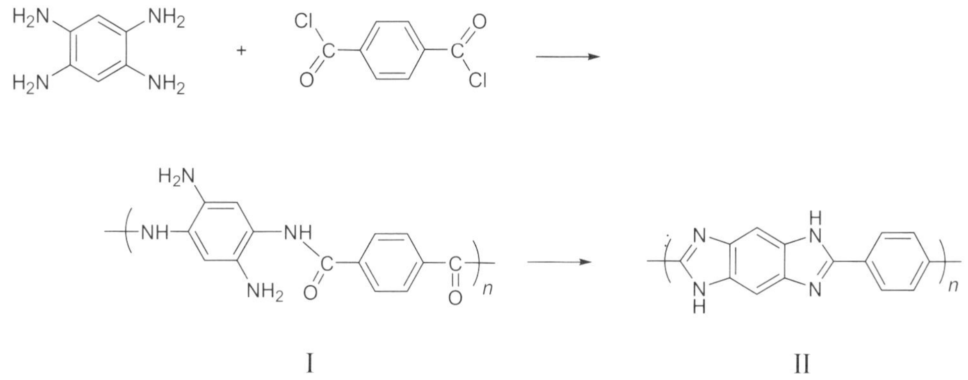

Generic source-based names:
  I. polyamide:[(terephthaloyl dichloride)-*alt*-benzene-1,2,4,5-tetramine]
  II. polybenzimidazole:[(terephthaloyl dichloride)-*alt*-benzene-1,2,4,5-tetramine]

Source-based name:
  poly[(terephthaloyl dichloride)-*alt*-benzene-1,2,4,5-tetramine]
    (I and II have the same source-based name.)

Structure-based names:
  I. poly[imino(2,5-diamino-1,4-phenylene)iminoterephthaloyl]
  II. poly[(1,5-dihydrobenzo[1,2-*d*:4,5-*d'*]diimidazole-2,6-diyl)-1,4-phenylene]

Example 2.4

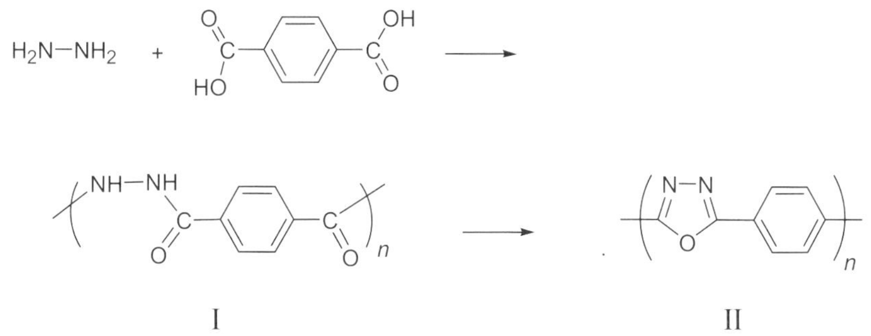

NOMENCLATURE

Generic source-based names:
I. polyhydrazide:[hydrazine-*alt*-(terephthalic acid)]
II. polyoxadiazole:[hydrazine-*alt*-(terephthalic acid)]

Source-based name:
poly[hydrazine-*alt*-(terephthalic acid)]
 (I and II have the same source-based name)

Structure-based names:
I. poly(hydrazine-1,2-diylterephthaloyl)
II. poly(1,3,4-oxadiazole-2,5-diyl-1,4-phenylene)

Example 2.5

$$\left[\text{O}-(\text{CH}_2)_4-\text{O}-\overset{\text{O}}{\underset{\parallel}{\text{C}}}-\text{NH}-(\text{CH}_2)_6-\text{NH}-\overset{\text{O}}{\underset{\parallel}{\text{C}}}\right]_p \left[\text{O}-(\text{CH}_2)_2-\text{O}-\overset{\text{O}}{\underset{\parallel}{\text{C}}}-\underset{}{\bigcirc}-\overset{\text{O}}{\underset{\parallel}{\text{C}}}\right]_r$$

Generic source-based name:
polyurethane:[butane-1,4-diol-*alt*-(hexane-1,6-diyl diisocyanate)]-*block*-polyester:
[(ethylene glycol)-*alt*-(terephthalic acid)]

Structure-based name (see *Note 5*):
poly(oxybutane-1,4-diyloxycarbonyliminohexane-1,6-diyliminocarbonyl)—
poly(oxyethane-1,2-diyloxyterephthaloyl)

Example 2.6

$$\left[\text{NH}-\overset{\text{O}}{\underset{\parallel}{\text{C}}}-(\text{CH}_2)_4-\overset{\text{O}}{\underset{\parallel}{\text{C}}}-\text{NH}-(\text{CH}_2)_6\right]_n + \underset{\text{O}}{\triangle}-\text{CH}_2\left[\text{O}-\text{CH}_2-\text{CH}_2\right]_p\text{O}-\text{CH}_3 \longrightarrow$$

$$\left[-\text{NH}-\overset{\text{O}}{\underset{\parallel}{\text{C}}}-(\text{CH}_2)_4-\overset{\text{O}}{\underset{\parallel}{\text{C}}}-\text{NH}-(\text{CH}_2)_6-\ /\ -\underset{\underset{\underset{\underset{\text{OH}}{|}}{\text{CH}-\text{CH}_2[\text{O}-\text{CH}_2-\text{CH}_2]_p\text{O}-\text{CH}_3}}{|}}{\text{N}}-\overset{\text{O}}{\underset{\parallel}{\text{C}}}-(\text{CH}_2)_4-\overset{\text{O}}{\underset{\parallel}{\text{C}}}-\text{NH}-(\text{CH}_2)_6\right]_n$$

Generic Source-based name:
polyamide:[hexane-1,6-diamine-*alt*-(adipic acid)]-*graft*-polyether:(ethylene oxide)

Rule 3
When more than one type of functional group or heterocyclic system is present in the polymer structure, names should be alphabetized – for example, poly(GG′):(A-*alt*-B).
Note: It is preferable, but not mandatory, to cite all generic classes.

GENERIC SOURCE-BASED NOMENCLATURE

Example 3.1

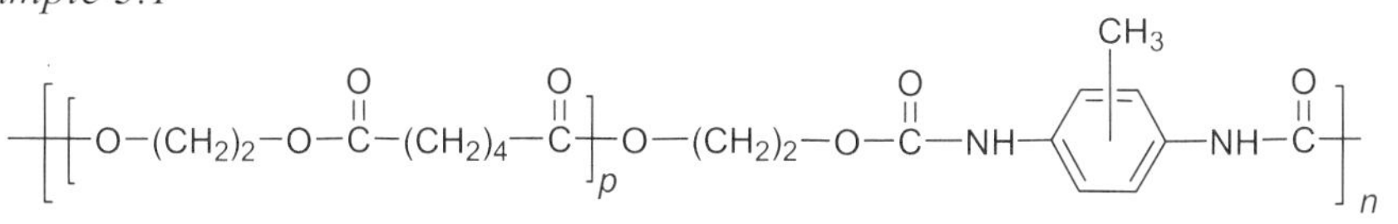

[chemical structure]

Generic source-based name:
  polyesterurethane:α,ω-dihydroxy{oligo[(ethylene glycol)-*alt*-(adipic acid)]-*alt*-(2,5-methyl-1,4-phenylene diisocyanate)}

Structure-based name:
  poly{[oligo(oxyethane-1,2-diyloxyhexanedioyl)]oxyethane-1,2-diyloxycarbonylimino(*x*-methyl-1,4-phenylene)iminocarbonyl}

Example 3.2

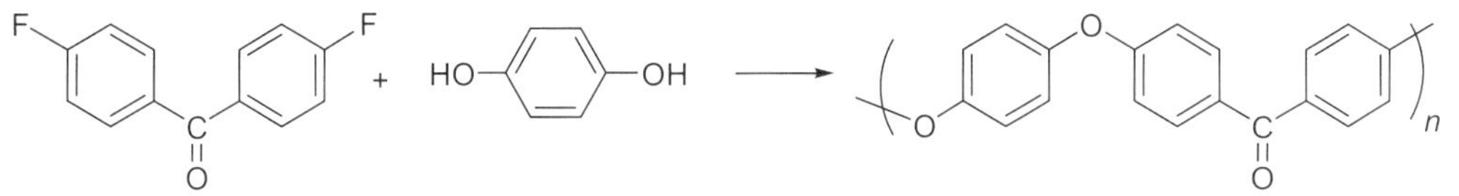

[chemical reaction scheme]

Generic source-based name:
  polyetherketone:(4,4′-difluorobenzophenone-*alt*-hydroquinone)

Structure-based name:
  poly(oxy-1,4-phenyleneoxy-1,4-phenylenecarbonyl-1,4-phenylene)

Rule 4
Polymer class names relevant only to the main chain are specified in the name; names of side-chain functional groups may also be included after a hyphen if they are formed during the polymerization reaction.

Example 4.1

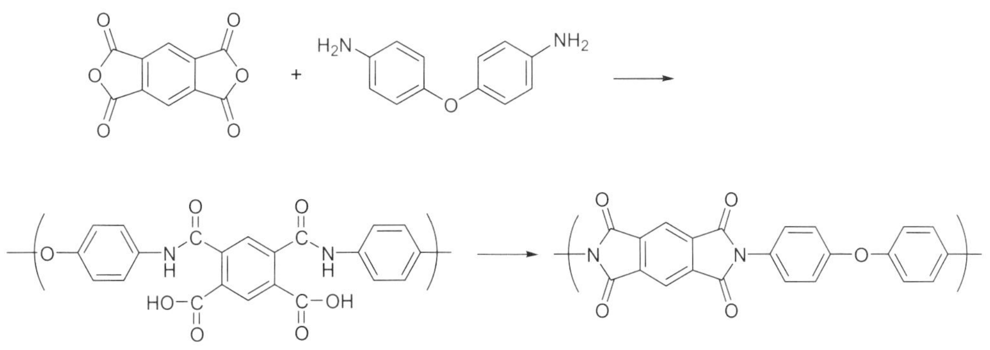

[chemical reaction scheme showing structures I and II]

I   II

Generic source-based names:
  I. poly(amide-acid):[(pyromellitic dianhydride)*-*alt*-(4,4′-oxydianiline)]
     (Both carboxy groups result from the polymerization reaction.)

---

* benzene-1,2,4,5-tetracarboxylic-1,2:4,5-dianhydride

NOMENCLATURE

II. polyimide:[(pyromellitic dianhydride)-*alt*-(4,4'-oxydianiline)]

Structure-based names:
I. poly[oxy-1,4-phenyleneiminocarbonyl(4,6-dicarboxy-1,3-phenylene)carbonylimino-1,4-phenylene]
II. poly[(1,3,5,7-tetraoxo-5,7-dihydrobenzo[1,2-*c*:4,5-*c*′]dipyrrole-2,6-(1H,3H)-diyl)-1,4-phenyleneoxy-1,4-phenylene]

Example 4.2

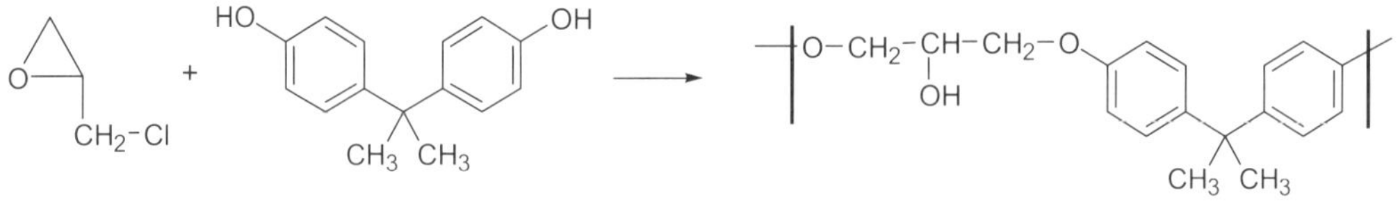

Generic source-based names:
poly(ether-alcohol):(epichlorohydrin*-*alt*-bisphenol A)

Structure-based name:
poly[oxy(2-hydroxypropane-1,3-diyl)oxy-1,4-phenylene(dimethylmethylene)-1,4-phenylene]

Rule 5
In the case of carbon-chain polymers such as vinyl polymers or diene polymers, the generic name is to be used only when different polymer structures may arise from a given monomeric system.

Example 5.1

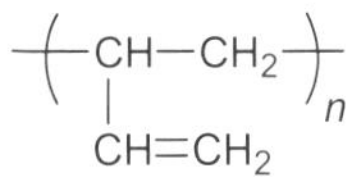

Generic source-based name: polyalkylene:buta-1,3-diene
Source-based name: poly(buta-1,3-diene)
Structure-based name: poly(1-vinylethane-1,2-diyl)

Example 5.2

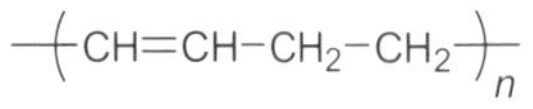

Generic source-based name: polyalkenylene:buta-1,3-diene
Source-based name: poly(buta-1,3-diene)
Structure-based name: poly(but-1-ene-1,4-diyl)

---

* 2-(chloromethyl)oxirane

GENERIC SOURCE-BASED NOMENCLATURE

Example 5.3

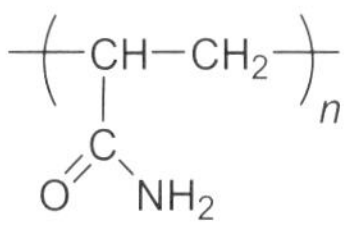

[structure: poly(CH-CH₂) with side group C(=O)NH₂]

Generic source-based name: polyalkylene:acrylamide
Structure-based name: poly(1-carbamoylethane-1,2-diyl)

Example 5.4

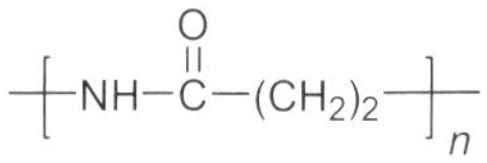

[structure: -[NH-C(=O)-(CH₂)₂]-ₙ]

Generic source-based name: polyamide:acrylamide
Structure-based name: poly[imino(1-oxopropane-1,3-diyl)]

Note: The terms polyalkylene and polyalkenylene refer to saturated carbon (main) chains and those with C=C bond(s), respectively.

4 FURTHER APPLICATIONS OF GENERIC SOURCE-BASED NAMES OF POLYMERS

Generic source-based nomenclature can be extended to more complicated polymers such as spiro and cyclic polymers and networks.

Example 6.1

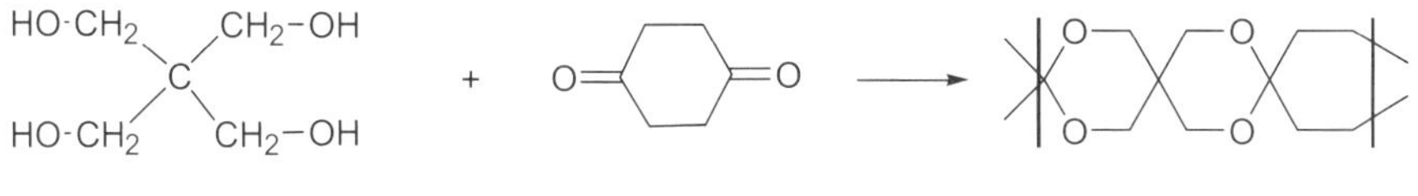

[structure: pentaerythritol + cyclohexane-1,4-dione → spiroketal polymer]

Generic source based name:
  polyspiroketal:{[2,2-bis(hydroxymethyl)propane-1,3-diol]-*alt*-cyclohexane-1,4-dione}
  or
  polyspiroketal:(pentaerythritol[*]-*alt*-cyclohexane-1,4-dione)

Structure based name:
  poly[2,4,8,10-tetraoxaspiro[5.5]undecane-3,3,9,9-tetrayl-9,9-bis(ethane-1,2-diyl)]

Example 6.2

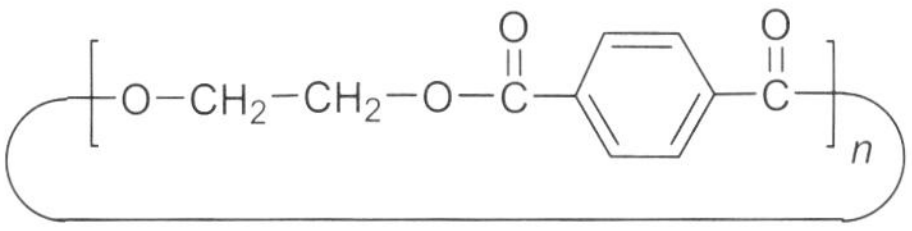

[structure: cyclic poly(ethylene terephthalate) -[O-CH₂-CH₂-O-C(=O)-C₆H₄-C(=O)]-ₙ]

---

[*] 2,2-bis(hydroxymethyl)

NOMENCLATURE

Generic source-based name:
  cyclic polyester:[(ethylene glycol)-*alt*-(terephthalic acid)]

Example 6.3

[structure: —O—C(=O)—CH=CH—C(=O)—O—(CH₂)₄— / —O—C(=O)—(o-C₆H₄)—C(=O)—O—(CH₂)₄—] + CH=CH₂–C₆H₅ ⟶ network

Generic source-based name:
  polyester:{butane-1,4-diol-*alt*-[(maleic anhydride);(phthalic anhydride)]}-*net*-polyalkylene: (maleic anhydride)-*co*-styrene]

5 REFERENCES

1. IUPAC. 'Nomenclature of regular single-strand organic polymers (IUPAC Recommendations 2002)', *Pure Appl. Chem.* **74**, 1921-1956 (2002). Reprinted as Chapter 15 this volume.
2. IUPAC. 'Nomenclature of regular double-strand (ladder and spiro) organic polymers (IUPAC Recommendations 1993)', *Pure Appl. Chem.* **65**, 1561-1580 (1993). Reprinted as Chapter 16 this volume.
3. IUPAC. 'Structure-based nomenclature for irregular single-strand organic polymers (IUPAC Recommendations 1994)', *Pure Appl. Chem.* **66**, 873-889 (1994). Reprinted as Chapter 17, this volume.
4. IUPAC. 'Source-based nomenclature for copolymers (Recommendations 1985)', *Pure Appl. Chem.*, **57**, 1427-1440 (1985). Reprinted as Chapter 19 this volume.
5. IUPAC. 'Source-based nomenclature for non-linear macromolecules and macromolecular assemblies (IUPAC Recommendations 1997)', *Pure Appl. Chem.* **69**, 2511-2521 (1997). Reprinted as Chapter 20 this volume.
6. IUPAC. 'A Guide to IUPAC Nomenclature of Organic Compounds.' R. Panico, W. H. Powell, and J-C. Richer (Eds.), Blackwell Scientific Publications, Oxford (1993).

22: ISO Abbreviations for Names of Polymeric Substances

PREAMBLE

The first recommendations on abbreviations for polymeric substances by the IUPAC Commission on Macromolecular Nomenclature were published in 1974 [1]. These were incorporated into an expanded list, published in 1986, by the International Organization for Standardization (ISO) [2]. The IUPAC recommendations of 1974 and the ISO list of 1986 were published by IUPAC in 1987 [3] and reproduced as Chapter 9 of the first edition of the Purple Book. More recently, ISO published a revised list in 2001 [4] and the present list of abbreviations is derived from that list.

Whilst the IUPAC Subcommittee on Polymer Terminology recognises that abbreviations have their uses, it realises the inherent difficulty of assigning systematic and unique abbreviations to polymeric structures. In this regard, it is to be noted that the ISO list uses nomenclature that is not necessarily in accord with IUPAC recommendations. The Subcommittee also reminds the reader of the IUPAC policy on the use of abbreviations in the chemical literature [5], which declares, in part, that: 'there are great advantages in defining all abbreviations,…, in a single conspicuous place in each paper. This is preferably done near the beginning of the paper in a single list'. The Subcommittee also urges that each abbreviation be fully defined the first time it appears in the text and that no abbreviation be used in titles and abstracts of publications.

List of abbreviations based on the 2001 International Standard ISO 1043-1:2001

AB	acrylonitrile-butadiene
ABAK	acrylonitrile-butadiene-acrylate[a]
ABS	acrylonitrile-butadiene-styrene
AEPDS	acrylonitrile-(ethene-propene-diene)-styrene[a]
AMMA	acrylonitrile-methyl methacrylate
ASA	acrylonitrile-styrene-acrylate
CA	cellulose acetate
CAB	cellulose acetate butyrate
CAP	cellulose acetate propionate
CEF	cellulose formaldehyde
CF	cresol-formaldehyde
CMC	(carboxy)methyl cellulose
CN	cellulose nitrate
COC	cyclo-olefin copolymer
CP	cellulose propionate
CTA	cellulose triacetate
EAA	ethylene-acrylic acid

Prepared by a Working Group consisting of W. V. Metanomski (USA), R. F. T. Stepto (UK) and E. S. Wilks (USA) using the ISO Standard of 2001[4].

NOMENCLATURE

EBAK	ethylene-butyl acrylate[a]
EC	ethyl cellulose
EEAK	ethylene-ethyl acrylate[a]
EMA	ethylene-methacrylic acid
EP	epoxide; epoxy
E/P	ethylene-propylene[a]
ETFE	ethylene-tetrafluoroethylene
EVAC	ethylene-vinyl acetate[a]
EVOH	ethylene-vinyl alcohol
FEP	perfluoro(ethylene-propylene)[a]
MABS	methyl methacrylate-acrylonitrile-butadiene-styrene
MBS	methyl methacrylate-butadiene-styrene
MC	methyl cellulose
MF	melamine-formaldehyde
MP	melamine-phenol
MPF	melamine-phenol-formaldehyde
MSAN	α-methylstyrene-acrylonitrile
PA	polyamide
PAA	poly(acrylic acid)
PAEK	polyaryletherketone
PAK	polyacrylate
PAN	polyacrylonitrile
PAR	polyarylate
PARA	poly(aryl amide)
PB	polybutene
PBAK	poly(butyl acrylate)
PBD	polybutadiene
PBN	poly(butylene naphthalate)
PBT	poly(butylene terephthalate)
PC	polycarbonate
PCL	polycaprolactone[*]
PCTFE	poly(chlorotrifluoroethene)
PDAP	poly(diallyl phthalate)
PE	polyethylene
PE-C	chlorinated polyethylene[a]
PE-HD	high-density polyethylene[a]
PE-LD	low-density polyethylene[a]
PE-LLD	linear low-density polyethylene[a]
PE-MD	medium-density polyethylene[a]
PE-UHMW	ultra-high-molecular-weight polyethylene[a]
PE-VLD	very low-density polyethylene[a]
PEC	polyestercarbonate
PEEK	polyetheretherketone
PEEST	polyetherester
PEI	polyetherimide
PEK	polyetherketone
PEN	poly(ethylene naphthalate)
PEOX	poly(ethylene oxide)

---

[*] caprolactone ≡ 1-oxaheptan-2-one

ABBREVIATIONS

PESTUR	polyesterurethane
PESU	polyethersulfone
PET	poly(ethylene terephthalate)
PEUR	polyetherurethane
PF	phenol-formaldehyde
PI	polyimide
PIB	poly(2-methylpropene); polyisobutylene
PIR	polyisocyanurate
PK	polyketone
PMMA	poly(methyl methacrylate)
PMPS	poly(methylphenylsiloxane)
PMS	poly(α-methylstyrene)
POM	poly(oxymethylene); polyformaldehyde
PP	polypropylene
PP-E	expandable polypropylene[a]
PP-HI	high-impact polypropylene[a]
PPE	poly(phenylene ether)
PPO	poly(1,4-phenylene oxide)
PPOX	poly(propylene oxide)
PPS	poly(phenylene sulfide)
PPSU	poly(phenylene sulfone)
PS	polystyrene
PS-E	expandable polystyrene[a]
PS-HI	high-impact polystyrene[a]
PSU	polysulfone
PTFE	polytetrafluoroethylene
PTT	poly(trimethylene terephthalate)
PUR	polyurethane
PVAC	poly(vinyl acetate)
PVAL	poly(vinyl alcohol)[a]
PVB	poly(vinyl butyral)
PVC	poly(vinyl chloride)
PVC-C	chlorinated poly(vinyl chloride)[a]
PVC-U	unplasticized poly(vinyl chloride)[a]
PVDC	poly(vinylidene dichloride)
PVDF	poly(vinylidene difluoride)
PVF	poly(vinyl fluoride)
PVFM	poly(vinyl formal)
PVK	poly(N-vinylcarbazole)
PVP	poly(N-vinylpyrrolidone)
SAN	styrene-acrylonitrile
SB	styrene-butadiene
SI	silicone
SMAH	styrene-maleic anhydride[a]
SMS	styrene-α-methylstyrene
UF	urea-formaldehyde
UP	unsaturated polyester
VCE	vinyl chloride-ethylene
VCEMAK	vinyl chloride-ethylene-methyl acrylate[a]

NOMENCLATURE

VCEVAC	vinyl chloride-ethylene-vinyl acetate
VCMAK	vinyl chloride-methyl acrylate[a]
VCMMA	vinyl chloride-methyl methacrylate
VCVAC	vinyl chloride-vinyl acetate
VCVDC	vinyl chloride-vinylidene chloride
VE	vinyl ester

[a] these polymeric substances had the following previous abbreviations and names [2,3] that are no longer recommended:

ABA	acrylonitrile-butadiene-acrylate
AEPDMS	acrylonitrile-(ethylene-propylene-diene)-styrene
CPE	chlorinated polyethylene
CPVC	chlorinated poly(vinyl chloride)
EBA	ethylene-butyl acrylate
EEA	ethylene-ethyl acrylate
EPM	ethylene-propylene
EPP	expandable polypropylene
EPS	expandable polystyrene
EVA	ethylene-vinyl acetate
HDPE	high-density polyethylene
HIPP	high-impact polypropylene
HIPS	high-impact polystyrene
LDPE	low-density polyethylene
LLDPE	linear low-density polyethylene
MDPE	medium-density polyethylene
PFEP	perfluoro(ethylene-propylene)
PVOH	poly(vinyl alcohol)
S/MA *or* SMA	styrene-maleic anhydride
UHMWPE	ultra-high-molecular-weight polyethylene
UPVC	unplasticized poly(vinyl chloride)
VCEMA	vinyl chloride-ethylene-methyl acrylate
VCMA	vinyl chloride-methyl acrylate
VLDPE	very low-density polyethylene

ABBREVIATIONS

REFERENCES

1. IUPAC. "List of standard abbreviations (symbols) for synthetic polymers and polymer materials 1974", *Pure Appl. Chem.* **40**, 473-476 (1974).
2. ISO. International Standard ISO 1043:1986. *Plastics – Symbols and Codes – Part 1: Symbols for basic polymers and their modifications, and for plasticizers.*
3. IUPAC. "Use of abbreviations for names of polymeric substances (Recommendations 1986)", *Pure Appl. Chem.* **59**, 691-693 (1987).
4. ISO. International Standard ISO 1043-1:2001. *Plastics - Symbols and Abbreviated Terms – Part 1: Basic polymers and their special characteristics.*
5. IUPAC. "Use of abbreviations in the chemical literature (Recommendations 1979)", *Pure Appl. Chem.* **52**, 2229-2232 (1980).

Appendix: Bibliography of Biopolymer-Related IUPAC-IUBMB Nomenclature Recommendations

CONTENTS

A Amino acids and peptides
B Nucleic acids and polynucleotides
C Oligo- and polysaccharides

A AMINO ACIDS AND PEPTIDES

1. IUPAC-IUBMB. 'Abbreviations and symbols for the description of the conformation of polypeptide chains (tentative rules 1969)'; *Pure Appl. Chem.* **40**, 291-308 (1974); IUBMB. *Biochemical Nomenclature and Related Documents*, 2nd ed., Portland Press Ltd., London (1992), pp. 73-81.

2. IUPAC-IUBMB. 'Abbreviated nomenclature of synthetic polypeptides (polymerized amino acids) (Recommendations 1971)', *Pure Appl. Chem.* **33**, 437-444 (1973); IUBMB.*Biochemical Nomenclature and Related Documents*, 2nd ed., Portland Press Ltd., London (1992), pp. 70-72.

3. IUPAC-IUBMB. 'Nomenclature and symbolism for amino acids and peptides (Recommendations 1983)', *Pure Appl. Chem.* **56**, 595-624 (1984); *Spec. Period. Rep.: Amino Acids, Pept., Proteins* **16**, 387-410 (1985). Corrections: *Eur. J. Biochem.* **152**, 1 (1985); IUBMB. *Biochemical Nomenclature and Related Documents*, 2nd ed., Portland Press Ltd., London (1992), pp. 39-67.

4. IUPAC-IUBMB. 'Nomenclature of glycoproteins, glycopeptides and peptidoglycans (Recommendations 1985)', *Pure Appl. Chem.* **60**, 1389-1394 (1988). (Modifies ref. 8 below by giving the short form for symbolizing oligosaccharide chains.); IUBMB. *Biochemical Nomenclature and Related Documents*, 2nd ed., Portland Press Ltd., London (1992), pp. 84-89.

B NUCLEIC ACIDS AND POLYNUCLEOTIDES

5. IUPAC-IUBMB. 'Abbreviations and symbols for nucleic acids, polynucleotides and their constituents (Recommendations 1970)', *Pure Appl. Chem.* **40**, 277-290 (1974); IUBMB. *Biochemical Nomenclature and Related Documents*, 2nd ed., Portland Press Ltd., London (1992), pp. 109-114.

6. IUPAC-IUBMB. 'Abbreviations and symbols for the description of conformations of polynucleotide chains (Recommendations 1982)', *Pure Appl. Chem.* **55**, 1273-1280

APPENDIX: BIOPOLYMER-RELATED NOMENCLATURE

(1983); IUBMB. *Biochemical Nomenclature and Related Documents*, 2nd ed., Portland Press Ltd., London (1992), pp. 115-121.

C OLIGO- AND POLYSACCHARIDES

7. IUPAC-IUBMB. 'Abbreviated terminology of oligosaccharide chains (Recommendations 1980)', (ref. 4 above gives an extension of these recommendations.); now incorporated into ref. 11. IUBMB. *Biochemical Nomenclature and Related Documents*, 2nd ed., Portland Press Ltd., London (1992), pp. 169-173.

8. IUPAC-IUBMB. 'Polysaccharide nomenclature (Recommendations 1980)', *Pure Appl. Chem.* **54**, 1523-1526 (1982); now incorporated into ref. 11. IUBMB. *Biochemical Nomenclature and Related Documents*, 2nd ed., Portland Press Ltd., London (1992), pp. 174-176.

9. IUPAC-IUBMB. 'Symbols for specifying the conformation of polysaccharide chains (Recommendations 1981)', *Pure Appl. Chem.* **55**, 1269-1272 (1983); IUBMB. *Biochemical Nomenclature and Related Documents*, 2nd ed., Portland Press Ltd., London (1992), pp. 177-179.

10. IUPAC-IUBMB. 'Nomenclature of carbohydrates (Recommendations 1996)', *Pure Appl. Chem.* **68**, 1919-2008 (1996).

INDEX

Terminology

ablation	13/1
ablator	13/2
abrasion	13/3
absolute compliance	8/5.18
absolute modulus	8/5.14
additive	9/1.30
adhering thread	7/4.9.3.1
adhesion	9/1.24
chemical adhesion	9/1.25
interfacial adhesion (tack)	9/1.26
adhesion promoter	9/1.37
adhesive strength	9/1.26
adjacent re-entry	6/4.8
aerosol hydrolysis	11/5.1
affine chain behaviour	11/4.2.1
agglomerate	9/1.42, 11/2.1, 11/2.2
agglomeration	9/1.41, 11/5.2, 11/5.3
aggregate	9/1.42, 11/2.3, 11/2.4
aggregation	9/1.41, 11/5.4, 11/5.5
aging	13/4
aqueous aging	13/4
biologically-induced aging	13/4
cosmic aging	13/4
gel aging	11/5.18
oxidative aging	13/4
photochemical aging	13/4
photo-oxidative aging	13/4
physical aging	13/4
thermal aging	13/4
thermo-oxidative aging	13/4
underground aging	13/4
alternating copolymer	1/2.11
alternating copolymerization	1/3.11
amphiphilic mesogen	7/2.11.1
amphitropic compound	7/2.4.4
ampholytic polymer (polyampholyte)	10/1
amphoteric polyelectrolyte	12/3.16
angle of observation	3/3.3.3
angular velocity (of a forced oscillation)	8/5.10
angular velocity of the resonance frequency	8/6.7
anion-exchange polymer	10/2
anionic polymer	10/3
anionic polymerization	1/3.19
anisotropy of physical properties	7/5.8
antagonism	13/5

INDEX - Terminology

anti-fatigue agent	13/6
antiferroelectric chiral smectic C mesophase	7/3.1.5.1.2
antimesophase	7/3.1.6.3
antioxidant	13/7
chain-breaking antioxidant	13/13
chain-terminating antioxidant	13/13
thermal antioxidant	13/7
photo-antioxidant	13/7
mechano-antioxidant	13/7
antiradiant	13/8
apparent molar mass	3/2.11
apparent molecular weight	3/2.11
apparent relative molecular mass	3/2.11
apparent viscosity	8/4.12
Archibald's method	3/3.2.15
ashing	13/9
asymmetric chirogenic polymerization	5/2
asymmetric liquid-crystal dimer	7/2.11.2.9
asymmetric helix-chirogenic polymerization	5/2
asymmetric enantiomer-differentiating polymerization	5/3
asymmetric polymerization	5/1
atactic block	2/1.20
atactic macromolecule	1/1.25
atactic polymer	1/2.21, 2/1.10
autoxidation	13/10
average degree of polymerization	3/2.12
Avrami equation	6/5.3
axialite	6/3.5
backbone	1/1.34, 15/2.5, 20/1.4
backflow	7/5.7
banana mesogen	7/2.11.2.10
banded texture (banded texture)	7/6.12
barotropic mesophase	7/2.4.2
Basic terms in polymer science	Ch. 1
molecules and molecular structure	Ch. 1, Sec. 1
substances	Ch. 1, Sec. 2
reactions	Ch. 1, Sec. 3
alphabetical index of terms	Ch. 1, Sec. 4
bâtonnet	7/4.10.1
bead-rod model	3/3.2.4
bead-spring model	3/3.2.5
bend deformation	7/5.2.2
biaxial mesophase	7/3.3
biaxial mesophase anisotropies	7/5.8.2
biaxial nematic mesophase (biaxial nematic)	7/3.3.1
biaxial smectic A mesophase	7/3.3.2
biforked mesogen	7/2.11.2.5
binodal (binodal curve)	9/2.6
biodegradable polymer	12/3.1
biodegradation	12/1.13, 13/11

INDEX - Terminology

bipolar droplet texture	7/4.9.1.1
bipolymer	1/2.5
bis-swallow-tailed mesogen	7/2.11.2.7
Blends, composites and multiphase polymeric materials	Ch. 9
basic terms in polymer mixtures	Ch. 9, Sec. 1
phase domain behaviour	Ch. 9, Sec. 2
domain morphologies	Ch. 9, Sec. 3
references	Ch. 9, Sec. 4
bibliography	Ch. 9, Sec. 5
alphabetical index of terms	Ch. 9, Sec. 6
block	1/1.62, 2/1.18
block copolymer	1/2.24
block macromolecule	1/1.26
block polymer	1/2.22
blooming	9/2.17
blue phase	7/3.1.4
board-shaped polymer	7/6.9
boojums	7/4.9.1.1
boundaries of Grandjean	7/4.10.4
bowlic mesogen	7/2.11.2.3
bowtie entanglement	11/4.2.3.1
branch (side-chain, pendant chain)	1/1.53, 20/1.6
branched chain	1/1.33, 20/1.3
branched polymer	1/2.34
branch point	1/1.54, 11/4.2.2, 20/1.8
branch unit	1/1.55, 20/1.7
branching index	3/1.25
brush macromolecule	1/1.52
bulk compliance	8/4.6
bulk compression	8/2.9
bulk compressive compliance	8/4.6
bulk compressive modulus	8/4.5
bulk compressive strain	8/2.9
bulk modulus (bulk compressive modulus)	8/4.5
bulk substance	9/3.12
butterfly entanglement	11/4.2.3.2
calamitic mesogen	7/2.11.2.1
calamitic mesophases	7/3.1
calcination (in polymer networks)	11/5.6
carbonization	13/12
carbo-reduction	11/5.7
cation-exchange polymer	10/4
cationic polymer	10/5
cationic polymerization	1/3.20
Cauchy tensor	8/1.8, 8/1.9
centered rectangular mesophase	7/3.1.6.3
ceramer	11/4.1.1
ceramic	11/4.1.2
ceramic-precursor	11/4.1.3
ceramic-reinforced polymer	11/4.1.4

INDEX - Terminology

ceramic yield	11/4.1.5
ceramization	11/5.8
chain	1/1.30, 20/1.1
chain axis	6/2.1
chain conformational repeating unit	6/2.3
chain entanglement	11/4.2.3
chain folding	6/4.3
(chain) identity period	6/2.2
chain-orientational disorder	6/2.10.2
chain polymerization	1/3.6, 4/1
(chain) repeating distance	6/2.2
chain scission	1/3.24
characteristic ratio	3/1.13
chelating polymer	12/2.1
chemical amplification	12/1.1
chemical functionality	11/2.5
chemical modification	12/1.2
chiral columnar oblique mesophase	7/3.2.2.3
chiral nematic mesophase (chiral nematic)	7/3.1.3
chiral nematogen	7/2.11
chiral smectic C mesophase	7/3.1.5.1.3
chiral smectic F mesophase	7/3.1.5.2.2
chiral smectic I mesophase	7/3.1.5.2.3
cholesteric mesophase (cholesteric)	7/3.1.3
cistactic polymer	2/1.16
class of helix	6/2.8
clearing point	7/2.6
clearing temperature	7/2.6
cloud point	9/2.9
cloud-point curve	9/2.10
cloud-point temperature	9/2.11
coagulation	11/5.9
coalescence	9/2.18
co-continuous phase domains	9/3.14
coefficient of viscosity	8/4.12
coexistence curve	9/2.6
coherent elastic scattering of radiation	3/3.3
colloid	11/2.6
colloidal	11/2.7
colloidal dispersion	11/2.8
colloidal processing	11/5.10
colloidal sol	11/2.9
colloidal suspension	11/2.10
columnar discotic mesophase (columnar discotic)	7/3.2.2
columnar hexagonal mesophase	7/3.2.2.1
columnar mesophase	7/3.2.2
columnar oblique mesophase	7/3.2.2.3
columnar rectangular mesophase	7/3.2.2.2
comb macromolecule	1/1.52, 20/1.13
comb polymer	1/2.33

combined liquid-crystalline polymer	7/6.6
comb-shaped (comb-like) polymer liquid-crystal	7/6.3
comb-shaped mesogen	7/2.11.2.3
compatibility	9/1.32
compatibilization	9/1.33
compatibilizer	9/1.36
complex compliance	8/5.19
complex modulus	8/5.15
complex rate of strain	8/5.22
complex strain	8/5.15, 8/5.19
complex stress	8/5.15, 8/5.19, 8/5.22
complex viscosity	8/5.22
compliance	8/4.4
composite	9/1.13, 11/4.1.6
polymer composite	9/1.14
nanocomposite	9/1.15, 11/4.1.20
compositional heterogeneity	3/2.1
compression ratio	8/2.3
compressive strain	8/2.4
compressive stress	8/3.3
condensation reaction	11/5.11
condensative chain polymerization	1/3.6, 4/1
conducting polymer	12/3.2
conducting polymer composite	
(solid polymer-electrolyte composite)	10/6, 12/3.2
ion-conducting polymer	12/3.2
photoconductive polymer	12/3.2
proton-conducting polymer	12/3.2
semiconducting polymer	12/3.2
configurational base unit	1/1.17, 2/1.2
configurational disorder	6/2.10.3
configurational homosequence	2/2.1.5
configurational repeating unit	1/1.18, 2/1.3
configurational sequence	1/1.64, 2/2.1.3
configurational unit	1/1.16, 2/1.1
conformational disorder	6/2.10.4
conformational repeating unit of a chain	6/2.3
conical mesogen	7/2.11.2.3
connectivity	11/2.11
constitutional heterogeneity	3/2.2
constitutional homosequence	2/2.1.2
constitutional repeating unit	1/1.15, 15/2.4, 16/2.3
constitutional sequence	1/1.63, 2/2.1.1
constitutional unit	1/1.14, 15/2.3, 16/2.2
constitutive equation for an elastic solid	8/4.1
constitutive equation for an incompressible viscoelastic liquid or solid	8/4.2
continuous distribution function	3/2.1.3
continuously curved chain	3/1.21
continuous phase domain	9/3.12

INDEX - Terminology

contour length	3/1.14
controlled-degradable polymer	13/20
co-oligomer	1/2.7
co-oligomerization	1/3.5
copolymer	1/2.11
copolymerization	1/3.4
copolymer micelle	3/1.28
core-shell morphology	9/3.15
co-solvency	3/3.1.15
coupling agent	9/1.37
covalent network	1/1.58, 11/4.1.21.3.3
cracking	13/14
chemical cracking	13/14
environmental stress cracking	13/19
oxidative cracking	13/14
ozone cracking	13/14
radiation cracking	13/14
solvent cracking	13/14
UV cracking	13/14
craze	9/1.29
crazing	13/15
creep	8/5.9, 11/4.1.7
creep compliance	8/5.9
creep function	8/5.9
critical concentration	11/5.11
critical ion-concentration in an ionomer	10/7
critical point	9/2.12
crosslink	1/1.59, 11/4.2.4, 20/1.9
permanent crosslink	11/4.2.4.1
transient crosslink	11/4.2.4.2
crosslink density	11/4.2.5
crosslinking	11/5.12, 12/1.3
crosslinking site	11/4.2.6
cross-over concentration	3/3.1.2
cross-section	3/3.3.7
cruciform polymer liquid-crystal	7/6.2
crystal B, E, G, H, J and K mesophases	7/3.1.5.3
crystalline polymer	6/1.2
Crystalline polymers	Ch. 6
general definitions	Ch. 6, Sec. 1
local conformation and structural aspects	Ch. 6, Sec. 2
morphological aspects	Ch. 6, Sec. 3
molecular conformation within polymer crystals	Ch. 6, Sec. 4
crystallization kinetics	Ch. 6, Sec. 5
crystallinity	6/1.1
cubic mesophase	7/3.1.9
curing	11/5.13, 12/1.4
electron-beam curing (EB curing)	11/5.13.2, 11/5.13.1, 12/1.4
photochemical curing (photo-curing)	11/5.13.3, 11/5.13.4,

INDEX - Terminology

	12/1.4
thermal curing	11/5.13.5, 12/1.4
cumulative distribution function	3/2.13
cybotactic groups	7/3.1.2
cyclization, polymer	12/1.12
cyclopolymerization	1/3.23
cylindrical morphology	9/3.16
damping curve	8/6.2
dashpot constant	8/5.3, 8/5.4
decad	1/1.63, 1/1.64
decay constant	8/6.3
decay frequency	8/6.4
decay time	7/5.21
defect (in liquid-crystalline polymers)	7/4.7
deflocculation	11/5.14
deformation gradients in an elastic solid	8/1.3
deformation gradients in a viscoelastic liquid or solid	8/1.6
deformation gradient in the orthogonal deformation of an elastic solid	8/2.1
deformation gradient tensor for an elastic solid	8/1.4
deformation gradient tensor for a viscoelastic liquid or solid	8/1.7
deformation of an elastic solid	8/1.3
deformation of a viscoelastic liquid or a solid	8/1.5
deformation ratio	8/2.3
deformation ratio in the orthogonal deformation of an elastic solid	8/2.1
degradation (polymer degradation)	12/1.13, 13/16
biodegradation	12/1.13, 13/11
oxidative degradation	13/16
photodegradation	13/16
photo-oxidative degradation	13/16
thermal degradation	13/16
thermochemical degradation	13/16
thermo-oxidative degradation	13/16
Degradation, aging, and related chemical transformations of polymers	Ch. 13
introduction	
terms	
reference	
alphabetical index of terms	
degree of compatibility	9/1.34
degree of crystallinity	2/4.3, 6/1.3
degree of incompatibility	9/1.34
degree of polymerization	1/1.13, 3/1.3
average degree of polymerization	3/2.12
degree of ripening	9/1.20
degrees of cistacticity and transtacticity	2/4.2
degrees of triad isotacticity, syndiotacticity and heterotacticity	2/4.1
delamination	9/1.18

INDEX - Terminology

denaturation	13/17
dendrite	6/3.6
densification	11/5.15
depolarization of scattered light	3/3.3.15
depolymerization	1/3.25, 12/1.5
diad	1/1.63, 1/1.64
differential distribution function	3/2.13
diisotactic polymer	2/1.14
dilute phase (polymer-poor phase)	3/3.4.2
dilute solution	3/3.1.1
director	7/3.1.1.1
disc-like mesogens (discotic mesophases, discotics)	7/3.2
disclination	7/4.7.2
disclination strength	7/4.9.2.2
discontinuous phase domain	9/3.13
discotic mesogen (discoid mesogen)	7/2.11.2.2
discotic nematic mesophase (discotic nematic)	7/3.2.1
discrete distribution function	3/2.13
discrete phase domain	9/3.13
dislocation	7/4.7.1
dispersed phase domain	9/3.13
dispersing agent (dispersing aid, dispersant)	9/1.40
dispersion	9/1.39
disruptor	7/6.5
dissymmetry of scattering	3/3.3.14
distortion in liquid crystals	7/5.2
distribution function	3/2.13
disyndiotactic polymer	2/1.15
ditactic polymer	2/1.12
divergence temperature	7/2.9
domain	7/4.1
domain boundary	9/3.4
domain interface	9/3.4
domain structure	9/3.5
dopant	10/8
doping	10/9
double-strand chain	1/1.40
double-strand copolymer	1/2.31
double-strand macromolecule	1/1.41
double-strand polymer	1/2.30, 16/2.6
drying control chemical additive	11/3.2
dual phase domain continuity	9/3.14
durability	13/18
dynamic-scattering mode	7/5.15
dynamic strain	8/5.1
dynamic stress	8/5.1
dynamic viscosity	8/5.20
elastic constants (elasticity moduli)	7/5.3
elastic modulus (modulus of elasticity)	8/4.3
elastically active network chain	11/4.2.7

INDEX - Terminology

elastomer	11/4.1.8
electrically conducting polymer	10/10
electroclinic effect	7/5.11
electrohydrodynamic instabilities	7/5.13
electroluminescent polymer	12/3.3
electron-exchange polymer	12/2.12
electro-optical polymer	12/3.10
elongational strain rate	8/2.12
elongational viscosity	8/4.9
elution volume	3/3.4.8
enantiotropic mesophase	7/2.4.1
end-group	1/1.35, 15/2.6
end-on fixed side-group polymer liquid-crystal	7/6.3
end-to-end distance	3/1.11
end-to-end vector	3/1.10
engineering strain	8/2.4
engineering stress	8/3.4
environmental stress cracking	13/19
environmentally degradable polymer	13/20
equilibrium sedimentation in a density gradient	3/3.2.16
equilibrium sedimentation (method)	3/3.2.13
equivalence postulate	6/2.5
equivalent chain	3/1.17
equivalent sphere	
thermodynamically equivalent sphere	3/1.4
hydrodynamically equivalent sphere	3/3.2.2
even-membered liquid-crystal dimer	7/2.11.2.9
excess scattering	3/3.3.8
excluded volume of a macromolecule	3/3.1.9
excluded volume of a segment	3/3.1.8
exfoliation	9/1.22, 11/5.16
expansion factor	3/3.1.10
extended-chain crystal	6/4.13
extender	9/1.43
extensional strain rate	8/2.12
extensional viscosity	8/4.9
extension ratio	8/2.3
extraction fractionation	3/3.4.5
fall time	7/5.22
fatigue	13/21
ferroelectric effects	7/5.9
ferroelectric polymer	12/3.4
ferromagnetic polymer	12/3.5
f-functional branch point	1/1.54, 11/4.2.2.1
f-functional branch unit	1/1.55
fibrillar morphology	9/3.17
fibrous crystal	6/3.7
fill factor	9/1.45
filler	9/1.44
Finger tensor	8/1.8, 8/1.11

INDEX - Terminology

fire retardancy (flame retardancy)	13/22
fire retardant (flame retardant)	13/22
first normal-stress coefficient	8/4.13
first normal-stress difference (first normal-stress function)	8/3.6
flexo-electric domain	7/5.17
flexo-electric effect	7/5.16
flexural deflection	8/6.9
flexural force	8/6.8
flexural modulus	8/6.10
flexural stress	8/6.8
flocculation	11/5.17
Flory constant	3/3.2.24
Flory distribution	3/2.17
Flory-Huggins theory (Flory-Huggins-Staverman theory)	3/3.1.11, 9/2.3
flow birefringence	3/3.2.9
focal-conic domain	7/4.10.2
focal-conic, fan-shaped texture	7/4.10.4
fold	6/4.4
fold domain	6/4.7
folded-chain crystal	6/4.11
fold plane	6/4.5
fold surface	6/4.6
forced flexural oscillation	8/6.7
forced oscillation	8/5.10
forced uniaxial extensional oscillation	8/6.6, 8/6.7
forked hemiphasmidic mesogen	7/2.11.2.5
fractal agglomerate	11/4.1.9
fractal dimension	11/4.1.10
fractionation	3/3.4.1
Frank constants	7/5.3
free oscillation	8/6.1
Fréedericksz transition	7/5.10
freely draining	3/3.2.6
freely jointed chain	3/1.16
freely rotating chain	3/1.19
free oscillation	8/6.1
frictional coefficient	3/3.2.1
friction coefficients (rotational viscosity coefficients)	7/5.6
fringed-micelle model	6/4.10
functional polymer	12/3.6
functionality (of a monomer)	11/2.12
fused twin mesogen	7/2.11.2.9
gel	11/3.1
aerogel	11/3.1.1
alcogel	11/3.1.2
aquagel	11/3.1.3
colloidal gel	11/3.1.4
humming gel	11/3.1.7
hydrogel	11/3.1.8
microgel (gel microparticle)	3/1.27, 11/3.1.9, 11/3.1.5

INDEX - Terminology

nanogel (gel nanoparticle)	11/3.1.10, 11/3.1.6
neutralized gel	11/3.1.11
particulate gel	11/3.1.12
polyelectrolyte gel	11/3.1.13
polymer gel	11/3.1.14
responsive gel	11/3.1.15
rheopexic gel	11/3.1.16
rheotropic gel	11/3.1.17
ringing gel	11/3.1.18
sonogel	11/3.1.19
thermoreversible gel	11/3.1.20
thixotropic gel	11/3.1.21
xerogel	11/3.1.22
gel fraction	11/4.1.11
gel point (gelation point)	11/3.3, 11/3.6
gel time (gelation time)	11/3.5, 11/3.8
gelation	11/5.19
gelation temperature (gel temperature)	11/3.7, 11/3.4
gel-permeation chromatography	3/3.4.6
general homogeneous deformation or flow of a viscoelastic liquid or solid	8/2.10
general homogeneous deformation of a elastic solid	8/2.10
geometrical equivalence	6/2.4
glassy mesophase	7/3.5
globular-chain crystal	6/4.14
graft copolymer	1/2.25
graft macromolecule	1/1.28
graft polymer	1/2.23
grafting	12/1.6
green body	11/4.1.12
Green tensor	8/1.8, 8/1.10
guest-host effect	7/5.23
guest polymer	9/3.13
Guinier plot	3/3.3.12
halatopolymer	10/11
halato-telechelic polymer	10/12
hard-segment phase domain	9/3.9
Hausdorff dimension	11/4.1.13
heat endurance	13/23
height equivalent to a theoretical plate	3/3.4.13
helix	6/2.6
helix residue	6/2.7
hemiphasmidic mesogens	7/2.11.2.5
Hencky strain	8/2.5
heptad	1/1.63, 1/1.64
hexad	1/1.63, 1/1.64
hexatic smectic mesophase	7/3.1.5.2
hipping	11/5.20
homeotropic alignment	7/4.3
homogeneous alignment	7/4.4

INDEX - Terminology

homogeneous deformation of an elastic solid	8/1.3
homogeneous deformation of a viscoelastic liquid or solid	8/1.5
homogeneous orthogonal deformation or flow of an incompressible viscoelastic liquid or solid	8/2.11
homogeneous polymer blend	9/1.3
homogeneous simple shear deformation or flow an incompressible viscoelastic liquid or solid	8/2.13
homopolymer	1/2.4
homopolymerization	1/3.3
host polymer	9/3.12
hot isostatic pressing	11/5.21
Huggins coefficient	3/3.2.23
Huggins equation	3/3.2.22
hybrid material	11/4.1.14
chemically bonded hybrid (material)	11/4.1.14.1
clay hybrid	11/4.1.14.2
polymer-clay composite	11/4.1.14.4
polymer-clay hybrid	11/4.1.14.5
hybrid polymer	11/4.1.14.3
hydrodynamic volume	3/3.2.3
hydrodynamically equivalent sphere	3/3.2.2
hydrolysis ratio	11/5.22
hydrolytic scission	12/1.8
hyperbranched polymer liquid-crystal	7/6.11
(chain) identity period ((chain) repeating distance)	6/2.2
immiscibility	9/1.11
impact-modified polymer	12/3.7
impregnation	9/1.19
induced mesophase	7/3.1.8
infinite-shear viscosity	8/4.12
inherent viscosity	3/3.2.20
inhibitor	13/24
inhomogeneous deformation of elastic solids	8/1.3
inorganic-organic polymer	11/4.1.15
inorganic polymer	11/4.1.15.1
insertion reaction	11/5.23
in-situ composite formation	11/5.24
integral distribution function	3/2.13
intercalated smectic mesophase	7/3.1.7
intercalation	9/1.21
intercalation reaction	11/5.23
interchange reaction	12/1.7
interdiffusion	9/2.16
interfacial agent	9/1.31
interfacial bonding	9/1.27
interfacial energy	9/1.26
interfacial fracture	9/1.28
interfacial region	9/3.6
interfacial-region thickness	9/3.8
interfacial tension	9/1.26

interfacial width	9/3.8
inter-junction molar mass	11/4.2.8
interphase	9/3.6
interphase elasticity	9/3.7
interphase thickness	9/3.8
intrinsically conducting polymer	10/13
intrinsic viscosity (limiting viscosity number, Staudinger index)	3/3.2.21
inverse hexagonal mesophase	7/3.2.2.1
inverse lamellar mesophase	7/3.1.5.1.1
ionene	10/15
ion-exchange membrane	10/16, 12/2.2
ion-exchange polymer	10/17, 12/2.2
ionic aggregates in an ionomer	10/18
ionic copolymerization	1/3.18
ionic polymer (ion-containing polymer)	10/19
ionic polymerization	1/3.17
ionomer	1/2.39, 10/20
ionomer cluster	10/21
ionomer molecule	1/1.66
ionomer multiplet	10/22
irregular macromolecule	1/1.5
irregular polymer	1/2.16
isopycnic	3/3.1.17
isorefractive	3/3.3.17
isostatic pressing	11/5.25
isotactic macromolecule	1/1.23
isotactic polymer	1/2.18
isotropic compression	8/2.9
isotropization temperature	7/2.6
junction point	1/1.54, 11/4.2.9
thermoreversible junction point	11/4.2.9.1
transient junction point	11/4.2.9.2
junction point density	11/4.2.10
junction unit	1/1.27
Kapustin domains	7/5.14
Kratky plot	3/3.3.13
Kuhn-Mark-Houwink-Sakurada equation	3/3.25
ladder chain	1/1.44
ladder macromolecule	1/1.45
ladder polymer	1/2.30
lamellar crystal	6/3.1
lamellar domain morphology	9/3.17
lamellar mesophase	7/3.1.5.1.1
lamina	9/1.16
laminate	9/1.16
lamination	9/1.17
large particle	3/3.3.2
lateral contraction ratio	8/2.6
lateral order	2/4.4

INDEX - Terminology

lateral strain	8/2.6
laterally branched mesogen	7/2.11.2.8
lath crystal	6/3.2
lattice distortion	6/2.10.1
length of the scattering vector	3/3.3.5
Leslie-Ericksen coefficients	7/5.4
lifetime	13/25
ligated twin mesogen	7/2.11.2.9
linear chain	1/1.32, 20/1.2
linear copolymer	1/2.28
linear macromolecule	1/1.6
linear polymer	1/2.27
linear viscoelastic behaviour	8/5.2
line repetition groups	6/2.9
liquid crystal	7/2.3
liquid-crystal dendrimer (dendrimeric liquid crystal, dendritic liquid crystal)	7/6.10
liquid-crystal dimer	7/2.11.2.9
liquid-crystal oligomer	7/2.11.2.9
liquid-crystalline phase	7/2.2.1
liquid-crystal polymer (polymer liquid-crystal, liquid-crystalline polymer)	7/6.1
Liquid crystals (low-molar-mass and polymer)	Ch. 7
preamble	Ch. 7, Sec. 1
general definitions	Ch. 7, Sec. 2
types of mesophase	Ch. 7, Sec. 3
textures and defects	Ch. 7, Sec. 4
physical characteristics of liquid crystals	Ch. 7, Sec. 5
liquid-crystal polymers	Ch. 7, Sec. 6
references	Ch. 7, Sec. 7
alphabetical index of terms	Ch. 7, Sec. 8
glossary of recommended abbreviations and symbols	Ch. 7, Sec. 9
liquid-crystal state (liquid crystalline state)	7/2.2
liquid-crystalline phase	7/2.2.1
living copolymerization	1/3.22
living polymer	12/2.3
living polymerization	1/3.21
local conformation	6/1.8.
logarithmic decrement	8/6.5
logarithmic normal distribution	3/2.20
logarithmic viscosity number	3/3.2.20
long chain	1/1.36
long-chain branch	1/1.53, 3/1.24
longitudinal order	2/4.5
long-range intramolecular interaction (long-range interaction)	3/1.6
long spacing	6/3.4
loose end	1/1.61, 11/4.2.11
loss angle of a forced oscillation	8/5.10
loss compliance	8/5.17
loss curve	8/5.11

INDEX - Terminology

loss factor	8/5.11
loss modulus	8/5.13
loss tangent	8/5.11
lower critical solution temperature	9/2.13
lyotropic liquid-crystalline polymer	12/3.8
lyotropic mesophase	7/2.4.3
macrocycle	1/1.57, 20/1.12
macrodispersion	9/1.39
macromolecular	1/1.1
macromolecular isomorphism	6/2.10.5
macromolecule	1/1.1
Macromolecules and dilute polymer solutions	Ch. 3
individual macromolecules	Ch. 3, Sec. 1
assemblies of macromolecules	Ch. 3, Sec. 2
dilute polymer solutions	Ch. 3, Sec. 3
macromonomer	1/2.35, 12/2.4
macromonomer molecule	1/1.9
macromonomeric unit (macromonomer unit)	1/1.12
macroporous-bead polymer support	12/3.23
macroporous polymer	12/3.9
macroradical	1/1.10
magnetic mesophase anisotropy	7/5.8.1
main chain	1/1.34, 15/2.5, 20/1.4
main-chain scission	12/1.8
main-chain polymer liquid crystal (main-chain liquid-crystalline polymer)	7/6.2
major biaxial mesophase anisotropy	7/5.8.2
marbled texture	7/4.9.4
Mark-Houwink equation	3/3.2.25
mass-average degree of polymerization	3/2.12
mass-average molar mass	3/2.7
mass-average relative molecular mass	3/2.7
mass-distribution function (weight-distribution function)	3/2.15
mass fractal dimension	11/4.1.16
matrix phase domain (matrix)	9/3.12
Maxwell element	8/5.3
Maxwell model	8/5.3
(Non-ultimate) Mechanical properties of polymers	Ch. 8
basic definitions	Ch. 8, Sec. 1
deformations used experimentally	Ch. 8, Sec. 2
stresses observed experimentally	Ch. 8, Sec. 3
quantities relating stress and deformation	Ch. 8, Sec. 4
linear viscoelastic behaviour	Ch. 8, Sec. 5
oscillatory deformations and stresses used experimentally for solids	Ch. 8, Sec. 6
references	Ch. 8, Sec. 7
alphabetical index of terms	Ch. 8, Sec. 8
glossary of symbols	Ch. 8, Sec. 9
mechanochemical reaction	12/1.9
mechanochemical scission	12/1.8

INDEX - Terminology

melted-grain boundary mesophase	7/3.6.3
mer	1/1.8
mesogen	7/2.11
mesogenic compound	7/2.11
mesogenic dimer	7/2.11.2.9
mesogenic group (mesogenic unit, mesogenic moiety)	7/2.10
mesogenic monomer	1/2.36
mesogenic oligomer	7/2.11.2.9
mesomorphic compound	7/2.1, 7/2.11
mesomorphic glass	7/2.1
mesomorphic state (mesomorphous state)	7/2.1
mesophase	7/2.4
mesophases of calamitic mesogens	7/3.1
mesophases of disc-like mesogens	7/3.2
mesoporous polymer	12/3.9
metal deactivator	13/26
metallomesogen	7/2.11.3
metastable miscibility	9/1.7
microdispersion	9/1.39
microdomain morphology	9/3.19
Mie scattering	3/3.3.18
Miesowicz coefficient	7/5.5
miscibility	9/1.2
metastable miscibility	9/1.7
miscibility gap	9/2.2
miscibility window	9/2.1
mixed ceramic	11/4.1.17
m,n-polycatenary mesogen	7/2.11.2.5
modulated smectic mesophase	7/3.1.6.3
modulus	8/4.3
molal refractive index increment	3/3.3.6
molar mass	3/1.2
molar-mass average	3/2.5
molar-mass exclusion limit	3/3.4.7
molecular conformation	6/1.7
molecular nucleation	6/5.2
molecular weight	3/1.1
molecular-weight average	3/2.5
molecular-weight exclusion limit	3/3.4.7
monodisperse polymer	1/2.13, 3/2.3
monodomain	7/4.2
monolith	11/4.1.18
monomer	1/2.1
monomer molecule	1/1.3
monomeric unit (monomer unit, mer)	1/1.8
monotropic mesophase	7/2.4.5
morphology	9/3.1
morphology coarsening (phase ripening)	9/2.19
most probable distribution (Schulz-Flory distribution, Flory distribution)	3/2.17

INDEX - Terminology

multicoat morphology	9/3.23
multilayer aggregate	6/3.3
multiphase copolymer	9/3.3, 11/4.1.19
multiple inclusion morphology	9/3.25
multi-strand chain	1/1.46
multi-strand macromolecule	1/1.47
nanodomain morphology	9/3.20
nematic	7/3.1.1
nematic droplet	7/4.9.1
nematic textures	7/4.9
nematogen	7/2.11
net shaping	11/5.26
network	1/1.58, 11/4.1.21, 20/1.10
bimodal network (bimodal polymer network)	11/4.1.21.3.1, 11/4.1.21.3.2
colloidal network	11/4.1.21.1
covalent network (covalent polymer network)	1/1.58, 11/4.1.21.3.3, 11/4.1.21.3.4
entanglement network	11/4.1.21.3.5
interpenetrating polymer network	1/2.43, 9/1.9, 11/4.1.21.3.6, 20/1.18
sequential interpenetrating polymer network	9/1.9, 11/4.1.21.3.6.1
simultaneous interpenetrating polymer network	9/1.9, 11/4.1.21.3.6.2
micronetwork	1/1.60, 11/4.1.21.3.7, 20/1.11
model network	11/4.1.21.3.8
oxide network	11/4.1.21.3.9
perfect network (perfect polymer etwork)	11/4.1.21.3.10, 11/4.1.21.3.11
physical network	1/1.58, 11/4.1.21.3.12
polymer network (network polymer)	1/2.41, 11/4.1.21.3, 11/4.1.21.2, 20/1.17
reversible network	11/4.1.21.3.13
semi-interpenetrating polymer network	1/2.42, 9/1.10, 11/4.1.21.3.14, 20/1.16
sequential semi-interpenetrating polymer network	9/1.10, 11/4.1.21.3.14.1
simultaneous semi-interpenetrating polymer network	9/1.10, 11/4.1.21.3.14.2
thermoreversible network	11/4.1.21.3.13.1
transient network	11/4.1.21.3.15
network-chain molar mass	11/4.2.12
network defect	11/4.2.13
neutral axis (neutral plane) (in forced flexural oscillation)	8/6.7
Newtonian liquid	8/4.2
nominal stress	8/3.4
nonad	1/1.63, 1/1.64
non-amphiphilic mesogen	7/2.11.2
non-draining	3/3.2.7
nonlinear-optical polymer	12/3.10
non-Newtonian liquid	8/4.2
non-uniform polymer (polydisperse polymer)	1/2.14, 3/2.4

INDEX - Terminology

normal stresses	8/3.5
n-star macromolecule	1/1.51
n-strand chain	1/1.46
n-strand macromolecule	1/1.47
nucleation	6/5.1
nucleation of phase separation	9/2.5
nucleus (in liquid crystals)	7/4.9.2.1
number-average degree of polymerization	3/2.12
number-average molar mass	3/2.6
number-average molecular weight	3/2.6
number-average relative molecular mass	3/2.6
number-distribution function	3/2.14
octad	1/1.63, 1/1.64
odd-membered liquid-crystal dimer	7/2.11.2.9
oligomer	1/2.3
oligomerization	1/3.2
oligomer molecule	1/1.2
onion morphology	9/3.21
optical texture	7/4.8
optically active polymer	12/3.11
ordered co-continuous double gyroid morphology	9/3.22
ordered sanidic phase	7/3.4.2
order parameter	7/5.1
organic-inorganic polymer	11/4.1.22
organically modified ceramic	11/4.1.23
organically modified silicate	11/4.1.24
organomodified ceramic	11/4.1.25
Ormocer	11/4.1.23
Ormosil	11/4.1.24
oscillatory (simple) shear flow	8/2.13
Oseen-Zocher-Frank constants	7/5.3
Ostwald ripening	11/5.27
out-of-phase viscosity	8/5.21
oxidation-reduction polymer	12/2.12
parabolic focal conic domain	7/4.10.2
parallel-chain crystal	6/4.12
partially draining	3/3.2.8
particle scattering factor	3/3.3.10
particle scattering function	3/3.3.10
pendant chain	1/1.53, 20/1.6
pendant group	1/1.56
pentad	1/1.63, 1/1.64
pentafunctional	1/1.54, 1/1.55
peptization	11/5.28
periodic copolymer	1/2.12
periodic copolymerization	1/3.12
peroxidation	13/27
peroxide decomposer	13/28
persistence length	3/1.22
perturbed dimensions	3/1.8

INDEX - Terminology

phantom chain behaviour	11/4.2.14
phase angle (of a forced oscillation)	8/5.10
phase domain	9/3.2
phase interaction	9/3.7
phase inversion	9/2.15
phase microdomain	9/3.2
phase nanodomain	9/3.2
phase ripening	9/2.19
phasmidic mesogen	7/2.11.2.5
photo-acid generator	12/1.1
photo-base generator	12/1.1
photochemical reaction	12/1.10, 12/1.18
photochromic polymer	12/3.14
photoelastic polymer	12/3.12
photoluminescent polymer	12/3.13
photopolymerization	12/1.10
photoreactive polymer	12/3.14
photorefractive polymer	12/3.10
photoresponsive polymer	12/3.14
photosensitive polymer	12/3.14
photosensitizer	13/29
photosensitization	13/29
piezoelectric polymer	12/3.15
pinning	9/2.19
Piola tensor	8/1.8
planar alignment (homogeneous alignment)	
(in liquid crystals)	7/4.4
uniform planar alignment (in liquid crystals)	7/4.5
plane strain	8/1.8
plane stress	8/1.2
plate height	3/3.4.14
plate number	3/3.4.13
Poisson distribution	3/2.18
Poisson's ratio	8/2.6
polyacid	10/23
polyaddition	1/3.8, 4/3
polyampholyte	10/1
polyanion	12/2.2
polybase	10/25
polybetaine	10/26
polycatenary mesogen	7/2.11.2.5
polycation	12/2.2
polycondensation	1/3.7, 4/2
polydisperse polymer	1/2.14, 3/2.4
polyelectrolyte (polymer electrolyte, polymeric electrolyte)	1/2.38, 12/3.16, 10/27, 10/30, 10/31
polyelectrolyte complex	10/28
polyelectrolyte molecule	1/1.65
polyelectrolyte network	10/29
polygonal texture	7/4.10.3

INDEX - Terminology

polymer	1/2.2
polymer absorbent	12/3.22
polymer acid	12/3.16
polymer adsorbent	12/3.22
polymer alloy	9/1.38, 11/4.1.26
polymer base	12/3.16
polymer blend	1/2.40, 9/1.1, 11/4.1.27, 20/1.15
compatible polymer blend	9/1.35, 11/4.1.27.1
homogeneous polymer blend (miscible polymer blend)	11/4.1.27.2, 9/1.3 11/4.1.27.3
homologous polymer blend	9/1.4
immiscible polymer blend (heterogeneous polymer blend)	9/1.12
isomorphic polymer blend	9/1.5
metastable miscible polymer blend	9/1.8
polymer catalyst	12/2.5
polymer compatibilizer	12/3.17
polymer complexation (polymer complex formation)	12/1.11
(polymer) crystal	6/1.4
(polymer) crystallite	6/1.5
polymer cyclization	12/1.12
polymer degradation	12/1.13
polymer derived ceramic	11/4.1.28
polymer drug	12/3.18
polymer functionalization	12/1.14
polymeric	1/1.1
polymeric inner salt	10/34
polymerization	1/3.1
Polymerization reactions, basic classification and definitions	Ch. 4
chain polymerization	Ch. 4, Sec. 1
polycondensation	Ch. 4, Sec. 2
polyaddition	Ch. 4, Sec. 3
summary	Ch. 4, Sec. 4
polymer liquid-crystal	7/6.1
polymer membrane	12/3.20
polymer-metal complex	12/2.6
polymer molecule	1/1.1
polymer phase-transfer catalyst	12/2.7
polymer-polymer complex	1/2.44, 9/1.6, 20/1.19
polymer-poor phase (dilute phase)	3/3.4.2
polymer reactant	12/2.9
polymer reaction	12/1.15
polymer reagent	12/2.9
polymer-rich phase (concentrated phase)	3/3.4.3
polymer solvent	12/3.21
polymer-solvent interaction	3/3.1.3
polymer sorbent	12/3.22
polymer support	12/3.23
polymer surfactant	12/3.24
polymer with mesogenic side-groups or side-chains	7/6.3

INDEX - Terminology

polymeric	1/1.1
Polymers containing ions and ionizable or ionic groups	Ch. 10
polymer-solvent interaction	3/3.1.3
polymer-supported catalyst	12/2.8
polymer-supported enzyme	12/2.8
polymer-supported reaction	12/1.16
polymer-supported reagent	12/2.9
polymolecularity correction	3/2.21
polymorphic modifications of strongly polar compounds	7/3.1.6
Porod-Kratky chain	3/1.21
pre-ceramic	11/4.1.29
pre-ceramic material	11/4.1.30
precipitation	11/5.29
precipitation fractionation	3/3.4.4
preferential sorption (selective sorption)	3/3.1.13
pre-gel regime	11/2.13
pre-gel state	11/2.14
prepolymer	1/2.37, 12/2.10
prepolymer molecule	1/1.11
prepreg	9/1.20
pre-tilted homeotropic alignment	7/4.3
pre-transitional temperature	7/2.9
primary crystallization	6/5.4
probability density function	3/2.13
protection of a reactive group	12/1.17
pseudo-co-oligomer	1/2.8
pseudo-copolymer	1/2.6
pure shear deformation or flow	8/3.1
pure shear of an elastic solid	8/2.7
pure shear stress	8/3.1
pyramidic mesogen	7/2.11.2.3
pyrolysis	11/5.30, 13/31
auto-pyrolysis	13/31
quality of solvent	3/3.1.4
quaterpolymer	1/2.5
racemate-forming chirogenic polymerization	5/2
racemate-forming enantiomer- differentiating polymerization	5/3
radial droplet texture	7/4.9.1.2
radiation reaction	12/1.18
radical copolymerization	1/3.16
radical polymerization	1/3.15
radius of gyration	3/1.9
random coil	3/1.15
random copolymer	1/2.10
random copolymerization	1/3.10
random-walk chain	3/1.16
rate-of-strain tensor	8/1.12
Rayleigh ratio	3/3.3.7
excess Rayleigh ratio	3/3.3.8
reaction injection moulding	11/5.31

INDEX - Terminology

reinforced reaction injection moulding	11/5.31.1
Reactions of polymers and functional polymeric materials	Ch. 12
introduction	
reactions involving polymers	Ch. 12, Sec. 1
polymer reactants and reactive polymeric materials	Ch. 12, Sec. 2
functional polymeric materials	Ch. 12, Sec. 3
references	
alphabetical index of terms	
reactive blending	12/1.19
reactive polymer	12/2.11
reactive polymer processing	11/5.32
recrystallization	6/5.7
rectangular sanidic mesophase	7/3.4.1
redox polymer	12/2.12
reduced viscosity	3/3.2.19
re-entrant mesophase	7/3.1.6.1
refractive index increment (refractive increment)	3/3.3.6
regular comb macromolecule	1/1.52
regular macromolecule	1/1.4
regular oligomer molecule	1/1.7
regular polymer	1/2.15, 15/2.1, 16/2.1
regular star macromolecule	1/1.51
relative biaxiality (of a biaxial mesophase)	7/5.8.2
relative molecular mass	3/1.1
relative-molecular-mass average	3/2.5
relative viscosity	3/3.2.17
relative viscosity increment	3/3.2.18
relaxation spectrum	8/5.6
relaxation time	8/5.6
reorganization	6/5.6
resin	12/2.13
resist polymer	12/3.25
electron-beam resist	12/3.25
ion-beam resist	12/3.25
negative-tone resist (negative resist)	12/3.25
photoresist	12/3.25
positive-tone resist (positive resist)	12/3.25
X-ray resist	12/3.25
resonance curve	8/6.11
width of the resonance curve	8/6.13
resonance frequency	8/6.7, 8/6.12
retardation spectrum	8/5.8
retardation time	8/5.8
retarder	13/32
retention volume	3/3.4.9
ribbon mesophase	7/3.1.6.3
rigid chain	7/6.7
ring-opening copolymerization	1/3.14
ring-opening polymerization	1/3.13
rise time	7/5.21

Rivlin-Ericksen tensors	8/1.14
rod-like morphology	9/3.24
root-mean-square end-to-end distance	3/1.12
rotational diffusion	3/3.2.10
rotational viscosity coefficients	7/5.6
salami-like morphology	9/3.25
sanidic mesogen	7/2.11.2.4
sanidic mesophase	7/3.4
scattering angle	3/3.3.3
scattering vector	3/3.3.4
length of the scattering vector	3/3.3.5
schlieren texture	7/4.9.2
Schulz-Flory distribution	3/2.17
Schulz-Zimm distribution	3/2.16
scission	
hydrolytic	12/1.8
main-chain	12/1.8
mechanochemical	12/1.8
oxidative	12/1.8
photochemical	12/1.8
thermal	12/1.8
secant modulus	8/4.7
secondary crystallization	6/5.5
second moment of area in a force flexural oscillation	8/6.7
second normal-stress coefficient	8/4.14
second normal-stress difference (second normal-stress function)	8/3.7
second virial corfficient	3/3.1.7
sedimentation	11/5.33
sedimentation coefficient	3/3.2.11
sedimentation equilibrium	3/3.2.12
sedimentation velocity method	3/3.2.14
segmented copolymer	9/3.11
segregated star macromolecule	1/1.51
segregation	6/5.8
selective sorption	3/3.1.13
selective solvent	3/3.1.14
semi-rigid chain	7/6.8
sensitizer	13/33
shape-memory polymer	12/3.26
shear	8/2.8, 8/2.13
shear compliance	8/4.11
shear modulus	8/4.10
shear rate	8/2.13
shear strain	8/2.8
shear stress	8/3.5
shear viscosity	8/4.12
shish-kebab structure	6/3.8
short chain	1/1.37
short-chain branch	1/1.53, 3/1.23

short-range intramolecular interaction	3/1.5
shrinkage	11/5.34
side-chain	1/1.53, 20/1.6
side-group or side-chain polymer liquid-crystal (side-group or side-chain liquid-crystalline polymer, polymer with mesogenic side-groups or side-chains, comb-shaped (comb-like) polymer liquid crystal)	7/6.3
side-group	1/1.56
side-on fixed side-group polymer liquid crystal	7/6.3
side-to-tail twin mesogen	7/2.11.2.9
simple shear of an elastic solid	8/2.8
single-strand chain	1/1.38
single-strand macromolecule	1/1.39
single-strand polymer	1/2.29, 15/2.2
sintering	11/5.35
size-exclusion chromatography	3/3.4.6
skeletal atom	1/1.49
skeletal bond	1/1.50
skeletal structure	1/1.48
slip	11/2.15
slip casting	11/5.36
small particle	3/3.3.1
small-strain tensor	8/1.10
smectic A_1, A_d, A_2, C_1, C_d, C_2 mesophases	7/3.1.6.2
smectic A mesophase	7/3.1.5.1.1
smectic B mesophase	7/3.1.5.2.1
smectic C mesophase	7/3.1.5.1.2
smectic F mesophase	7/3.1.5.2.2
smectic I mesophase	7/3.1.5.2.3
smectic mesophase	7/3.1.5
smectic mesophases with unstructured layers: SmA and SmC	7/3.1.5.1
smectic textures	7/4.10
smectogen	7/2.11
soft-segment phase domain	9/3.10
solid polymer electrolyte	10/33
sol	11/2.16
aerosol	11/2.16.1
particulate sol	11/2.26.2
polymeric sol	11/2.26.3
sonosol	11/2.26.4
sol fraction	11/2.17
sol-gel coating	11/ 4.1.31.1
sol-gel critical concentration	11/5.37
sol-gel material	11/4.1.31
sol-gel metal oxide	11/4.1.31.2
sol-gel process	11/5.38
sol-gel silica	11/4.1.31.3
sol-gel transition	11/5.39
Sols, gels, networks and inorganic-organic hybrid materials	Ch. 11
introduction	Ch. 11, Sec. 1

INDEX - Terminology

precursors	Ch. 11, Sec. 2
gels	Ch. 11, Sec. 3
solids	Ch. 11, Sec. 4
processes	Ch. 11, Sec. 5
references	Ch. 11, Sec. 6
alphabetical index of terms	Ch. 11, Sec. 7
solubility parameter	3/3.1.16
spacer	7/6.4
specific refractive index increment	3/3.3.6
spherulite	6/3.9
spinodal decomposition	9/2.8
spinodal phase-demixing	9/2.8
spinodal (spinodal curve)	9/2.7
spiro chain	1/1.42
spiro macromolecule	1/1.43
spiro polymer	1/2.30
splay deformation	7/5.2.1
spreading function	3/3.4.11
spring constant	8/5.3, 8/5.4
stability	13/34
biological stability	13/34
chemical stability	13/34
oxidative stability	13/34
photostability	13/34
photo-oxidative stability	13/34
radiation stability	13/34
thermal stability	13/34
thermostability	13/34
thermo-oxidative stability	13/34
stabilization	13/35
chemical stabilization	13/35
physical stabilization	13/35
stabilizer	13/36
light stabilizer	13/30
photoprotective agent	13/30
photostabilizer	13/30
thermal stabilizer	13/36
UV stabilizer	13/36
standard linear viscoelastic solid	8/5.5
star macromolecule	1/1.51, 20/1.14
star polymer	1/2.32
star polymer liquid-crystal	7/6.2
statistical coil	3/1.15
statistical copolymer	1/2.9
statistical copolymerization	1/3.9
statistical pseudo-copolymer	1/2.6
statistical segment	3/1.18
Staudinger index	3/3.2.21
steady (simple) shear flow	8/2.13
steady uniaxial homogeneous elongational deformation or	

INDEX - Terminology

flow of an incompressible viscoelastic liquid or solid	8/2.12
stem	6/4.2
stereoblock	2/1.21
stereoblock macromolecule	1/1.29
stereoblock polymer	1/2.26, 2/1.23
Stereochemical definitions and notations relating to polymers	Ch. 2
basic definitions	Ch. 2, Sec. 1
constitutional and configurational sequences	Ch. 2, Sec. 2.1
description of relative configurations	Ch. 2, Sec. 2.2
designation of conformation of polymer molecules	Ch. 2, Sec. 3.1
specific terminology for crystalline polymers	Ch. 2, Sec. 3.2
supplementary definitions	Ch. 2, Sec. 4
Stereochemically asymmetric polymerizations	Ch. 5
asymmetric polymerization	Ch. 5, Sec. 1
asymmetric chirogenic polymerization	Ch. 5, Sec. 2
asymmetric enantiomer-differentiating polymerization	Ch. 5 Sec. 3
stereohomosequence	2/2.1.6
stereoregular macromolecule	1/1.22
stereoregular polymer	1/2.20, 2/1.9
stereorepeating unit	1/1.19, 2/1.4
stereoselctive polymerization	5/3
stereosequence	2/2.1.4
stereospecific polymerization	2/1.11
steric factor	3/1.20
storage compliance	8/5.16
storage modulus	8/5.12
stored energy function	8/4.1
strain amplitude (of a forced oscillation)	8/5.10
strain tensor	8/1.8
streaming birefringence	3/3.2.9
stress	8/1.2
stress amplitude (of a forced oscillation)	8/5.10
stress relaxation	8/5.7
stress tensor (stress)	8/1.2
stress tensor resulting from orthogonal deformation or flow	8/3.1
stress tensor resulting from a simple shear deformation or flow	8/3.5
stress vector	8/1.1
structural disorder	6/2.10
subchain	1/1.31, 20/1.5
superabsorbent polymer	12/3.27
supercritical drying of a gel	11/5.40
surface disclination line	7/4.9.3.1
surface fractal dimension	11/4.1.32
surface grafting	12/1.21
surface pre-tilt	7/4.3
swallow-tailed mesogen	7/2.11.2.6
swelling	11/5.41
swelling agent	11/3.9
switchboard model	6/4.9

INDEX - Terminology

swollen-gel-bead polymer support	12/3.23
syndiotactic macromolecule	1/1.24
syndiotactic polymer	1/2.19, 2/1.8
syneresis	11/5.42
microsyneresis	11/5.42.1
tack	9/1.26
tactic block	2/1.19
tactic block polymer	2/1.22
tacticity	1/1.20, 2/1.6
tactic macromolecule	1/1.21
tactic polymer	1/2.17, 2/1.5
tail-to-tail twin mesogen	7/2.11.2.9
tangent modulus	8/4.7
telechelic molecule	1/1.11
telechelic oligomer	12/2.14
telechelic polymer	12/2.14
telomer	1/2.3
telomerization	1/3.2
tensile compliance	8/4.8
tensile modulus	8/4.7
tensile strain	8/2.4
tensile stress	8/3.2
terpolymer	1/2.5
tetrad	1/1.63, 1/1.64
tetrafunctional	1/1.54, 1/1.55
thermodynamic quality of solvent	3/3.1.4
thermodynamically equivalent sphere	3/1.4
thermolysis	11/5.43, 13/31
thermo-oxidative aging	13/4
thermo-oxidative degradation	13/16
thermo-oxidative stability	13/34
thermoplastic elastomer	9/1.46, 11/4.1.8.1
thermoset	12/2.15
thermosetting polymer	12/2.15
thermostability	13/34
thermotropic liquid-crystalline polymer	12/3.8
thermotropic mesophase	7/2.4.1
theta solvent (θ solvent)	3/3.1.5
theta state (θ state)	3/3.1.5
theta temperature (θ temperature)	3/3.1.6
threaded texture	7/4.9.3
three-point bending	8/6.7
three-point flexure	8/6.7
threshold fields	7/5.12
threshold electric field	7/5.12
threshold magnetic induction	7/5.12
tie molecule	6/4.1
tilted smectic mesophase	7/3.1.5
'time-off' of the electro-optical effect	7/5.20

INDEX - Terminology

'time-on' of the electro-optical effect	7/5.19
torsion pendulum	8/6.2
traction (stress vector)	8/1.1
transesterification	12/1.7
transitional entropy	7/2.8
transition temperature	7/2.5
transtactic polymer	2/1.17
triad	1/1.63, 1/1.64
trifunctional	1/1.54, 1/1.55
trimethylsilylation	12/1.17
triphase catalyst	12/2.7
tritactic polymer	2/1.13
true stress	8/1.2
Tung distribution	3/2.19
turbidimetric titration	3/3.3.16
turbidity	3/3.3.9
turn-off time	7/5.19
turn-on time	7/5.18
twin mesogen	7/2.11.2.9
twist alignment	7/4.6
twist deformation	7/5.2.3
twisted-nematic cell	7/5.18
twist grain-boundary mesophase	7/3.6
twist grain-boundary A* mesophase	7/3.6.1
twist grain-boundary C* mesophase	7/3.6.2
twist viscosity	7/5.6
undecad	1/1.63, 1/1.64
underground aging	13/4
uniaxial compliance	8/4.8
uniaxial strain (engineering strain)	8/2.4
(uniaxial) compressive strain	8/2.4
(uniaxial) tensile strain	8/2.4
uniaxial deformation of an elastic solid	8/2.2
uniaxial deformation or flow of an incompressible viscoelastic liquid or solid	8/2.11
uniaxial deformation ratio	8/2.3
uniaxial mesophase anisotropy	7/5.8.1
uniaxial nematic mesophase	7/3.1.1
uniaxial orthogonal deformation or flow	8/3.1
uniaxial pressing	11/5.44
uniform planar alignment	7/4.5
uniform polymer (monodisperse polymer)	1/2.13, 3/2.3
unit cell	6/1.6
universal calibration	3/3.4.10
unperturbed dimensions	3/1.7
unzipping	1/3.25
upper critical solution temperature	9/2.14
variegated star macromolecule	1/1.51
virial coefficients of the chemical potential	3/3.1.7
virtual transition temperature	7/2.7

INDEX - Terminology

viscoelasticity	8/5.1
viscosity	8/4.12
viscosity-average degree of polymerization	3/2.12
viscosity-average molar mass	3/2.10
viscosity-average molecular weight	3/2.10
viscosity-average relative molecular mass	3/2.10
viscosity function	3/3.2.24
viscosity number	3/3.2.19
viscosity ratio	3/3.2.17
viscous flow sintering	11/5.45
viscous sintering	11/5.46
Voigt-Kelvin model (Voigt-Kelvin element, Voigt model, Voigt element)	8/5.4
volume compression	8/2.9
vorticity tensor	8/1.13
vulcanization	11/5.47, 12/1.22
weak link	13/37
wear	13/38
abrasive wear	13/38
weatherability	13/39
weathering	13/39
accelerated weathering	13/39
artificial weathering	13/39
weight-average degree of polymerization	3/2.12
weight-average molecular weight	3/2.7
weight-distribution function	3/2.15
wetting	9/1.23
width of the resonance curve	8/6.13
Williams domains	7/5.14
worm-like chain	3/1.21
Young's modulus	8/4.7
z-average degree of polymerization	3/2.12
z-average molar mass	3/2.8
z-average molecular weight	3/2.8
z-average relative molecular mass	3/2.8
$(z+1)$-average degree of polymerization	3/2.12
$(z+1)$-average molar mass	3/2.9
$(z+1)$-average molecular weight	3/2.9
$(z+1)$-average relative molecular mass	3/2.9
zero-shear viscosity	8/4.12
Zimm plot	3/3.3.11
zwitterionic polymer	10/34
α-scission	12/1.8
β-scission	12/1.8
π-line disclination	7/4.9.2.2
2π-line disclination	7/4.9.2.2
χ interaction parameter (χ parameter)	3/3.1.12, 9/2.4

Nomenclature

Abbreviations for names of polymeric substances	Ch. 22
backbone	1/1.34, 15/2.5, 20/1.4
Biopolymer-related IUPAC-IUB nomenclature recommendations, bibliography of	Appendix
branch (side-chain, pendant chain)	1/1.53, 20/1.6
branched chain	1/1.33, 20/1.3
branch point	1/1.54, 11/4.2.2, 20/1.8
branch unit	1/1.55, 20/1.7
chain	1/1.30, 20/1.1
comb macromolecule	1/1.52, 20/1.13
constitutional repeating unit	1/1.15, 15/2.4, 16/2.3
constitutional unit	1/1.14, 15/2.3, 16/2.2
Copolymers, source-based nomenclature for	Ch. 19
preamble	
basic concept	
classification and definition of copolymers	
copolymers with an unspecified arrangement of monomeric units	Ch. 19, Sec. 1
statistical copolymers	Ch. 19, Sec. 2
alternating copolymers	Ch. 19, Sec. 3
other types of periodic copolymers	Ch. 19, Sec. 4
block copolymers	Ch. 19, Sec. 5
graft copolymers	Ch. 19, Sec. 6
copolymers made by condensation polymerization or related polymerization	Ch. 19, Sec. 7
specification with regard to mass fractions, mole fractions, molar masses and degrees of poymerization	Ch. 19, Sec. 8
appendix: alternative nomenclature for copolymers	
references	
crosslink	1/1.59, 11/4.2.4, 20/1.9
Double-strand (ladder and spiro) organic polymers	Ch. 16
introduction	Ch. 16, Sec. 1
definitions	Ch. 16, Sec. 2
structure-based nomenclature	Ch. 16, Sec. 3
fundamental principles	Ch. 16, Sec. 3.1
identification of the preferred constitutional repeating unit	Ch. 16, Sec. 3.2
orientation of the constitutional repeating unit	Ch. 16, Sec. 3.3
naming of the preferred constitutional unit	Ch. 16, Sec. 3.4
naming of the polymer	Ch. 16, Sec. 3.5
polymers constituted of repeatedly fused or spiro carbocycles	Ch. 16, Sec. 3.6
polymers constituted of repeatedly fused or spiro carbocyclic systems	Ch. 16, Sec. 3.7

INDEX - Nomenclature

polymers constituted of repeatedly fused or spiro heterocycles	Ch. 16, Sec. 3.8
polymers constituted of repeatedly fused or spiro heterocyclic systems	Ch. 16, Sec. 3.9
substituents	Ch. 16, Sec.3.10
end-groups	Ch. 16, Sec. 3.11
sourced-based nomenclature	Ch. 16, Sec. 4
acknowledgement	Ch. 16, Sec. 5
references	Ch. 16, Sec. 6
double-strand polymer	1/2.30, 16/2.6
end-group	1/1.35, 15/2.6
Generic source-based nomenclature for polymers	Ch. 21
introduction	Ch. 21, Sec. 1
source-based nomenclature for homopolymers	Ch. 21, Sec. 2
generic nomenclature	Ch. 21, Sec. 3
fundamental principles	Ch. 21, Sec. 3.1
general rules	Ch. 21, Sec. 3.2
further applications of generic names	Ch. 21, Sec. 4
references	Ch. 21, Sec. 5
Graphic representations (chemical formulae) of macromolecules	Ch. 18
preamble	
general rules	Ch. 18, Sec. 1
regular polymers	Ch. 18, Sec. 2
irregular polymers	Ch. 18, Sec. 3
copolymers	Ch. 18, Sec. 4
references	
interpenetrating polymer network	1/2.43, 9/1.9, 11/4.1.21.3.6, 20/1.18
Introduction to polymer nomenclature	Ch. 14
preamble	Ch. 14, Sec. 1
the principles of source-based nomenclature	Ch. 14, Sec. 1.1
the principles of structure-based nomenclature	Ch. 14, Sec. 1.2
IUPAC nomenclature	Ch. 14, Sec. 2
source-based nomenclature	Ch. 14, Sec. 2.1
homopolymers	Ch. 14, Sec. 2.1.1
copolymers	Ch. 14, Sec. 2.1.2
structure-based nomenclature	Ch. 14, Sec. 2.2
regular single-strand organic polymers	Ch. 14, Sec. 2.2.1
inorganic and coordination polymers	Ch. 14, Sec. 2.2.2
stereochemical definitions and notations	Ch. 14, Sec. 2.2.3
regular double-strand (ladder and spiro) organic polymers	Ch. 14, Sec. 2.2.4
irregular single-strand organic polymers	Ch. 14, Sec. 2.2.5
trade names and abbreviations	Ch. 14, Sec. 3
references	Ch. 14, Sec. 4
Irregular single-strand organic polymers, structure-based nomenclature	Ch. 17
preamble	
basic principles	

INDEX - Nomenclature

rules for naming irregular polymers	
additional examples	
references	
linear chain	1/1.32, 20/1.2
locant	15/2.10
macrocycle	1/1.57, 20/1.12
main chain (backbone)	1/1.34, 15/2.5, 20/1.4
micronetwork	1/1.60, 11/4.1.21.3.7, 20/1.11
network	1/1.58, 11/4.1.21, 20/1.10
network polymer	1/2.41, 11/4.1.21.3, 11/4.1.21.2, 20/1.17
Non-linear macromolecules and macromolecular assemblies, source-based nomenclature	Ch. 20
preamble	
definitions	Ch. 20, Sec. 1
general principles	Ch. 20, Sec. 2
non-linear homopolymer molecules	Ch. 20, Sec. 3
non-linear copolymer molecules	Ch. 20, Sec. 4
copolymer molecules comprising a single species of linear chain	Ch. 20, Sec. 4.1
copolymer molecules comprising a variety of species of chains	Ch. 20, Sec. 4.2
macromolecular assemblies	Ch. 20, Sec. 5
quantitative specifications	Ch. 20, Sec. 6
references	Ch. 20, Sec. 7
path length	15/2.8
pendant chain	1/1.53, 20/1.6
polymer blend	1/2.40, 9/1.1, 11/4.1.27, 20/1.15
polymer-polymer complex	1/2.44, 9/1.6, 20/1.19
regular polymer	1/2.15, 15/2.1, 16/2.1
Regular single-strand organic polymers, nomenclature of	Ch. 15
preamble	Ch. 15, Sec 1
definitions	Ch. 15, Sec 2
fundamental principles	Ch. 15, Sec 3
seniority of subunits	Ch. 15, Sec 4
heterocyclic rings and ring systems	Ch. 15, Sec 4.1
heteroatom chains	Ch. 15, Sec 4.2
carbocyclic rings and ring systems	Ch. 15, Sec 4.3
acyclic carbon chains	Ch. 15, Sec 4.4
selection of the preferred constitutional repeating unit (CRU)	Ch. 15, Sec 5
simple CRUs	Ch. 15, Sec 5.1
complex CRUs	Ch. 15, Sec 5.2
naming the preferred constitutional repeating unit (CRU)	Ch. 15, Sec 6

naming subunits	Ch. 15, Sec 6.1
naming the preferred CRU	Ch. 15, Sec 6.2
naming the polymer	Ch. 15, Sec 7
polymer chain as a substituent	Ch. 15, Sec 8
examples of polymer names	Ch. 15, Sec 9
references	Ch. 15, Sec 10
appendix	Ch. 15, Sec 11
list of names of common subunits	Ch. 15, Sec 11.1
structure- and source-based names for common polymers	Ch. 15, Sec 11.2
ring	16/2.4
ring system	16/2.5
semi-interpenetrating polymer network	1/2.42, 9/1.10, 11/4.1.21.3.14, 20/1.16
seniority	15/2.9
side-chain (branch, pendant chain)	1/1.53, 20/1.6
single-strand polymer	1/2.29, 15/2.2
star macromolecule	1/1.51, 20/1.14
subchain	1/1.31, 20/1.5
subunit	15/2.7